Analyzing Qualitative/Categorical Data

Analyzing Qualitative/Categorical Data

Log-Linear Models and Latent-Structure Analysis

Leo A. Goodman
University of Chicago

Jay Magidson, Editor

Abt Books
Cambridge, Massachusetts

Printed in the United States of America.

ISBN: 0-89011-513-3

Foreword

The need for this text became apparent from the response to a workshop on the analysis of qualitative variables held at Abt Associates Inc. during the summer of 1977. The participants were college and university teachers, applied researchers, and social scientists who came from 18 states and Canada to learn from Leo Goodman, James Davis, and Jay Magidson about the analysis of qualitative data using log-linear models and latent structure analysis. The four-day session proved successful in teaching the participants a sophisticated methodology without drawing on more than an elementary background in statistics.

The workshop's general approach was to break down any barriers posed by the seemingly complex notation. This was done by gradually developing the notation through the use of examples and analogies to the well-known linear regression model. Actual data were used to illustrate the techniques and provide concrete interpretations. Informal comments from the participants and formal evaluation questionnaires attested to the success of this approach.

This volume contains a collection of the materials used at that workshop. Its goal, like that of the workshop, is to provide a thorough understanding of the logic underlying this unified approach for formulating hypotheses, building and testing models, and reformulating ideas so that social scientists and others can apply these techniques to their own data.

In addition to being a valuable reference for applied researchers, this book should prove a useful text or supplementary text for courses in survey analysis and research methods for graduate students and advanced undergraduates. I expect that this book will create exciting learning experiences for you as it did for the workshop participants.

Clark C. Abt, President
Abt Associates Inc.
Cambridge, Mass.

Acknowledgments

I wish to thank Jay Magidson for his great help in preparing this volume for publication. He wrote the preface and general introduction, the introductions to the sections, and one of the chapters. He also did whatever editorial work needed to be done. I also wish to thank James Davis for permission to include his well-known exegesis of some of my work.

Leo A. Goodman

Chicago, Illinois
February 1978

Preface

Traditional methods for summarizing cross-tabulations have proven inadequate for answering many of the questions asked by social scientists. On the other hand, the recent development of appropriate statistical methods, together with the preparation of associated computer programs, has made it relatively easy to apply these methods to the analysis of qualitative or categorical data. The readings on log-linear models and latent-structure analysis found here address the pressing need of social researchers and others for a unified and systematic approach to the analysis of qualitative data.

One of the leading contributors to the development of these log-linear methods is Leo Goodman. In a series of recent publications he describes and applies a systematic procedure for building models and testing hypotheses pertaining to qualitative data. These articles are reprinted for the first time in this text together with some introductory-level expositions of these methods by James Davis and me. The book should serve as a useful reference for researchers in sociology, psychology, education, political science, marketing, and in many other areas where data from surveys are analyzed.

Attitudes (favorable, neutral, unfavorable), occurrence of crime, purchase behavior, death rate, race, sex, social class, and marital status are examples of the variables for which these methods are designed. The general approach has proven valuable for identifying relevant interaction effects and for revealing relatively simple structures underlying seemingly complex relationships among the variables.

The scope of applications is virtually unlimited. The methods can be used by the sociologist studying the relationship between poverty and crime; the educational researcher examining the reliability and validity of a set of test items; the psychometrician creating a new scale; the market researcher analyzing purchase behavior in different market segments; the political scientist studying voter behavior; and the medical researcher trying to identify factors associated with breast cancer.

I would like to express my deep gratitude to Leo Goodman and Jim Davis for their support and encouragement in assembling this book. The book owes much to Professor Goodman for laying its foundation with his many imaginative articles.

I wish to thank Clark Abt, Donald Muse, and Wendy Peter Abt for their support of the workshop, and the participants for their many helpful comments.

I am indebted to the editors and publishers of the journals and books in which the articles originally appeared for their permission to reprint. The journals and books are *American Journal of Sociology, American Sociological Review, Journal of the American Statistical Association, Technometrics, Biometrika,* and *Sociological Methodology.*

Jay Magidson

Cambridge, Massachusetts
March 1978

Table of Contents

Introduction

For a long time now, social science statistics has been split between the "correlators" and the "crosstabbers." The correlators seem to come mainly from the fields of economics, biometrics, or psychology, where interval measures abound and where practitioners have learned to dummycode variables that are not interval. The crosstabbers seem to be mainly sociologists who have had to contend with predicting political party from race, religion, gender, and region of the country or who have faced some similar problem dealing with two or more variables, each categorical with two or more categories (dichotomous or polytomous). Anthropologists and political scientists do not seem to have chosen sides. This volume should effect some rapprochement in the split.

Qualitative variables are the building blocks of theory construction in many of the social sciences. Unlike quantitative variables, they pertain to classifications rather than to measurements. They include nominal variables such as marital status, for which the categories (married, single, divorced, widowed) are unordered. They also include ordinal variables such as attitude toward the legalization of marijuana (favorable, neutral, unfavorable) and amount of education (less than high school, high school, college, post college), in which the categories are ordered. Numbers are associated with the "values" of a quantitative variable; categories or classes are associated with the "levels" of a qualitative variable.

The usual linear regression model is typically used to analyze and predict a quantitative (dependent) variable as a function of other quantitative explanatory (independent) variables. In a series of recent publications Leo Goodman (1970, 1971, 1972a, 1972b, 1973a, 1973b) presents a unified and systematic approach for analyzing qualitative variables using log-linear models. Goodman's approach is in a sense more general than the usual quantitative regression approach since with the former approach variables can be designated dependent on others; or each variable can be analyzed simultaneously as a function of all others without designating dependent and independent variables. His approach can be extended to quantitative explanatory variables or to a mixture of qualitative and quantitative explanatory variables (for example, see Haberman, 1974; Nerlove and Press, 1973).

Goodman's approach to the analysis of qualitative data is more general in at least two other respects as well. For example, interaction effects are an integral part of his log-linear system. A "saturated" model includes all possible interactions. Quantitative models on the other hand are typically assumed to be linear with no interactions or linear with only a small number of interactions added. Interaction terms are sometimes included in quantitative models as products of the explanatory variables. However, when interaction terms are included, the results are often difficult to interpret because they depend on whether the explanatory variables are standardized, along with other related issues (Mosteller and Tukey, 1977). Although interpretative difficulties are encountered with quantitative variables, Goodman's symmetric interactions appear to have a more natural interpretation within the context of his hierarchical system.

Second, normality is typically assumed in the analysis of a quantitative dependent variable using the usual regression methods. In practice the dependent variable may or may not be normally distributed around a regression line. Goodman's approach for qualitative data depends less on such distributional assumptions.

In addition to these models hypothesizing relationships among observed variables, Goodman (1974a, 1974b, 1975) shows how unobserved or "latent" variables can be analyzed and how maximum likelihood estimates can be calculated using models that contain both observed and unobserved qualitative variables. These latent-structure models are somewhat analogous to quantitative factor analytic models and models hypothesizing linear structural relations (Jöreskog, 1969, 1971, 1974; Jöreskog and Sörbom, 1976).

This book is divided into five parts. Some preliminary comments on each part may be useful at this point.

Part I: The Logit Model. This is the simplest log-linear model. It is somewhat analogous to the quantitative regression model. Applications are illustrated for the simplest case of a single dichotomous dependent variable.

Part II: The General Log-Linear Model. The general model is developed and applied to a variety of data sets.

Part III: Davis on Goodman's Approach. James Davis discusses the Goodman system in nontechnical terms. (Some readers may prefer to begin here.)

Part IV: Latent-Structure and Scaling Models. A wide class of latent-structure and scaling models is considered and applied to a variety of data sets.

Part V: Some Extensions to the Goodman System. Goodman's general unified approach is extended to causal systems of qualitative variables and to multiple populations.

The chapters are discussed in greater detail in the introduction to each part of the book.

References

Goodman, L.A. 1970. The multivariate analysis of qualitative data: Interactions among multiple classifications. *Journal of the American Statistical Association* 65:225-56.

. 1971. The analysis of multidimensional contingency tables: Stepwise procedures and direct estimation methods for building models for multiple classifications. *Technometrics* 13:33-61.

. 1972a. A modified multiple regression approach to the analysis of dichotomous variables. *American Sociological Review* 37:28-46.

. 1972b. A general model for the analysis of surveys. *American Journal of Sociology* 77:1035-86.

. 1973a. Causal analysis of data from panel studies and other kinds of surveys. *American Journal of Sociology* 78:1135-91.

. 1973b. The analysis of multidimensional contingency tables when some variables are posterior to others: A modified path analysis approach. *Biometrika* 60:179-92.

. 1974a. Exploratory latent structure analysis using both identifiable and unidentifiable models. *Biometrika* 61:215-31.

. 1974b. The analysis of systems of qualitative variables when some of the variables are unobservable. Part I: A modified latent structure approach. *American Journal of Sociology* 79:1179-1259.

. 1975. A new model for scaling response patterns: An application of the quasi-independence concept. *Journal of the American Statistical Association* 70:755-68.

Haberman, S.J. 1974. *The analysis of frequency data.* Chicago: University of Chicago Press.

Jöreskog, K.G. 1969. A general approach to confirmatory maximum likelihood analysis. *Psychometrika* 34:183-202.

. 1971. Simultaneous factor analysis in several populations. *Psychometrika* 36:409-26.

. 1974. Analyzing psychological data by structural analysis of covariance matrices. In *Contemporary developments in mathematical psychology,* vol. II. Edited by R.C. Atkinson, D.H. Krantz, and R.D. Suppes, pp. 1-56. San Francisco: W.H. Freeman and Company.

Jöreskog, K.G., and Sörbom, A. 1976. Some models and estimation methods for analysis of longitudinal data. In *Latent variables in socioeconomic models.* Edited by D.J. Aigner and A.S. Goldberger, pp. 285-325. Amsterdam: North Holland Publishing Co.

Mosteller, F., and Tukey, J.W. 1977. *Data analysis and regression.* Reading, Mass.: Addison-Wesley, Chapter 13.

Nerlove, M., and Press, S.J. 1973. *Univariate and multivariate log-linear and logistic models.* Rand Report R-1306-EPA/NIH. Santa Monica, Ca.: Rand.

Part One
The Logit Model

Part I focuses on a simple regression in which the dependent variable and all the explanatory variables are dichotomous. The traditional assumptions underlying quantitative regression analysis are violated when the dependent variable is dichotomous; and the usual least squares methods of estimation in quantitative regression analysis, if used when the dependent variable is dichotomous, can lead to estimates of probabilities outside the meaningful 0-1 range.

In chapter 1 Goodman proposes a modified multiple regression approach for analyzing data of this kind. The modified approach is to use a log-linear model (the logit model) in place of the usual linear model and to use a maximum likelihood estimation procedure instead of least squares. The methods are illustrated using data from the well-known study *The American Soldier* by Samuel Stouffer et al.

Although this modified multiple regression approach is designed especially for use with qualitative data, researchers more familiar with the traditional regression approach may hesitate to use the newer methods. Chapter 2 attempts to motivate the use of the newer methods by describing similarities between the parameters (regression coefficients) of a dummy variable regression model and the corresponding parameters from the multiplicative version of the logit model. The former parameters are shown to express changes in probabilities; the latter, changes in odds. Both sets of parameters are readily interpreted.

Chapter 2 also compares the results of applying both models to the data from *The American Soldier.* The traditional regression approach yields interpretations similar to those of the modified approach, but the traditional regression model does not fit the data as well as the corresponding logit model. This chapter also indicates when the two approaches yield different conclusions.

Chapter 1

A Modified Multiple Regression Approach to the Analysis of Dichotomous Variables

To illustrate the models and methods of the present article, we shall reanalyze those data in the famous study of The American Soldier *by Stouffer et al. (1949), subsequently analyzed by Coleman (1964), Zeisel (1968), and Theil (1970). The methods we present reveal how the odds pertaining to a given dichotomized variable (e.g., the odds that a soldier would prefer a Northern to a Southern Camp assignment) are related to other dichotomized variables (e.g., (a) the soldier's race, (b) his region of origin, (c) his present camp location). The usual regression analysis methods do not suit the case considered here, where the dependent variable is the odds pertaining to a given dichotomous variable. Nor do the usual methods suit the case where the dependent variable is a proportion pertaining to the dichotomous variable. This article presents some relatively elementary models and methods suitable for analyzing the odds (or a proportion) pertaining to the given dichotomous dependent variable. Applying these models and methods to the data referred to above, new insights are obtained.*

LET us begin by describing the data that will be analyzed here for illustrative purposes. These data, which are based on the earlier data first presented by Stouffer et al. (1949), appear in Table 1 below. This four-way table cross-classifies soldiers by the following four dichotomous variables: (A) race (Negro or white); (B) region of origin (North or South); (C) location of present camp (North or South); and (D) preference as to camp location (North or South).

Table 1 shows that for say, a Negro Northerner in a Northern camp the odds are 387 to 36 that he will prefer a Northern camp. This table also shows that for, say, a white Southerner in a Southern camp, the odds are 91 to 869 that he will prefer a Northern camp. In the next section, we present a model that describes quantitatively how these and the other odds in Table 1 are affected by (A) race, (B) region of origin, and (C) present camp location. We show how to test whether the model fits the data, and we measure how well the model fits using an index that is analogous to the usual multiple correlation coefficient of regression analysis. We also show how to assess the statistical significance of the contribution made by certain parameters in the model, and we measure the contribution's magnitude with indices that are analogous to the usual partial and multiple-partial correlation coefficients of regression analysis.

With the model that will be described in the next section, and with the more general model that follows it, we can estimate how the odds for preferring a Northern camp are changed by the "main effects" of race, region or origin, and present camp location, as well as by certain "interaction effects" among these variables. With each model considered here we can also estimate what the expected frequencies in Table 1 would be if the model were true. We can compare these estimated expected frequencies with the corresponding observed frequencies to determine if the model fits the data. For the model that will be described in the next section, Table 2 gives the expected frequencies estimated under the assumption that the model is true. Under this model, the estimated odds are 390.64 to 32.36 that the Negro Northerner in a Northern camp will prefer a Northern camp; and the estimated odds are 91.74 to 868.26 that the white Southerner in a Southern camp will prefer a Northern camp. When we compare the corresponding entries in Tables 1 and 2

This research was supported in part by Research Contract No. NSF GS 2818 from the Division of the Social Sciences of the National Science Foundation. For helpful comments, the author is indebted to R. D. Bock, S. Haberman, P. F. Lazarsfeld and A. Stinchcombe.

Table 1. Cross-Classification of Soldiers With Respect to Four Dichotomized Variables: (A) Race, (B) Region of Origin, (C) Location of Present Camp, and (D) Preference as to Camp Location

Variable A Race	Variable B Region of Origin	Variable C Location of Present Camp	Variable D Number of Soldiers Preferring Camp* In North	In South
Negro	North	North	387	36
Negro	North	South	876	250
Negro	South	North	383	270
Negro	South	South	381	1712
White	North	North	955	162
White	North	South	874	510
White	South	North	104	176
White	South	South	91	869

*The numbers in this table were recalculated from the percentage table in Stouffer et al. (1949, p. 553). These numbers are consistent with the percentages in the 1949 table, but they may differ somewhat from the actual observed frequencies due to rounding of the percentages. A related percentage table was given also by Coleman (1964, p. 198) and Theil (1970, p. 104). "Preference for camp in North" includes (a) those who prefer to move to a specific camp located in the North, and (b) those whose present camp is in the North and who prefer to stay there. Similarly, "preference for camp in South" includes (a) those who prefer to move to a specific camp located in the South, and (b) those whose present camp is in the South and who prefer to stay there.

(by methods that will be described later herein), we find that the model fits the data well.

Using the expected frequencies estimated under the given model (see Table 2), we can also estimate the expected *proportion* preferring a Northern camp (as well as the *odds* referred to above), for the individuals in each row in Table 2, under the assumption that the model is true. The models considered herein, which describe how the *odds* for preferring a Northern camp are changed by certain specified "main effects" and "interaction effects," can also be used to describe how the *proportion* preferring a Northern camp is changed by these effects.

In order to understand how the dependent variable (preference as to camp location) is related to the other three variables (race, region of origin, and present camp location), we began with the four-way table (Table 1). Is it necessary to use a four-way table to describe how the dependent variable is related to the other three variables, or can this relationship be summarized adequately using the information contained in tables of smaller dimension (e.g., two-way and/or three-way tables)? The methods presented in the present article can be used to answer this question. To estimate the relationship between the dependent variable and the other three variables, the model that will be presented in the next section will actually use only the information contained in (1) the two-way table describing the relationship between the dependent variable and race, and (2) the three-way table describing the relationship between the dependent variable, region of origin and present camp location. Since that model fits the data well, we find that the relationship between the dependent variable and the other three variables can be summarized adequately using only the information contained in the particular two-way and three-way tables noted above. This topic will be discussed more fully later herein when Table 5 is presented.

We propose to analyze the data in Table 1 by methods quite different from those used in earlier analyses of these data. Our model

Table 2. Estimate of the Expected Frequencies in the Four-Way Contingency Table (Table 1), Under the Model in Which the Odds for Preferring a Northern Camp Depend on Race, Region of Origin, Location of Present Camp, and on the Interaction Between Region of Origin and Location of Present Camp

Variable A Race	Variable B Region of Origin	Variable C Location of Present Camp	Variable D Number of Soldiers Preferring Camp In North	In South
Negro	North	North	390.64	32.36
Negro	North	South	879.31	246.69
Negro	South	North	376.79	276.21
Negro	South	South	380.26	1712.74
White	North	North	951.36	165.64
White	North	South	870.69	513.31
White	South	North	110.21	169.79
White	South	South	91.74	868.26

fits the data better than Coleman's (1964) and we present a more parsimonious explanation of these data than he does. With the estimated parameters in our model, we can explain, in a more comprehensive and compact way, various interesting features of these data noted by Zeisel (1968). Some of the models considered in the present article are related to those in Theil (1970), but the methods we use are easier to apply than his. In the final section herein, we shall compare more fully ours with earlier methods.

A MODEL FOR ANALYZING THE ODDS

The symbols A, B, C, and D denote the four dichotomized variables in the four-way table (Table 1): (A) race, (B) region of origin, (C) location of present camp, and (D) preference as to camp location. For variable A, we use numbers 1 and 2 to denote Negro and white. For variables B, C, and D, we use numbers 1 and 2 to denote North and South. Each of Table 1's sixteen cells can be designated (i, j, k, l), where i = 1 or 2; j = 1 or 2; k = 1 or 2; l = 1 or 2. For example, entry 387 is in cell (1, 1, 1, 1), and is a case where variables A, B, C, D all take on value 1; entry 36 is in the cell (1, 1, 1, 2), with variable A, B, and C taking on value 1 and variable D value 2; entry 876 is in cell (1, 1, 2, 1) with variables A, B, and D taking on value 1 and variable C value 2; entry 250 is in cell (1, 1, 2, 2) with variables A and B taking on value 1 and variables C and D value 2.

Let f_{ijkl} denote the observed frequency in cell (i, j, k, l) of Table 1. For example, $f_{1111} = 387$, $f_{1112} = 36$, $f_{1121} = 876$, $f_{1122} = 250$, etc. Note that each row of Table 1 can be described by the triplet (i, j, k). For example, the first row is (1, 1, 1); the second (1, 1, 2); etc. Let n_{ijk} denote the total observed frequency in a row (i, j, k). In other words, we can write n_{ijk} as

$$n_{ijk} = f_{ijk1} + f_{ijk2}. \quad (1)$$

For example, $n_{111} = 423$, $n_{112} = 1126$, etc.

For those in row (i, j, k) the observed odds in favor of a preference for a Northern camp (i.e., the odds that variable D will take on value 1) can be written as

$$\omega_{ijk} = f_{ijk1}/f_{ijk2}. \quad (2)$$

For example, $\omega_{111} = 10.75$, $\omega_{112} = 3.50$, etc. In other words, when variables A, B, and C all take on value 1, the odds are 10.75 to 1 that variable D will take on that value. When variables A and B take on value 1 and variable C value 2, the odds are 3.50 to 1 that variable D will take on value 1.

For row (i, j, k) in Table 1, let p_{ijk} denote that row's observed proportion of observations for which variable D takes on value 1. In other words, we can write p_{ijk} as

$$p_{ijk} = f_{ijk1}/n_{ijk}. \quad (3)$$

For example, $p_{111} = .91$, $p_{112} = .78$, etc. We also let q_{ijk} denote the observed proportion of observations in row (i, j, k) for which variable D takes on value 2. Thus,

$$q_{ijk} = f_{ijk2}/n_{ijk} = 1 - p_{ijk}. \qquad (4)$$

From (2)–(4), we see that

$$\omega_{ijk} = p_{ijk}/q_{ijk}. \qquad (5)$$

We can also express the p_{ijk} and q_{ijk} in terms of the observed odds ω_{ijk}:

$$p_{ijk} = \omega_{ijk}/(1 + \omega_{ijk}),$$
$$q_{ijk} = 1/(1 + \omega_{ijk}). \qquad (6)$$

From (3) and (6) we see that the observed frequencies f_{ijkl} can be expressed in terms of the observed odds ω_{ijk} and the n_{ijk}:

$$f_{ijk1} = n_{ijk}\omega_{ijk}/(1 + \omega_{ijk})$$
$$f_{ijk2} = n_{ijk}/(1 + \omega_{ijk}). \qquad (7)$$

Let F_{ijkl} denote the expected frequency in cell (i, j, k, l) under some specified model. For example, for the model referred to at the end of the preceding section, we see from Table 2 that F_{1111} and F_{1112} are estimated as 390.64 and 32.36, respectively. (The calculation of the entries in Table 2 will be commented upon later herein after we have presented the material in Table 5.) Letting Ω_{ijk} denote the odds based on the expected frequencies, we see that

$$\Omega_{ijk} = F_{ijk1}/F_{ijk2}. \qquad (8)$$

Formula (8) corresponds to (2). In addition, corresponding to (7), we have the following:

$$F_{ijk1} = n_{ijk}\Omega_{ijk}/(1 + \Omega_{ijk})$$
$$F_{ijk2} = n_{ijk}/(1 + \Omega_{ijk}). \qquad (9)$$

(For a related matter, see (44) later herein.) Thus, from the F_{ijkl}, we can calculate the "expected odds" Ω_{ijk}; and from the Ω_{ijk} and n_{ijk}, we can calculate the F_{ijkl}.

Our models will express the Ω_{ijk} in terms of a set of parameters that describe the "main effects" of variables A, B, and C, and certain "interaction effects" among these variables, in a way that is somewhat analogous to the corresponding effects in the usual analysis of variance model. In the present section, we shall present a particular model that fits the data (Table 1) well; and in the next section, we shall present a more general model, namely, a "saturated model" for analyzing the odds, that can help determine the various "unsaturated models" that should be examined further.[1]

Our analysis of this saturated model led us to the particular unsaturated model that we shall present now. For expository reasons we present the unsaturated model first.

Consider the following model:

$$\Omega_{ijk} = \gamma\, \gamma^{A}{}_{i}\gamma^{B}{}_{j}\gamma^{C}{}_{k}\gamma^{BC}{}_{jk} \qquad (10)$$

where

$$\gamma^{A}{}_{1} = 1/\gamma^{A}{}_{2},\ \gamma^{B}{}_{1} = 1/\gamma^{B}{}_{2},\ \gamma^{C}{}_{1} = 1/\gamma^{C}{}_{2},$$
$$\gamma^{BC}{}_{11} = \gamma^{BC}{}_{22} = 1/\gamma^{BC}{}_{12} = 1/\gamma^{BC}{}_{21}. \qquad (11)$$

Parameters γ, $\gamma^{A}{}_{1}$, $\gamma^{B}{}_{1}$, and $\gamma^{C}{}_{1}$ describe the "main effects" on Ω_{ijk} of the general mean[2] and variables A, B, and C, respectively; and parameter $\gamma^{BC}{}_{11}$ describes the "interaction effect" of variables B and C on Ω_{ijk}.[3]

Formula (10) describes the effects of the parameters on Ω_{ijk}, expressing Ω_{ijk} explicitly in terms of the model's parameters. These parameters can also be explicitly expressed in terms of Ω_{ijk}. From (10)–(11), we obtain the following expressions for the parameters in terms of Ω_{ijk}:

$$\gamma = \left[\prod_{i=1}^{2}\prod_{j=1}^{2}\prod_{k=1}^{2}\Omega_{ijk}\right]^{1/8}, \qquad (12)$$

$$\gamma^{A}{}_{1} = [\Omega_{1jk}/\Omega_{2jk}]^{1/2} \text{ (for } j = 1, 2;\ k = 1, 2)$$
$$= \left[\prod_{j=1}^{2}\prod_{k=1}^{2}(\Omega_{1jk}/\Omega_{2jk})\right]^{1/8}, \qquad (13)$$

$$\gamma^{B}{}_{1} = \left[\prod_{i=1}^{2}\prod_{k=1}^{2}(\Omega_{i1k}/\Omega_{i2k})\right]^{1/8}, \qquad (14)$$

$$\gamma^{C}{}_{1} = \left[\prod_{i=1}^{2}\prod_{j=1}^{2}(\Omega_{ij1}/\Omega_{ij2})\right]^{1/8}, \qquad (15)$$

$$\gamma^{BC}{}_{11} = [(\Omega_{i11}\Omega_{i22})/(\Omega_{i12}\Omega_{i21})]^{1/4} \text{ (for } i = 1, 2)$$
$$= \left[\prod_{i=1}^{2}[(\Omega_{i11}\Omega_{i22})/(\Omega_{i12}\Omega_{i21})]\right]^{1/8}. \qquad (16)$$

---

[1] The saturated model, which we present in the next section, can also be described as a full model or an unrestricted model. The unsaturated models can also be described as restricted models. The various models we consider, which assume that the expected odds Ω_{ijk} are subject to certain multiplicative main and interaction effects (see, e.g., formulas (10) and (29)), are quite different from models of the kind appearing in, for example, Coleman (1964) and Boudon (1968). For further comment, see the final section of the present article.

[2] Since γ is somewhat analogous to the main effect of the general mean in the usual model for the analysis of variance (i.e., the constant term in that model), we refer to γ as the main effect of the general mean on the Ω_{ijk}. γ actually equals the *geometric* mean of the Ω_{ijk} corresponding to the eight possible values of (i, j, k) obtained when i=1, 2; j=1, 2; k=1, 2. For further details, see formula (12) below.

[3] The relationship between the model described above by (10)–(11) and the usual model for the analysis of variance will be clarified when we discuss formulas (20)–(22).

From (12), we see that γ is actually the geometric mean of the eight Ω_{ijk}. From (13), we see that γ^{A}_{1} is the square-root of the odds-ratio $\Omega_{1jk}/\Omega_{2jk}$. From (12)–(16), we see that all the γ parameters can be expressed in terms of Ω_{ijk}.[4] Since Table 2 presents the estimated values of F_{ijkl} (under model (10)), we can use these to estimate first, Ω_{ijk} (see (8)) and second, the γ parameters (see (12)–(16)). Table 3 gives the estimated values of the γ parameters.

To emphasize the fact that odds Ω_{ijk} pertain to variable D, and that the γ parameters describe the main and interaction effects on these odds, we could replace the symbols Ω_{ijk}, γ, γ^{A}_{i}, γ^{B}_{j}, γ^{C}_{k}, γ^{BC}_{jk} in (10) (16) by $\Omega^{\bar{D}}_{ijk}$, $\gamma^{\bar{D}}$, $\gamma^{A\bar{D}}_{i}$, $\gamma^{B\bar{D}}_{j}$, $\gamma^{C\bar{D}}_{k}$, $\gamma^{BC\bar{D}}_{jk}$, respectively. This notation was used in Table 3 and later in Table 4, where each of the above parameters is identified by its superscript. From Table 3, we see that the estimated main effect of each variable (A, B, C) is positive (i.e., the estimates of $\gamma^{A\bar{D}}_{1}$, $\gamma^{B\bar{D}}_{1}$, $\gamma^{C\bar{D}}_{1}$ are all larger than 1); but the estimated interaction effect between variables B and C is negative (i.e., the estimate of $\gamma^{BC\bar{D}}_{11}$ is less than 1). This means, among other things, that the estimated effect on $\Omega^{\bar{D}}_{ijk}$ of being a Northerner in a Northern camp is less positive (due to the multiplicative factor of 0.86 pertaining to $\gamma^{BC\bar{D}}_{11}$) than might be surmised simply by combining the main effect of being a Northerner with the main effect of his being in a Northern camp. More precisely, after taking account of the model's various main effects, we must multiply the estimate of $\Omega^{\bar{D}}_{ijk}$ by the factor 0.86 for a Northerner located in a Northern camp, to account for the interaction effect $\gamma^{BC\bar{D}}_{11}$ between variables B and C (i.e., the effect on $\Omega^{\bar{D}}_{ijk}$ of the interaction between region of origin and present camp location).[5] By applying the numerical values of Table 3 to formula (10), we see, for example, that

Table 3. Estimate of the Main Effects and Interaction Effects of the Three Variables (A,B,C) on the Odds Ω_{ijk} Pertaining to Variable D in the Four-Way Contingency Table (Table 1), Under Models (10) and (20)

Variable	γ Effects in Model (10)	β Effects in Model (20)
D	1.31	.27
AD	1.45	.37
BD	3.45	1.24
CD	2.14	.76
BCD	0.86	-.15

$$\Omega^{\bar{D}}_{111} = (1.31)(1.45)(3.45)(2.14)(0.86) = 12.07. \qquad (17a)$$

(All calculations in this paper were carried out to more significant digits than are reported here.)

Further insight into the meaning of the γ parameters can be gained by noting how the estimated value of these parameters affect the estimate of $\Omega^{\bar{D}}_{ijk}$, for $i = 1, 2$; $j = 1, 2$; $k = 1, 2$. By applying the numerical values of Table 3 to formulas (10)–(11), we find that the $\Omega^{\bar{D}}_{ijk}$ can be estimated by (17a) and as follows:

$$\Omega^{\bar{D}}_{211} = (1.31)\left(\frac{1}{1.45}\right)(3.45)(2.14)(0.86) = 5.74, \qquad (17b)$$

$$\Omega^{\bar{D}}_{121} = (1.31)(1.45)\left(\frac{1}{3.45}\right)(2.14)\left(\frac{1}{0.86}\right) = 1.36, \qquad (17c)$$

$$\Omega^{\bar{D}}_{112} = (1.31)(1.45)(3.45)\left(\frac{1}{2.14}\right)\left(\frac{1}{0.86}\right) = 3.56, \qquad (17d)$$

et cetera. Comparing (17a) with (17b), we

[4] The relationship between formulas (12)–(16) and certain formulas in the analysis of variance will be clarified when we discuss formulas (23)–(27).

[5] From the relationship between γ^{BC}_{11}, γ^{BC}_{22}, γ^{BC}_{12} and γ^{BC}_{21} described by formula (11), we see that, after taking account of the various main effects in the model, the estimate of the expected odds $\Omega^{\bar{D}}_{1jk}$ favoring a preference for a Northern camp must be multiplied by the factor 0.86 for those whose region of origin is the same as their present camp location (viz., Northerners in a Northern and Southerners in a Southern camp); and it must be divided by the factor 0.86 for those whose region of origin differs from their present camp location (viz., Northerners in a Southern and Southerners in a Northern camp). For further comments, see the final section herein.

see the effect of $\gamma^{A\bar{D}}_1$. Comparing (17a) with (17c), we see the effect of $\gamma^{B\bar{D}}_1$ and $\gamma^{BC\bar{D}}_{11}$. Comparing (17a) with (17d), we see the effect of $\gamma^{C\bar{D}}_1$ and $\gamma^{BC\bar{D}}_{11}$.

We used the superscript $\bar{D}$ in the preceding two paragraphs to emphasize the fact that the odds Ω_{ijk} pertain to variable D, and that the γ parameters describe the main and interaction effects on these odds. To simplify notation we will delete this superscript hereafter, in all but one section of the paper.

Formula (10) expresses Ω_{ijk} as a product of certain main and interaction effect parameters. This formula can also be expressed in an additive form via logarithms. First, corresponding to Ω_{ijk}, we let Φ_{ijk} denote the natural logarithm of Ω_{ijk}; i.e., we define Φ_{ijk} as

$$\Phi_{ijk} = \log \Omega_{ijk}, \qquad (18)$$

where "log" denotes the natural logarithm. Second, corresponding to formula (10)'s set of parameters (γ, γ^A_i, γ^B_j, γ^C_k, γ^{BC}_{jk}), we define a new set as follows:

$$\beta = \log \gamma,\ \beta^A_i = \log \gamma^A_i,$$
$$\beta^B_j = \log \gamma^B_j,\ \text{etc.} \qquad (19)$$

Then from (10) and (18)–(19) we see that

$$\Phi_{ijk} = \beta + \beta^A_i + \beta^B_j + \beta^C_k + \beta^{BC}_{jk}. \quad (20)$$

From (11) and (19) we see that

$$\beta^A_1 = -\beta^A_2,\quad \beta^B_1 = -\beta^B_2,\quad \beta^C_1 = -\beta^C_2,$$
$$\beta^{BC}_{11} = \beta^{BC}_{22} = -\beta^{BC}_{12} = -\beta^{BC}_{21}, \quad (21)$$

which can also be expressed as follows:

$$\sum_{i=1}^{2} \beta^A_i = 0,\quad \sum_{j=1}^{2} \beta^B_j = 0,\quad \sum_{k=1}^{2} \beta^C_k = 0,$$

$$\sum_{j=1}^{2} \beta^{BC}_{jk} = 0 \ (\text{for } k = 1, 2), \qquad (22)$$

$$\sum_{k=1}^{2} \beta^{BC}_{jk} = 0 \ (\text{for } j = 1, 2).$$

Parameters β, β^A_1, β^B_1, and β^C_1 describe the main effects on Φ_{ijk} of the general mean [6] and variables A, B, and C; and parameter β^{BC}_{11} describes the interaction effect of variables B and C on Φ_{ijk}. The model described by formula (20), which expresses Φ_{ijk} in terms of five parameters (β, β^A_1, β^B_1, β^C_1, β^{BC}_{11}), is equivalent to that described by formula (10), which expresses the corresponding Ω_{ijk} in terms of the corresponding five parameters (γ, γ^A_1, γ^B_1, γ^C_1, γ^{BC}_{11}).

We noted earlier that model (10)'s parameters could be expressed explicitly in terms of Ω_{ijk} (see formulas (12)–(16)). Similarly, the parameters in model (20) can be expressed explicitly in terms of Φ_{ijk}. From (20)–(22), we obtain the following expressions for these parameters in terms of Φ_{ijk}:

$$\beta = \left[\sum_{i=1}^{2} \sum_{j=1}^{2} \sum_{k=1}^{2} \Phi_{ijk} \right] / 8, \qquad (23)$$

$$\beta^A_1 = [\Phi_{1jk} - \Phi_{2jk}]/2 \ (\text{for } j = 1, 2;\ k = 1, 2) \qquad (24)$$

$$= \left[\sum_{j=1}^{2} \sum_{k=1}^{2} (\Phi_{1jk} - \Phi_{2jk}) \right] / 8,$$

$$\beta^B_1 = \left[\sum_{j=1}^{2} \sum_{k=1}^{2} (\Phi_{i1k} - \Phi_{i2k}) \right] / 8, \quad (25)$$

$$\beta^C_1 = \left[\sum_{i=1}^{2} \sum_{j=1}^{2} (\Phi_{ij1} - \Phi_{ij2}) \right] / 8, \quad (26)$$

$$\beta^{BC}_{11} = [\Phi_{i11} + \Phi_{i22} - \Phi_{i12} - \Phi_{i21}]/4 \ (\text{for } i = 1, 2) \qquad (27)$$

$$= \left[\sum_{i=1}^{2} [\Phi_{i11} + \Phi_{i22} - \Phi_{i12} - \Phi_{i21}] \right] / 8.$$

Formulas (23)–(27) are equivalent to the corresponding formulas (12)–(16).[7] Formula (23) states that β is the arithmetic mean of Φ_{ijk} (corresponding to the eight possible values of (i, j, k)). Formula (24) states that β^A_1 can be expressed both as one-half the difference $\Phi_{1jk} - \Phi_{2jk}$ (for j = 1, 2; k = 1, 2), and as one-half the arithmetic mean of the differences $\Phi_{1jk} - \Phi_{2jk}$ corresponding to the four possible values of (j, k) obtained when j = 1, 2; k = 1, 2. Formula (25) states that β^B_1 equals one-half the arithmetic mean of the differences $\Phi_{i1k} - \Phi_{i2k}$ corresponding

[6] Since β is somewhat analogous to the main effect of the general mean in the usual model for the analysis of variance (i.e., the constant term in that model), we refer to β as the main effect of the general mean on Φ_{ijk}. β actually equals the *arithmetic* mean of the Φ_{ijk} corresponding to the eight possible values of (i, j, k) obtained when i=1, 2; j=1, 2; and k=1, 2. For further details, see formula (23) below.

[7] Indeed, instead of obtaining (23)–(27) from formulas (20)–(22), we could also have obtained (23)–(27) from formulas (12)–(16) and (18)–(19). Similarly, we could have obtained (12)–(16) from formulas (23)–(27), making use of formula (28) below and the fact that $\Omega_{ijk} = \exp \Phi_{ijk}$.

to the four possible values of (i, k) obtained when i = 1, 2; k = 1, 2. Formula (26) can be similarly expressed, and formula (27) also has a somewhat similar interpretation.

Since Table 2 presents the estimated values of F_{ijkl} (under model (10) or the equivalent model (20)), we can use these values to estimate first, Ω_{ijk} and Φ_{ijk} (see (8) and (18)) and second, the β parameters (see (23)–(27)). Table 3 includes the β parameters' estimated values. We can also use these values to calculate the γ parameters' estimated values since the relationship between the β and γ parameters can be expressed by (19) or by the following equivalent set of formulas:[8]

$$\gamma = \exp \beta,\ \gamma^{A}{}_{i} = \exp \beta^{A}{}_{i},\ \gamma^{B}{}_{j} = \exp \beta^{B}{}_{j}, \qquad (28)$$

etc., where "exp" denotes the exponential function.[9]

Earlier we discussed Table 3's γ parameter estimated values. Now let us examine the estimated values of β, $\beta^{A}{}_{1}$, $\beta^{B}{}_{1}$, $\beta^{C}{}_{1}$, $\beta^{BC}{}_{11}$, also given in Table 3. In line with our earlier discussion of Table 3, in examining the estimated β parameters, we note that the estimated main effect of each variable (A, B, C) is positive (i.e., the estimates of the $\beta^{A}{}_{1}$, $\beta^{B}{}_{1}$, $\beta^{C}{}_{1}$, which could have been written as $\beta^{A\bar{D}}{}_{1}$, $\beta^{B\bar{D}}{}_{1}$, $\beta^{C\bar{D}}{}_{1}$, are all positive); but the estimated interaction effect between variables B and C is negative (i.e., the estimate of $\beta^{BC}{}_{11}$, which could have been written as $\beta^{BC\bar{D}}{}_{11}$, is negative).[10]

Later we shall show how to assess the statistical significance of the contribution made by certain parameters (e.g., $\gamma^{BC}{}_{11}$) in model (10), and by certain parameters (e.g., $\beta^{BC}{}_{11}$) in model (20), and we also show how to measure this contribution's magnitude.

Our models express the expected odds Ω_{ijk} in terms of the γ parameters (see (10) and also (29) below); or they express the expected log-odds Φ_{ijk} in terms of the β parameters (see (20) and also (35) below). These two forms of expression are equivalent. Since the expected frequencies F_{ijkl} can be expressed in terms of the Ω_{ijk} (see (9)), our models can also be used to express F_{ijkl} in terms of the γ parameters. In addition, letting P_{ijk} and Q_{ijk} denote the expected proportions F_{ijk1}/n_{ijk} and F_{ijk2}/n_{ijk}, respectively (see (3)–(4)), note that our models can also be used to express P_{ijk} (and Q_{ijk}) in terms of these γ parameters.

Before closing this section, we should note the relationship between model (20) and the usual models for (a) the analysis of variance and (b) the analysis of the "logit" pertaining to variable D.

Model (20) and the usual model for the three-way analysis of variance may be compared in several ways. (Note that the Φ_{ijk} in (20) can be presented in a three-way array, while the F_{ijkl} are presented in a four-way table.) In the usual three-way analysis of variance, one must assume homoscedasticity, i.e., that each observation in the three-way table has the same variance. On the other hand, for our kind of data, the homoscedasticity assumption would be contradicted in a way that could not be ignored.[11] Our data also contradict the assumption in the usual analysis of variance that each observation has a normal distribution.[12]

---

[8] For expository purposes, we discussed the γ before the β parameters. Since Table 3 already provided the γ parameters' estimated values, we could have used them in turn to estimate the β parameters (see (19)). Actually, rather than calculate the estimated β from the estimated γ parameters, calculated earlier from the estimated Ω_{ijk} (see (12)–(16)), it is easier to calculate the estimated γ from the corresponding estimated β parameters (see (28)), which can be calculated from the estimated Φ_{ijk} (see (23)–(27)).

[9] The exponential function is the inverse of the natural logarithm. Comparison of (19) and (28) should make this point clear. For example, for a given γ value, we can calculate β from (19) using a table of natural logarithms; and for a given β value, we can calculate γ either from (19), with the natural-logarithm table used now in so to speak, inverted order, or equivalently from (28) using a table of the exponential function.

[10] The fact that the estimate of $\beta^{BC}{}_{11}$ is negative corresponds to the fact that the estimate of $\gamma^{BC}{}_{11}$ is less than 1 (see (19) and (28)). We interpreted this fact in footnote (5). For further comment, see the final section herein.

[11] In the present context, we note that the variance of the observed proportion p_{ijk} (see (3)) will depend both on the magnitude of n_{ijk} (see (1)) and the expected proportion $P_{ijk}=F_{ijk1}/n_{ijk}$. A similar remark applies to the variance of the observed odds ω_{ijk} (see (2) and (5)) and the variance of the logarithm of ω_{ijk}. (The logarithm of the ω_{ijk} is of interest here since it corresponds to Φ_{ijk} in the same sense that ω_{ijk} corresponds to Ω_{ijk}; see (2), (8), (18).)

[12] On the other hand, when n_{ijk} is large, the

Note also that formulas (20)–(22) and (23)–(27) are similar to formulas appearing in the usual analysis of variance. However, to estimate the β parameters under model (20), we use the estimated values of the expected frequencies F_{ijkl} under the model (see Table 2) to estimate first Ω_{ijk} and Φ_{ijk} (see (8) and (18)); and then we use these estimated values of Φ_{ijk} in (23)–(27) to estimate the β parameters. In contrast, in the usual analysis of variance (assuming homoscedasticity), the quantity corresponding to the estimated Φ_{ijk} in formulas (23)–(27) is replaced by the observation in cell (i, j, k); and formulas (24) and (27) are replaced simply by the corresponding expressions on the second line of these two formulas.

Now let us consider the usual model for analyzing the logit pertaining to variable D. This logit is usually defined as being $\Phi_{ijk}/2$ (see, e.g., Fisher and Yates 1963). Model (20) states that this logit (multiplied by 2) can be expressed as a sum of parameters β, $\beta^A{}_i$, $\beta^B{}_j$, $\beta^C{}_k$, $\beta^{BC}{}_{jk}$ (i.e., the main effects of the general mean and of variables A, B, C, and the interaction effect between variables B and C). We can rewrite this model as a regression model expressing variable D's logit as a linear function of dummy variables pertaining to the main effects of variables A, B, C and the interaction effect between variables B and C; but homoscedasticity can not be assumed in this model. Later we shall test the statistical significance of the contribution made by certain parameters in this model, and we shall measure the contributions' magnitude by applying methods proposed in Goodman (1970, 1971a). For some related material, see also Dyke and Patterson (1952), Bishop (1969), Theil (1970), and the final section below.

A GENERAL MODEL FOR ANALYZING THE ODDS

Model (10) included the main effect on Ω_{ijk} of all three variables (A, B, C), but only one of three possible two-factor interaction effects (viz., $\gamma^{BC}{}_{jk}$); and it did not include the three-factor interaction effect (viz., $\gamma^{ABC}{}_{ijk}$). This model assumed that $\gamma^{AB}{}_{ij}$, $\gamma^{AC}{}_{ik}$, and $\gamma^{ABC}{}_{ijk}$ all equal 1. We shall now consider the model that includes all possible main and interaction effects and that makes no assumptions about which (if any) of these effects equals 1. Instead of model (10), we now have the following "saturated" model:

$$\Omega_{ijk} = \gamma\,\gamma^A{}_i\gamma^B{}_j\gamma^C{}_k\gamma^{AB}{}_{ij}\gamma^{AC}{}_{ik}\gamma^{BC}{}_{jk}\gamma^{ABC}{}_{ijk}, \tag{29}$$

where

$$\gamma^A{}_1 = 1/\gamma^A{}_2, \ldots, \gamma^{AB}{}_{11} = \gamma^{AB}{}_{22} = 1/\gamma^{AB}{}_{12} = 1/\gamma^{AB}{}_{21}, \ldots,$$

$$\gamma^{ABC}{}_{111} = \gamma^{ABC}{}_{221} = \gamma^{ABC}{}_{212} = \gamma^{ABC}{}_{122} = 1/\gamma^{ABC}{}_{112} = 1/\gamma^{ABC}{}_{121} = 1/\gamma^{ABC}{}_{211} = 1/\gamma^{ABC}{}_{222}. \tag{30}$$

Formula (29) describes the effects of the γ parameters on Ω_{ijk}. It expresses Ω_{ijk} explicitly in terms of the model's γ parameters. These parameters can also be expressed explicitly in terms of Ω_{ijk}. From (29)–(30), we obtain the following expressions for the parameters in terms of the Ω_{ijk}:[13]

$$\gamma = \left[\prod_{i=1}^{2}\prod_{j=1}^{2}\prod_{k=1}^{2}\Omega_{ijk}\right]^{1/8}, \tag{31}$$

$$\gamma^A{}_1 = \left[\prod_{j=1}^{2}\prod_{k=1}^{2}(\Omega_{1jk}/\Omega_{2jk})\right]^{1/8}, \tag{32}$$

$$\ldots,$$

$$\gamma^{AB}{}_{11} = \left[\prod_{k=1}^{2}(\Omega_{11k}\Omega_{22k})/(\Omega_{12k}\Omega_{21k})\right]^{1/8}, \ldots, \tag{33}$$

$$\gamma^{ABC}{}_{111} = [(\Omega_{111}\Omega_{221}\Omega_{212}\Omega_{122})/(\Omega_{112}\Omega_{121}\Omega_{211}\Omega_{222})]^{1/8}. \tag{34}$$

For the saturated model (29)–(30), we can estimate the γ parameters by formulas (31)–(34), replacing the expected odds Ω_{ijk} in these formulas by the corresponding observed odds ω_{ijk}.[14] With the saturated

observed proportion p_{ijk} will be approximately normally distributed (as long as the expected proportion P_{ijk} differs sufficiently from the extreme values of 0 and 1). A similar remark applies to the observed odds ω_{ijk} and the logarithm of ω_{ijk}.

[13] Formulas (31)–(34) for the saturated model (29)–(30) correspond to formulas (12)–(16) for the unsaturated model (10)–(11).

[14] In contrast to this procedure for the saturated model, note that for an unsaturated model (e.g., model (10)) we use the estimated values of the expected frequencies F_{ijkl} under the model (see Table 2) to estimate Ω_{ijk} (see (8)); then we can use these estimated values of the Ω_{ijk} in (31)–(34) to estimate the γ parameters. (When these estimated values of the Ω_{ijk} are used in (31)–(34), we

Table 4. Estimate of the Main Effects and Interaction Effects of the Three Variables (A,B,C) on the Odds Ω_{ijk} Pertaining to Variable D in the Four-Way Contingency Table (Table 1), Under the Saturated Models (29) and (35)

Variable	γ Effects in Model (29)	β Effects in Model (35)	Standardized Value
D	1.28	.25	6.96
AD	1.44	.37	10.21
BD	3.43	1.23	34.36
CD	2.10	.74	20.65
ABD	0.96	-.04	-1.11
ACD	1.00	.00	0.00
BCD	0.86	-.15	-4.31
ABCD	0.97	-.03	-0.86

model's γ parameters thus estimated, the observed data fit perfectly. (For further comments on this point, see footnote 19 later herein.) Based on Table 1's data, the γ parameters' estimated values are given in Table 4. Note that, for Table 1's data, Table 4's estimated γ's are quite similar to the corresponding quantities of Table 3.

Having replaced the unsaturated (10) with the saturated model (29), we can also replace the unsaturated (20) with the following saturated model:

$$\Phi_{ijk} = \beta + \beta^{A}{}_{i} + \beta^{B}{}_{j} + \beta^{C}{}_{k} + \beta^{AB}{}_{ij} + \beta^{AC}{}_{ik} + \beta^{BC}{}_{jk} + \beta^{ABC}{}_{ijk}, \quad (35)$$

where

$$\beta^{A}{}_{1} = -\beta^{B}{}_{2}, \; \ldots, \; \beta^{AB}{}_{11} = \beta^{AB}{}_{22} = -\beta^{AB}{}_{12} = -\beta^{AB}{}_{21}, \ldots,$$
$$\beta^{ABC}{}_{111} = \beta^{ABC}{}_{221} = \beta^{ABC}{}_{212} = \beta^{ABC}{}_{122} = -\beta^{ABC}{}_{112} = -\beta^{ABC}{}_{121} = -\beta^{ABC}{}_{211} = -\beta^{ABC}{}_{222}. \quad (36)$$

obtain the same results as when they are used in (12)–(16).) For an unsaturated model (e.g., model (10)), the entries in Table 2 are the maximum-likelihood estimates of the F_{ijkl} under the model, and they are calculated by an iterative procedure which we shall comment upon later herein after we have presented the material in Table 5. The observed frequencies f_{ijkl} are the maximum-likelihood estimates of the F_{ijkl} under the saturated model, but *not* under an unsaturated model. Similarly, the observed odds ω_{ijk} are the maximum-likelihood estimates of the Ω_{ijk} under the saturated model, but *not* under an unsaturated model.

Model (35)–(36) is, of course, equivalent to model (29)–(30). Similarly, formulas (31)–(34) are equivalent to the following set of formulas:

$$\beta = \left[\sum_{i=1}^{2} \sum_{j=1}^{2} \sum_{k=1}^{2} \Phi_{ijk} \right] / 8, \quad (37)$$

$$\beta^{A}{}_{1} = \left[\sum_{j=1}^{2} \sum_{k=1}^{2} (\Phi_{1jk} - \Phi_{2jk}) \right] / 8, \quad (38)$$

$$\ldots,$$

$$\beta^{AB}{}_{11} = \left[\sum_{k=1}^{2} (\Phi_{11k} + \Phi_{22k} - \Phi_{12k} - \Phi_{21k}) \right] / 8, \quad (39)$$

$$\ldots,$$

$$\beta^{ABC}{}_{111} = [\Phi_{111} + \Phi_{221} + \Phi_{212} + \Phi_{122} - \Phi_{112} - \Phi_{121} - \Phi_{211} - \Phi_{222}]/8. \quad (40)$$

For the saturated model (35)–(36) we can estimate the β parameters by formulas (37)–(40), replacing the "expected log-odds" Φ_{ijk} in these formulas by the corresponding log ω_{ijk}.[15] In addition, the variance of the estimated β parameters can be estimated by the following formula:[16]

$$S^2_{\hat{\beta}} = \left[\sum_{i=1}^{2} \sum_{j=1}^{2} \sum_{k=1}^{2} \sum_{l=1}^{2} (1/f_{ijkl}) \right] / 64. \quad (41)$$

By dividing each estimated β parameter by its estimated standard deviation $S_{\hat{\beta}}$, we obtain the corresponding "standardized value" of the estimate. Each standardized value can be used to test whether the corresponding β parameter is nil.[17] Table 4 in-

[15] Remarks similar to those in footnotes 8 and 14 would apply here as well.

[16] Note should be taken of the fact that the estimation method presented herein for the saturated model can be improved upon by replacing f_{ijkl} in (41) by $f_{ijkl}+\frac{1}{2}$, and replacing the ω_{ijk} that are used in (31)–(34) (or in (37)–(40)) by $\omega_{ijk}=(f_{ijk1}+\frac{1}{2})/(f_{ijk2}+\frac{1}{2})$. It should also be noted that formula (41) and some of the other results presented herein are applicable both in the case where the observed four-way table (Table 1) describes results obtained for a random sample of individuals cross-classified with respect to the four variables (A, B, C, D), and also in the case where the f_{ijk1} and f_{ijk2} in row (i, j, k) of Table 1 describe results obtained with respect to variable D for a random sample of n_{ijk} individuals at levels i, j, k on variables A, B, C, respectively. For further details, see Goodman (1970) and Haberman (1970).

[17] The term "standardized value" of a statistic is used here to mean the ratio of the statistic and

cludes the β parameters' estimated values, and their corresponding standardized values.

By examining the magnitudes of Table 4's standardized values, we find that the model in which β^{AB}_{ij}, β^{AC}_{ik}, β^{ABC}_{ijk} are set equal to zero in (35) should merit consideration. (Recall that these three parameters could also have been written as $\beta^{AB\overline{D}}_{ij}$, $\beta^{AC\overline{D}}_{ik}$, $\beta^{ABC\overline{D}}_{ijk}$, respectively.) But the model obtained when these particular parameters are set equal to zero in (35) is equivalent to model (20). Thus, for Table 1's data, examining the saturated models (29) and (35) leads to models (10) and (20).

HOW TO TEST WHETHER A MODEL FOR THE ODDS FITS THE DATA

To test whether the hypothesis H described by model (10) fits Table 1's data, we first estimate the expected frequencies F_{ijkl} under the hypothesis H (see Table 2), and then compare the observed frequency f_{ijkl} in Table 1 with the corresponding estimate of the F_{ijkl} in Table 2, by calculating either the usual chi-square goodness-of-fit statistic

$$\sum_{i=1}^{2}\sum_{j=1}^{2}\sum_{k=1}^{2}\sum_{l=1}^{2}(f_{ijkl}-F_{ijkl})^2/F_{ijkl}, \qquad (42)$$

or the corresponding chi-square based on the likelihood-ratio statistic; viz.,

$$2\sum_{i=1}^{2}\sum_{j=1}^{2}\sum_{k=1}^{2}\sum_{l=1}^{2} f_{ijkl}\log\,[f_{ijkl}/F_{ijkl}]. \qquad (43)$$

The chi-square value obtained from (42) or (43) can be assessed by comparing its numerical value with the percentiles of the tabulated chi-square distribution. The degrees of freedom for testing hypothesis H will be $8-5=3$ (since (a) there are eight observed odds in Table 1, and (b) there are five γ parameters estimated in model (10)).

Using (42), we obtain a goodness-of-fit chi-square value of 1.46, and using (43), a likelihood-ratio chi-square value of 1.45. Since there were three degrees of freedom under H, the model fits the data well.

Model (10) is obtained from the saturated model (29) by making a specific set of its γ parameters equal to one.[18] Model (20) is obtained from the saturated model (35) by making a specific set of its β parameters equal to zero. Models obtained this way from saturated models we call "unsaturated."[19] Of course, all unsaturated models obtained from (29) or (35) are models for the odds Ω_{ijk} (or the log-odds Φ_{ijk}) pertaining to variable D. Thus, all these unsaturated models view the four-way table (Table 1) asymmetrically. In the four-way table for variables (A, B, C, D), we treated variable D as the dependent variable; i.e., we viewed the odds (or log odds) pertaining to variable D as depending on the level of variables (A, B, C).

When each individual in a sample is classified by four dichotomous variables (e.g., A, B, C, D), we obtain a four-way table (e.g., Table 1); and, in some contexts, any one of the four variables might be viewed as the dependent variable. For the four-way table, the expected frequencies estimated under a given unsaturated model that treats variable D as the dependent variable (see, e.g., Table 2) will usually differ from the corresponding expected frequencies estimated under a model that treats one of the other variables as the dependent variable.

In some contexts, the research worker will know which variable should be treated as the dependent variable; in others, any one of the four might be treated so. In still others, a different point of view would be appropriate. We could, for example, con-

its estimated standard deviation. The same or similar words have also been used by other writers to denote other things with which the usage here should not be confused.

If a particular β parameter is nil, then the standardized value of the corresponding estimated β will be approximately normally distributed with zero mean and unit variance (when the sample size is large). For comments on related matters, see Goodman (1970, 1971a).

[18] Indeed, the three degrees of freedom used above to test model (10) correspond to the three γ parameters (viz., γ^{AB}_{11}, γ^{AC}_{11}, γ^{ABC}_{111}) in (29) that are set equal to one under model (10).

[19] The number of degrees of freedom used to test a given unsaturated model will equal the number of γ parameters in (29) that are set equal to one under the unsaturated model. Since none of the γ parameters in (29) are set equal to one under that model (i.e., the number of γ parameters set equal to one is zero), there will be zero degrees of freedom under the saturated model. This corresponds to the fact that the observed data fit perfectly under the saturated model, since it includes all possible main and interaction effects (i.e., all possible γ parameters).

sider the case where none of the variables is the dependent variable, but where all are mutually related in some sense (see, e.g., Goodman, 1970). For the four-way table, Goodman (1970) described in his Table 4 a large class of models that would include as special cases models like our (10) and (20) which treat one variable as the dependent variable, as well as "unsaturated" models of a different kind where none of the variables is treated as the dependent variable but where some or all may be mutually related variables. Goodman's Table 4 (1970) contains fifty-three different "unsaturated models" in which one of the four variables is treated as the dependent variable and 113 different "unsaturated" models in which none of the four variables is viewed as the dependent variable. For the case where a given variable (say variable D) is the dependent variable, Goodman's Table 4 (1970) lists nineteen different unsaturated models.

Earlier herein we considered the case where a given variable is the dependent variable. Our models well suit this case (see (10), (20), (29), (35)). Many readers will find our exposition of this case easier to understand than the exposition of the more general case in Goodman (1970). Nevertheless, the more general models and methods of the earlier article also apply to the special case we considered. For each unsaturated model of the kind considered herein, and also for other kinds of "unsaturated" models, Goodman's Table 4 (1970) gave the corresponding degrees of freedom when each variable in the contingency table is dichotomous. He also described ways to calculate the degrees of freedom when some variables are polytomous but not necessarily dichotomous. A single computer program can be used to calculate the estimate of the F_{ijkl}, and the corresponding chi-square values (42) and (43), for any set of "unsaturated" models of the kinds considered herein and in Goodman (1970). For related material dealing with such models see, e.g., Bishop (1969), Goodman (1970, 1971a, 1972).

Let us reconsider model (10), which we obtained from the saturated model (29) by making some of its γ parameters equal to one. We can describe this unsaturated model in any of the following equivalent ways: (1) By listing the γ parameters that are included in model (10); viz., $\gamma^{\bar{D}}$ $\gamma^{A\bar{D}}_{i}$, $\gamma^{B\bar{D}}_{j}$, $\gamma^{C\bar{D}}_{k}$, $\gamma^{BC\bar{D}}_{jk}$.[20] (2) By listing the γ parameters in (29) that are set equal to one under the model; viz., $\gamma^{AB\bar{D}}_{ij}$, $\gamma^{AC\bar{D}}_{ik}$, $\gamma^{ABC\bar{D}}_{ijk}$. (3) By listing the particular marginal tables that are fitted under the model—a topic we shall now discuss.

From our Table 1 we can determine n_{ijk} as defined by formula (1). In all unsaturated models obtained from the saturated model (29), the n_{ijk} are considered fixed; thus in these models the expected frequencies F_{ijkl} (under the model) will satisfy the following condition:

$$F_{ijk1} + F_{ijk2} = n_{ijk}. \qquad (44)$$

By comparing the n_{ijk} from Table 1 with the estimated value of $F_{ijk1} + F_{ijk2}$ from Table 2, we see that condition (44) is satisfied. Since the n_{ijk} describe the three-way marginal table pertaining to variables (A, B, C), we shall use the symbol {ABC} to denote this table. Condition (44) states that the marginal table {ABC} is fitted under the model.

In addition to the marginal table {ABC}, two other marginal tables are fitted under model (10); viz., the two-way marginal table {AD} and the three-way marginal table {BCD}. Table 5 gives the three marginal tables fitted under model (10).[21] In the preceding paragraph, we explained why the marginal table {ABC} was fitted. Under model (10), we also fit the marginal tables {AD} and {BCD} because it includes the parameters $\gamma^{A\bar{D}}_{i}$ and $\gamma^{BC\bar{D}}_{jk}$, which pertain to the relationship between variables A and D (as displayed in the marginal table {AD})

[20] We return now to the notation used earlier where the letter $\bar{D}$ was included in the superscript of each γ parameter to emphasize the fact that the γ parameters describe the main and interaction effects on the odds pertaining to variable D. This notation will facilitate some of our present exposition. This notation's utility will become clearer two paragraphs below.

[21] The four-way contingency table of observed data (Table 1) can be displayed as a 2 x 2 x 2 x 2 table, or an 8 x 2 table (as in Table 1); similarly the three-way marginal table {ABC} can be displayed as a 2 x 2 x 2 table, or a 4 x 2 table (as in Table 5), or as a 8 x 1 table (as we would obtain if we present it as the marginal of the 8 x 2 table displayed in Table 1).

Table 5. The Three Marginal Tables That are Fitted When Models (10) and (20) are Applied to the Four-Way Contingency Table (Table 1)

I. Table {ABC}

Variable A	Variable B	Variable C North	Variable C South
Negro	North	423	1126
Negro	South	653	2093
White	North	1117	1384
White	South	280	960

II. Table {AD}

Variable A	Variable D North	Variable D South
Negro	2027	2268
White	2024	1717

III. Table {BCD}

Variable B	Variable C	Variable D North	Variable D South
North	North	1342	198
North	South	1750	760
South	North	487	446
South	South	472	2581

and to the joint relationship among variables B, C, and D (as displayed in the marginal table {BCD}).[22]

The reader will find that the entries in the three marginal tables in Table 5, which were calculated from Table 1's data, equal the corresponding entries in the three marginal tables calculated from Table 2's estimated F_{ijkl}. The computer program, to which we referred in the fourth paragraph preceding this one, calculated the estimated values in Table 2 (viz., the maximum-likelihood estimates of the expected frequencies F_{ikjl} under model (10)) by an iterative procedure which insured that the three marginal tables given in Table 5 would be fitted when Table 2's estimated F_{ijkl} are used. For further details about the computing procedure, see, for example, the literature cited in the paragraph referred to above.

Although the three marginal tables (viz., {ABC}, {AD}, {BCD}) in Table 5 are fitted under model (10), we noted earlier that the reason for fitting {ABC} in the present context is somewhat different from the reason for fitting {AD} and {BCD}. The marginal table {ABC} is considered to be fixed under model (10); i.e., the n_{ijk} in (1) and (44) are viewed as constants. Aside from the n_{ijk} constants, to estimate the F_{ijkl} under model (10), we use only the information contained in the observed marginals tables {AD} and {BCD}.

The above remarks pertain to model (10), but they can be extended in a straightforward way to a wide range of unsaturated models obtained from the saturated model (29) by setting certain specified γ parameters in (29) equal to one. Now let's apply several such unsaturated models to Table 1's data. Table 6 lists the chi-square values (42) and (43) obtained in testing these models. We include both chi-square values (42) and (43) in Table 6; but, in the present context, (43) has some advantages (see, e.g., Goodman 1968, 1970). In the remaining discussion, we shall use only the chi-square value based on (43).

Each model in Table 6 is described there by listing the marginal tables fitted under the model. For the sake of brevity, we actually list in Table 6 the "minimal set" of marginal tables fitted under the model, rather than the entire set of marginal tables that will in fact be fitted (see footnote 22 herein and Goodman 1970). For example, for model (10), which is presented as H_1 in Table 6, we list in Table 6 the following "minimal set" of marginal tables fitted under the model: {ABC}, {AD}, {BCD}. From this "minimal set" of marginal tables,

[22] When the three-way marginal table {BCD} is fitted, then the following two-way marginal tables will fit automatically: {BC}, {BD}, {CD}. Similarly, when a two-way marginal table, say, {AD} is fitted, then the two one-way marginals, {A} and {D}, will fit automatically. Corresponding to the superscript of each γ parameter in model (10) (with the letter $\overline{D}$ added to each superscript), a marginal table pertaining to that superscript will be included in the set of marginal tables fitted under the model. Under model (10), it will suffice to include {AD} and {BCD} (in addition to table {ABC}) in the set of fitted marginal tables, since then all the marginal tables corresponding to the model's γ parameters (viz., {D}, {AD}, {BD}, {CD}, {BCD}) will actually be fitted.

Table 6. Chi-Square Values for Some Models Pertaining to Table 1

Model	Fitted Marginals	Degrees of Freedom	Likelihood-Ratio Chi-Square	Goodness-of-Fit Chi-Square	γ Parameters Included in the Model
H_1	{ABC},{AD},{BCD}	3	1.45	1.46	[D],[AD],[BD],[CD],[BCD]
H_2	{ABC},{AD},{BD},{CD}	4	24.96	25.73	[D],[AD],[BD],[CD]
H_3	{ABC},{BCD}	4	152.65	147.59	[D],[BD],[CD],[BCD]
H_4	{ABC},{BD},{CD}	5	186.36	180.26	[D],[BD],[CD]
H_5	{ABC},{AD},{CD}	5	2286.83	2187.71	[D],[AD],[CD]
H_6	{ABC},{AD},{BD}	5	695.01	727.16	[D],[AD],[BD]
H_7	{ABC},{D}	7	3111.47	2812.64	[D]
H_8	{ABC},{ACD},{BCD}	2	1.32	1.34	[D],[AD],[BD],[CD],[ACD],[BCD]
H_9	{ABC},{ABD},{BCD}	2	0.68	0.69	[D],[AD],[BD],[CD],[ABD],[BCD]
H_{10}	{ABC},{ABD},{ACD}	2	17.29	18.73	[D],[AD],[BD],[CD],[ABD],[ACD]
H_{11}	{ABD},{ACD},{BCD}	2	24.79	25.11	*
H_{12}	None	15	5469.88	5989.11	*

*Models H_{11} and H_{12} cannot be expressed in terms of the γ parameters. See related discussion in the present article.

we find that the following marginal tables will in fact be fitted: {D}, {AD}, {BD}, {CD}, {BCD} as well as {ABC} and all the marginal tables formed from {ABC}. Variable D is included in five of the marginal tables listed above, and the model under consideration (i.e., model H_1 of Table 6) will include the following five γ parameters corresponding to these five marginal tables: $\gamma^{\bar{D}}$, $\gamma^{A\bar{D}}_{i}$, $\gamma^{B\bar{D}}_{j}$, $\gamma^{C\bar{D}}_{k}$, $\gamma^{BC\bar{D}}_{jk}$.

Consider now hypothesis H_2 in Table 6. Since this model fits the marginals {ABC}, {AD}, {BD}, {CD}, it will include the following γ parameters: $\gamma^{\bar{D}}$, $\gamma^{A\bar{D}}_{i}$, $\gamma^{B\bar{D}}_{j}$, $\gamma^{C\bar{D}}_{k}$.[23] Similarly, hypothesis H_3 in Table 6 fits the marginals {ABC}, {BCD}; and thus that model includes the following γ parameters: $\gamma^{\bar{D}}$, $\gamma^{B\bar{D}}_{j}$, $\gamma^{C\bar{D}}_{k}$, $\gamma^{BC\bar{D}}_{jk}$. Hypothesis H_4 in Table 6 fits the marginals {ABC}, {BD}, {CD}; and thus that model includes the following γ parameters: $\gamma^{\bar{D}}$, $\gamma^{B\bar{D}}_{j}$, $\gamma^{C\bar{D}}_{k}$. Let us discuss these and other models in Table 6 further.

As we have already noted, model (10) is listed as H_1 of Table 6. If we now make $\gamma^{BC\bar{D}}_{jk}$ equal to 1 in model (10), we get H_2 of Table 6. In model H_2, the odds pertaining to variable D are expressed in terms of the parameters $\gamma^{\bar{D}}$, $\gamma^{A\bar{D}}_{i}$, $\gamma^{B\bar{D}}_{j}$, $\gamma^{C\bar{D}}_{k}$, i.e., the main effects of the general mean and variables A, B, and C. To test whether the parameter $\gamma^{BC\bar{D}}_{jk}$ in model (10) contributes in a statistically significant way, we can use the difference between the corresponding chi-square values for H_2 and H_1 as a chi-square statistic with one degree of freedom. (We get the one degree of freedom by subtracting the corresponding degrees of freedom for H_2 and H_1; i.e., $4 - 3 = 1$.) From Table 6's chi-square values for H_2 and H_1, we see that $\gamma^{BC\bar{D}}_{jk}$ does contribute to model (10) in a statistically significant way.

[23] For the reader who has difficulty determining which γ parameters are included in the model from the description of the model in terms of the marginal tables that are fitted, we include this information in Table 6's final column. In that column, we use the symbols [D], [AD], [BD], . . . , to denote $\gamma^{\bar{D}}$, $\gamma^{A\bar{D}}_{i}$, $\gamma^{B\bar{D}}_{j}$, . . . , respectively.

If we set $\gamma^{A\bar{D}}{}_{i}$ equal to 1 in model (10), we get H_3 of Table 6. To test whether the parameter $\gamma^{A\bar{D}}{}_{i}$ in model (10) contributes in a statistically significant way, we can use the difference between the corresponding chi-square values for H_3 and H_1 as a chi-square statistic with one degree of freedom. From Table 6's chi-square values for H_3 and H_1, we see that $\gamma^{A\bar{D}}{}_{i}$ does contribute to model (10) in a statistically significant way.

If we set $\gamma^{A\bar{D}}{}_{i}$ and $\gamma^{BC\bar{D}}{}_{jk}$ equal to 1 in model (10), we get H_4 of Table 6. If we set $\gamma^{B\bar{D}}{}_{j}$ and $\gamma^{BC\bar{D}}{}_{jk}$ equal to 1 in model (10), we get H_5. If we set $\gamma^{C\bar{D}}{}_{k}$ and $\gamma^{BC\bar{D}}{}_{jk}$ equal to 1 in model (10), we get H_6. Comparing the magnitudes of Table 6's corresponding three chi-square values, we see that the worst fitting model was H_5, the next worst H_6, and the least worst H_4. In other words, by comparing the three models obtained from H_2 by deleting the main effect of one of the variables (A, B, C), we see that $\gamma^{B\bar{D}}{}_{j}$ contributes the most.

If we set $\gamma^{A\bar{D}}{}_{i}$, $\gamma^{B\bar{D}}{}_{j}$, $\gamma^{C\bar{D}}{}_{k}$, and $\gamma^{BC\bar{D}}{}_{jk}$ equal to 1 in model (10), we get H_7 of Table 6. In model H_7, the odds pertaining to variable D depend on $\gamma^{\bar{D}}$ (the main effect of the general mean), but are unaffected by the level of variables A, B, and C. In other words, model H_7 states that variable D is independent of the joint variable A, B, C. From the chi-square value for H_7 in Table 6, we see that the data contradict this model.

If we set $\gamma^{AB\bar{D}}{}_{ij}$ and $\gamma^{ABC\bar{D}}{}_{ijk}$ equal to 1 in model (29), we get H_8 of Table 6. If we set $\gamma^{AC\bar{D}}{}_{ik}$ and $\gamma^{ABC\bar{D}}{}_{ijk}$ equal to 1 in model (29), we get H_9. If we set $\gamma^{BC\bar{D}}{}_{jk}$ and $\gamma^{ABC\bar{D}}{}_{ijk}$ equal to 1 in model (29), we get H_{10}. Table 6 shows that H_8 and H_9 fit the data well, but H_{10} does not.

From the above description of H_8 and H_9, we can express model H_1 as follows: Model H_1 states both that H_8 is true *and* that $\gamma^{AC\bar{D}}{}_{ik}$ in H_8 equals 1. Model H_1 also states both that H_9 is true *and* that $\gamma^{AB\bar{D}}{}_{ij}$ in H_9 equals 1. Thus, if H_1 is true, then H_8 and H_9 will also be true; but H_8 and H_9 can be true in cases where H_1 is not. H_1 implies models H_8 and H_9.

Models H_{11} and H_{12} of Table 6 differ from H_1 to H_{10} in an important respect. These last two models do not include the marginal {ABC} among the marginals that are fitted under the model. Therefore, the expected frequencies F_{ijkl} estimated under models H_{11} and H_{12} will *not* satisfy condition (44) (except in some special cases). These two models cannot be expressed as unsaturated models obtained from the saturated model (29), except in cases where condition (44) is satisfied.

Model H_{12} of Table 6 is easier to describe than H_{11}, so I will describe it first. Model H_{12} states that the sixteen cells of Table 1 are equiprobable. From the chi-square value for H_{12} in Table 6, we see that the data contradict this model.

Now let us consider H_{11} of Table 6. As we noted above, this model cannot be expressed as one in which variable D is the dependent variable, since table {ABC} is not fitted under it. However, since the other three-way marginal tables (viz., {ABD}, {ACD}, {BCD}) are fitted under model H_{11}, we see that H_{11} is a model in which any one of the other variables (C, B, or A) can be viewed as the dependent variable. (Note that three of the four possible three-way marginal tables are fitted under H_{11}, and also under H_8, and H_9, and H_{10}.) From the chi-square value for H_{11} in Table 6, we see that the data contradict this model.

We noted earlier that Goodman's Table 4 (1970) included models in which one of the variables is treated as the dependent variable, and it included other kinds of models as well. To test whether any of these other kinds of models might fit the data in our four-way table, we would first consider H_{11} of Table 6; for this model assumes only that the F_{ijkl} are not affected by the three factor "interaction effect" among the three variables A, B, C, nor by the four-factor "interaction effect" among variables A, B, D, C.[24] If this particular model does not fit

[24] Under model H_{11}, the only tables not fitted to the data are the three-way marginal table {ABC} and the four-way table {ABCD}; so the F_{ijkl} (under model H_{11}) are not affected by the three-factor "interaction effect" among variables A, B, C (as displayed in the marginal table {ABC}) nor by the four-factor "interaction effect" among variables A, B, C, D (as displayed in the four-way table). We use the term "interaction effect" in the preceding sentence, and in the sentence to which this footnote applies, in a way

Table 7. Analysis of the Variation in the Odds Pertaining to Variable D in the Four-Way Contingency Table (Table 1)

Source of Variation	Degrees of Freedom	Chi-Square	Numerical Value
1. Total variation due to the "main effects" of variables A,B,C and "interaction effects" among these variables	7	$X^2(H_7)$	3111.47
1a. Due to variation unexplained by model H_1	3	$X^2(H_1)$	1.45
1b. Due to variation explained by model H_1	4	$X^2(H_7)-X^2(H_1)$	3110.02
Partition of (1a)			
1a.1. Due to variation unexplained by model H_9	2	$X^2(H_9)$	0.68
1a.2. Due to variation explained by the $\gamma_{ij}^{AB\bar{D}}$ parameter in model H_9	1	$X^2(H_1)-X^2(H_9)$	0.77

the data, then the data will also contradict any of the other kinds of "unsaturated" models that do not treat variable D as the dependent variable.[25] For examples of data that do not contradict these other kinds of "unsaturated" models, see, Goodman (1970, 1971a).

Before closing this section, we should note that some of the material discussed above could be presented in summary form in tables that are somewhat analogous to the usual analysis of variance tables. Table 7 is an example.

MULTIPLE AND PARTIAL CORRELATION COEFFICIENTS FOR MODELS FOR THE ODDS

In the usual multiple regression analysis for quantitative variables (predicting variable Y from, say, variables X_1 and X_2), the quantity $R^2_{Y \cdot X_1X_2}$, which is the square of the multiple correlation coefficient, can be interpreted as follows: It is the relative decrease in Y's "unexplained variation" obtained when comparing the case where X_1 and X_2 are not used to predict Y with the case where both are used. Similarly, the quantity $r^2_{YX_1 \cdot X_2}$, which is the square of the partial correlation coefficient, can be interpreted as follows: It is the relative decrease in Y's unexplained variation obtained when comparing the case where X_2 but not X_1 is used to predict Y with the case where both are used. The quantity $R^2_{Y \cdot X_1X_2}$ is sometimes referred to as the coefficient of multiple determination, and the quantity $r^2_{YX_1 \cdot X_2}$ can be called the coefficient of partial determination. Goodman (1970, 1971a) introduced coefficients that are somewhat analogous to the usual coefficients of multiple and partial determination for analyzing the odds pertaining to a given variable in the four-way contingency table. We shall now illustrate their calculation.

For a given model in Table 6 (say model H_i, for $i = 1, 2, \ldots, 12$), we shall use the symbol $X^2(H_i)$ to denote its chi-square value. In the preceding section, we noted, among other things, that the statistic $X^2(H_2) - X^2(H_1)$ could be used to test whether the parameter γ^{BC}_{jk} in H_1 contributed in a statistically significant way.[26] To

related to but different from the way we used it earlier. Earlier the term referred to the interaction effects of certain variables on the expected odds Ω_{ijk} pertaining to variable D; whereas above the term refers to the "interaction effects" *among* certain variables in the four-way table. For further details, see Goodman (1970, 1971a).

[25] Except for model H_{11}, any other unsaturated model that does not treat variable D as the dependent variable can be viewed as a model that states both that H_{11} is true *and* that some additional "interaction effects" (in addition to the particular three and four-factor "interaction effects" noted in sentence one of footnote 24) can be set equal to one. Thus, if any other unsaturated model (of the above kind) is true, then H_{11} will also be true. If H_{11} is not true, then none of the other unsaturated models (of the above kind) can be true. For related matters, see Goodman (1970, 1971a).

[26] To facilitate exposition in the preceding section, we included the letter $\bar{D}$ in the superscript of

measure the contribution's magnitude, we recommend the following coefficient, which we shall call the coefficient of partial determination between the odds ω_{ijk} and the parameter γ^{BC}_{jk}, when the other γ's in model H_1 are taken into account:[27]

$$r^2_{\omega\ \gamma_{BC}\cdot H_1} = [X^2(H_2) - X^2(H_1)]/X^2(H_2). \quad (45)$$

From Table 6, we see that this coefficient equals .94 for Table 1's data.

We also noted in the preceding section that the statistic $X^2(H_3) - X^2(H_1)$ could be used to test whether the parameter γ^A_i in H_1 contributed in a statistically significant way. As in the preceding paragraph, we shall measure this contribution's magnitude by the following coefficient of partial determination:[28]

$$r^2_{\omega\ \gamma_A\cdot H_1} = [X^2(H_3) - X^2(H_1)]/X^2(H_3). \quad (46)$$

From Table 6, we see that this coefficient equals .99 for Table 1's data.

To measure how well model H_1 fits the data, we consider the following coefficient, which we call the coefficient of multiple determination between ω and the γ parameters in model H_1:

$$R^2_{\omega\cdot H_1} = [X^2(H_7) - X^2(H_1)]/X^2(H_7). \quad (47)$$

From Table 6, we see that this coefficient equals 1.00 (to two decimal places) for Table 1's data.

We might also consider the following coefficient, which we shall call the coefficient of multiple-partial determination between ω and the parameters γ^A_i and γ^{BC}_{jk} in model H_1, when H_1's other γ parameters (viz., γ, γ^B_j, γ^C_k) are taken into account.

$$R^2_{\omega\ (\gamma_A,\ \gamma_{BC})\cdot\gamma,\ \gamma_B,\ \gamma_C} = [X^2(H_4) - X^2(H_1)]/X^2(H_4). \quad (48)$$

From Table 6 we see that the coefficient equals .99 (to two decimal places) for Table 1's data. Similarly, we can measure the contribution of γ^B_j and γ^{BC}_{jk} (using H_5 rather than H_4 in (48)) or the contribution of γ^C_k and γ^{BC}_{jk} (using H_6 rather than H_4 in (48)).

We can also use the above coefficients to measure the magnitude of the contribution made by the parameters in other models in Table 6. For example, to measure the magnitude of γ^{AC}_{ik} in model H_8, we use the following coefficient of partial determination:

$$r^2_{\omega\ \gamma_{AC}\cdot H_8} = [X^2(H_1) - X^2(H_8)]/X^2(H_1). \quad (49)$$

From Table 6, we see that this coefficient equals .09 for Table 1's data.

All of the r^2 and the R^2 coefficients given by (45)–(49) above took the general form

$$R^2 = [X^2(H'') - X^2(H')]/X^2(H''), \quad (50)$$

where the γ parameters in model H'' are also included among the γ parameters in model H'. We could also write each coefficient as follows:

$$\frac{\sum_{i=1}^{2}\sum_{j=1}^{2}\sum_{k=1}^{2}\sum_{l=1}^{2} F'_{ijkl} \log [F'_{ijkl}/F''_{ijkl}]}{\sum_{i=1}^{2}\sum_{j=1}^{2}\sum_{k=1}^{2}\sum_{l=1}^{2} f_{ijkl} \log [f_{ijkl}/F''_{ijkl}]} \quad (51)$$

where F'_{ijkl} and F''_{ijkl} denote the expected frequencies estimated under model H' and model H'', respectively. The expression of the coefficient in the form (51) is somewhat analogous to the expression of the coefficients of multiple and partial determination, in the usual multiple regression analysis, as

---

each γ parameter; e.g., γ^{BC}_{jk} in (10) became $\gamma^{BC\bar{D}}_{jk}$. In the present section, we have no need for this more cumbersome notation and will not include the letter $\bar{D}$ in the superscript. The reader should, of course, keep in mind that, say, γ^{BC}_{jk} here has the same meaning as $\gamma^{BC\bar{D}}_{jk}$ earlier, and that the various γ parameters describe the main and interaction effects on the odds pertaining to variable D.

[27] In the subscript of r^2 in (45), we changed the γ^{BC} notation to the γ_{BC} notation because of typographical considerations. This simple notational change should not confuse the reader. Similar notational changes will be made in other formulas in this section.

Since model H_1 includes the γ parameters γ, γ^A_i, γ^B_j, γ^C_k, γ^{BC}_{jk}, we could let

$$r^2_{\omega\gamma_{BC}\cdot\gamma,\ \gamma_A,\ \gamma_B,\ \gamma_C}$$

denote the coefficient defined by (45). To test whether this coefficient differs significantly from zero, we use the statistic $X^2(H_2)$—$X^2(H_1)$ as noted earlier.

[28] Remarks like those in the second paragraph of footnote 27 can be applied to the coefficients defined by (46)–(49). For example, for (46), we could let

$$r^2_{\omega\gamma_A\cdot\gamma,\ \gamma_B,\ \gamma_C,\ \gamma_{BC}}$$

denote this coefficient, and we could assess the statistical significance of this coefficient using the statistic $X^2(H_3)$—$X^2(H_1)$ noted earlier.

a ratio of the "explained variation" (when model H' is used to "explain" the variation that was not explained by H") to the "unexplained variation" (when model H" is used).[29]

COMMENTS ON SOME RELATED WORK

As we noted earlier, Coleman's model and methods differ from ours in several ways. His model does not fit the data as well, and his explanation of the data is less parsimonious. Indeed, he observes in his book (1964) that his model differs from the actual data in certain systematic ways, attributing these deviations to a supposed interaction for Negroes (but not whites) between their region of origin and present camp location.[30] In contrast, we find that (a) model (10) in the present article fits the data very well, (b) it does not require an ostensible interaction for Negroes (but not whites) of the kind considered by Coleman, (c) it includes an interaction effect γ^{BC}_{jk} between region of origin and present camp location, which applies equally to both Negroes and whites, (d) this interaction effect is statistically significant, and (e) it both reduces the expected odds favoring a preference for a Northern camp for those whose region of origin is the same as their present camp location, and increases these expected odds for those whose region of origin differs from their present camp location, after the various main effects in the model have been taken into account.[31]

Coleman's article did not show how to test whether his model fit the actual data, nor was he able to measure how well it fit. Furthermore, he did not show how to test the statistical significance of the contribution made by the various parameters in the model, nor could he measure their contribution's magnitude. In addition, the variance of Coleman's estimates of the main effects in his model was larger than it would have been had he used more efficient estimation methods (e.g., maximum-likelihood estimation methods); and his estimates are biased to the extent that his model excluded relevant interaction effects.[32]

Coleman's model states that the effects on the expected proportions P_{ijk} are linear.[33] Applying Coleman's estimation methods to his model, it is possible to obtain clearly incorrect estimates of the P_{ijk} under the model; e.g., estimates of the expected *proportions* P_{ijk} that are negative or larger than one.[34] Furthermore, his model and methods do not take into account the fact that the variance of the observed proportion p_{ijk} will depend on the magnitude of P_{ijk}.[35]

Some of the limitations of Coleman's approach apply to the usual multiple regression model (and analysis of variance model) if used in the present context. For data of the kind considered in the present article, the assumption of homoscedasticity made in the usual multiple regression model (and in

---

[29] When H" is taken as H_7 of Table 6, then the denominator in (51) (i.e., the "unexplained variation" when model H_7 is used) corresponds to the "total variation" in the denominator of the usual coefficient of multiple determination in multiple regression analysis. For related matters, see Goodman (1970).

[30] Coleman (1964) did not provide methods for including interaction effects in his models, and so could not measure the magnitude of the ostensible interaction to which he referred, nor could he judge whether introducing the ostensible interaction would improve the fit of his model.

[31] For further details, see footnote 5. In addition to the effect of γ^{BC}_{jk} described there, the effect can also be described as follows: For the estimate of the expected odds $\Omega^{\overline{D}}_{ijk}$ in favor of a Northern camp, the effect on the estimated $\Omega^{\overline{D}}_{ijk}$ of being at present in a Northern rather than a Southern camp is less for those from the North than from the South. Similarly, the effect on the estimated $\Omega^{\overline{D}}_{ijk}$ of being a Northerner rather than a Southerner is less for those presently in a Northern rather than a Southern camp.

[32] The remarks above apply to Coleman's (1964) and Boudon's (1968) articles, except that Boudon's model did allow for interaction effects.

[33] Recall that $P_{ijk} = F_{ijk1}/n_{ijk}$, using our notation. In contrast to Coleman's, our model states that the expected odds Ω_{ijk} can be expressed in terms of multiplicative effects. In many substantive contexts, it will be more useful to consider multiplicative rather than additive effects. Furthermore, from the point of view of statistical theory, there are a number of reasons for preferring multiplicative models of our kind for analyzing data of the kind presented in Table 1. We shall not pursue these matters further here.

[34] Since P_{ijk} denotes an expected *proportion,* it should not be negative nor larger than one. Therefore, it is undesirable to use models and methods that can lead to *estimates* of the P_{ijk} that are negative or that are larger than one.

[35] In other words, Coleman implicitly assumes homoscedasticity when, on the contrary, his data violate this assumption.

the usual analysis of variance model) would be contradicted in a way that could not be ignored. In addition, as with Coleman's analysis, if one applied the usual multiple regression methods to the model in which the effects on the expected proportions P_{ijk} are linear, one could obtain clearly incorrect estimates of the P_{ijk} under the model, in the sense described above.[36]

We noted earlier that our data were also analyzed by Zeisel (1968) and Theil (1970). Zeisel described various interesting features of these data. These features can be explained, in a more comprehensive and compact way, in terms of the estimated parameters in model (10) of the present article. For example, from the estimates for model (10) presented in Table 3 herein, we find that the estimated product of the parameters γ and γ^{BC}_{11} is approximately one (more precisely, this product is 1.13), and this single fact can be used to explain the following features of the data: (a) the preference for a Northern camp location among Negro Northerners in Northern camps is approximately equal to the preference for a Southern camp location among white Southerners in Southern camps; and (b) the preference for a Northern camp location among white Northerners in Northern camps is approximately equal to the preference for a Southern camp location among Negro Southerners in Southern camps. (In order to see that this single fact explains features (a) and (b), insert the estimated values of the parameters in model (10).) The other interesting features noted by Zeisel can also be explained in similar terms, with one exception. This exception pertains to Zeisel's mention of a supposed effect on camp preference due to the interaction between race and region of origin, among those in Northern camps. Applying the methods of the present article, we find that this ostensible interaction effect is not statistically significant, and there is no need to include it in our model (10).

We comment next on the article by Theil (1970). He used the logit model corresponding to (20), but his estimation method and his analysis differed from ours. Theil (1970) used a weighted least-squares procedure, as did Grizzle, Starmer, and Koch (1969) in the same context; whereas, all the estimates presented in the present article are maximum-likelihood estimates. In commenting on the weighted least-squares procedure, the Grizzle-Starmer-Koch article notes that estimates obtained by their procedure have a somewhat larger variance than maximum-likelihood estimates (see also Rao 1965); similarly Theil's estimates have a somewhat larger variance than our maximum-likelihood estimates. We also find that it is harder to use the methods proposed by Theil, and by Grizzle, Starmer, and Koch than the methods proposed in the present article, when studying the kinds of hypotheses we have discussed for the four-way contingency table (or in Goodman 1970, for the five-way table).[37]

Before closing, we remind the reader that our methods were for the case where a given variable (e.g., variable D) can be viewed as the dependent variable which is affected by the other variables under consideration. Where this is not the case, we refer the reader to the more general techniques presented in, for example, Goodman (1970, 1972).

[36] The comments in footnotes 33 and 34 are relevant here.

[37] For further details, see Goodman 1971a.

REFERENCES

Bishop, Y. Y. M.
1969 "Full contingency tables, logits, and split contingency tables." Biometrics 25:383–400.

Boudon, R.
1968 "A new look at correlation analysis." In H. M. Blalock, Jr. and A. Blalock (eds.), Methodology in Social Research. New York: McGraw-Hill.

Coleman, J. S.
1964 Introduction to Mathematical Sociology. New York: Free Press.

Dyke, G. V. and H. D. Patterson
1952 "Analysis of factorial arrangements when the data are proportions." Biometrics 8:1–12.

Fisher, R. A. and F. Yates
1963 Statistical Tables for Biological, Agricultural and Medical Research. Sixth Edition, New York: Hafner Publishing Co., Inc.

Goodman, L. A.
1968 "The analysis of cross-classified data: Independence, quasi-independence and interactions in contingency tables with or with-

out missing entries." Journal of the American Statistical Association 63:1091–1131.
1970 "The multivariate analysis of qualitative data: Interactions among multiple classifications." Journal of the American Statistical Association 65:226–256.
1971a "The analysis of multidimensional contingency tables: Stepwise procedures and direct estimation methods for building models for multiple classifications." Technometrics 13:33–61.
1971b "Partitioning of chi-square, analysis of marginal contingency tables, and estimation of expected frequencies in multidimensional contingency tables." Journal of the American Statistical Association 66: 339–344.
1972 "A general model for the analysis of surveys." American Journal of Sociology 77 (in press).

Grizzle, J. E., C. F. Starmer, and G. G. Koch
1969 "Analysis of categorical data by linear models." Biometrics 25:489–504.

Haberman, S. J.
1970 "The general log-linear model." Ph.D. Thesis. University of Chicago.

Lazarsfeld, P. F.
1971 "Regression analysis with dichotomous attributes." Unpublished manuscript.

Rao, C. R.
1965 "Criteria of estimation in large samples." Pp. 345–362 in Contributions to Statistics. New York: Pergamon Press.

Stouffer, S. A., E. A. Suchman, L. C. Devinney, S. A. Star, and R. M. Williams, Jr.
1949 The American Soldier: Adjustment during Army Life. Studies in Social Psychology in World War II, Vol. 1. Princeton, N.J.: Princeton University Press.

Theil, H.
1970 "On the estimation of relationships involving qualitative variables." American Journal of Sociology 76:103–154.

Zeisel, H.
1968 Say It with Figures. Fifth Edition, Revised. New York: Harper & Row.

Leo A. Goodman

Chapter 2

An Illustrative Comparison of Goodman's Approach to Logit Analysis with Dummy Variable Regression Analysis

Jay Magidson

In regression analysis a designated dependent variable is predicted as a function of one or more explanatory variables. The dependent variable is typically assumed to be continuous, following a normal distribution around the regression line. Instead, suppose that the dependent variable is dichotomous (pass-fail, live-die, buy-no buy) or polytomous (buy, no buy, uncertain; favorable, neutral, unfavorable). In these cases the normality assumption inherent in classical linear regression analysis is not appropriate. The regression equation is not necessarily a linear function of the explanatory variables, nor is it necessarily additive (interaction terms may occur).

This chapter develops two different regression models in which the dependent variable and all the explanatory variables are dichotomous. (For some related results for polytomous variables see Goodman, 1971.) Our criteria for preferring a model are the following:

1. Its parameters should represent substantively meaningful quantities.
2. The parameters should be expressed in units that are readily interpretable.
3. Statistical tests associated with the estimates of its parameters should be available.
4. It should provide a parsimonious explanation of data.

For concreteness, we develop our formulations using data from *The American Soldier* by Stouffer et al. (1949) where we wish to predict camp preference (North or South) as a function of three dichotomous explanatory variables. For algebraic simplicity, we first formulate models that describe the relationships between camp preference and the two explanatory variables (B) region of origin (North or South) and (C) present location (North or South). Later we expand our models to include (A) race (black or white) as a third explanatory variable.

We begin by using criteria 1 and 2 to develop two models. The first model, the additive probability model, is identical to a dummy variable regression model; the second, a multiplicative odds model, is equivalent to the logit model described in the literature on log-linear models (Goodman, 1970; Bishop, Fienberg, and Holland, 1975) and recommended by Goodman (1972). We compare the parameters of these models at an elementary technical level.

Author's Note: An earlier version of this paper was presented at the 1978 annual meetings of the American Educational Research Association.

The Data

Table 1 presents the frequency distribution for 8036 soldiers on the dependent variable (*D*) camp preference and on the two explanatory variables (*B*) region of origin and (*C*) present location. These three dichotomies classify each soldier into one of eight groupings.

Since we wish to predict camp preference as a function of region of origin and present location, it is useful to express these eight frequencies in the form of a two-way table. The columns denote the two categories of the dependent variable, and the rows denote the four joint levels of the explanatory variables.

Table 1. The Observed Frequencies of American Soldiers Classified by *(B)* Region of Origin, *(C)* Present Location of Camp, and *(D)* Preference of Camp Location

j	k	l	*(B)* Region of Origin	(C) Present Location	*(D)* Camp Preference	Observed Frequencies f_{jkl}
1	1	1	North	North	North	1342
1	1	2	North	North	South	198
1	2	1	North	South	North	1750
1	2	2	North	South	South	760
2	1	1	South	North	North	487
2	1	2	South	North	South	446
2	2	1	South	South	North	472
2	2	2	South	South	South	2581

Note: Total observations: 8036; 50.4% born in North, 30.8% presently located in North, 50.4% prefer North.

Table 2 displays the desired array and presents calculated values for the observed conditional proportions and conditional odds associated, respectively, with a northern and a southern camp preference. The proportions (probabilities) in each row are calculated by dividing the observed frequency of soldiers preferring the North by the total number of soldiers in that row ($p_{11} = 1342/1540 \cong 0.871$). Odds are computed by dividing the observed frequency of soldiers preferring the North by the observed frequency of soldiers preferring the South

Table 2. Probabilities and Odds Associated with Camp Preference (North or South) for *American Soldier* Data

j	k	*(B)* Region of Origin	*(C)* Present Location	*(D)* Camp Preference: North ($\ell = 1$) f_{jk1}	North ($\ell = 1$) p_{jk}	North ($\ell = 1$) ω_{jk}	South ($\ell = 2$) f_{jk2}	South ($\ell = 2$) $1 - p_{jk}$	South ($\ell = 2$) $1/\omega_{jk}$	Total n_{jk}
1	1	North	North	1342	0.871	6.778	198	0.129	0.148	1540
1	2	North	South	1750	0.697	2.303	760	0.303	0.434	2510
2	1	South	North	487	0.522	1.092	446	0.478	0.916	933
2	2	South	South	472	0.155	0.183	2581	0.845	5.468	3053
		Total		4051	0.504	1.02	3985	0.496	0.984	8036

Notes: For Northern camp preference

P_{jk} (or P_{jk}^{BC}) $\equiv$ Prob (Preferring North $| B = B_j$ and $C = C_k) = f_{jk1}/n_{jk}$

ω_{jk} (or ω_{jk}^{BC}) $\equiv$ Odds (Preferring North$| B = B_j$ and $C = C_k) = f_{jk1}/f_{jk2}$

where $j, k = 1, 2$

Conversion formulas:

Odds = Prob/ (1 − Prob)

Prob = Odds/ (1 + Odds)

(ω_{11} = 1342/198 ≅ 6.78). Probabilities are converted to odds by dividing the corresponding probability by one minus the probability; and odds are converted to probabilities by dividing the corresponding odds by one plus the odds. These conversion formulas are presented in the lower right-hand corner of table 2.

While 87.1% of the northern-born soldiers presently located in the North state a preference for a northern camp, only 15.5% of the southern-born soldiers presently located in the South prefer a northern camp. Expressed in terms of odds, these figures are 6.78 (or 6.78:1) in favor of a northern camp preference for northern-born soldiers situated in the North, while only 0.18:1 for southern-born soldiers presently living in the South. When the second figure is inverted to convert to odds favoring a southern camp preference, the odds are 1:0.18 or 5.47:1 in favor of a southern camp preference for this latter group of soldiers.

Given the proportion preferring the North, we can find the proportion preferring the South. Similarly, if we know the odds in favor of a northern camp preference, we can calculate the odds in favor of a southern camp preference. *Thus the four probabilities or the four odds, presented in table 2 in the column pertaining to northern camp preference, capture all the relevant information in the table about the relationships between the dependent variable (the columns) and the explanatory variables (the rows).* Since our purpose is to formulate a model with parameters that describe the relationship between camp preference and the two explanatory variables (region of origin and present location), we can focus either on these four probabilities or these four odds.

Table 3 displays the four conditional probabilities and the four conditional odds in the form of a two-factor ANOVA with one observation per cell. We have two possibilities for the number in each cell, the odds or the probabilities. Since both numbers are easily interpreted, we will formulate two models, one based on probabilities, the other on odds.

Table 3 also presents unweighted averages for each of the two rows and each of the two columns. For the probabilities the averages are simple arithmetic means. For the odds we take simple geometric means. Probabilities are *additive* quantities. We speak of increasing the probability by 0.3 from 0.2 to 0.5 or from 0.5 to 0.8. Odds, on the other hand, are *multiplicative* quantities. We speak of doubling or tripling odds.

Probabilities and odds are different units for expressing the same quantity, just as Fahrenheit and Centigrade scales are different units for measuring temperatures. There are simple formulas for going back and forth from probabilities to odds, just as there are simple formulas for converting from Centigrade to Fahrenheit and from Fahrenheit to Centigrade. However, there is no simple formula for converting between additive *changes* in probabilities and multiplicative *changes* in odds.

For example, odds of 1:1 convert to a probability of 0.50. If the odds triple to 3:1, the probability increases to 0.75, an increase of 0.25. Here the odds increase 3-fold, and the probability increases 0.25. There is no way to get from the

Table 3. The Conditional Odds and Conditional Probabilities Associated with a Northern Camp Preference (Conditional Probabilities in Parentheses)

(B)		(C) Present Location		
		North	South	
Region of Origin	j \ k	1	2	Average*
North	1	6.778 (0.871)	2.303 (0.697)	3.95 (0.784)
South	2	1.092 (0.522)	0.183 (0.155)	0.45 (0.338)
Average*		2.72 (0.697)	0.65 (0.426)	1.33 (0.561)

*Calculation of "Unweighted" Averages:

Average $[\text{Prob}_1, \text{Prob}_2] = \frac{1}{2}[\text{Prob}_1 + \text{Prob}_2]$

Average $[\text{Odds}_1, \text{Odds}_2] = [\text{Odds}_1 \times \text{Odds}_2]^{1/2}$

Average $[\log(\text{Odds}_1), \log(\text{Odds}_2)] = \frac{1}{2}[\log(\text{Odds}_1) + \log(\text{Odds}_2)]$
$= \log[\text{Average}(\text{Odds}_1, \text{Odds}_2)]$

number 3 to the number 0.25 without using another piece of information such as the initial probability of 0.50. To see why this is the case, suppose that a probability increases from 0.01 to 0.26, again an increase in probability of 0.25. But here the odds increase from 1:99 (approximately 0.01) to 26:74 (approximately 0.35), a 35-fold increase in odds. Thus in one case the additive increase in probability of 0.25 corresponds to a 3-fold increase in odds, while in the other case it converts to a 35-fold increase in odds. The next section shows how this lack of comparability makes the additive probability model substantively different from the multiplicative odds model.

Thus far we have considered only criterion 2. Both probabilities and odds are reasonable candidates for the units of the parameters.

We now develop some substantively meaningful parameters for criterion 1.

Main Effects and Interaction Effects

To calculate some estimates of meaningful parameters for our models, we ask

1. What is the main effect of region of origin on camp preference? More specifically, what is the average increase in the odds (probability) of a northern preference associated with a northern versus a southern region of origin?
2. What is the main effect of present location on camp preference? More specifically, what is the average increase in the odds (probability) of a northern preference associated with a northern versus a southern present location?
3. Is there an interaction effect? In other words, does the effect of present location on camp preference depend on region of origin?

We wish to calculate a (partial) regression coefficient to answer each of these questions.

The main and interaction effects formulated in the analysis of variance (ANOVA) are regression coefficients (Cohen, 1968) and can be interpreted as partial measures of association, controlling for (holding constant) the other explanatory variables. We thus rely on the ANOVA effects to define our parameters. However, we shall illustrate the calculations somewhat differently than is generally done to emphasize the interpretation of the parameters as (average) changes in probabilities or changes in odds. (Readers familiar with the ANOVA formulation for the effects may wish to skim the remainder of this section.)

The calculations of these parameters are based on the proportions and odds reported in table 3 and summarized in tables 4, 5, and 6. We begin with the main effect for (*B*) region of origin. We estimate this main effect by computing the change in probabilities (odds) of preferring the north associated with a northern versus southern region of origin. This change is computed separately for soldiers presently located in the North and those presently located in the South. The main effect of region of origin on camp preference is the average of these two changes.

For soldiers located in the North (col. 1 of table 3), the proportion preferring the North is (0.871 – 0.522 = 0.349) greater for northern-born soldiers than for southern-born soldiers. Similarly, we calculate the corresponding change in proportions for soldiers presently located in the South. Here the proportion preferring the North (col. 2 of table 3) is (0.697 – 0.155 = 0.542) greater for northern-born soldiers than for southern-born soldiers. The average change in probability units is 0.446, the arithmetic mean of 0.349 and 0.542.

Alternatively, we can calculate the change in odds. For soldiers located in the North the odds in favor of the North are (6.78/1.09 = 6.21) times greater for northern-born soldiers than for southern-born soldiers. For soldiers located in the South the odds in favor of the North are (2.30/0.18 = 12.59) times greater for northern-born soldiers than for southern-born soldiers. The average change in odds is 8.84, the geometric mean of 6.21 and 12.59.

Table 4. Estimation of the Main Effect of *(B)* Region of Origin on *(D)* Camp Preference

Additive probability model

$$\tfrac{1}{2}\,[(0.871 - 0.522) + (0.697 - 0.155)]$$
$$= \tfrac{1}{2}\,[0.349 + 0.542]$$
$$= 0.446 > 0$$

(Divide by 2 for ANOVA formulation) **0.223**

Multiplicative odds model

$$\left[\left(\frac{6.778}{1.092}\right) \times \left(\frac{2.303}{0.183}\right)\right]^{1/2}$$
$$= [(6.207) \times (12.585)]^{1/2}$$
$$= 8.838 > 1$$

(Take square root for ANOVA formulation; Goodman 1970) **2.937**

Logit model

$$\tfrac{1}{2}\,[(\log 6.78 - \log 1.09) + (\log 2.30 - \log 0.18)]$$
$$= \tfrac{1}{2}[\log 6.21 + \log 12.59]$$
$$= \log 8.84 > 0$$

(Divide by 2 for ANOVA formulation **1.09**

Note: The last two models are different formulations of the same model and therefore represent equivalent models. The first model represents a substantively different model which may yield different conclusions about a data set.

The average change of 0.446 in probabilities is greater than 0, indicating that the main effect for region of origin on camp preference is positive for the additive probability model. The average change of 8.84 in odds is greater than 1, indicating that the main effect for region of origin on camp preference is also positive for the multiplicative odds models. Thus the two models agree that the main effect of *B* is positive in this case; that is, on the average, northern-born soldiers are more likely to prefer the North than southern-born soldiers in the same location. How much more likely? The answer is 0.446 more likely in terms of probabilities or 8.84 times more likely in terms of odds.

Goodman (1976) shows that these two models yield similar interpretations of the data when all the conditional probabilities are between 0.25 and 0.75. If some proportions are outside this range, the two models may yield different conclusions about any set of data. Examples in a later section illustrate how different interpretations can result for some data, depending on whether the additive probability model or the multiplicative odds model is used. Since these two models define effects differently, different interpretations of a given set of data are possible.

The calculations for the additive probability model and the multiplicative odds model are summarized in table 4. By taking the natural logarithm of the odds, we can convert the multiplicative odds model to an additive logit model. The conclusion that the odds model yields a positive main effect for region of origin can be obtained from the logit model by noting that the logarithm of 8.84 (2.18) is greater than 0.

The multiplicative odds model and the additive logit model are just different formulations of the same model. They are equivalent models; and they are different from the additive probability model.

Table 5. Estimation of the Main Effect of *(C)* Present Location on *(D)* Camp Preference

Additive probability model

$$\tfrac{1}{2}\,[(0.871 - 0.697) + (0.522 - 0.155)]$$
$$= \tfrac{1}{2}\,[0.174 + 0.367]$$
$$= 0.271 > 0$$

(Divide by 2 for ANOVA formulation) **0.135**

Multiplicative odds model

$$\left[\left(\frac{6.778}{2.303}\right)\times\left(\frac{1.092}{0.183}\right)\right]^{1/2}$$
$$= [(2.943) \times (5.967)]^{1/2}$$
$$= 4.191 > 1$$

(Take square root for ANOVA formulation) **2.047**

Logit model

$$\log 4.19 > 0$$

(Divide by 2 for ANOVA formulation) **0.72**

Table 5 presents calculations for the main effect for (*C*) present location on (*D*) camp preference. The additive probability model and the multiplicative odds model (or the equivalent additive logit model) agree that there is a positive main effect. Soldiers presently in the North are on the average 0.271 more likely in terms of probabilities and 4.19 times more likely in terms of odds to prefer the North than those soldiers presently located in the South who have the same region of origin.

Our third parameter is the interaction effect. For the effect of (*C*) present location, we computed increases in probability of 0.174 and 0.367 for the two levels of the other explanatory variable, region of origin. For the multiplicative odds model the corresponding increases in odds are 2.94 and 5.97. Table 6 calculates the interaction effects for the two models. Both models agree that the estimated interaction effect is negative. Based on different definitions for the effects, both models imply that the effect of present location on camp preference depends on region of origin, the main effect of present location being larger for southern-born soldiers than for northern-born soldiers.

Table 6. Estimation of the *BC* Interaction Effect on *(D)* Camp Preference

Additive probabality model

$$\tfrac{1}{2}\,[(0.871 - 0.522) - (0.697 - 0.155)]$$

$$= \tfrac{1}{2}\,[(0.871 - 0.697) - (0.522 - 0.155)]$$

$$= -\,0.097 < 0$$

(Divide by 2 for ANOVA formulation) **−0.048**

Multiplicative odds model

$$\left[\left(\frac{6.778}{1.092}\right)\times\left(\frac{2.303}{0.183}\right)\right]^{1/2}$$

$$= \left[\left(\frac{6.778}{2.303}\right)\times\left(\frac{1.092}{0.183}\right)\right]^{1/2}$$

$$= 0.702 < 1$$

(Take square root for ANOVA formulation) **0.838**

Logit model

$$\log 0.70 < 0$$

(Divide by 2 for ANOVA formulation) **−0.18**

Table 7. Verification that the ANOVA Effects Equal One Half of the Effects Defined Earlier for the Additive Probability Model

Estimated Main Effect of $B \equiv p_{1.} - p_{..}$

$= \frac{1}{2}(0.871 + 0.697) - \frac{1}{4}(0.871 + 0.697 + 0.522 + 0.155)$

$= \frac{1}{4}(0.871 + 0.697 - 0.522 - 0.155)$

$= \frac{1}{2}[(0.871 - 0.522) + (0.697 - 0.155)]/2$

Estimated Main Effect of $C \equiv p_{.1} - p_{..}$

$= \frac{1}{2}(0.871 + 0.522) - \frac{1}{4}(0.871 + 0.697 + 0.522 + 0.155)$

$= \frac{1}{4}(0.871 + 0.522 - 0.697 - 0.155)$

$= \frac{1}{2}[(0.871 - 0.697) + (0.522 - 0.155)]/2$

Estimated BC Interaction Effect $\equiv p_{11} - p_{1.} - p_{.1} + p_{..}$

$= 0.871 - \frac{1}{2}(0.871 + 0.697) - \frac{1}{2}(0.871 + 0.522)$

$+ \frac{1}{4}(0.871 + 0.697 + 0.522 + 0.155)$

$= \frac{1}{4}(0.871 - 0.522 - 0.697 + 0.155)$

$= \frac{1}{2}[(0.871 - 0.522) - (0.697 - 0.155)]/2$

$= \frac{1}{2}[(0.871 - 0.697) - (0.522 - 0.155)]/2$

The Saturated Model

One of our major considerations in formulating a model to predict camp preference as a function of region of origin and present location is that the parameters represent substantively meaningful quantities. We have just calculated two sets of substantively meaningful quantities, one based on probabilities, the other based on odds. We now formulate the additive probability model and the multiplicative odds model.

The ANOVA formulation (Scheffé, 1959) for main effects and interaction effects yields estimates in the saturated model that equal our estimates divided by two. It is convenient to divide our estimates of substantively meaningful quantities by two because we can then use the symmetric ANOVA representation to express our models. In this section we first show that the ANOVA effects are in fact equal to our effects divided by two. We then formulate the saturated models using the ANOVA formulation.

In table 3 we displayed the proportions (and odds) in the form of a two-way factor layout with one observation per cell. We now introduce the following notation for the unweighted averages calculated in that table.

$$p_{1.} \equiv \tfrac{1}{2}(p_{11} + p_{12}) = 0.784$$

$$p_{.1} \equiv \tfrac{1}{2}(p_{11} + p_{21}) = 0.697$$

Let $p_{..}$ denote the grand mean of the conditional proportions.

$$p_{..} \equiv \tfrac{1}{4}(p_{11} + p_{12} + p_{21} + p_{22}) = 0.561$$

The estimates for the ANOVA effects are given in table 7. Note that these estimates equal the estimates for the additive probability model divided by two. As a multiplicative analog for the ANOVA calculations, we take the square root of our estimated effects in the multiplicative model (or equivalently divide the estimates from the logit model by two) for our corresponding multiplicative ANOVA formulation.

In table 3 four numbers conveyed all the information about the association between the dependent variable camp preference and the explanatory variables region of origin and present location. These four conditional probabilities can alternatively be expressed as conditional odds. Using the information provided by the conditional probabilities (conditional odds), we calculated two sets of three parameters that were more substantively meaningful for our purposes than the raw probabilities or raw odds. We will now show that the four probabilities can be perfectly reproduced from the three effect parameters and a constant in the additive probability model. Similarly, the four odds in table 3 can be perfectly reproduced from the three effect parameters and a constant in the multiplicative odds model (or the additive logit model). This means that the estimated effect parameters include all the information conveyed by the observed proportions and odds. Hence the effects are alternative ways of representing the observed relationships among the variables.

Table 8 shows how (1) the four probabilities are exactly reproduced in terms of the parameters of the additive probability model; (2) the four conditional odds are exactly reproduced in terms of the parameters from the multiplicative odds models; (3) the four logits are exactly reproduced in terms of the parameters from the additive logit model.

Let us first consider the additive probability model. The first row of numbers corresponds to the first conditional probability in table 3 (0.871). To reproduce this number we simply sum the three effect parameters from the additive probability model and add the constant 0.561, which is the overall average of the probabilities ($p_{..}$). (Any differences here and in similar calculations are due solely to rounding error.) That the four estimates denoted $\hat{a}$, $\hat{b}$, $\hat{c}$, and $\hat{d}$ are summed is designated by +1 in columns a through d. The second probability is reproduced exactly by summing the first two estimates $\hat{a}$ and $\hat{b}$ and subtracting $\hat{c}$ and $\hat{d}$. This is designated by +1 in columns a and b and by −1 in columns

Table 8. Verification that the Effects for the Saturated Regression-Type Models Perfectly Reproduce the Observed Conditional Probabilities and Conditional Odds

					Indicator Variables			
B	*C*	Prob	Odds	Logit*	(a) Constant	(b) Main Effect for *B*	(c) Main Effect for *C*	(d) *BC* Interaction Effect
North	North	0.871	6.778	1.91	+1	+1	+1	+1
North	South	0.697	2.303	0.83	+1	+1	–1	–1
South	North	0.522	1.092	0.04	+1	–1	+1	–1
South	South	0.155	0.183	–0.85	+1	–1	–1	+1
		Prob			0.561	0.223	0.135	–0.048
		Odds			1.329	2.973	2.047	0.838 (gammas)
		Logit*			0.28	1.09	0.72	–0.18 (betas)

* The logit is the natural logarithm of the odds. Thus, Odds ≡ e^{Logit}

c and d. Similarly, the third probability is reproduced exactly by adding $\hat{a}$ and $\hat{d}$ and subtracting $\hat{b}$ and $\hat{c}$. The final sample probability is exactly reproduced by adding a and d and subtracting b and c.

These relationships can be summarized using indicator variables.

$$p_{jk} = \hat{a} + \hat{b}\,X^B + \hat{c}X^C + \hat{d}X^{BC}, \tag{1}$$

where

$$X^B = \begin{cases} +1 & \text{if region of origin is North} \\ -1 & \text{otherwise} \end{cases}$$

$$X^C = \begin{cases} +1 & \text{if present location is North} \\ -1 & \text{otherwise} \end{cases}$$

$$X^{BC} = X^B X^C = \begin{cases} +1 & \text{if region of origin and present location are the same (both North or both South)} \\ -1 & \text{otherwise} \end{cases}$$

For example, for northern-born soldiers located in the North, all three indicator variables X^B, X^C, and X^{BC} equal +1 so that p_{11} equals the sum of all four parameters.

Now let us formulate the multiplicative odds model. Denoting the parameters by γ's (gammas), we have

$$\omega_{jk} = \hat{\gamma}\,(\hat{\gamma}^B)^{X^B}\,(\hat{\gamma}^C)^{X^C}\,(\hat{\gamma}^{BC})^{X^{BC}} \tag{2}$$

For northern-born soldiers located in the North, all three indicator variables equal +1 so that the corresponding conditional odds of preferring the North are exactly reproduced by the product of the four estimates $\hat{\gamma}$, $\hat{\gamma}^B$, $\hat{\gamma}^C$, and $\hat{\gamma}^{BC}$. This equality can be verified from table 8. Similarly, for northern-born soldiers located in the South, X^B equals +1 while X^C and X^{BC} each equal −1. The observed odds of preferring the North for northern-born soldiers located in the south (ω_{12}) are therefore reproduced by dividing the product of $\hat{\gamma}$ and $\hat{\gamma}^B$ by the product of $\hat{\gamma}^C$ and $\hat{\gamma}^{BC}$.

Alternatively, we can convert model (2) to the additive logit formulation by taking natural logarithms.

$$\begin{aligned} \log \omega_{jk} &= (\log \hat{\gamma}) + (\log \hat{\gamma}^B)X^B + (\log \hat{\gamma}^C)X^C + (\log \hat{\gamma}^{BC})X^{BC} \\ &= \hat{\beta} + \hat{\beta}^B X^B + \hat{\beta}^C X^C + \hat{\beta}^{BC} X^{BC} \end{aligned} \tag{3}$$

The parameters are now denoted by β's (betas), where the β's are the natural logarithms of the corresponding γ's. The logarithm of the odds (the logits) are also given in table 8 together with the β estimates. From table 8 we can also verify that model (3) is correct.

All three models are saturated in parameters, so the equalities *always* hold. They are tautologies. We now show how to express these models as saturated regression models where the predicted variable is camp preference (North or South).

The Saturated Regression Models

Regression models predict the value of the dependent variable to be its conditional expectation expressed as a function of the explanatory variables. The additive probability model can be expressed as a regression model since the population conditional probabilities (the *P*'s) are the conditional expectations of the indicator variable Z^D, coded 1 for soldiers with a northern preference and 0 for soldiers with a southern preference. The additive saturated regression model for the probabilities is thus

$$Z^D = a + bX^B + cX^C + dX^{BC} + \epsilon, \tag{4}$$

where ϵ is the error term representing the fact that camp preference (North or South) cannot be perfectly predicted solely on the basis of region of origin and present camp location.

This model is a saturated regression model for the prediction of *D* from *B* and *C* because all information about the variables is included in the model. We cannot introduce any more explanatory variables associated with *B* and *C* because any other variable associated with *B* and *C* can be expressed as a linear function of X^B, X^C, and X^{BC} and can thus be taken into account (absorbed) by these indicator variables in the additive model. Therefore no new parameters (regression coefficients) can be added to the saturated regression model. It is already saturated in parameters.

The predictions for the dependent variable Z^D are conditional expectations or probabilities. An infinite number of saturated regression models could be formulated to predict the population probabilities as a function of *B* and *C*. Let *P* denote the conditional probabilities in the population expressed as a function of *B* and *C*.

$$P = f(B, C) \tag{5}$$

where *f* denotes the explicit functional form for some saturated regression model.

For the (saturated) additive probability model (4) we have

$$f(B, C) = a + bX^B + cX^C + dX^{BC}. \tag{6}$$

The estimates for these parameters given earlier for this saturated model are maximum likelihood estimates. The resulting estimates for the unknown population probabilities are the observed sample proportions.

The second saturated model that we considered is the multiplicative odds model or the equivalent additive logit model. Denoting the population odds by Ω, we noted that the relationships between the probability P and the odds can be expressed as

$$\Omega = P/(1 - P) \text{ and } P = \Omega/(1 + \Omega); \tag{7}$$

and we can write the logarithm of the odds (the logit) as $\log \Omega$. The logit regression model is given by

$$\log \Omega = g(B, C), \tag{8}$$

where

$$g(B, C) = \beta + \beta^B X^B + \beta^C X^C + \beta^{BC} X^{BC}. \tag{9}$$

This model is equivalent to the following multiplicative model:

$$\Omega = e^{g(B, C)} \tag{10}$$

$$= \gamma (\gamma^B)^{X^B} (\gamma^C)^{X^C} (\gamma^{BC})^{X^{BC}}, \tag{11}$$

where the γ's and β's are related as in formula (3). Model (10)–(11), or the equivalent (8)–(9), states that the conditional probability P can be written as

$$P = e^{g(B, C)}/(1 + e^{g(B, C)}), \tag{12}$$

where $g(B, C)$ is defined by (9). Model (12) for the conditional probability P is equivalent to model (10)–(11) and to model (8)–(9), and it is different from model (5)–(6).

The estimates for the parameters of the saturated multiplicative odds model or the equivalent additive logit model are maximum likelihood estimates. The resulting estimates for the (unknown) population odds of preferring the North are the observed sample odds, and therefore the estimated population probabilities in this model are the observed sample probabilities.

The estimates for the main and interaction effects from the additive probability saturated model might be quite different from the corresponding estimates from the multiplicative saturated model for the odds, but these two saturated models both estimate the population probabilities to be the sample proportions. The parameters in both models represent substantively meaningful quantities and are expressed in units that are easy to interpret (probabilities and odds).

We next consider criterion 3. We are particularly interested in testing whether the parameters satisfy certain restrictions corresponding to (1) no interaction effect, (2) no row effects, (3) no column effects, and (4) no row or column effects, which are of interest in the analysis of variance. We wish to test the goodness of fit for unsaturated or restricted regression models that omit some parameters (set them to 0 or 1).

With the additive probability model, both ordinary least squares and weighted least squares yield the maximum likelihood estimates for the saturated model; but the estimates for unsaturated (restricted) models obtained using either procedure are not maximum likelihood estimates and they are not efficient estimates. The usual *t*, *F*, and χ^2 statistics are not justified in this case (see Goodman, 1976).

The next section describes some statistical tests associated with the parameters in the saturated and various unsaturated multiplicative odds models. For this discussion we follow Goodman's (1970) hierarchical model notation. We then expand the additive probability model and the multiplicative odds model to include race as a third explanatory variable. We consider criterion 4 as we compare results given by these two models for some unsaturated formulations, including those estimated by Goodman (1972) using the logit model.

Statistical Tests for Hierarchical Logit Models

In the usual two-way analysis of variance, we might be interested in the saturated model (which includes row effects, column effects, and row × column interaction effects) and four different unsaturated models. Table 9 lists the corresponding logit models for predicting variable *D* using the predictor variables *B* and *C*.

A variable *B* effect corresponds to a row effect, a variable *C* effect corresponds to a column effect, and a *BC* interaction effect corresponds to a row X column interaction. For each model in table 9 the maximum likelihood estimate of the expected frequencies can be obtained using available computer programs, such as Goodman's ECTA program (Everyman's Contingency Table Analyzer). For the saturated model the estimated probabilities (and the corresponding estimated expected frequencies) are equal to the corresponding observed quantities, so this model can be conveniently described by noting that the estimated expected frequencies "fit" the observed three-way table *(BCD).* Similarly, each of the models can be conveniently described in terms of the marginal tables that are fitted by the maximum likelihood estimates obtained from the computer program. The computer program also provides other relevant information for each model tested: the chi-square goodness-of-fit statistic comparing the observed and estimated expected frequencies, a corresponding chi square based on the likelihood-ratio statistic, the number of degrees of freedom associated with the chi-square statistic (the number of parameters omitted from the model), and the maximum likelihood estimate of the parameters in the model.

Table 9. Hierarchical Logit Models for Predicting (D) Location Preference from (B) Region of Origin and (C)) Present Location

Model	Description	Effects Included in the Model	Effects Omitted from Model	Degrees of Freedom for Testing Fit of Model	Fitted Marginal Tables
L_0	Saturated model	B, C, and BC interaction	None	0	$\{BCD\}$
L_1	Main effects only	B and C	BC interaction	1	$\{BC\}$, $\{BD\}$, $\{CD\}$
L_2	Main effect for B only $[D \otimes C\ B]$ *	B	C and BC interaction	2	$\{BC\}$, $\{BD\}$
L_3	Main effect for C only $[D \otimes B\ C]$ **	C	B and BC interaction	2	$\{BC\}$, $\{CD\}$
L_4	No effects $[D \otimes BC]$ ***	None	All 3 effects	3	$\{BC\}$, $\{D\}$

* Read "D is independent of C given B."

** Read "D is independent of B given C."

*** Read "D is independent of the joint variable BC." [This model is equivalent to total independence in the 4×2 table (recall table 2) and implies that D is independent of both B and C.]

The unsaturated models are labeled L_1, L_2, L_3, and L_4 in table 9. Model L_1, the main-effects-only model, specifies no interaction effect of region of origin and present location on camp preference. That is, it hypothesizes that the ratio of odds of preferring the North for soldiers located in the North to the corresponding odds for soldiers located in the South is the same for northern-born soldiers as it is for southern-born soldiers. Our calculations in table 4 for the saturated model showed that these odds ratios are 6.207 for northern-born soldiers and 12.585 for southern-born soldiers.

The statistical test for the equality of the two odds ratios is equivalent to the test of goodness of fit for the expected frequencies estimated from model L_1. This test has one degree of freedom, the number of effects omitted from the saturated model. The precise details of the statistical tests are given in section 4 of Goodman (1972). We leave it as an exercise to show that model L_1 is rejected in favor of model L_0 at the 0.05 level of significance. Models L_2, L_3, and L_4 also hypothesize no interaction effect and are also rejected in favor of the saturated model.

Some Logit Models with Three Explanatory Variables

For illustrative purposes we have limited the discussion so far to two explanatory variables. We now expand the model to include (A) race (black or white) as a third explanatory variable. Table 10 summarizes the new $2^3 \times 2$ cross-classification, dividing the soldiers into eight groups.

The estimated effects for our new saturated additive probability model are obtained by averaging *four* differences for each effect (instead of two differences, as in tables 4, 5, and 6). Similarly, the estimated effects for the saturated multiplicative odds model generalize by averaging four ratios (instead of two). For example, the main effect of B in the saturated multiplicative odds model becomes 11.8, the average conditional odds ratio in favor of preferring a northern camp for northern- versus southern-born soldiers.

$$\left[\left(\frac{10.750}{1.419}\right)\times\left(\frac{5.895}{0.591}\right)\times\left(\frac{3.504}{0.223}\right)\times\left(\frac{1.714}{0.105}\right)\right]^{1/4} \cong 11.8 \tag{13}$$

As in tables 4, 5, and 6, we take the square root of 11.8 to convert to the corresponding estimate of 3.4 for the ANOVA formulation. (See also equation 14 in Goodman, 1972.) Calculation of the other parameter estimates in the saturated model is left as an exercise.

Models (1) and (2) are easily generalized to three explanatory variables by defining X^A, X^{AB}, X^{AC}, and X^{ABC} as X^B, X^C and X^{BC} were defined. These variables are displayed in table 11.

The reader should verify the following relationships:

$$X^{ABC} = X^A X^{BC} = X^B X^{AC} = X^C X^{AB} = X^A X^B X^C. \tag{14}$$

Table 10. Cross-Classification of Soldiers with respect to Four Dichotomized Variables: *(A)* Race, *(B)* Region of Origin, *(C)* Location of Present Camp, and *(D)* Preference as to Camp Location

(B) Region of Origin	*(C)* Location of Present Camp	*(A)* Race	Camp Preference North Freq.	North Prob.	North Odds	South Freq.	South Prob.	South Odds	Total
North	North	Black	387	0.915	10.750	36	0.085	0.093	423
North	North	White	955	0.855	5.895	162	0.145	0.170	1117
North	South	Black	876	0.778	3.504	250	0.222	0.285	1126
North	South	White	874	0.632	1.714	510	0.368	0.584	1384
South	North	Black	383	0.587	1.419	270	0.413	0.705	653
South	North	White	104	0.371	0.591	176	0.629	1.692	280
South	South	Black	381	0.182	0.223	1712	0.818	4.493	2093
South	South	White	91	0.095	0.105	869	0.905	9.549	960
			4051	0.504	1.017	3985	0.496	0.984	8036

Table 11. The Variables for the Saturated Logit Model

i	*j*	*k*	*Constant*	X^A	X^B	X^C	X^{AB}	X^{AC}	X^{BC}	X^{ABC}
1	1	1	+1	+1	+1	+1	+1	+1	+1	+1
1	1	2	+1	+1	+1	−1	+1	−1	−1	−1
1	2	1	+1	+1	−1	+1	−1	+1	−1	−1
1	2	2	+1	+1	−1	−1	−1	−1	+1	+1
2	1	1	+1	−1	+1	+1	−1	−1	+1	−1
2	1	2	+1	−1	+1	−1	−1	+1	−1	+1
2	2	1	+1	−1	−1	+1	+1	−1	−1	+1
2	2	2	+1	−1	−1	−1	+1	+1	+1	−1

Table 12. The Results for Some Logit Models for the Prediction of *(D)* Location Preference

Model	Effects Included in the Model	Degrees of Freedom	Likelihood Ratio Chi-Square	Goodness of Fit Chi-Square
H_0	ALL (A, B, C, AB, AC, BC, ABC)	0	0	0
H_9	A, B, C, AB, BC	2	0.68	0.69
H_8	A, B, C, AC, BC	2	1.32	1.34
H_1	A, B, C, BC	3	1.45	1.46
H_{10}	A, B, C, AB, AC	2	17.29	18.73
H_2	A, B, C	4	24.96	25.73
H_3	B, C, BC	4	152.65	147.59
H_4	B, C	5	186.36	180.26
H_6	A, B	5	695.01	727.16
H_5	A, C	5	2286.83	2187.71
H_7	None	7	3111.47	2812.64

Table 12 summarizes the results for some logit models for the prediction of *(D)* location preference. Model H_1 in table 12 is the most parsimonious model consistent with these data (Goodman, 1972). It hypothesizes only a single interaction term, the *BC* term. Letting Ω denote expected odds, we can express model H_1 as

$$\Omega = (\gamma^A)^{X^A}(\gamma^B)^{X^B}(\gamma^C)^{X^C}(\gamma^{BC})^{X^{BC}} \tag{15}$$

Since there are no interaction terms associated with (*A*) race in model H_1, this model hypothesizes that the black/white odds ratio for preferring the North is constant over the four joint categories of region of origin/present location. Thus the model states that black soldiers are about 2.1 times more likely to prefer the North than white soldiers having the same region of origin and the same present location. The constant 2.1 is the maximum likelihood estimate of the effect of race $(\gamma^A)^2$ under this model (see table 13).

From (15) we have

$$\frac{(\Omega \mid X^A = 1, B = B_j, C = C_k)}{(\Omega \mid X^A = -1, B = B_j, C = C_k)} = \frac{\gamma^A}{(\gamma^A)^{-1}} = (\gamma^A)^2 \tag{16}$$

$$\text{for } j = 1, 2, \ k = 1, 2.$$

Table 13. Maximum Likelihood Estimates for the Parameters in Models H_0 and H_1

	Saturated Model H_0		Model H_1	
ANOVA Parameters	γ	γ^2	γ	γ^2
γ	1.28	1.6	1.31	1.7
γ^A	1.44	2.1	1.45	2.1
γ^B	3.43	11.8	3.45	11.9
γ^C	2.10	4.4	2.14	4.6
γ^{AB}	0.96	0.9	1.00†	1.0†
γ^{AC}	1.00	1.0	1.00†	1.0†
γ^{BC}	0.86	0.7	0.86	0.7
γ^{ABC}	0.97	0.8	1.00†	1.0†

† Restricted to be zero effect ($\gamma = 1$).

Table 13 compares all the estimates of effects in the saturated model H_0 with those of model H_1. (These estimates are also reported in tables 3 and 4 of Goodman (1972), and their meaning is discussed. A complete discussion of statistical tests is also presented in that article.)

A Comparison of Results from the Logit Model and the Additive Probability Model

The results for the logit models presented in table 12 show that the most parsimonious model consistent with the data is model H_1. This model contains four terms, one term associated with each of the three explanatory variables *A, B,* and *C* and a single interaction term associated with *B* and *C.* The best-fitting *three-term* model is the main-effects-only model (H_2), but the chi-square value for this model is significantly large at the 0.01 level, so this model is rejected in favor of model H_1.

To compare the logit models with the additive probability models for these data, the saturated additive probability model was formulated and one term at a time was eliminated until only four terms remained. At each step the regression coefficient with the smallest magnitude was set to 0. The resulting model included the three terms associated with the main variables *A, B,* and *C* together with the *BC* interaction term (model H_1). This was true for both the ordinary least squares (OLS) and weighted least squares (WLS) method of estimation. Thus both the logit model and the additive probability model selected H_1 as the best model containing four terms.

At the next step the *BC* interaction term was deleted, yielding the main-effects-only model H_2 as the best three-variable model. Again this was true using both OLS and WLS. Thus the additive probability model also agrees with the logit model that H_2 is the best model containing three terms.

Table 14 compares the logit model with the additive probability model with respect to (a) the saturated model H_0, (b) model H_1, and (c) the main-effects-only model H_2. The estimates are all identical in sign and similar in relative magnitude. When comparing the absolute magnitudes of estimates from the logit model with those of the additive probability model, note that they are in different metrics and that estimates from the additive probability models are bounded between –0.5 and +0.5. Also notice that the OLS and WLS estimates are almost identical for these models. (They are always exactly equal for any saturated model.)

The major difference between these models is how well they fit the data. The logit models fit these data much better than do the corresponding additive probability models, as indicated by the chi-square values in table 15. While the logit model H_1 fits these data exceptionally well, the corresponding additive probability models $H_1{}^*$ and $H_1{}^{**}$ do not, as indicated by the large chi-square values. A similar comparison of model H_2 shows the same pattern; the logit model again fits the data much better.

Table 14. Comparison of Estimates under the Logit Models H_0, H_1, and H_2, with the Corresponding Estimates from Additive Probability Models based on OLS (H_0^*, H_1^*, H_2^*) and WLS (H_0^*, H_1^{**}, H_2^{**}) Estimation

	Saturated Model							
Effect	Logit (H_0)	Additive Probability (H_0^*)	H_1	H_1^*	H_1^{**}	H_2	H_2^*	H_2^{**}
B	1.23	0.24	1.24	0.24	0.24	1.31	0.26	0.26
C	0.74	0.13	0.76	0.14	0.14	0.78	0.13	0.12
A	0.37	0.06	0.37	0.06	0.05	0.38	0.06	0.06
BC	–0.15	–0.04	–0.15	–0.04	–0.04	0†	0†	0†
AB	–0.04	–0.01	0†	0†	0†	0†	0†	0†
AC	0.00	0.01	0†	0†	0†	0†	0†	0†
ABC	–0.03	–0.03	0†	0†	0†	0†	0†	0†
χ^2_G	0	0	1.46	37	25	26	–	129
Degrees of freedom	0	0	3	3	3	4	4	4

† Restricted to zero.

Table 15. The Probability of Preferring a Northern Camp and the Associated Residuals (+ or –) Estimated under Models H_0, H_1, H_2, H_1^*, H_1^{**}, H_2^*, and H_2^{**}

B	C	A	H_0 (Observed)	H_1	H_1^*	H_1^{**}	H_2	H_2^*	H_2^{**}
1	1	1	.915	.923 (–)	.956 (–)	.941 (–)	.940 (–)	1.007 (–)	.974 (–)
1	1	2	.855	.851 (+)	.839 (+)	.839 (+)	.880 (–)	.882 (–)	.864 (–)
1	2	1	.778	.781 (–)	.762 (+)	.757 (+)	.770 (+)	.738 (+)	.729 (+)
1	2	2	.632	.629 (+)	.645 (–)	.655 (–)	.610 (+)	.613 (+)	.619 (+)
2	1	1	.587	.577 (+)	.557 (+)	.552 (+)	.536 (+)	.485 (+)	.453 (+)
2	1	2	.371	.394 (–)	.440 (–)	.450 (–)	.350 (+)	.360 (+)	.342 (+)
2	2	1	.182	.182 (+)	.191 (–)	.188 (–)	.197 (–)	.216 (–)	.208 (–)
2	2	2	.095	.096 (–)	.074 (+)	.087 (+)	.103 (–)	.092 (+)	.098 (–)
χ^2_G			0	1.46	37	25	26	–	129
Degrees of freedom			0	3	3	3	4	4	4

Table 15 shows how these models translate into estimated probabilities of preferring a northern camp. The estimates based on the logit model are generally closer to the observed proportions than the corresponding estimates from the additive probability models. Notice also that the pattern of the residuals for main-effects-only models H_2, $H_2{}^*$ and $H_2{}^{**}$ indicate that the *BC* interaction term should be added to the model. Finally, note that one of the proportions is estimated to be greater than 1.0 under $H_2{}^*$. Although none of the proportions was estimated by WLS to be outside the 0-1 range, this can occur with some data.

Although we used the chi-square value as a measure of goodness of fit for the additive probability models based on OLS and WLS estimation, this use is not strictly appropriate because the OLS and WLS estimates are not maximum likelihood estimates and they are not efficient estimates.

Some Examples of Extreme Cases

In this example the additive probability model and the multiplicative odds model give similar interpretations for the data. The models agree that there are positive main effects and a negative *BC* interaction effect. Now let us briefly consider some hypothetical examples showing that the additive probability model and the multiplicative odds model can yield different conclusions for some data.

Goodman (1976) showed that when all the observed proportions are between 0.25 and 0.75, the additive probability model and the multiplicative odds model yield essentially the same conclusions. If some proportions are very small or very large, however, the two models can yield quite different conclusions. To see why this happens, let us consider the four hypothetical examples in table 16, which are designed to show how *interaction* effects can be quite different under the two models.

In example 1 of table 16 the four proportions in the table are all between 0.20 and 0.51. Both the probability model and the odds model for this example agree that there is a very small positive effect associated with *B*, a moderate sized positive effect associated with *C*, and a very small interaction effect which is absolute zero for the additive probability model.

In example 2 the additive probability model indicates no interaction effect. The increase in probability from 0.50 to 0.51 is equal to the increase in probability from 0.001 to 0.011. The odds model, however, yields a large negative interaction effect. In terms of odds the increase in probability from 0.001 to 0.011 is much greater than the probability increase from 0.50 to 0.51. The odds increase 11-fold in the former case while only slightly (from 1 to 1.04) in the latter case.

Example 3 shows the reverse situation where the odds model indicates no interaction while the probability model detects a large positive interaction effect. The increase in probability from 0.5 to 0.917 is an 11-fold increase in odds

Table 16. Comparison of Effects Based on Probabilities with Effects Based on Odds for Four Hypothetical Examples

Example 1

	C_1	C_2		Effects: B	C	BC
B_1	1.04 (0.51)	0.27 (0.21)	Prob	Very small		
			Model	+	+	0
B_2	1.00 (0.50)	0.25 (0.20)	Odds	Very small		Very small
			Model	+	+	–

Example 2

	C_1	C_2		Effects: B	C	BC
B_1	1.04 (0.51)	0.011 (0.011)	Prob	Very small	Large	
			Model	+	+	0
B_2	1.00 (0.50)	0.001 (0.001)	Odds	Large	Very large	Large
			Model	+	+	–

Example 3

	C_1	C_2		Effects: B	C	BC
B_1	11.0 (0.917)	0.011 (0.011)	Prob		Very large	Large
			Model	+	+	+
B_2	1.0 (0.500)	0.001 (0.001)	Odds	Large	Extremely large	
			Model	+	+	0

Example 4

	C_1	C_2
B_1	9.0 (0.9)	X
B_2	0.25 (0.2)	4.0 (0.8)

For every value of X the additive probability model shows a positive interaction effect. For $X = 144$ (odds) or 0.993 (probability) the odds model shows no interaction effect.

(from 1:1 to 11:1). For the probability model this increase is much greater than the 0.01 increase in probability from 0.001 to 0.011. For the odds model the increases have the same magnitude.

Finally, example 4 raises a possible problem with the formulation of the additive probability model. For any increase in probability from B_2 to B_1 in column C_2, the probability model detects a positive interaction effect. In the C_1 category the increase from 0.2 to 0.9 is substantial and cannot be matched by a corresponding increase at the C_2 level because the initial B_2 probability is already at 0.8. On the other hand, in the odds model an increase in probability from 0.8 to 0.993 results in no interaction effect. This final example reminds us that the estimated probabilities can fall outside the 0-1 range for the additive probability model under the traditional least squares methods of estimation.

Summary

We have considered two different models for the prediction of a dichotomous variable based on two or three dichotomous explanatory variables. The parameters of both models represent substantively meaningful quantities. The first model, an additive probability model, can be formulated as a dummy variable regression model. The second, a multiplicative odds model, is equivalent to a logit model. The models can yield different substantive conclusions about some data sets. Maximum likelihood techniques have been developed for estimating and testing various logit models, and a unified approach for estimating and testing such models is given by Goodman (1970, 1972). Using data from *The American Soldier* by Samuel Stouffer et al., we compared results obtained for the logit model with corresponding results obtained for the additive probability models and found that logit models fit the data substantially better than the corresponding additive probability models.

References

Bishop, Y.M.M. 1969. Full contingency tables, logits and split contingency tables. *Biometrics* 25:383-400.

Bishop, Y.M.M.; Fienberg, S.E.; and Holland, P.W. 1975. *Discrete multivariate analysis:Theory and practice.* Cambridge, Mass.: MIT Press.

Cohen J. 1968. Multiple regression as a general data-analytic system. *Psychological Bulletin* 70:426-443.

Goodman, L.A. 1970. The multivariate analysis of qualitative data: Interactions among multiple classifications. *Journal of the American Statistical Association* 65:225-56.

. 1971. The analysis of multidimensional contingency tables: Stepwise procedures and direct estimation methods for building models for multiple classifications. *Technometrics* 13:33-61.

. 1972. A modified multiple regression approach to the analysis of dichotomous variables. *American Sociological Review* 37:28-46.

. 1975. The relationship between modified and usual multiple-regression approaches to the analysis of dichotomous variables. In *Sociological Methodology 1976.* Edited by David R. Heise, pp. 83-110. San Francisco: Jossey-Bass, Inc.

Scheffé, H. 1959. *The analysis of variance.* New York: John Wiley and Sons.

Stouffer, S.A., et al. 1949. *The American Soldier: Adjustment during army life.* Studies in social psychology in World War II, vol. 1. Princeton, N.J.: Princeton University Press.

Part Two
The General Log-Linear Model

The maximum likelihood methods described in part I for the logit model form the basis of a unified system for analyzing qualitative data. Part II shows how the logit model is formulated for two or more dependent variables and how all logit models are subsumed under a more general log-linear model. The general log-linear formulation incorporates all "interactions" in the multidimensional contingency table without the need to designate some of the variables as dependent on the others.

Applications to five sets of data are presented in chapters 3-6. Chapter 3 shows how recursive and nonrecursive path diagrams and more general causal systems can be formulated and tested using the general log-linear model. The approach is illustrated in a reanalysis of data from the Wilner, Walkley, and Cook (1955) study of the contact hypothesis, which states that the close proximity of whites and blacks in integrated housing increases favorable interracial sentiments through the intervening variables of contact and local norms.

Chapter 4 further develops the log-linear model and describes the general approach to estimation and hypothesis testing for two, three, four, and five variables. It also reanalyzes data pertaining to five dichotomized variables that classify individuals according to whether they (1) read newspapers, (2) listen to radio, (3) read books and magazines, (4) attend lectures, and (5) have a good or poor knowledge of cancer. The chapter models that fit the data better than those obtained in earlier published analyses of the same data.

In chapter 5 Goodman proposes and illustrates stepwise procedures and direct estimation methods for building models for multiple classifications. These procedures are somewhat analogous to the classical stepwise regression methods for adding terms (forward selection) and deleting terms (backwards elimination) for quantitative models. The methods are illustrated with a classification of 1008 consumers according to their preference for laundry detergents, previous use of brand M, water temperature, and water softness.

The methods are further illustrated in chapter 6, in which three sets of data are reanalyzed and many new insights result. The first two data sets are (a) Lazarsfeld's famous two-attribute turnover table (the 16-fold table) on voting intention and (b) a two-attribute turnover table on student attitudes analyzed earlier by Coleman. Models that fit these data are somewhat similar to the models developed in chapter 3 for the Wilner, Walkley, and Cook (1955) data on the contact hypothesis. The data in chapter 3 are analyzed further in chapter 6. Latent variables are introduced into one of the analyses, and the topic of latent variables and latent-structure analysis is further explored in part IV.

The notation for the log-linear model in part II is slightly more general than that for the logit model in part I. We now provide the algebraic relationship between the logit model and the log-linear model for the simple case of a dichotomous dependent variable D and two dichotomous explanatory variables B and C. If we express the expected odds for the variable D as the ratio of expected frequencies, the multiplicative version for the logit model is

$$\frac{F_{jk1}}{F_{jk2}} = \gamma^{D}\gamma_{j}^{BD}\gamma_{k}^{CD}\gamma_{jk}^{BCD}, \text{ for } j, k = 1, 2. \tag{1}$$

The multiplicative version for the log-linear model expresses the expected frequencies (rather than the expected odds) as the product of parameters. The general model is

$$F_{jkl} = \tau\ \tau_{j}^{B}\ \tau_{k}^{C}\ \tau_{l}^{D}\ \tau_{jk}^{BC}\ \tau_{jl}^{BD}\ \tau_{kl}^{CD}\ \tau_{jkl}^{BCD}, \text{ for } j, k, l = 1, 2.$$

Thus F_{jk1} and F_{jk2} are given by

$$F_{jk1} = \tau\,\tau_{j}^{B}\ \tau_{k}^{C}\ \tau_{1}^{D}\ \tau_{jk}^{BC}\ \tau_{j1}^{BD}\ \tau_{k1}^{CD}\ \tau_{jk1}^{BCD}, \tag{2}$$

$$F_{jk2} = \tau\,\tau_{j}^{B}\ \tau_{k}^{C}\ \tau_{2}^{D}\ \tau_{jk}^{BC}\ \tau_{j2}^{BD}\ \tau_{k2}^{CD}\ \tau_{jk2}^{BCD}. \tag{3}$$

Dividing equation (2) by equation (3) cancels four τ parameters, yielding the logit model (1).

$$\begin{aligned}\frac{F_{jkl}}{F_{jk2}} = &\left(\frac{\tau_{1}^{D}}{\tau_{2}^{D}}\right)\times\left(\frac{\tau_{j1}^{BD}}{\tau_{j2}^{BD}}\right)\times\left(\frac{\tau_{k1}^{CD}}{\tau_{k2}^{CD}}\right)\times\left(\frac{\tau_{jk1}^{BCD}}{\tau_{jk2}^{BCD}}\right)\\ &\left(\tau_{1}^{D}\right)^{2}\ \left(\tau_{j1}^{BD}\right)^{2}\ \left(\tau_{k1}^{CD}\right)^{2}\ \left(\tau_{jk1}^{BCD}\right)^{2}\\ &\gamma^{D}\qquad \gamma_{j}^{BD}\qquad \gamma_{k}^{CD}\qquad \gamma_{jk}^{BCD}\end{aligned} \tag{4}$$

Thus the logit model is a special case of the general log-linear model where the parameters τ^{B}, τ^{C}, and τ^{BC} associated with the explanatory variables B and C are considered fixed. Like the classical quantitative regression model, the logit model expresses a conditional relationship between the response variable and fixed values of the explanatory variables B and C. On the other hand, the general log-linear model can also be formulated when all variables are responses and hence all parameters are free to vary.

Chapter 3

A General Model for the Analysis of Surveys[1]

This article shows how the combined use of direct estimation methods and indirect testing procedures, which was advocated by Goodman (1970, 1971*a*), can be applied in survey analysis. This approach to survey analysis replaces a "causal system" diagram (of the kind used, e.g., by Davis 1971) by a more quantitatively explicit model of the system, which then leads to a more direct and more unified assessment of the relationships among the variables in the system. The methods presented in the present article can also help the survey analyst to determine whether his survey data support or negate a given hypothesized causal system; and in some cases these methods can be used to determine alternative causal systems that provide better descriptions of the phenomena under investigation. To illustrate the application of these methods, we shall reanalyze data that were analyzed recently by Davis (1971) and earlier by Wilner, Walkley, and Cook (1955) in their well-known study of the "contact hypothesis." We also include in the present article some new results on how the relationship between two given dichotomous variables is affected by the introduction of additional variables.

INTRODUCTION

Let us begin by describing the survey that will be reanalyzed here for illustrative purposes. This survey, by Wilner and his associates (1955), was concerned with the relationships among the following four variables: Proximity (P), Contact (C), Norms (N), and Sentiment (S). Wilner was looking at the effect of segregated and integrated public housing on racial attitudes. His main hypothesis was that the close proximity (P) of whites and Negroes in integrated housing would increase favorable interracial sentiments (S) through the intervening variables of contact (C) and local norms (N). With respect to variables P, C, and S, Wilner hypothesized that the close proximity of whites and Negroes would lead to their having more contact with each other; the contact would lead to liking, and liking to contact. With respect to variables P, N, and S, Wilner hypothesized that the integrated housing project would tend to develop local norms favoring positive interracial sentiments; the people would tend to adopt the sentiments favored by local norms, and their sentiments would rein-

[1] This research was supported in part by Research Contract NSF GS 2818 from the Division of the Social Sciences of the National Science Foundation. For helpful comments, the author is indebted to J. A. Davis, O. D. Duncan, R. Fay, W. H. Kruskal, P. F. Lazarsfeld, D. McFarland, S. Schooler, and A. Stinchcombe.

force the local norms. In addition, with respect to variables *C* and *N*, it was assumed (by Davis 1971) that these two variables would have a mutual positive relationship. The "causal system" for the four variables (*P, C, N, S*), which we described above, can be summarized by the diagram in figure 1, which was presented by Davis (1971).[2]

To investigate the relationships among the four variables (*P, C, N, S*), Wilner collected data on these variables in interviews with 608 white women living in public housing projects having varying degrees of spatial segregation. These data are summarized in table 1.[3] (Note that the data in table 1 are limited to the study of white women. Different results might have been obtained for white men or for Negro men and/or women.)

TABLE 1

OBSERVED CROSS-CLASSIFICATION OF 608 WHITE WOMEN LIVING IN PUBLIC HOUSING PROJECTS, WITH RESPECT TO FOUR DICHOTOMIZED VARIABLES: (1) PROXIMITY TO A NEGRO FAMILY, (2) FREQUENCY OF CONTACTS WITH NEGROES, (3) FAVORABLENESS OF LOCAL NORMS TOWARD NEGROES, (4) FAVORABLENESS OF RESPONDENT'S ATTITUDES (SENTIMENTS) TOWARD NEGROES IN GENERAL

			SENTIMENT	
PROXIMITY	CONTACT	NORMS	+	−
+	+	+	77*	32*
+	+	−	30*	36*
+	−	+	14	19
+	−	−	15	27
−	+	+	43	20
−	+	−	36	37
−	−	+	27	36
−	−	−	41	118

SOURCE.—The observed frequencies in the above table were recalculated from the percentage table in Wilner, Walkley, and Cook (1955, p. 106). The asterisked numbers differ somewhat from the corresponding numbers in Davis (1971, p. 149), since his numbers do not agree with those calculated from the 1955 percentage table.

[2] Figure 1 served in Davis (1971) as a diagrammatic representation of the various hypothesized relationships among variables *P, C, N, S,* which were described in the above paragraph. Later herein we shall discuss how this kind of diagram is related to the more usual kinds of diagrammatic representations of the relationships among quantitative variables (see, e.g., Blalock 1964, and Duncan 1970).

[3] Although the data are displayed asymmetrically (as regards the *S* variable) in table 1, they could just as well have been displayed in a more symmetrical way as a four-way cross-classification. The asymmetric display might appear, at first sight, to suggest that variable *S* is the "dependent" variable; but the causal system considered will actually view variable *S* as both a "dependent" variable (in a certain sense) and also as a variable that can affect some of the other "dependent" variables in the system. This point will be discussed more fully later in the section on diagrams of causality and systems of simultaneous equations.

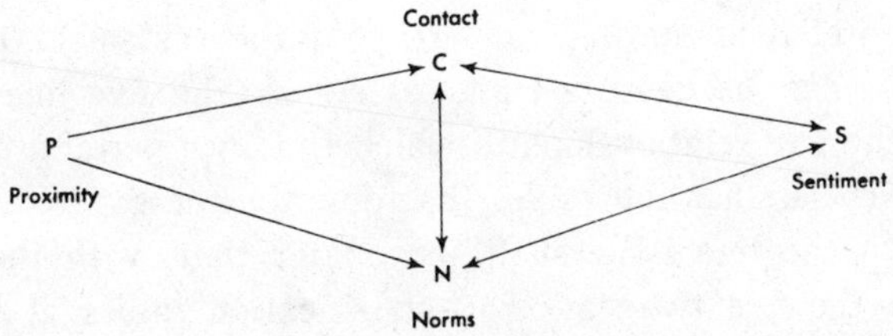

FIG. 1.—Diagram of the causal system among the four variables (P, C, N, S), as hypothesized by Wilner, Walkley, and Cook (1955), and Davis (1971).

In the present article, we shall replace the causal system diagram (fig. 1) by a more quantitatively explicit model of the system, and we shall show how to use this model to estimate, in a more direct and unified way, the relationships among the variables in the system. In addition, we shall show how to test whether the survey data (table 1) support or negate the hypothesized causal system. We shall also introduce alternative causal systems, and shall show that they, too, can be tested to determine whether they are supported or negated by the data.

Most of the methods presented in the present article are not necessarily limited to the study of the relationships among four variables. They can be used to study the relationships among any given number (say, 3, 4, 5, . . .) of variables. Also, most of the methods are not necessarily limited to the study of a set of dichotomous variables. They can be applied (with appropriate modifications) to the study of a set of polytomous variables as well.[4]

A CAUSAL SYSTEM

Let us first consider the causal system as described in figure 1. The five lines in this diagram pertain to the relationship between the following pairs of variables: (1) variables P and C; (2) variables P and N; (3) variables C and N; (4) variables C and S; (5) variables N and S. (Note that there is no line in figure 1 corresponding to the relationship between variables P and S.) These five relationships can be described by the following two-way cross-classification tables: (1) the two-way table describing the relationship between variables P and C; (2) the two-way table describing the relationship between variables P and N; etc. Corresponding to the five lines in figure 1, there are the five two-way cross-classification tables noted above. (The fact that some lines in fig. 1 have two arrowheads and some have only one will be discussed later.) Figure 1

[4] Of course, some surveys collect information pertaining to a set of quantitative (continuous) variables in addition to (or instead of) information pertaining to a set of dichotomous (or polytomous) variables. The methods presented in the present article are limited to the analysis of polytomous or polytomized variables, which would of course include the analysis of dichotomous or dichotomized variables.

can be interpreted as stating, in part, that the system of four variables (*P*, *C*, *N*, *S*) can be described adequately by the five lines in figure 1, that is, by the five relationships to which the lines pertain, or by the five two-way cross-classification tables that describe the five relationships corresponding to the lines. We shall show later that, with the information contained in the five two-way cross-classification tables, it is possible to describe very well the entire system of four variables (*P*, *C*, *N*, *S*) as represented by the observed distribution of individuals within the four-way cross-classification table (table 1). In other words, we shall show that the observed distribution of individuals within the four-way table can be fitted very well using only the information contained in the five two-way cross-classification tables.

From Wilner's survey of 608 women, the two-way cross-classification table describing the relationship between variables *P* and *C* can be obtained from the data in table 1 by ignoring the other two variables (*N* and *S*); that is, by "summing over" or "collapsing" the other two variables. We shall refer to this two-way table as a "two-way marginal table" since it can be obtained from the four-way table (table 1) by ignoring two of the variables in the four-way table. We shall use the symbol $\{PC\}$ to denote this particular two-way marginal table. The five two-way marginal tables, to which we referred in the preceding paragraph, can now be described by the symbols $\{PC\}$, $\{PN\}$, $\{CN\}$, $\{CS\}$, and $\{NS\}$.

As we have already noted, figure 1 can be interpreted as describing, in part, the hypothesis that the information contained in the set of five marginal tables ($\{PC\}$, $\{PN\}$, $\{CN\}$, $\{CS\}$, $\{NS\}$) is sufficient to describe the entire system of four variables as represented by the four-way cross-classification table (table 1).[5] We shall use the symbol H to denote this hypothesis. Later we shall describe H in more quantitatively explicit terms, which will then lead to the calculation of the "expected distribution" of individuals within the four-way table under the assumption that hypothesis H is true. For the reader who is anxious to see the numerical results, we give in table 2 the "expected frequencies" in the four-way table, estimated under the assumption that H is true. By comparing tables 1 and 2, we see that the hypothesis H is supported by the data in table 1.[6]

[5] This hypothesis also states that the information contained in the two-way marginal table $\{PS\}$, and the additional information contained in the four three-way marginal tables ($\{PCN\}$, $\{PCS\}$, $\{PNS\}$, $\{CNS\}$), are not needed at all for this purpose. (Since the hypothesis stated that the set of five marginal tables [$\{PC\}$, $\{PN\}$, $\{CN\}$, $\{CS\}$, $\{NS\}$] is sufficient for this purpose, it follows that the other marginal tables [$\{PS\}$ and the four three-way marginal tables] are not needed.)

[6] Later we shall provide a quantitative assessment of how well hypothesis H fits the data.

TABLE 2

ESTIMATE OF EXPECTED FREQUENCIES IN THE FOUR-WAY CONTINGENCY TABLE (TABLE 1), TAKING INTO ACCOUNT OBSERVED TWO-WAY CROSS-CLASSIFICATION FOR FOLLOWING PAIRS OF VARIABLES: (1) PROXIMITY AND CONTACT, (2) PROXIMITY AND NORMS, (3) CONTACT AND NORMS, (4) CONTACT AND SENTIMENT, (5) NORMS AND SENTIMENT

			SENTIMENT	
PROXIMITY	CONTACT	NORMS	+	—
+	+	+	74.80	34.48
+	+	—	32.28	33.45
+	—	+	14.75	17.98
+	—	—	11.30	30.97
—	+	+	42.94	19.79
—	+	—	35.99	37.29
—	—	+	28.51	34.76
—	—	—	42.44	116.29

NOTE.—The observed two-way cross-classification for variables P and S (Proximity and Sentiment) was not taken into account in the calculation of the expected frequencies in the four-way table.

THE MODEL THAT CORRESPONDS TO THE CAUSAL SYSTEM

In table 1 we have used the usual symbols + and — to denote the two categories of each dichotomous variable represented in the table. (For example, with respect to the proximity variable P, the individuals who live close to a Negro family are placed in the + category for P, and the individuals who live less close to a Negro family are placed in the — category for P.) Now it will be convenient here to replace the + and — symbols by the numbers 1 and 2, respectively.[7] Viewing table 1 as a four-way cross-classification table, we let f_{ijkl} denote the observed frequency in cell (i,j,k,l) of the table (where $i = 1,2$; $j = 1,2$; $k = 1,2$; $l = 1,2$). (For example, corresponding to cell (1,1,1,1) in table 1, the f_{1111} value is 77; corresponding to cell (1,1,1,2), the f_{1112} value is 32; etc.) For some specified hypothesis about the system of variables in table 1 (e.g., for the hypothesis H discussed in the preceding section), we shall use the symbol F_{ijkl} to denote the expected frequency in cell (i,j,k,l) of table 1, calculated under the asumption that the specified hypothesis is true.[8]

Let n denote the total sample size (e.g., $n = 608$ in table 1). From the definition of the f_{ijkl}, we see that

[7] This change in notation will facilitate the exposition and presentation of formulae, and it will also facilitate the extension of results from the case where variables are dichotomous to the case where variables are polytomous.

[8] In other words, the symbol F_{ijkl} will denote the *expected value* of the observed frequency in cell (i,j,k,l) of table 1, under the specified hypothesis. We shall use the term "expected" to refer to the expected value F_{ijkl} under some specified hypothesis and also to certain quantities that are based upon the F_{ijkl}.

$$\sum_{i=1}^{2}\sum_{j=1}^{2}\sum_{k=1}^{2}\sum_{l=1}^{2} f_{ijkl} = n. \tag{1}$$

Because equation (1) is satisfied for the observed frequencies f_{ijkl}, we also impose the same condition on the expected frequencies F_{ijkl}. That is, we also require that

$$\sum_{i=1}^{2}\sum_{j=1}^{2}\sum_{k=1}^{2}\sum_{l=1}^{2} F_{ijkl} = n. \tag{2}$$

The two-way marginal table $\{PC\}$, which we discussed in the preceding section, can be calculated from the four-way table in the following way. Letting f^{PC}_{ij} denote the observed frequency in cell (i,j) of the two-way table $\{PC\}$ (where $i = 1,2$; $j = 1,2$), we can calculate f^{PC}_{ij} by the formula

$$f^{PC}{}_{ij} = \sum_{k=1}^{2}\sum_{l=1}^{2} f_{ijkl}. \tag{3}$$

Similar formulae can be applied in order to calculate the other two-way marginal tables. Calculated from the data in the four-way table (table 1), we give in table 3 the two-way marginal table $\{PC\}$, and also the marginal tables $\{PN\}$, $\{CN\}$, $\{CS\}$, $\{NS\}$ to which we referred in the preceding section.

From the two-way table $\{PC\}$, we can calculate the one-way marginals corresponding to the "row" marginal totals and "column" marginal totals of the two-way table. We shall use the symbols $\{P\}$ and $\{C\}$ to denote the one-way marginal frequency arrays corresponding to the P and C variables, respectively. Letting f^P_i and f^C_j denote the observed frequency in cell (i) and cell (j), respectively, of the one-way marginals $\{P\}$ and $\{C\}$ (where $i = 1,2$; $j = 1,2$), we can calculate f^P_i and f^C_j by the following formulae:

$$\begin{aligned} f^P{}_i &= \sum_{j=1}^{2} f^{PC}{}_{ij}, \\ f^C{}_j &= \sum_{i=1}^{2} f^{PC}{}_{ij}. \end{aligned} \tag{4}$$

Consider now the set of five two-way marginal tables discussed in the preceding section, namely $\{PC\}$, $\{PN\}$, $\{CN\}$, $\{CS\}$, $\{NS\}$. (This set of five tables, which we calculated from the data in table 1, is presented in table 3 below.) From the information contained in this set of tables, we can obtain the one-way marginals $\{P\}$, $\{C\}$, $\{N\}$, and $\{S\}$. The hypothesis H of the preceding section can be expressed by stating that the set of nine

TABLE 3

THE TWO-WAY MARGINAL TABLE FOR THE FOLLOWING PAIRS OF VARIABLES, CALCULATED FROM DATA IN TABLE 1: (I) PROXIMITY AND CONTACT, (II) PROXIMITY AND NORMS, (III) CONTACT AND NORMS, (IV) CONTACT AND SENTIMENT, (V) NORMS AND SENTIMENT

Table I
Proximity and Contact

		Contact +	Contact −
Proximity	+	175	75
	−	136	222

Table II
Proximity and Norms

		Norms +	Norms −
Proximity	+	142	108
	−	126	232

Table III
Contact and Norms

		Norms +	Norms −
Contact	+	172	139
	−	96	201

Table IV
Contact and Sentiment

		Sentiment +	Sentiment −
Contact	+	186	125
	−	97	200

Table V
Norms and Sentiment

		Sentiment +	Sentiment −
Norms	+	161	107
	−	122	218

marginals (namely, $\{P\}$, $\{C\}$, $\{N\}$, $\{S\}$, $\{PC\}$, $\{PN\}$, $\{CN\}$, $\{CS\}$, $\{NS\}$) is sufficient to estimate the expected distribution of individuals within the four-way table.

The hypothesis H can be made quantitatively explicit in different ways. We shall now describe one of these ways. Corresponding to the set of nine marginals listed in the preceding paragraph, we now introduce a set of nine parameters which we shall denote as τ^P, τ^C, τ^N, τ^S, τ^{PC}, τ^{PN}, τ^{CN}, τ^{CS}, τ^{NS}. The parameter τ^P can be viewed as pertaining to the "main effect" of the variable P on the expected frequency F_{ijkl} in cell (i,j,k,l) of table 1; the parameter τ^{PC} can be viewed as pertaining to the "interaction effect" between the variables P and C on the F_{ijkl}, etc.[9] The hypothesis H states that the expected frequencies F_{ijkl} can be calculated from these main effects and interaction effects. More explicitly, the F_{ijkl} under H can be expressed as

[9] The meaning of the τ parameters will become clearer later when we present and discuss formulae (7) and (14)–(17).

$$F_{ijkl} = \eta\tau^{P}{}_{i}\tau^{C}{}_{j}\tau^{N}{}_{k}\tau^{S}{}_{l}\tau^{PC}{}_{ij}\tau^{PN}{}_{ik}\tau^{CN}{}_{jk}\tau^{CS}{}_{jl}\tau^{NS}{}_{kl}, \tag{5}$$

where

$$\tau^{P}{}_{1} = \tau^{P}, \tau^{P}{}_{2} = 1/\tau^{P}, \tau^{PC}{}_{11} = \tau^{PC}{}_{22} = \tau^{PC}, \tau^{PC}{}_{12} = \tau^{PC}{}_{21} = 1/\tau^{PC}, \text{etc.}, \tag{6}$$

and where η is a constant that is introduced to insure that the F_{ijkl} satisfy condition (2). From formulae (5)–(6) we see that the effect of, say, the τ^P parameter upon the F_{ijkl} is to introduce the multiplicative factor τ^P when variable P is at level 1, and the multiplicative factor $1/\tau^P$ when variable P is at level 2. Similarly, we see that the effect of, say, the τ^{PC} parameter upon the F_{ijkl} is to introduce the multiplicative factor τ^{PC} when both variable P and variable C are at the same level, and the multiplicative factor $1/\tau^{PC}$ when variable P and variable C are at different levels. Et cetera.

From formulae (5)–(6), we also find that

$$[(F_{11kl}F_{22kl})/(F_{12kl}F_{21kl})] = (\tau^{PC})^4. \tag{7}$$

The quantity on the left of the equality sign in (7) is called the "cross-product ratio" or the "odds-ratio" in the 2×2 table describing the expected relationship between variables P and C when the levels of the remaining variables N and S are set at k and l, respectively.[10] From (7) we see that, when the remaining variables (N,S) are "held constant" by setting them at a given level (k,l), then the expected odds-ratio pertaining to the relationship between variables P and C will have the same value at each of the four possible levels (for $k = 1,2$; $l = 1,2$) under model (5)–(6), and this expected odds-ratio will be equal to the fourth power of τ^{PC}. In other words, this expected odds-ratio (under model [5]–[6]) is independent of the level (k,l) of the remaining variables (N,S), and from (7) we can obtain an explicit expression for τ^{PC} in terms of the expected odds-ratio. This provides a clearer meaning for the parameter τ^{PC} which we introduced in the preceding paragraph. A similar kind of interpretation can be obtained for the other τ parameters introduced in the preceding paragraph.

The F_{ijkl} can be expressed in terms of the τ parameters (as in [5]), and the τ parameters can be expressed in terms of the F_{ijkl} (as in [7]). Since we have already presented in table 2 the estimated values of the F_{ijkl}, we can use these values in turn to estimate the τ parameters.[11] The

[10] Since the quantity on the left of the equality sign in (7) is based upon the expected frequencies F_{ijkl} (under model [5]–[6]), we shall call this quantity the "expected odds-ratio." When the expected odds-ratio pertaining to the relationship between variables P and C is calculated at a given level (k,l) of the remaining variables (N,S), all the τ parameters except τ^{PC} cancel out under model (5)–(6).

[11] Later we shall comment further upon the calculation of the estimated F_{ijkl} and the estimated τ parameters (see [14]–[17] and the Appendix).

estimated values of the τ parameters are given in table 4. Recall that the estimated values of the F_{ijkl} in table 2 were calculated from the informa-

TABLE 4

ESTIMATE OF MAIN EFFECTS AND INTERACTION EFFECTS AMONG FOUR VARIABLES (P,C,N,S) IN FOUR-WAY CONTINGENCY TABLE (TABLE 1), UNDER MODELS (5) AND (8)

Variable	τ Effects in Model (5)	λ Effects in Model (8)
P	0.83	−.19
C	1.13	.12
N	0.91	−.09
S	0.94	−.06
PC	1.35	.30
PN	1.18	.17
PS	1.00	.00
CN	1.15	.14
CS	1.28	.24
NS	1.22	.20

tion contained in the set of five two-way marginals (namely, $\{PC\}$, $\{PN\}$, $\{CN\}$, $\{CS\}$, $\{NS\}$—the set of five tables presented in table 3), and similarly the τ parameters can be estimated directly from this information. Note the relative magnitudes of the τ parameters in table 4. (Since the parameter τ^{PS} is not included in model [5], its numerical value in table 4 is given as 1.00, which is equivalent to stating that there is no τ^{PS} effect in the multiplicative model [5].)

Formula (5) expresses the F_{ijkl} as a product of certain main effect and interaction effect parameters. This formula can also be expressed in an additive form via logarithms. Letting G_{ijkl} denote the natural logarithm of F_{ijkl} (i.e., $G_{ijkl} = \log F_{ijkl}$, where log denotes the natural logarithm), we see from (5) that the G_{ijkl} can be expressed as follows:[12]

$$G_{ijkl} = \theta + \lambda^{P}{}_{i} + \lambda^{C}{}_{j} + \lambda^{N}{}_{k} + \lambda^{S}{}_{l} + \lambda^{PC}{}_{ij} + \lambda^{PN}{}_{ik} + \lambda^{CN}{}_{jk} + \lambda^{CS}{}_{jl} + \lambda^{NS}{}_{kl}, \qquad (8)$$

where

$$\theta = \log \eta, \lambda^{P}{}_{i} = \log \tau^{P}{}_{i}, \lambda^{C}{}_{j} = \log \tau^{C}{}_{j}, \text{etc.} \qquad (9)$$

Corresponding to the τ parameters discussed earlier in this section, the λ parameters are defined as

$$\lambda^{P} = \log \tau^{P}, \lambda^{PC} = \log \tau^{PC}, \text{etc.} \qquad (10)$$

12 The model described by formula (5) is, of course, equivalent to the model described by formula (8).

From (6), (9), and (10), we find that

$$\lambda^{P} = \lambda^{P}{}_{1} = -\lambda^{P}{}_{2}, \lambda^{PC} = \lambda^{PC}{}_{11} = \lambda^{PC}{}_{22} = -\lambda^{PC}{}_{12} = -\lambda^{PC}{}_{21}, \text{etc.} \tag{11}$$

Estimates of the λ parameters are included in table 4.[13] (Since the parameter λ^{PS} is not included in model [8], its numerical value in table 4 is given as .00, which corresponds to a λ effect that is nil in this model.)

We shall show later how to assess the statistical significance of the contribution made by a given τ parameter (e.g., τ^{PC}) in model (5), or by a given λ parameter (e.g., λ^{PC}) in model (8). We shall also show how to measure the magnitude of this contribution, using coefficients that are somewhat analogous to the usual partial correlation coefficients of regression analysis. In addition, we shall discuss how to test whether model (5) (or the corresponding model [8]) fits the data, and we shall measure how well the model fits using a coefficient that is somewhat analogous to the usual multiple correlation coefficient.

A GENERAL MODEL

In the preceding discussion, we began with a particular hypothesized causal system (fig. 1), we then replaced this system by a model (equation [5] or [8]) that served as a possible explication of it, we checked (informally) whether the survey data supported the model (comparing tables 1 and 2), and we then used the data to estimate the parameters in the model (see table 4). In the section before the preceding one, we found that the data actually did support the model (comparing tables 1 and 2), but in many surveys we might expect to find that the hypothesized causal system is not supported by the observed data. In such cases, we would consider alternative causal systems that might fit the observed data better. We shall now provide a general model that can be used for this purpose. This model can also be used in the analysis of surveys that were conducted without the benefit of a hypothesized causal system.

Corresponding to the set of nine marginals listed earlier in the preceding section (namely, $\{P\}$, $\{C\}$, $\{N\}$, $\{S\}$, $\{PC\}$, $\{PN\}$, $\{CN\}$, $\{CS\}$, $\{NS\}$), we introduced a set of nine parameters (τ^{P}, τ^{C}, τ^{N}, τ^{S}, τ^{PC}, τ^{PN}, τ^{CN}, τ^{CS}, τ^{NS}), which were used to describe the main effects and interaction effects in model (5), and which in turn could be described in terms of the F_{ijkl} (see [7]). There were four one-way marginals and five two-way marginals in the set of nine marginals; and, similarly, there were four one-factor τ's

[13] The calculation of the estimated λ parameters will be discussed later, after model (18)–(19) has been presented (see also n. 20).

and five two-factor τ's in the set of nine τ's.[14] (We noted earlier that a sixth two-factor τ [τ^{PS}] was not included in model [5], since the causal system diagram [fig. 1] did not include a line pertaining to the corresponding two-way marginal $\{PS\}$.) Now instead of introducing only the nine τ parameters, let us introduce all possible τ parameters: the four one-factor τ's (τ^P, τ^C, τ^N, τ^S), the six two-factor τ's (τ^{PC}, τ^{PN}, τ^{PS}, τ^{CN}, τ^{CS}, τ^{NS}), the four three-factor τ's (τ^{PCN}, τ^{PCS}, τ^{PNS}, τ^{CNS}), and the one four-factor τ (i.e., τ^{PCNS}). Instead of model (5), we now have the following "saturated" model:[15]

$$F_{ijkl} = \eta \tau^P{}_i \tau^C{}_j \tau^N{}_k \tau^S{}_l \tau^{PC}{}_{ij} \tau^{PN}{}_{ik} \tau^{PS}{}_{il} \tau^{CN}{}_{jk} \tau^{CS}{}_{jl} \tau^{NS}{}_{kl} \tau^{PCN}{}_{ijk} \tau^{PCS}{}_{ijl} \tau^{PNS}{}_{ikl} \tau^{CNS}{}_{jkl} \tau^{PCNS}{}_{ijkl}, \qquad (12)$$

where

$$\tau^P{}_1 = \tau^P, \tau^P{}_2 = 1/\tau^P, \tau^{PC}{}_{11} = \tau^{PC}{}_{22} = \tau^{PC}, \tau^{PC}{}_{12} = \tau^{PC}{}_{21} = 1/\tau^{PC},$$

$$\tau^{PCN}{}_{111} = \tau^{PCN}{}_{221} = \tau^{PCN}{}_{212} = \tau^{PCN}{}_{122} = \tau^{PCN},$$

$$\tau^{PCN}{}_{211} = \tau^{PCN}{}_{121} = \tau^{PCN}{}_{112} = \tau^{PCN}{}_{222} = 1/\tau^{PCN}, \text{etc.} \qquad (13)$$

Model (5) can be obtained from the saturated model (12) by assuming that certain τ parameters in (12) (τ^{PS}, τ^{PCN}, τ^{PCS}, τ^{PNS}, τ^{CNS}, τ^{PCNS}) are set equal to one. Models that are obtained from (12) in this way we shall call "unsaturated" models.[16]

Model (12) describes the F_{ijkl} in terms of the τ parameters, and the τ parameters in this model can in turn be expressed in terms of the F_{ijkl}. (Recall that formula [7] described a τ parameter in model [5] in terms of the F_{ijkl} under that model.) From formulae (12)–(13), we obtain the fol-

[14] A one-factor τ (say, τ^P) could also be described as a one-variable τ, since this parameter pertains to the effect of a single variable (say, variable P) on the F_{ijkl}. A two-factor τ (say, τ^{PC}) could also be described as a two-variable τ, since this parameter pertains to the interaction effect between two variables (say, variables P and C) on the F_{ijkl}. Etc.

[15] The saturated model, which is given by formulae (12)–(13), can also be described as an "unrestricted" model or as a "full" model. With this model, we can represent the F_{ijkl} in terms of the multiplicative main effects (τ^P, τ^C, τ^N, τ^S) and interaction effects (τ^{PC}, τ^{PN}, . . . , τ^{PCNS}). This representation is related to, but quite different from, the Lazarsfeld representation as presented in Lazarsfeld (1961, 1968) and Bahadur (1961). The Lazarsfeld representation expresses the F_{ijkl} in terms of a set of parameters that are quite different from the parameters used in the present article; but it is related to the present representation ([12]–[13]) in the sense that they are both saturated models. For a further comment on this matter, see n. 16 below.

[16] An unsaturated model can also be described as a "restricted" model. The unsaturated models that are obtained from (12) in this way are quite different from the models that can be obtained from the Lazarsfeld representation (see n. 15 above) by setting certain parameters in that representation equal to zero.

lowing explicit formulae for the τ parameters expressed in terms of the F_{ijkl}:[17]

$$\tau^{P} = \left[\prod_{j=1}^{2}\prod_{k=1}^{2}\prod_{l=1}^{2}(F_{1jkl}/F_{2jkl})\right]^{1/16}, \ldots, \quad (14)$$

$$\tau^{PC} = \left[\prod_{k=1}^{2}\prod_{l=1}^{2}(F_{11kl}F_{22kl})/(F_{12kl}F_{21kl})\right]^{1/16}, \ldots, \quad (15)$$

$$\tau^{PCN} = \left[\prod_{l=1}^{2}(F_{111l}F_{221l}F_{212l}F_{122l})/(F_{112l}F_{121l}F_{211l}F_{222l})\right]^{1/16}, \ldots, \quad (16)$$

$$\tau^{PCNS} = \left[\frac{F_{1111}F_{2211}F_{2121}F_{1221}F_{2112}F_{1212}F_{1122}F_{2222}}{F_{1112}F_{1121}F_{1211}F_{2111}F_{2221}F_{2212}F_{2122}F_{1222}}\right]^{1/16} \quad (17)$$

Formula (7) stated that $(\tau^{PC})^4$ is equal to the expected odds-ratio pertaining to the relationship between variables P and C, at level (k,l) of the remaining variables (N,S), under model (5); while formula (15) states that $(\tau^{PC})^4$ is equal to the geometric mean of these expected odds-ratios, over the four possible levels (k,l) (for $k = 1,2$; $l = 1,2$) of the remaining variables (N,S). Similarly, formula (14) states that $(\tau^{P})^2$ is equal to the geometric mean of the "expected odds" F_{1jkl}/F_{2jkl} pertaining to variable P, over the eight possible levels (j,k,l) (for $j = 1,2$; $k = 1,2$; $l = 1,2$) of the remaining variables (C,N,S). (For related material, see formula [52] below.) Formulae (16) and (17) can be interpreted in a similar way.

Since the saturated model (12) includes all possible τ parameters (i.e., all possible main effects and interaction effects), the expected frequencies F_{ijkl} under the model can be estimated by the observed frequencies f_{ijkl}, and these estimates of the F_{ijkl} can be inserted in (14)–(17) in order to obtain estimates of the τ parameters.[18] With the τ parameters in the

[17] Formulae (14)–(17) can be applied to model (12), and they can also be applied to model (5) discussed earlier. Although formula (15) is different from the corresponding formula (7) presented earlier for model (5), use of either formula (7) or (15) will give the same numerical value for τ^{PC}, if these formulae are applied to the expected frequency F_{ijkl} under model (5). For a related comment, see n. 18 below.

[18] In contrast to this procedure for the saturated model (12), it is important to note that, with respect to the unsaturated model (5), in order to estimate the τ parameters under the model, formulae (14)–(17) can be used with the F_{ijkl} in these formulae replaced by the estimated expected frequencies under the model, which were given in table 2. Under the unsaturated model, the expected frequencies F_{ijkl} can *not* be replaced by the observed frequencies f_{ijkl}. Since the estimated expected frequencies F_{ijkl} under the unsaturated model (5) (see table 2) are somewhat different from the estimated F_{ijkl} under the saturated model (12) (namely, the f_{ijkl} in table 1), the

saturated model estimated as indicated above, the observed f_{ijkl} are fitted perfectly. (For further comments on this point, see the second paragraph of n. 23 below.) Based upon the data given in table 1, the estimated values of the τ parameters are given in table 5. Note that, for the data in

TABLE 5

ESTIMATE OF MAIN EFFECTS AND INTERACTION EFFECTS AMONG FOUR VARIABLES (P, C, N, S) IN FOUR-WAY CONTINGENCY TABLE (TABLE 1), UNDER SATURATED MODELS (12) AND (18)

Variable	τ Effects in Model (12)	λ Effects in Model (18)	Standardized Value of Estimated λ
P	0.84	−.18	−3.89
C	1.12	.11	2.37
N	0.90	−.10	−2.18
S	0.95	−.05	−1.06
PC	1.33	.28	6.14
PN	1.16	.15	3.28
PS	1.03	.03	0.58
CN	1.16	.15	3.23
CS	1.25	.23	4.89
NS	1.20	.18	3.88
PCN	1.00	.00	0.09
PCS	0.97	−.03	−0.67
PNS	0.99	−.01	−0.30
CNS	1.05	.05	1.06
$PCNS$	1.05	.05	1.03

table 1, the estimated τ's in table 5 are quite similar to the corresponding quantities obtained in table 4.[19]

We noted earlier that model (5) could be replaced by the equivalent model (8); and similarly the saturated model (12) can be replaced by the following equivalent model:

$$G_{ijkl} = \theta + \lambda^{P}{}_{i} + \lambda^{C}{}_{j} + \lambda^{N}{}_{k} + \lambda^{S}{}_{l} + \lambda^{PC}{}_{ij} + \lambda^{PN}{}_{ik} + \lambda^{PS}{}_{il} + \lambda^{CN}{}_{jk} + \lambda^{CS}{}_{jl} + \lambda^{NS}{}_{kl}$$

estimated τ parameters under model (5) will also be somewhat different from the corresponding quantities under model (12) (compare tables 4 and 5).

[19] We did not include in tables 4 and 5 estimates of the constant η for models (5) and (12) because this constant is of no real substantive interest. As we noted earlier, the constant η in (5) and (12) was introduced there to insure that the F_{ijkl} satisfy condition (2). For the reader who wishes to estimate this constant, we note that it is possible to obtain (from [2] and [12]) a simple explicit expression for η as a ratio of the sample size n and a function of the τ's (see, e.g., Goodman 1971*a*, p. 36). Thus, using the estimated values of the τ's, we can also estimate η. This estimate can also be obtained by calculating the geometric mean of the estimated expected frequencies.

$$+ \lambda^{PCN}{}_{ijk} + \lambda^{PCS}{}_{ijl} + \lambda^{PNS}{}_{ikl} + \lambda^{CNS}{}_{jkl} + \lambda^{PCNS}{}_{ijkl}, \quad (18)$$

where θ and the λ's are defined as in (9). Corresponding to the τ parameters, we now have the λ parameters defined as in (10), and the relationships described by (11) can be written as

$$\lambda^{P} = \lambda^{P}{}_{1} = -\lambda^{P}{}_{2}, \lambda^{PC} = \lambda^{PC}{}_{11} = \lambda^{PC}{}_{22} = -\lambda^{PC}{}_{12} = -\lambda^{PC}{}_{21},$$

$$\lambda^{PCN} = \lambda^{PCN}{}_{111} = \lambda^{PCN}{}_{221} = \lambda^{PCN}{}_{212} = \lambda^{PCN}{}_{122}$$

$$= -\lambda^{PCN}{}_{211} = -\lambda^{PCN}{}_{121} = -\lambda^{PCN}{}_{112} = -\lambda^{PCN}{}_{222}, \text{etc.} \quad (19)$$

Since we have already provided estimates of the τ parameters in Table 5, we see from (10) that the estimated λ parameters can be calculated as the logarithm of the corresponding estimated τ parameters.[20] In addition, the variance of the estimated λ parameters can be estimated by the following formula:[21]

$$S^2_{\hat{\lambda}} = \Big[\sum_{i=1}^{2}\sum_{j=1}^{2}\sum_{k=1}^{2}\sum_{l=1}^{2}(1/f_{ijkl})\Big]\Big/(16)^2. \quad (20)$$

By dividing each estimated λ parameter by its estimated standard deviation $S_{\hat{\lambda}}$, we obtain the corresponding "standardized value" of the estimate. Each standardized value can be used to test whether the corresponding λ is nil.[22] Table 5 includes the estimated values of the λ parameters and their corresponding standardized values.

By an examination of the relative magnitudes of the standardized

[20] For example, the estimated λ^P can be calculated from the estimated τ^P, which in turn can be calculated from equation (14), with F_{ijkl} (under model [18]) replaced by f_{ijkl} in this equation. In actual practice, it is simpler to calculate first the logarithm of the right side of equation (14) with F_{ijkl} (under model [18]) replaced by f_{ijkl} in this equation, which provides then the estimated λ^P. The estimated λ^P can then be used to calculate the estimated τ^P, since $\tau^P = \exp \lambda^P$, where exp denotes the exponential function (i.e., the inverse of the logarithmic function). The above remarks apply to the saturated model (18). With respect to the unsaturated model (8), the above remarks apply except for the fact that the F_{ijkl} can *not* be replaced by the f_{ijkl} (see n. 18 above).

[21] Note should be taken of the fact that the estimation method presented for the saturated model can be improved upon by replacing the f_{ijkl} in (20) by $f_{ijkl} + \frac{1}{2}$, and by making the same kind of replacement in the f_{ijkl} that are used in (14)–(17). It should also be noted that formula (20) and some of the other results presented are applicable in the case where the observed four-way table (table 1) describes results obtained for a random sample of individuals cross-classified with respect to the four variables (P, C, N, S), and also in some other cases as well. For further details see Goodman (1970) and Haberman (1970).

[22] If a particular λ parameter is nil, then the standardized value of the corresponding estimated λ will be approximately normally distributed with zero mean and unit variance (when the sample size is large). For comments on related matters, see Goodman (1970, 1971*a*).

values for the two-factor, three-factor, and four-factor interactions in table 5, we find that the model in which the following parameters are set equal to zero should be worthy of consideration: λ^{PS}, λ^{PCN}, λ^{PCS}, λ^{PNS}, λ^{CNS}, λ^{PCNS}. But the model that is obtained when these particular parameters are set equal to zero in (18) is equivalent to model (8). (None of the one-factor λ's are set equal to zero in [18] for reasons that will be explained at the end of the Appendix below.) Thus, for the data in table 1, the examination of the saturated models (12) and (18) actually leads to the hypothesized models (5) and (8), which were based upon the particular hypothesized causal system.

In cases where the survey data do not support the model corresponding to the hypothesized causal system, and in cases where the survey data have been collected without the benefit of a hypothesized causal system, an examination of the standardized values corresponding to the λ parameters in the saturated model (see table 5) can be used by the survey analyst to assist him in his search for models that fit the data. For some cases where the standardized values were used to do this, see Goodman (1970, 1971*a*, 1972*b*).

Before closing this section, we take note of the fact that the standard deviation of each estimated λ parameter in the saturated model (see table 5) was estimated by $S^{*}_{\hat{\lambda}} = .046$, using formula (20) (see n. 21). This estimated standard deviation can be applied to each estimated λ parameter in table 5 for the saturated model, and it also can serve as an estimate of an upper bound for the standard deviation of each estimated λ parameter in table 4 for the unsaturated model (8).

HOW TO TEST WHETHER A MODEL FITS THE DATA

To test whether the hypothesized H described by model (5) fits the data in table 1, we first estimate the expected frequencies F_{ijkl} under the hypothesis H (see table 2), and we then compare each observed frequency f_{ijkl} in table 1 with the corresponding estimate of the F_{ijkl} in table 2, by calculating either the usual χ^2 goodness-of-fit statistic

$$\sum_{i=1}^{2}\sum_{j=1}^{2}\sum_{k=1}^{2}\sum_{l=1}^{2} (f_{ijkl} - F_{ijkl})^2/F_{ijkl}, \tag{21}$$

or the corresponding χ^2 based upon the likelihood-ratio statistic; namely,

$$2\sum_{i=1}^{2}\sum_{j=1}^{2}\sum_{k=1}^{2}\sum_{l=1}^{2} f_{ijkl} \log (f_{ijkl}/F_{ijkl}). \tag{22}$$

The χ^2 value obtained from (21) or (22) can be assessed by comparing its numerical value with the percentiles of the tabulated χ^2 distribution.

The degrees of freedom for testing hypothesis H will be $16 - 1 - 9 = 6$ (since [*a*] there are 16 cells in table 1, [*b*] there is one restriction of the form [2] to which the F_{ijkl} are subject, and [*c*] there are nine τ parameters that are estimated in model [5]).

Using (21), we obtain a goodness-of-fit χ^2 value of 2.62; and using (22), we obtain a likelihood-ratio χ^2 value of 2.53. Since there were 6 df under H, the model fits the data very well indeed.

Model (5) is a particular example of a model that is obtained from the saturated model (12) by setting a specified set of τ parameters in (12) equal to one.[23] The corresponding model (8) is a particular example of a model that is obtained from the saturated model (18) by setting a specified set of λ parameters in (18) equal to zero. For the four-way cross-classification table, Goodman (1970) described in his table 4 a large class of models (there are 166 different models in this class) that can be obtained from the saturated model by setting a certain specified set of λ parameters in (18) equal to zero (or by setting the corresponding set of τ parameters in [12] equal to one). For each of the models in this class (namely, the class of "hierarchial" models), table 4 in Goodman (1970) gave the corresponding degrees of freedom when the four-way table is a 2^4 table, and that article also described various ways of calculating the degrees of freedom when some of the variables in the four-way table are not necessarily dichotomous. A single computer program can be used to calculate the estimate of the F_{ijkl}, and the corresponding χ^2 values (21) or (22), for any set of models in this class.[24] For related material, see, for example, Birch (1963), Ku and Kullback (1968), Bishop (1969), Goodman (1969*b*, 1970, 1971*a*), and the Appendix below.

We shall now consider a number of different models pertaining to table 1 and shall list in table 6 the χ^2 values obtained in testing these models. We include both the χ^2 values (21) and (22) in table 6; but,

[23] Indeed, the 6 df that were used to test model (5) correspond to the six τ parameters (τ^{PS}, τ^{PCN}, τ^{PCS}, τ^{PNS}, τ^{CNS}, τ^{PCNS}) in (12) that are set equal to one under model (5).

More generally, the number of degrees of freedom used to test a given unsaturated model will be equal to the number of τ parameters in (12) that are set equal to one under the model. Since the saturated model can be described by saying that none of the τ parameters in (12) are set equal to one under that model (i.e., the number of τ parameters that are set equal to one is zero), there will be 0 df under the saturated model. This corresponds to the fact that the observed data are fitted perfectly under the saturated model, since this model includes all possible main effects and interaction effects (i.e., all possible τ parameters).

[24] This computer program calculated, e.g., the entries in table 2 above (namely, the maximum-likelihood estimates of the F_{ijkl} under model [5]), and the corresponding χ^2 values. In order to calculate these maximum-likelihood estimates, the computer program uses an iterative procedure that insures that the table 2 entries will fit the five marginal tables given in table 3. For further details about the computing procedure, see, e.g., the Appendix below.

TABLE 6

CHI-SQUARE VALUES FOR SOME MODELS PERTAINING TO TABLE 1

Model	Fitted Marginals	Degrees of Freedom	Likelihood-Ratio χ^2	Goodness-of-Fit χ^2
H_1	$\{PC\},\{PN\},\{CN\},\{CS\},\{NS\}$	6	2.53	2.62
H_2	$\{PC\},\{PN\},\{CN\},\{CS\}$	7	24.22	24.21
H_3	$\{PC\},\{PN\},\{CN\},\{NS\}$	7	34.31	34.31
H_4	$\{PC\},\{PN\},\{CS\},\{NS\}$	7	12.37	13.04
H_5	$\{PC\},\{CN\},\{CS\},\{NS\}$	7	16.48	16.46
H_6	$\{PN\},\{CN\},\{CS\},\{NS\}$	7	50.21	49.87
H_7	$\{PC\},\{PN\},\{PS\},\{CN\},\{CS\},\{NS\}$	5	2.25	2.29
H_8	$\{PC\},\{PN\},\{PS\},\{CN\},\{CS\}$	6	22.71	22.92
H_9	$\{PC\},\{PN\},\{PS\},\{CN\},\{NS\}$	6	30.00	30.52
H_{10}	$\{PC\},\{PN\},\{PS\},\{CS\},\{NS\}$	6	12.31	12.86
H_{11}	$\{PC\},\{PS\},\{CN\},\{CS\},\{NS\}$	6	14.97	15.10
H_{12}	$\{PN\},\{PS\},\{CN\},\{CS\},\{NS\}$	6	45.91	46.44
H_{13}	$\{PCN\},\{CNS\}$	4	1.89	1.95
H_{14}	$\{CNS\},\{PC\},\{PN\},\{PS\}$	4	1.63	1.65
H_{15}	$\{PCN\},\{PS\},\{CS\},\{NS\}$	4	2.24	2.29
H_{16}	$\{CNS\},\{PC\},\{PN\}$	5	1.90	1.95
H_{17}	$\{PCN\},\{CS\},\{NS\}$	5	2.52	2.62
H_{18}	$\{P\},\{C\},\{N\},\{S\}$	11	178.37	226.07
H_{19}	None	15	209.43	265.79

in the present context, (22) had some advantages (see, e.g., Goodman 1968, 1970). In the remaining discussion, we shall use only the χ^2 value based upon (22).

The hypothesis H described by model (8) (or model [5]) is listed as H_1 in table 6. In this model, there are five two-factor λ's—λ^{PC}, λ^{PN}, λ^{CN}, λ^{CS}, λ^{NS}. We next consider the models H_2–H_6 in table 6. In each of these models, there are four two-factor λ's. One of the five two-factor λ's in H_1 has been deleted in each of the models H_2–H_6.

In order to test whether the parameter λ^{PC} in H_1 contributes in a statistically significant way, the difference between the corresponding χ^2s for H_6 and H_1 can be used as a χ^2 statistic with 1 df. (The 1 df is obtained as the difference between the corresponding degrees of freedom for H_6 and H_1; i.e., $7 - 6 = 1$.) Similarly, each of the other four two-factor λ's in H_1 can be tested. From H_1 to H_6 of table 6, we see that each of the five two-factor λ's in H_1 contributes in a statistically significant way.[25] (Note

[25] For example, comparing H_4 with H_1, we see that λ^{CN} in H_1 contributes in a statistically significant way. On the other hand, it should also be noted that, when H_4 is tested by the data, the χ^2 value obtained is significant at the 10% level, but not at the 5% level (see table 6). (For some cautionary remarks on the interpretation of

that, for the data in table 1, the χ^2 values in table 6 for H_2–H_6 agree in their rank order with the standardized values of the corresponding λ's in table 5.)

Recall that the two-factor λ^{PS} was not included in H_1, but the other five two-factor λ's were. We next consider the model H_7 in which all six two-factor λ's are included. In order to test whether the parameter λ^{PS} in H_7 contributes in a statistically significant way, the difference between the corresponding χ^2s for H_1 and H_7 can be used as a χ^2 statistic with 1 df. (The 1 df is obtained as the difference between corresponding degrees of freedom for H_1 and H_7; i.e., $6 - 5 = 1$.) Similarly, each of the other five two-factor λ's in H_7 can be tested. From H_1 and H_7 to H_{12} of table 6, we see that λ^{PS} in H_7 does not contribute in a statistically significant way, but the other five two-factor λ's do. (Note that, for the data in table 1, the χ^2 values in table 6 for H_1 and H_8 to H_{12} agree in their rank order with the standardized values of the corresponding λ's in table 5.)

Next consider H_{13} of table 6. This model differs from H_1 to H_{12} in that it includes two three-factor λ's (λ^{PCN} and λ^{CNS}), whereas the other models included only two-factor λ's. Model H_{13} is fitted using the corresponding two three-way marginal tables $\{PCN\}$ and $\{CNS\}$. The F_{ijkl} for this model are particularly easy to calculate since this model states only that there is independence in the 2×2 table describing the expected relationship between variables P and S when the level of the two remaining variables, C and N, is given (see Goodman 1970). Corresponding to each of the four possible levels [i.e., (1,1), (1,2), (2,1), (2,2)] of the remaining variables (C,N), we have a single 2×2 table describing the relationship between variables P and S; and the χ^2 for testing independence in the 2×2 table will have 1 df. The χ^2 for H_{13} is equal to the sum of the four χ^2 values (corresponding to the four possible levels of the remaining variables $[C,N]$), and it will therefore have 4 df. From table 6, we see that model H_{13} fits the data well.

As with (3) and (4), we now find that, from the information contained in the three-way table $\{PCN\}$, we can obtain the three two-way marginal tables $\{PC\}$, $\{PN\}$, $\{CN\}$, and we can also obtain the three one-way marginals $\{P\}$, $\{C\}$, $\{N\}$. Similarly, from $\{CNS\}$ we can obtain the three two-way marginal tables $\{CN\}$, $\{CS\}$, $\{NS\}$, and the three-one way marginals $\{C\}$, $\{N\}$, $\{S\}$. From the set of marginal tables fitted under H_{13} ($\{PCN\}$ and $\{CNS\}$), we can obtain the four one-way marginals ($\{P\}$, $\{C\}$, $\{N\}$, $\{S\}$), and five two-way marginals ($\{PC\}$, $\{PN\}$, $\{CN\}$, $\{CS\}$, $\{NS\}$). Corresponding to this set of nine marginals, we have the λ parameters λ^P, λ^C, λ^N, λ^S, λ^{PC}, λ^{PN}, λ^{CN}, λ^{CS}, λ^{NS}; and corresponding to

---

these levels of significance, see, e.g., Goodman 1970.) The λ^{CN} parameter is not included in model H_4, but it is included in model H_1.

the two three-way tables we have λ^{PCN} and λ^{CNS}. There are a total of 11 λ's here. These 11 λ's are the ones that are included in model H_{13}.

From the remarks in the preceding paragraph, we now see that model H_1 can be expressed as follows. Model H_1 states both that model H_{13} is true *and* that the parameters λ^{PCN} and λ^{CNS} in H_{13} are equal to zero. Thus, if H_1 is true, then H_{13} will also be true; but H_{13} can be true in cases where H_1 is not true. Model H_{13} is implied by model H_1.

Consider next H_{14} of table 6. This model differs from H_{13} in the following two respects: The λ^{PCN} in H_{13} is set equal to zero in H_{14}; and the λ^{PS} in H_{14} is set equal to zero in H_{13}. Both of these models have the same number of degrees of freedom (namely, 4); and from table 6 we see that both models fit the data well.

Model H_{14} includes the four one-factor λ's, the six two-factor λ's, and also the three-factor λ^{CNS}. Because of this, we now see that model H_7 can be expressed as follows. Model H_7 states both that model H_{14} is true *and* that the parameter λ^{CNS} in H_{14} is equal to zero. Thus, if H_7 is true, then H_{14} will also be true; but H_{14} can be true in cases where H_7 is not true. Model H_{14} is implied by model H_7.

Model H_{15} of table 6 is similar in form to model H_{14}. The remarks in the preceding two paragraphs can be applied to model H_{15} simply by interchanging the letters *P* and *S* in the above statements.

Model H_{16} of table 6 is obtained from H_{14} by setting λ^{PS} in H_{14} equal to zero. From table 6, we see that H_{16} also fits the data well. A comparison of H_{16} and H_7 in table 6 indicates that H_{16} provides a slightly better fit to the data. (Models H_{16} and H_7 differ in the following two respects: the λ^{CNS} in H_{16} is set equal to zero in H_7; and the λ^{PS} in H_7 is set equal to zero in H_{16}.)

Model H_{17} of table 6 is similar in form to model H_{16}. This model also fits the data well, but not quite as well as H_{16} and H_7. Remarks similar to those presented in the preceding paragraph can also be applied to model H_{17}.

Models H_{13} to H_{17} are actually equivalent to models that arise in "logit analysis," but the other models of table 6 can *not* be expressed in that form. The class of models considered in the present article is more general than the class of models that arise in the usual logit analysis. In addition, the techniques proposed here are easier to apply than the corresponding techniques of logit analysis. For further details on these matters, see Bishop (1969) and Goodman (1970, 1971*a*, 1972*b*).

We noted earlier that model H_{13} was implied by model H_1. Similarly, we find that models H_7 and H_{14}–H_{17} are also implied by H_1; and that models H_{13}–H_{15} are implied by H_7.

Consider now models H_{18} and H_{19} of table 6. The former model states that the four variables in table 1 are mutually independent; and the latter model states that the 16 cells of the four-way cross-classification table (table 1) are equiprobable. From table 6, we see that these two models do not fit the data at all.

Before closing this section, we should point out that some of the material that we discussed above could be presented in summary form in tables that are somewhat analogous to the usual analysis of variance tables. We shall next present two examples of such tables. Many others could also be presented.

With respect to a given model in table 6 (say model H_i, for $i = 1,2, \ldots, 19$), we shall use the symbol $X^2(H_i)$ to denote its χ^2 value. We can view $X^2(H_i)$ as a measure of the "unexplained variation" in the f_{ijkl} when model H_i is used to "explain" the variation.[26] Having noted earlier in this section that the model of mutual independence among the four variables (model H_{18}) does not fit the data, we now present in table 7 a partitioning

TABLE 7

ANALYSIS OF ASSOCIATION IN TABLE 1, USING MODELS H_{18}, H_1, AND THEN H_7

Source of Variation	Degrees of Freedom	χ^2	Numerical Value
1. Total mutual dependence among variables P,C,N,S	11	$X^2(H_{18})$	178.37
1a. Due to mutual dependence unexplained by model H_1	6	$X^2(H_1)$	2.53
1b. Due to mutual dependence explained by model H_1	5	$X^2(H_{18}) - X^2(H_1)$	175.84
Partition of (1a)			
1a.1. Due to mutual dependence unexplained by model H_7	5	$X^2(H_7)$	2.25
1a.2. Due to mutual dependence explained by λ^{PS} in model H_7	1	$X^2(H_1) - X^2(H_7)$	0.28

of the unexplained variation $X^2(H_{18})$ (i.e., the total variation due to the mutual *dependence* among the four variables) into the following two components: (1a) The variation due to mutual dependence that remains unexplained after model H_1 has been introduced to explain the variation (i.e., after the five parameters λ^{PC}, λ^{PN}, λ^{CN}, λ^{CS}, λ^{NS}, have been

[26] The meaning of the term "unexplained variation" here is related to, but different from, the analogous term in the usual analysis of variance and in the usual multiple regression analysis.

included with the λ parameters of model H_{18}); and (1b) the variation due to mutual dependence that is explained by model H_1 (i.e., by the inclusion of the five two-factor λ's listed above with the λ parameters of model H_{18}). As noted in table 7, component (1a) is measured by $X^2(H_1)$, and component (1b) by $X^2(H_{18}) - X^2(H_1)$. Table 7 also partitions (1a) into the following two components: (1a.1) The variation due to mutual dependence that remains unexplained after model H_7 has been introduced (i.e., after the λ^{PS} parameter has been included with the λ parameters of model H_1); and (1a.2) the variation due to mutual dependence that is explained by the inclusion of λ^{PS} in the model. As noted in table 7, component (1a.1) is measured by $X^2(H_7)$, and component (1a.2) by $X^2(H_1) - X^2(H_7)$. We see here that model H_1 explains a great deal, and that there is no need to include the λ^{PS} parameter with the λ parameters of model H_1.

Table 7 provided examples of the partitioning of unexplained variation, namely, $X^2(H_{18})$ into (1a) and (1b); and $X^2(H_1)$ into (1a.1) and (1a.2). In table 8 we shall provide an example of the partitioning of "explained variation."

TABLE 8

ANALYSIS OF ASSOCIATION IN TABLE 1, USING MODELS H_{18}, H_7, AND THEN H_1

Source of Variation	Degrees of Freedom	χ^2	Numerical Value
1. Total mutual dependence among variables *P,C,N,S*	11	$X^2(H_{18})$	178.37
1a. Due to mutual dependence unexplained by model H_7	5	$X^2(H_7)$	2.25
1b. Due to mutual dependence explained by model H_7	6	$X^2(H_{18}) - X^2(H_7)$	176.12
Partition of (1b)			
1b.1. Due to mutual dependence explained by λ^{PS} in model H_7	1	$X^2(H_1) - X^2(H_7)$	0.28
1b.2. Due to mutual dependence explained by model H_1 (i.e., by the five two-factor λ's in that model)	5	$X^2(H_{18}) - X^2(H_1)$	175.84

If we had not started the analysis herein with a hypothesized causal system represented by figure 1 (i.e., by model H_1), after noting that model H_{18} did not fit the data, it would then seem natural to see how much of the unexplained variation $X^2(H_{18})$ could be explained by model

H_7 (i.e., by the inclusion of all six two-factor λ's with the λ parameters of model H_{18}). This is done on lines (1a) and (1b) of table 8. We next partition the variation due to mutual dependence that is *explained* by model H_7 into the following two components: (1b.1) the variation due to mutual dependence that is explained by the inclusion of λ^{PS} in the model; and (1b.2) the variation due to mutual dependence that is explained by the five two-factor λ's in model H_1. As noted in table 8, component (1b.1) is measured by $X^2(H_1) - X^2(H_7)$, and component (1b.2) by $X^2(H_{18}) - X^2(H_1)$. (Note that [1a], [1b.1], and [1b.2] of table 8 correspond to [1a.1], [1a.2], and [1b] of table 7.) We see here that model H_7 explains a great deal, but that the λ^{PS} parameter in that model could be deleted, thus yielding model H_1.

For additional examples of the use of tables like tables 7 and 8 (i.e., tables that are analogous to the usual analysis of variance tables) for the analysis of association in contingency tables, see, for example, Goodman (1970, 1971*a*).

MULTIPLE AND PARTIAL CORRELATION COEFFICIENTS FOR MULTIPLE CLASSIFICATIONS

In the usual multiple regression analysis for quantitative variables (predicting Y from, say, variables X_1 and X_2), the quantity $R^2_{Y \cdot X_1 X_2}$, which is the square of the multiple correlation coefficient, can be interpreted as follows: it is the relative decrease in the unexplained variation in Y that is obtained when comparing the case where X_1 and X_2 are not used to predict Y with the case where both X_1 and X_2 are used. Similarly, the quantity $r^2_{YX_1 \cdot X_2}$, which is the square of the partial correlation coefficient, can be interpreted as follows: it is the relative decrease in the unexplained variation in Y that is obtained when comparing the case where X_2 is used to predict Y (but X_1 is not used to predict Y) with the case where both X_1 and X_2 are used. The quantity $R^2_{Y \cdot X_1 X_2}$ is sometimes referred to as the coefficient of multiple determination, and the quantity $r^2_{YX_1 \cdot X_2}$ can be called the coefficient of partial determination. Coefficients that are somewhat analogous to the usual coefficients of multiple and partial determination were introduced by Goodman (1970, 1971*a*) for the analysis of qualitative variables such as the variables P, C, N, S considered herein. We shall now illustrate their calculation.

In the preceding section, we noted, among other things, that the statistic $X^2(H_6) - X^2(H_1)$ could be used to test whether the parameter λ^{PC} in H_1 contributed in a statistically significant way. To measure the magnitude of this contribution, we recommend the following coefficient, which we shall call the coefficient of partial determination between f and λ^{PC}, when the other λ's in model H_1 are "taken into account":[27]

$$r^2_{f\lambda_{PC}\cdot H_1} = [X^2(H_6) - X^2(H_1)]/X^2(H_6). \tag{23}$$

The coefficient defined by (23) measures the relative decrease in the unexplained variation in the f_{ijkl} that takes place when model H_1 is used to explain the variation, over the case when model H_6 is used for this purpose.[28] In other words, this coefficient measures the relative decrease in the unexplained variation that takes place when λ^{PC} is used (together with the λ's in model H_6) to explain the variation, over the case when only the λ's in model H_6 are used. From table 6, we see that this coefficient is equal to .95 for the data in table 1. In a similar way, we can measure the contribution made by each of the other four two-factor λ's in H_1.

In the preceding section, we also noted that the statistic

$$X^2(H_1) - X^2(H_7)$$

could be used to test whether the parameter λ^{PS} in H_7 contributed in a statistically significant way. As in the preceding paragraph, we shall measure the magnitude of this contribution by the following coefficient of partial determination:

$$r^2_{f\lambda_{PS}\cdot H_7} = [X^2(H_1) - X^2(H_7)]/X^2(H_1). \tag{24}$$

From table 6, we see that this coefficient is equal to .11 for the data in table 1. In a similar way, we can measure the contribution made by each of the other five two-factor λ's in H_7.

To measure how well model H_1 fits the data, we consider the following coefficient, which we shall call the coefficient of multiple determination between f and the λ parameters in model H_1:

$$R^2_{f\cdot H_1} = [X^2(H_{19}) - X^2(H_1)]/X^2(H_{19}). \tag{25}$$

From table 6, we see that this coefficient is equal to .99 for the data in table 1. It is also of interest to consider the following coefficient, which we shall call the coefficient of multiple-partial determination between f and the two-factor λ parameters in model H_1, when the one-factor λ parameters (λ^P, λ^C, λ^N, λ^S) in H_1 are taken into account:

$$R^2_{fH_1\cdot\lambda_P,\lambda_C,\lambda_N,\lambda_S} = [X^2(H_{18}) - X^2(H_1)]/X^2(H_{18}). \tag{26}$$

27 In the subscript of r^2 in (23), we changed the λ^{PC} notation to the λ_{PC} notation because of typographical considerations. This simple change in notation should not confuse the reader. Similar notational changes are also made later in this section.

28 Recall that the λ's in model H_6 are the same as in model H_1 except that λ^{PC} is not included in H_6.

From table 6, we see that this coefficient is also equal to .99 (to two decimal places) for the data in table 1.

In the preceding paragraph, we measured how well model H_1 fits the data. In a similar way, we can measure how well any of the other models in table 6 fit the data.

Before closing this section, we take note of the fact that, as in the usual multiple regression analysis, the coefficients of partial, multiple, and multiple-partial determination do not include adjustments for the fact that the two terms in the numerator of the coefficient (see, e.g., [26]) have different degrees of freedom. If such adjustments are desired, then each X^2 term in both the numerator and denominator (of, e.g., [26]) can be divided by its degrees of freedom.[29]

ON DIAGRAMS OF CAUSALITY AND SYSTEMS OF SIMULTANEOUS EQUATIONS

Earlier we introduced model (5) in order to describe in quantitative terms the causal system represented by figure 1. Now we shall show that model (5) can also be expressed as a system of simultaneous equations that describe how each "dependent" variable in the system is affected (in a certain sense) by the other variables in the system. In order to do this, we first introduce the notation presented in the next paragraph.

With respect to those individuals who are at levels *i, j,* and *k* on variables *P*, *C*, and *N*, respectively, the observed odds in favor of a + response on variable *S* (i.e., the odds that variable *S* will take on the value 1) can be written as

$$\omega^{\bar{S}}_{ijk\cdot} = f_{ijk1}/f_{ijk2}.$$

(For example, corresponding to level (1,1,1) on the joint variable (*P*,*C*,*N*) in table 1, the value of $\omega^{\bar{S}}_{111\cdot}$ is $77/32 = 2.41$.) With respect to those individuals who are at levels *i, j,* and *l* on variables *P*, *C*, and *S*, respectively, the observed odds in favor of a + response on variable *N* (i.e., the odds that variable *N* will take on the value 1) can be written as

$$\omega^{\bar{N}}_{ij\cdot l} = f_{ij1l}/f_{ij2l}. \qquad (27)$$

Similarly, we define the following odds:

$$\omega^{\bar{C}}_{i\cdot kl} = f_{i1kl}/f_{i2kl},$$
$$\omega^{\bar{P}}_{\cdot jkl} = f_{1jkl}/f_{2jkl}. \qquad (28)$$

[29] For the corresponding adjustment in the multiple regression context, see, e.g., Goldberger (1964) and Wiorkowski (1970). The numerical results reported here do not include this adjustment.

Corresponding to these observed odds, we now define the following "expected odds":

$$\Omega^{\bar{S}}{}_{ijk\cdot} = F_{ijk1}/F_{ijk2}, \tag{29}$$

$$\Omega^{\bar{N}}{}_{ij\cdot l} = F_{ij1l}/F_{ij2l}, \tag{30}$$

$$\Omega^{\bar{C}}{}_{i\cdot kl} = F_{i1kl}/F_{i2kl}, \tag{31}$$

$$\Omega^{\bar{P}}{}_{\cdot jkl} = F_{1jkl}/F_{2jkl}. \tag{32}$$

We next define a new set of parameters (the γ parameters) in terms of the τ parameters as follows:

$$\gamma^{\bar{S}} = (\tau^{S}{}_1)^2, \gamma^{C\bar{S}}{}_j = (\tau^{CS}{}_{j1})^2, \gamma^{N\bar{S}}{}_k = (\tau^{NS}{}_{k1})^2. \tag{33}$$

Applying now (29) and (33), we see from model (5) that

$$\Omega^{\bar{S}}{}_{ijk\cdot} = \gamma^{\bar{S}}\, \gamma^{C\bar{S}}{}_j\, \gamma^{N\bar{S}}{}_k. \tag{34}$$

Formula (34) states that the expected odds $\Omega^{\bar{S}}{}_{ijk\cdot}$ pertaining to variable S are dependent (in a certain sense) upon variables C and N, but not upon variable P.[30] Similarly, from model (5) we obtain the following additional formulae:

$$\Omega^{\bar{N}}{}_{ij\cdot l} = \gamma^{\bar{N}}\, \gamma^{P\bar{N}}{}_i\, \gamma^{C\bar{N}}{}_j\, \gamma^{S\bar{N}}{}_l, \tag{35}$$

$$\Omega^{\bar{C}}{}_{i\cdot kl} = \gamma^{\bar{C}}\, \gamma^{P\bar{C}}{}_i\, \gamma^{N\bar{C}}{}_k\, \gamma^{S\bar{C}}{}_l, \tag{36}$$

$$\Omega^{\bar{P}}{}_{\cdot jkl} = \gamma^{\bar{P}}\, \gamma^{C\bar{P}}{}_j\, \gamma^{N\bar{P}}{}_k, \tag{37}$$

where

$$\gamma^{\bar{N}} = (\tau^{N}{}_1)^2, \gamma^{\bar{C}} = (\tau^{C}{}_1)^2, \gamma^{\bar{P}} = (\tau^{P}{}_1)^2,$$
$$\gamma^{P\bar{N}}{}_i = (\tau^{PN}{}_{i1})^2, \gamma^{P\bar{C}}{}_i = (\tau^{PC}{}_{i1})^2, \text{etc.} \tag{38}$$

Note should be taken of the following relationships:

$$\gamma^{C\bar{S}}{}_1 = \gamma^{S\bar{C}}{}_1, \gamma^{N\bar{S}}{}_1 = \gamma^{S\bar{N}}{}_1, \gamma^{P\bar{N}}{}_1 = \gamma^{N\bar{P}}{}_1, \text{etc.} \tag{39}$$

Figure 1 can be viewed as a diagrammatic representation of the system of equations (34)–(36), in which condition (39) is satisfied.[31] On the other

[30] The parameters $\gamma^{\bar{S}}$, $\gamma^{C\bar{S}}{}_j$, and $\gamma^{N\bar{S}}{}_k$ in (34) describe the main effects on $\Omega^{\bar{S}}{}_{ijk\cdot}$ of the "general mean" and the variables C and N, respectively. Formula (34) states that $\Omega^{\bar{S}}{}_{ijk\cdot}$ can be expressed in terms of the main effects of the "general mean" and the variables C and N, and that there is no main effect of variable P upon $\Omega^{\bar{S}}{}_{ijk\cdot}$. For further discussion of formulas of this kind, see, e.g., Goodman (1972*b*).

[31] The double-headed arrow between, say, N and S in fig. 1 signifies that variable S is affected by variable N in equation (34), that variable N is affected by variable S

hand, the system of equations (34)–(37), with condition (39) satisfied, can be represented by figure 2. Both figures 1 and 2 can be described in

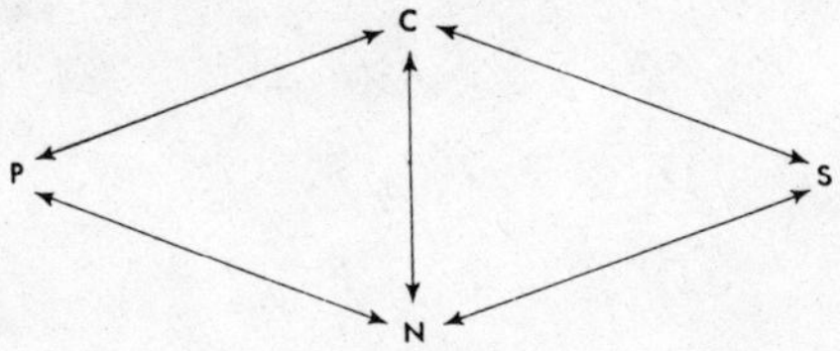

FIG. 2.—Diagram of the causal system described by the system of equations (34)–(37)

quantitative terms by model (5). To distinguish between these two causal systems (fig. 1 or 2), we must rely upon additional kinds of information that would help us to determine whether or not variables C and N are partial "causes" of variable P.

Equation (34) is equivalent to the unsaturated model obtained from (12) by setting the following τ parameters in (12) equal to one:[32] τ^{PS}, τ^{PCS}, τ^{PNS}, τ^{CNS}, τ^{PCNS}. We have listed these τ parameters in table 9, using

TABLE 9

THE τ PARAMETERS IN (12) THAT ARE SET EQUAL TO ONE, WHEN EQUATIONS PERTAINING TO EXPECTED ODDS ARE EXPRESSED AS UNSATURATED MODELS OBTAINED FROM (12)

Equation for Expected Odds	Equation No.	Parameters Set at One
Ω^S	(34)	[PS], [PCS], [PNS], [CNS], [$PCNS$]
Ω^N	(35)	[PCN], [PNS], [CNS], [$PCNS$]
Ω^C	(36)	[PCN], [PCS], [CNS], [$PCNS$]
Ω^P	(37)	[PS], [PCN], [PCS], [PNS], [$PCNS$]

in equation (35), and that the parameter $\gamma^{N\bar{S}}{}_1$ in (34) is equal to the parameter $\gamma^{S\bar{N}}{}_1$ in (35) (see condition [39]). We draw a double-headed arrow between N and S, rather than two single-headed arrows (one from N to S and one from S to N), as a way of taking note of the fact that the parameter $\gamma^{N\bar{S}}{}_1$ and $\gamma^{S\bar{N}}{}_1$ in (34) and (35), respectively, are equal. The single-headed arrow from, say, P to C in fig. 1 signifies that variable C is affected by variable P in equation (36). (Variable P is *not* affected directly by variable C in the system of equations [34]–[36]; we do *not* draw a double-headed arrow between P and C in fig. 1.) The absence of any kind of arrow between S and P in fig. 1 signifies that variable S is *not* affected directly by variable P (see eq. [34]) and that variable P is *not* affected directly by variable S in the system of equations (34)–(36). The above remarks should help to clarify the meaning of the double-headed and single-headed arrows in fig. 1, and in the other figures presented later.

[32] For related results, see, e.g., Bishop (1969) and Goodman (1970, 1971*a*).

the symbols $[PS]$, $[PCS]$, . . . , to denote τ^{PS}, τ^{PCS}, . . . , respectively. Recall now that model (5) was equivalent to the unsaturated model obtained by setting the following τ parameters in (12) equal to one: τ^{PS}, τ^{PCN}, τ^{PCS}, τ^{PNS}, τ^{CNS}, τ^{PCNS}. Thus, model (5) states both that equation (34) is true *and* that τ^{PCN} in (12) is set equal to one.

From table 9, we see that equation (35) is equivalent to the unsaturated model obtained from (12) by setting the following τ parameters in (12) equal to one: τ^{PCN}, τ^{PNS}, τ^{CNS}, τ^{PCNS}. Thus, model (5) states both that equation (35) is true *and* that τ^{PS} and τ^{PCS} in (12) are set equal to one. We also see that model (5) is equivalent to the statement that both equations (34) and (35) are true. A diagram of the causal system described by equations (34) and (35) is presented as figure 3*A*.

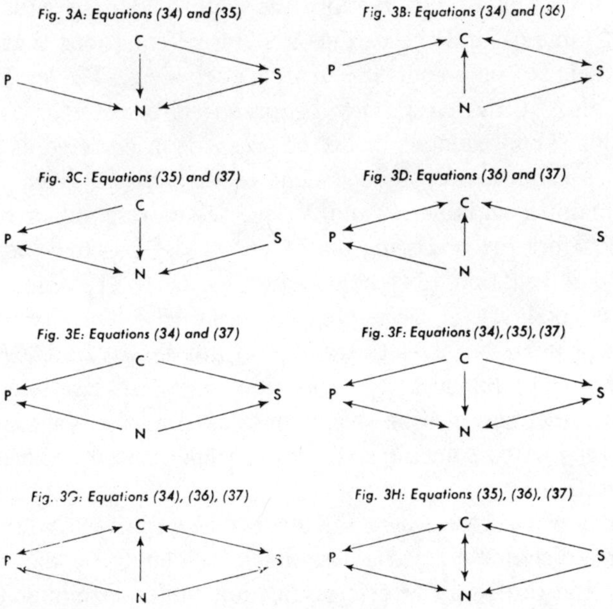

FIG. 3.—Diagrams of some additional causal systems described by model (5)

From table 9 we also find that model (5) is equivalent to the statement that both equations (34) and (36) are true, or that both equations (35) and (37) are true, or that both equations (36) and (37) are true, or that both equations (34) and (37) are true. Diagrams of the causal systems described by each of these pairs of equations are presented as figures 3*B*–3*E*.[33] Some additional causal systems that are also equivalent to model (5)

[33] Note that the causal system in fig. 3*E* is more parsimonious (in a certain sense) than the causal systems in figs. 3*A*–3*D*, which are in turn more parsimonious than the other causal systems in figs. 1–3. If we restrict consideration only to systems in which

are presented as figures 3*F*–3*H*. As we noted earlier, in order to distinguish among the various causal systems in figures 1–3, we must rely upon additional kinds of information pertaining to causality within the system of variables.[34]

Note that although there is no line between *P* and *C* in figure 3*A*, the model describing this causal system (model [5]) included the τ^{PC} parameter, since this parameter is not included among the τ parameters that are set equal to one in the system of equations (34)–(35) (see table 9). A similar kind of remark applies in figures 3*B*–3*E*. To further illustrate this point, we shall consider below what might happen in a simpler context, namely, when the system of variables under study consists of three dichotomous variables rather than four.

Before moving to the simpler context, we remind the reader that the preceding discussion in this section was concerned only with model (5) (i.e., H_1 of table 6) and the various systems of equations that correspond to that model. In some contexts, other models (e.g., H_{13} or H_{17} of table 6), which also fit the data, may be preferred because of their ease of interpretation. (For example, model H_{13} may be interpreted as stating that variable *S* is conditionally independent of variable *P*, given the level of the two remaining variables *C* and *N* [see the corresponding comments in the section before the preceding one]; and model H_{17} may be interpreted as stating that equation [34] holds true [see table 9], which means that the expected odds $\Omega^{\bar{S}}_{ijk\cdot}$ pertaining to variable *S* are dependent [in a certain sense] upon variables *C* and *N*, but not upon variable *P* [see n. 30 above].) The diagrammatic representation of models that are easy to interpret will sometimes also be easy (consider, e.g., the diagram representing model [34]), and sometimes not. (For some related comments, see the following section.)

Consider now the case where the system of variables under study consists of three dichotomous variables, which we denote by the letters *A, B,* and *C*. For the three-way cross-classification table pertaining to variables *A, B,* and *C*, we let f_{ijk} denote the observed frequency in cell (i,j,k) of the table (where $i = 1,2$; $j = 1,2$; $k = 1,2$), and we let F_{ijk} denote the cor-

---

variable *P* is an antecedent variable (see figs. 1, 3*A*, 3*B*), then the causal systems in figs. 3*A* and 3*B* are more parsimonious than the causal system in fig. 1.

[34] We shall discuss in a separate report how to distinguish among the causal systems that can be described by a given unsaturated model. When the unsaturated model describes a number of *different* possible causal systems, the model then does not serve as a complete description of a given causal system, since the system will not be uniquely determined by the model in that case. We shall state that the model describes the causal system even when it does not determine it uniquely. It might have been preferable, in that case, to say that the model serves as a partial description of the causal system, or that the model is consistent with the causal system. The abbreviated terminology used should cause no confusion.

responding expected frequency under some specified hypothesis. As in (12), we can write F_{ijk} as follows for the saturated model:

$$F_{ijk} = \eta \tau^{A}{}_{i} \tau^{B}{}_{j} \tau^{C}{}_{k} \tau^{AB}{}_{ij} \tau^{AC}{}_{ik} \tau^{BC}{}_{jk} \tau^{ABC}{}_{ijk}, \qquad (40)$$

where

$$\tau^{A}{}_{1} = \tau^{A}, \tau^{A}{}_{2} = 1/\tau^{A},$$
$$\tau^{AB}{}_{11} = \tau^{AB}{}_{22} = \tau^{AB}, \tau^{AB}{}_{12} = \tau^{AB}{}_{21} = 1/\tau^{AB},$$
$$\tau^{ABC}{}_{111} = \tau^{ABC}{}_{221} = \tau^{ABC}{}_{212} = \tau^{ABC}{}_{122} = \tau^{ABC},$$
$$\tau^{ABC}{}_{211} = \tau^{ABC}{}_{121} = \tau^{ABC}{}_{112} = \tau^{ABC}{}_{222} = 1/\tau^{ABC}, \text{etc.} \qquad (41)$$

Consider now the unsaturated model obtained from (40) by assuming that $\tau^{ABC} = 1$. In this case, the F_{ijk} under the model can be expressed as follows:

$$F_{ijk} = \eta \tau^{A}{}_{i} \tau^{B}{}_{j} \tau^{C}{}_{k} \tau^{AB}{}_{ij} \tau^{AC}{}_{ik} \tau^{BC}{}_{jk}. \qquad (42)$$

Applying the notation used earlier in the present section, we let $\Omega^{\bar{C}}{}_{ij.}$ denote the "expected odds" pertaining to variable C, when variables A and B are at levels i and j, respectively; that is,

$$\Omega^{\bar{C}}{}_{ij.} = F_{ij1}/F_{ij2}. \qquad (43)$$

From (42)–(43), we find that

$$\Omega^{\bar{C}}{}_{ij.} = \gamma^{\bar{C}} \gamma^{A\bar{C}}{}_{i} \gamma^{B\bar{C}}{}_{j}, \qquad (44)$$

where the γ's are defined as in (33) and (38). The model described by (44) is equivalent to the model described by (42).[35] Formula (44) states that the expected odds $\Omega^{\bar{C}}{}_{ij.}$ pertaining to variable C can be expressed in terms of the main effects of the "general mean" and the variables A and B.[36]

Consider now the causal system described in figure 4. This diagram is supposed to tell us that variable C is a "consequent variable," and that variables A and B are (partial) causes of variable C. This causal system

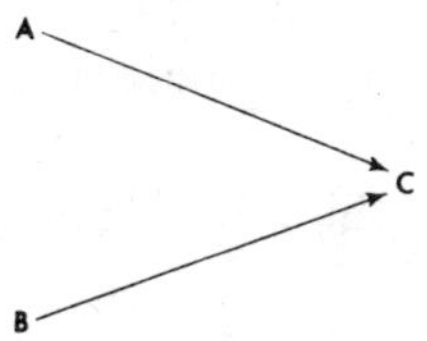

FIG. 4.—Diagram of a causal system among the three variables A, B, and C, in which variables A and B are (partial) causes of variable C.

[35] See references in n. 32 above.

[36] See comments in n. 30 above.

can be made quantitatively explicit in various ways. One of these ways would be to use the model described by formula (44). Since the model described by (44) is equivalent to the model described by (42), the causal system in figure 4 could also be described by model (42); that is, the unsaturated model obtained from (40) by setting τ^{ABC} equal to one.

It is interesting to note that model (42) includes the τ^{AB} parameter in it, but the causal system (fig. 4) which this model describes does ***not*** include a line between variables A and B. (A similar remark was made earlier when discussing model [5] and the causal systems in figs. 3*A*–3*E*.) Indeed, with respect to the causal system in figure 4, if there is statistical independence between variables A and B in the 2×2 marginal table for these two variables,[37] then the numerical value of the τ^{AB} parameter must be less than one (under certain conditions to be specified in the final section below).

The fact that the τ^{AB} parameter is included in model (42) could be noted in figure 4 by the introduction of, say, a wiggly double-headed arrow between A and B. Similarly, if there is not statistical independence between variables A and B in the 2×2 marginal table for these variables, we could draw attention to this fact by the use of, say, a curved double-headed arrow between A and B in figure 4 when neither variable is prior to the other (see, e.g., Duncan 1970). In addition, other symbols could be introduced into figure 4 to draw attention to the "variation" that still remains unexplained by the model. (This variation was discussed in the preceding two sections.) Various symbols of the kind noted above could help to make the diagrammatic representation of the causal system more informative. For the sake of simplicity and brevity, we shall not pursue this matter further here (except for a passing reference to it later when figs. 6 and 7 are presented).

The τ^{AB} parameter in model (42) pertains to the relationship between variables A and B, when the level of the remaining variable C is given. (See formula [7] for a corresponding result for the system of four variables.) In other words, the parameter τ^{AB} pertains to the relationship between variables A and B ***within*** the system of three variables. It does ***not*** pertain directly to the relationship between variables A and B in the 2×2 marginal table for these two variables; but, in a more indirect way, the τ^{AB} parameter together with some of the other τ parameters in the model will affect this relationship. We shall discuss this point in more detail in our final section.

In the paragraph before the preceding one, we noted that figure 4 did not include a line between variables A and B, but the unsaturated model

[37] In other words, we consider for the moment the case where the two-way marginal table $\{AB\}$ pertaining to variables A and B exhibits statistical independence between the "row variable" and the "column variable."

describing this causal system, model (42), did include the τ^{AB} parameter. Now we shall present a different causal system, which also does not include a line between variables A and B, but in this case the model describing it does *not* include the τ^{AB} parameter. This causal system is presented in figure 5, which is supposed to tell us that variable C is an "antecedent variable," and that it affects both variables A and B.

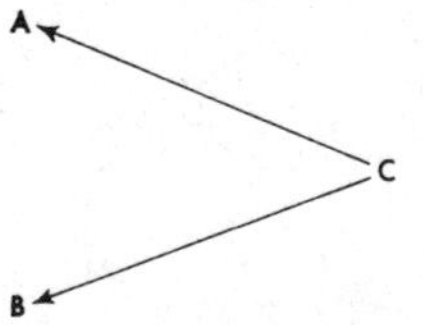

FIG. 5.—Diagram of a causal system among the three variables A, B, and C, in which variable C affects both variables A and B.

In order to obtain a more quantitative description of the causal system in figure 5, consider now the unsaturated model obtained from (40) by setting τ^{AB} and τ^{ABC} equal to one. In this case, the F_{ijk} under the model can be expressed as follows:

$$F_{ijk} = \eta\tau^{A}{}_{i}\tau^{B}{}_{j}\tau^{C}{}_{k}\tau^{AC}{}_{ik}\tau^{BC}{}_{jk}. \qquad (45)$$

Letting $\Omega^{\bar{A}}{}_{\cdot jk}$ and $\Omega^{\bar{B}}{}_{i\cdot k}$ denote the "expected odds" pertaining to variables A and B, respectively, that is,

$$\Omega^{\bar{A}}{}_{\cdot jk} = F_{1jk}/F_{2jk},\ \Omega^{\bar{B}}{}_{i\cdot k} = F_{i1k}/F_{i2k}, \qquad (46)$$

we find that model (45) is equivalent to the model described by the following equations:[38]

$$\Omega^{\bar{A}}{}_{\cdot jk} = \gamma^{\bar{A}}\,\gamma^{C\bar{A}}{}_{k}, \qquad (47)$$

$$\Omega^{\bar{B}}{}_{i\cdot k} = \gamma^{\bar{B}}\,\gamma^{C\bar{B}}{}_{k}, \qquad (48)$$

where the γ's are defined as in (33) and (38). Formula (47) states that the expected odds $\Omega^{\bar{A}}{}_{\cdot jk}$ pertaining to variable A can be expressed in terms of the main effect of the "general mean" and the variable C, and that there is no main effect of variable B upon $\Omega^{\bar{A}}{}_{\cdot jk}$. Formula (48) states that the expected odds $\Omega^{\bar{B}}{}_{i\cdot k}$ pertaining to variable B can be expressed in terms of the main effect of the "general mean" and the variable C, and that there is no main effect of variable A upon $\Omega^{\bar{B}}{}_{i\cdot k}$.

The causal system in figure 5 can be made quantitatively explicit in various ways. One of these ways would be to use the model described by the system of equations (47)–(48). Since the model described by (47)–(48) is equivalent to the model described by (45), the causal system in

[38] Model (45) is actually equivalent to equation (47) alone, to equation (48) alone, and to the system of equations (47)–(48).

figure 5 could also be described by model (45); that is, by the unsaturated model obtained from (40) by setting τ^{AB} and τ^{ABC} equal to one.

As we noted earlier, a given unsaturated model can serve as a description of a number of different causal systems (see figs. 1–3). In figure 6, we present some causal systems, aside from the one in figure 4, that can be described by model (42); and in figure 7, we present some causal systems, aside from the one in figure 5, that can be described by model (45).

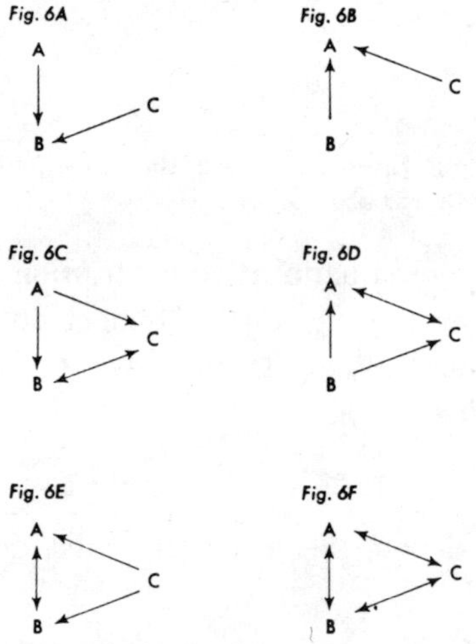

FIG. 6.—Diagrams of some additional causal systems described by model (42)

The four figures near the bottom of figure 7 require special comment. We included figures *7D.a* and *7D.b* in figure 7 in order to take note of the fact that figure *7D* can represent model (47) when either (*a*) variables *B* and *C* are mutually dependent in the 2 × 2 marginal table for these variables (see my earlier comment about curved double-headed arrows), or (*b*) variable *B* is a (partial) cause of variable *C* in the 2 × 2 marginal table, or (*c*) variables *B* and *C* are statistically independent in the marginal table.[39] A similar comment applies to figures *7E.a* and *7E.b*.[40]

[39] The arrow from *B* to *C* in fig. *7D.b* (and the arrow from *A* to *C* in fig. *7E.b*) do not have the same kind of meaning as the other arrows used previously in this paper, since that arrow in fig. *7D.b* (and the one in fig. *7E.b*) pertain to certain relationships within *marginal* tables whereas the other arrows used pertain to relationships within the *full* cross-classification table. (For example, the arrow from *C* to *A* in fig. *7D*, and in figs. *7D.a* and *7D.b*, refers to equation (47) for the expected odds $\Omega^{\bar{A}}_{\cdot jk}$ rather than for the corresponding expected odds $\Omega^{\bar{A}}_{\cdot k}$ pertaining to variable *A* obtained from the 2 × 2 marginal table for variables *A* and *C*. Although the value of the parameter $\gamma^{C\bar{A}}_{k}$ in equation (47) will remain the same under model (47) for $\Omega^{\bar{A}}_{\cdot jk}$ and under the corresponding model for $\Omega^{\bar{A}}_{\cdot k}$, in general the parameter values and the models themselves will change—see, e.g., Goodman 1971*b*.)

[40] It should also be noted that figs. *6A* and *6B* (and fig. 4) can be supplemented in a similar way to the supplementation of figs. *7D* and *7E* by *7D.a*, *7D.b*, *7E.a*, *7E.b*.

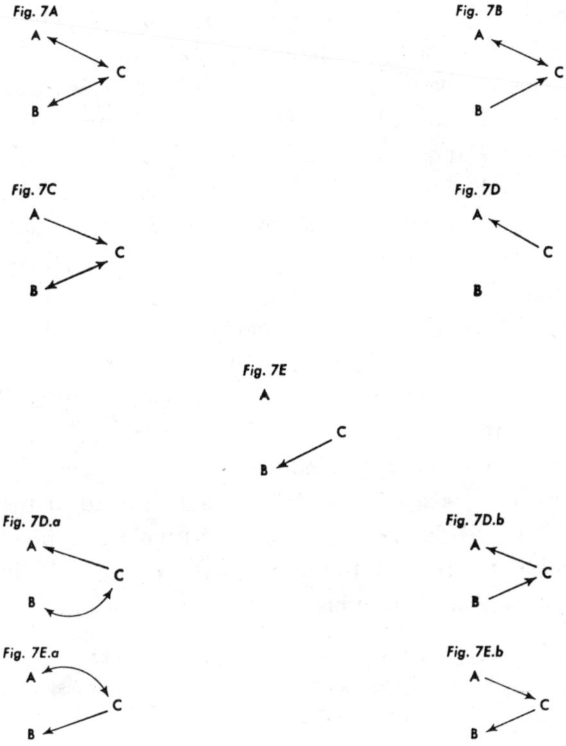

FIG. 7.—Diagrams of some additional causal systems described by model (45)

The methods presented in the earlier sections can be applied in order to decide whether the observed data support model (42) or model (45); that is, whether τ^{ABC} can be set equal to one or whether τ^{AB} and τ^{ABC} can be set equal to one. Similarly, when a causal system in figure 4 or 6 is being compared with a causal system in figures 5 or 7, the methods referred to above can be used to distinguish between these causal systems.

If we were to limit consideration to causal systems in which there is no reciprocal causation, as is often done (see, e.g., Blalock 1964 and Duncan 1970), then model (42) yields only figures 4, 6*A*, and 6*B* (and their supplementary forms);[41] model (45) yields only figures 5, 7*D*, and 7*E* (and their supplementary forms); and model (5) yields only figure 3*E*. On the other hand, if we also consider relationships within certain marginal tables (in a way that would be analogous to what is done in the analysis of recursive systems), then additional causal systems are obtained. For example, if instead of using equations (34) and (35) for $\Omega^{\bar{S}}_{ijk\cdot}$ and $\Omega^{\bar{N}}_{ij\cdot l}$ (as in fig. 3*A*), we replaced (35) by the corresponding equation for the expected odds $\Omega^{\bar{N}}_{ij\cdot}$ pertaining to variable N in the marginal table

[41] See n. 40 above.

$\{PCN\}$,[42] then the only change in figure 3*A* would be the replacement of the double-headed arrow by a single-headed arrow from *N* to *S*.[43] Similar kinds of changes can be made in figures 3*B*–3*D* and 3*F*–3*H*, and in figures 1 and 2.[44] In addition, figures 3*A*–3*E* can be supplemented as we did earlier to figure 7*D* and 7*E*.

It should also be noted that, as in the usual analysis of recursive systems, by making certain assumptions as to which variables are prior to which other variables, the number of possible causal systems corresponding to a given model can be reduced. For example, if the order of priority is from variable *P* to *N* to *C* to *S*, then model (5) yields only *one* causal system; namely, the changed figure 3*B*, with the double-headed arrow replaced by a single-headed arrow from *C* to *S*, and with the insertion of a single-headed arrow from *P* to *N*. On the other hand, if neither variable *P* nor *N* is prior to the other, but both are prior to *C*, which is prior to *S*, then the above remark applies with the single-headed arrow from *P* to *N* replaced by a curved double-headed arrow. Remarks similar to the above ones can also be applied with the four letters (*P*, *N*, *C*, *S*) interchanged in the ordering, as long as *S* or *P* has the final position in the ordering.[45]

[42] Under model (5), we find that, for the marginal table $\{PCN\}$ obtained when variable *S* is collapsed, the expected odds $\Omega^{\bar{N}}_{ij\cdot}$ can be expressed by the following equation:

$$\Omega^{\bar{N}}_{ij\cdot} = \gamma^{\bar{N}}\gamma^{P\bar{N}}_{i}\gamma^{C\bar{N}}_{j},$$

where the value of the parameters $\gamma^{\bar{N}}$ and $\gamma^{C\bar{N}}_{j}$ may differ from the corresponding values in equation (35).

[43] The arrows from *P* to *N* and from *C* to *N* will not have the same kind of meaning as the arrows that pertain to certain relationships within the *full* cross-classification table, rather than within the marginal table $\{PCN\}$. For related comments, see n.39 above.

[44] Because of the possible changes that may occur in the relationships among the variables when one of the variables is collapsed, caution must be exercised here. For example, consider the case where equation (34) for $\Omega^{\bar{S}}_{ijk\cdot}$ is replaced by the equation obtained for the expected odds $\Omega^{\bar{S}}_{ij\cdot}$ pertaining to variable *S* in the marginal table $\{PCS\}$ obtained when variable *N* is collapsed. Under model (5), we find that the following equation is obtained:

$$\Omega^{\bar{S}}_{ij\cdot} = \gamma^{\bar{S}}\gamma^{P\bar{S}}_{i}\gamma^{C\bar{S}}_{j}\gamma^{PC\bar{S}}_{ij},$$

where the value of the parameters $\gamma^{\bar{S}}$ and $\gamma^{C\bar{S}}_{i}$ in this equation may differ from the corresponding values in equation (34), and where the new parameters $\gamma^{P\bar{S}}_{i}$ and $\gamma^{PC\bar{S}}_{ij}$ are introduced. Because the above equation introduces these new parameters, various complications can arise. For example, if instead of using equations (34) and (35) for $\Omega^{\bar{S}}_{ijk\cdot}$ *and* $\Omega^{\bar{N}}_{ij\cdot l}$ (as in fig. 3*A*), we replaced (34) by the above equation for $\Omega^{\bar{S}}_{ij\cdot}$, then the system of equations thus obtained could *not* be represented by a simple modification of fig. 3*A* (i.e., the replacement of the double-headed arrow by a single-headed arrow from *S* to *N* is *not* sufficient for this purpose), and this system of equations would *not* be equivalent to model (5).

[45] The material considered in the preceding two paragraphs will be discussed in more detail in Goodman (1972*a*).

Before closing this section, we remind the reader that the data analyzed here in the cross-classification (table 1) describe how variables vary together, and they can not be used to establish causation unless certain kinds of assumptions are made. For example, we must assume that the system under study is closed in the sense that, if variables outside the system (e.g., variables other than *P, C, N, S* of table 1) have any effect on the variables in the system, it is to produce stochastic disturbances corresponding to the usual random variation of the observed frequency f_{ijkl} from its expected value F_{ijkl} (when sampling is random). In addition, we must make other assumptions (e.g., assumptions as to which variables are prior to which other variables) in order to select one causal system from among the various causal systems that are consistent with a given model.

ON DIAGRAMS OF CAUSALITY AND THREE-FACTOR INTERACTIONS IN THE MODELS

In analyzing a $2 \times 2 \times 2$ cross-classification table describing the relationship between variables *A, B,* and *C*, let us suppose that there was a strong positive relationship between variables *A* and *B* when variable *C* was at level 1, and that there was a less strong positive relationship (or possibly a zero or a negative relationship) between variables *A* and *B* when variable *C* was at level 2. This situation is an example of the case where there is three-factor interaction in the three-way table (i.e., where the interaction between variables *A* and *B* depends upon the level of variable *C*). This case can occur frequently in practice (see, e.g., Goodman 1964, 1965). On the other hand, the usual diagrams of causal systems (and also diagrams of the kind used in this paper) are not suited to describing this case, since the arrow from, say, *A* to *B* in the diagram does not usually indicate whether it pertains to the case where variable *C* is at level 1 or at level 2.

In the analysis of table 1, we noted that it actually was not necessary to include the three-factor (or four-factor) interactions in the model that fit the data well. (For some examples of cases where the inclusion of some three-factor interactions did improve the fit, see Goodman [1970, 1972*b*].) The models presented (see, e.g., [12]) can include three-factor interactions (and also interactions pertaining to more than three factors), and the methods of analyzing these models will be the same as those used in the present article. On the other hand, the survey analyst who considers only causal systems that can be described by the usual kinds of diagrams is limited to the case where only two-factor interactions are considered (and the interactions pertaining to three or more factors are assumed to be nil). In this respect, he will appreciate the fact that the models and methods presented here will allow him to study a wider range of systems of polytomous variables than he could otherwise do.

All of the methods of analysis presented in the preceding sections could be applied to systems of polytomous variables (see, e.g., Goodman 1970, 1971*a*). Although table 1 dealt with dichotomous variables, these methods of analysis could be applied more generally. In the next (final) section we shall deal only with the case where the variables are dichotomous.

HOW THE RELATIONSHIP BETWEEN TWO GIVEN DICHOTOMOUS VARIABLES IS AFFECTED BY THE INTRODUCTION OF ADDITIONAL VARIABLES

Consider the situation where a research worker begins his study by examining the relationship between two dichotomous variables, say, variables A and B. He does this by examining a 2×2 cross-classification table pertaining to variables A and B. Let us suppose that the research worker then introduces a third dichotomous variable, say, variable C, into his study in order to see what the relationship between variables A and B is at each level of variable C. He then obtains a $2 \times 2 \times 2$ cross-classification table pertaining to variables A, B, and C, which can be viewed as a pair of 2×2 tables, namely, a 2×2 table pertaining to the relationship between variables A and B when variable C is at level 1, and a corresponding 2×2 table when variable C is at level 2. In the present section, we shall show how the relationship between variables A and B, which the research worker examined first (before he introduced variable C into his study), is affected by the introduction of variable C. We shall see how the relationship in the 2×2 table pertaining to variables A and B, which the research worker examined first, is related to the corresponding relationship obtained when variable C is at level 1 and also when variable C is at level 2.

This topic was studied earlier, for example, by Lazarsfeld (1961) and Davis (1971) (see also Yule and Kendall [1950] and Kendall and Lazarsfeld [1950]). Lazarsfeld (1961) showed how the "cross-product difference" between two given dichotomous variables can be affected by the introduction of a third variable, but he also noted (as did Davis 1971) that the cross-product difference is not usually a satisfactory measure of the relationship between the two given dichotomous variables. In its place, Davis used Yule's Q, but then he found that the effect on Q of the introduction of the third variable was more complicated. He presented a "rule of thumb" concerning the possible effect on Q, but this rule can give incorrect results. (As Davis [1971] has noted, his "rule of thumb . . . has no mathematical necessity.") In the present section, we shall present some new results on this topic. The point of view developed in the earlier sections herein will lead us to these new results.

From (7) we see that the expected odds-ratio defined there is closely related to the corresponding two-factor τ parameter in our models. The odds-ratio is also closely related to Yule's Q, and to some other measures of association as well (see, e.g., Goodman 1965). For the sake of simplicity, the results which we shall present in this section will be expressed in terms of the odds-ratio, but some of these results can be directly translated into corresponding results about the other measures of association (such as Yule's Q) that are monotonic functions of the odds-ratio.

Consider now the three-way cross-classification table pertaining to the dichotomous variables A,B, and C. In discussing this table, we shall use the notation developed earlier when presenting formulae (40)–(41). We shall also use the symbol $\zeta_{AB \cdot k}$ to denote the conditional expected odds-ratio between variables A and B when variable C is at level k (for $k = 1,2$), that is,

$$\zeta_{AB \cdot k} = (F_{11k}F_{22k})/(F_{12k}F_{21k}). \tag{49}$$

Consider now the unsaturated model (42) obtained from (40) by setting τ^{ABC} equal to one. From (41), (42), and (49), we find that

$$\zeta_{AB \cdot 1} = \zeta_{AB \cdot 2} = (\tau^{AB})^4. \tag{50}$$

Formula (50) tells us that under model (42) the conditional expected odds-ratio pertaining to variables A and B when variable C is at level 1 is equal to the corresponding quantity when variable C is at level 2;[46] and that this quantity is equal to $(\tau^{AB})^4$.

In the more general case where τ^{ABC} need not be equal to one,[47] then

$$\sqrt{\zeta_{AB \cdot 1}\, \zeta_{AB \cdot 2}} = (\tau^{AB})^4. \tag{51}$$

We shall use the symbol $\zeta_{AB \cdot C}$ to denote the geometric mean of the two expected odds-ratio on the left side of (51), and we shall call this quantity the *partial* "expected odds-ratio" between variables A and B when variable C is held constant. Formula (51) can be written as

[46] In other words, under the unsaturated model (42) (i.e., when $\tau^{ABC} = 1$), the introduction of variable C does not produce a so-called specification effect on the relationship between variables A and B.

[47] In this more general case, the odds-ratio between variables A and B when variable C is at level 1 need not be equal to the corresponding quantity when variable C is at level 2. In other words, the introduction of variable C here can produce a so-called specification effect on the relationship between variables A and B. It should also be noted that the more general case (namely, when τ^{ABC} need not be equal to one) refers to the saturated model (40); and the estimate of the expected frequency F_{ijk} under the saturated model will equal to the corresponding observed frequency f_{ijk}. Thus, in the more general case, any quantity based upon the estimated F_{ijk} (e.g., the estimate of the conditional *expected* odds-ratios $\zeta_{AB \cdot 1}$ and $\zeta_{AB \cdot 2}$) will be equal to the corresponding quantity based upon the observed f_{ijk} (e.g., the conditional *observed* odds-ratios).

$$\zeta_{AB\cdot C} = (\tau^{AB})^4. \tag{52}$$

Returning now to the case where $\tau^{ABC} = 1$, we see from (50) and (52) that

$$\zeta_{AB\cdot C} = \zeta_{AB\cdot 1} = \zeta_{AB\cdot 2} \tag{53}$$

in this special case. Formula (53) states that the *partial* ζ is equal to each of the two *conditional* ζ's in the case considered here.

Letting F^{AB}_{ij} denote the expected frequency in cell (i,j) in the marginal 2×2 table $\{AB\}$, we shall use the symbol υ_{AB} to denote the expected odds-ratio between variables A and B in this table,[48] that is,

$$\upsilon_{AB} = (F^{AB}{}_{11}\, F^{AB}{}_{22})/(F^{AB}{}_{12}\, F^{AB}{}_{21}). \tag{54}$$

In the case where $\tau^{ABC} = 1$, we then find

$$\upsilon_{AB} \begin{matrix} > \\ = \\ < \end{matrix} \zeta_{AB\cdot C} \tag{55}$$

according as

$$(1 - \zeta_{AC\cdot B})(1 - \zeta_{BC\cdot A}) \begin{matrix} > \\ = \\ < \end{matrix} 0, \tag{56}$$

where

$$\zeta_{AC\cdot B} = (\tau^{AC})^4, \zeta_{BC\cdot A} = (\tau^{BC})^4. \tag{57}$$

This result states that, when $\tau^{ABC} = 1$, the expected odds-ratio υ_{AB} pertaining to variables A and B will be larger than the corresponding *partial* expected odds-ratio $\zeta_{AB\cdot C}$, when the other two *partial* expected odds-ratios, $\zeta_{AC\cdot B}$ and $\zeta_{BC\cdot A}$ are both larger than one or when they are both less than one. Formulae (55)–(56) also state that υ_{AB} will be less than $\zeta_{AB\cdot C}$ when $\zeta_{AC\cdot B} > 1$ and $\zeta_{BC\cdot A} < 1$ or when $\zeta_{AC\cdot B} < 1$ and $\zeta_{BC\cdot A} > 1$. Furthermore, these formulae state that υ_{AB} will be equal to $\zeta_{AB\cdot C}$ when $\zeta_{AC\cdot B} = 1$ or when $\zeta_{BC\cdot A} = 1$.

In the case considered above (i.e., where $\tau^{ABC} = 1$), we see from (53)

[48] The estimate of the expected frequency F^{AB}_{ij} in the marginal table $\{AB\}$ will be equal to the corresponding observed frequency f^{AB}_{ij} under the saturated model (40) (see the last two sentences of n. 47 above); and this will also be the case under the unsaturated model obtained when $\tau^{ABC} = 1$, since the marginal table $\{AB\}$ is included among the marginal tables fitted under that model. Thus, both in the more general case and in the case where $\tau^{ABC} = 1$, any quantity based upon the estimated F^{AB}_{ij} (e.g., the estimate of the *expected* odds-ratio υ_{AB}) will be equal to the corresponding quantity based upon the observed f^{AB}_{ij} (e.g., the *observed* odds-ratio).

that $\zeta_{AB\cdot C}$ can be expressed simply as the conditional expected odds-ratio $\zeta_{AB\cdot 1}$ or $\zeta_{AB\cdot 2}$. Similarly, in this case we find that

$$\zeta_{AC\cdot B} = \zeta_{AC\cdot 1} = \zeta_{AC\cdot 2},$$

$$\zeta_{BC\cdot A} = \zeta_{BC\cdot 1} = \zeta_{BC\cdot 2}. \tag{58}$$

When $\zeta_{AB\cdot C}$ is larger than one, we say that the *partial* relationship between variables A and B is positive; when $\zeta_{AB\cdot C}$ is less than one, we say that this relationship is negative; and when $\zeta_{AB\cdot C}$ is equal to one, we say that this relationship is nil.[49] A similar remark applies to $\zeta_{AC\cdot B}$ and $\zeta_{BC\cdot A}$. Thus, formulae (55)–(56) state that $v_{AB} > \zeta_{AB\cdot C}$ when both the *partial* relationship between variables A and C and the *partial* relationship between variables B and C have the same sign (i.e., both positive or both negative). Formulae (55)–(56) also state that $v_{AB} < \zeta_{AB\cdot C}$ when the *partial* relationship between variables A and C and the *partial* relationship between variables B and C have opposite signs (one positive and one negative). These formulae also state that $v_{AB} = \zeta_{AB\cdot C}$ when at least one of the other two *partial* relationships is nil.

The application of formulae (55)–(56) might be illustrated as follows: consider the relationship between variables A and B described in table 10.

TABLE 10

HYPOTHETICAL 2 × 2 CROSS-CLASSIFICATION TABLE DESCRIBING RELATIONSHIP BETWEEN VARIABLES A AND B

	B	$\bar{B}$
A	19	5
$\bar{A}$	7	2

For the data in table 10, the estimated odds-ratio v_{AB} is 38/35. The relationship between variables A and B is positive, since the estimated v_{AB} is greater than one. Now can we find a variable C that will "explain" the relationship between variables A and B? In other words, can we find a

[49] Note that the *partial* relationship between variables A and B is positive, negative, or nil according as $\zeta_{AB\cdot C} - 1$ is positive, negative, or nil. In other words, the sign of the *partial* relationship between variables A and B is the same as the sign of $\zeta_{AB\cdot C} - 1$ In the case considered above (viz., where $\tau^{ABC} = 1$), the sign of $\zeta_{AB\cdot C} - 1$ will also be the same as the sign of the corresponding *partial* Q coefficient. Thus, in the case considered above, a comment about the sign of the *partial* relationship between variables A and B (i.e., a comment about the sign of $\zeta_{AB\cdot C} - 1$) can be viewed also as a comment about the sign of the corresponding *partial* Q coefficient.

variable C which is such that the *partial* relationship between variables A and B becomes nil when variable C is taken into account, that is, $\zeta_{AB \cdot C} = 1$. From formulae (55)–(56), we see that, in the case where variable C does not have a specification effect,[50] in order for variable C to explain the relationship between variables A and B, both the *partial* relationship between A and C and the *partial* relationship between B and C must have the same sign (both positive or both negative).[51]

As an illustration of the above remarks, suppose that when variable C is introduced, the 2×2 table (table 10), which describes the relationship between variables A and B, becomes the $2 \times 2 \times 2$ tables presented as table 11. For the data in table 11, variable C actually explains the relationship

TABLE 11

HYPOTHETICAL $2 \times 2 \times 2$ CROSS-CLASSIFICATION TABLE DESCRIBING RELATIONSHIP BETWEEN VARIABLES A, B, AND C

	C: B	C: $\bar{B}$		$\bar{C}$: B	$\bar{C}$: $\bar{B}$
A	4	2	A	15	3
$\bar{A}$	2	1	$\bar{A}$	5	1

between variables A and B, since the estimated $\zeta_{AB \cdot C}$ is equal to one. (Note also that variable C does not have a specification effect, since the estimated $\zeta_{AB \cdot 1}$ and $\zeta_{AB \cdot 2}$ are equal.) The $2 \times 2 \times 2$ table (table 11) can also be rewritten in the two equivalent forms presented as table 12. From table 12 we see that the *partial* relationship between variables A and C is negative (since the estimated $\zeta_{AC \cdot B}$ is equal to $2/3$) and the *partial* relationship between variables B and C is also negative (since the $\zeta_{BC \cdot A}$ is equal to $2/5$). If one of these two *partial* relationships had been positive and the other negative, then from formulae (55)–(56) we would know that variable C could *not* explain the relationship between variables A and B.

The above remarks applied to the case when $v_{AB} > 1$ (see table 10). In this case, we have noted that in order to explain the relationship between variables A and B, we must seek a variable C for which both the

50 Recall that variable C will not have a specification effect when $\zeta_{AB \cdot 1} = \zeta_{AB \cdot 2}$; i.e., when $\tau^{ABC} = 1$ (see formula [50] and n. 46 above).

51 When both $1 - \zeta_{AC \cdot B}$ and $1 - \zeta_{BC \cdot A}$ have the same sign, we can be sure that $v_{AB} > \zeta_{AB \cdot C}$ (see formulae [55]–[56]); but we can not be sure that $\zeta_{AB \cdot C} = 1$. In the case where $v_{AB} > 1$, the condition described above (that both $1 - \zeta_{AC \cdot B}$ and $1 - \zeta_{BC \cdot A}$ have the same sign) is a necessary condition in order that $\zeta_{AB \cdot C} = 1$; but it is not a sufficient condition to insure that $\zeta_{AB \cdot C} = 1$.

TABLE 12

RELATIONSHIP BETWEEN VARIABLES A, B, AND C PRESENTED IN TABLE 11, BUT EXPRESSED IN TWO OTHER EQUIVALENT FORMS

Table 12A

	B			$\bar{B}$	
	C	$\bar{C}$		C	$\bar{C}$
A	4	15	A	2	3
$\bar{A}$	2	5	$\bar{A}$	1	1

Table 12B

	A			$\bar{A}$	
	C	$\bar{C}$		C	$\bar{C}$
B	4	15	B	2	5
$\bar{B}$	2	3	B	1	1

partial relationship between A and C and the *partial* relationship between B and C must have the same sign. From formulae (55)–(56), we also see that, when $\upsilon_{AB} < 1$, in order to explain the relationship between variables A and B, we must seek a variable C for which the *partial* relationship between A and C and the *partial* relationship between B and C have opposite signs (one positive and one negative).

We have been concerned above with the search for a variable C that is an "explanatory" variable. Formulae (55)–(56) can also be applied to obtain further insight in the case where variable C is a "suppressor" variable or where variable C has "no effect." With respect to a suppressor variable, recall that $\upsilon_{AB} < \zeta_{AB \cdot C}$ (so that a positive relationship between variables A and B becomes more positive when variable C is taken into account) when the *partial* relationship between variables A and C and the *partial* relationship between variables B and C have opposite signs; and also that $\upsilon_{AB} > \zeta_{AB \cdot C}$ (so that a negative relationship between variables A and B becomes more negative when variable C is taken into account) when both the *partial* relationship between variables A and C and the *partial* relationship between variables B and C have the same sign. With respect to a variable that has no effect, recall that $\upsilon_{AB} = \zeta_{AB \cdot C}$ when at least one of the other two *partial* relationships is nil.[52]

[52] In all of the above applications of formulae (55)–(56), we consider only the case where $\tau^{ABC} = 1$, i.e., where variable C has no specification effect.

Now let us move from the case where $\tau^{ABC} = 1$ to the more general case where τ^{ABC} need not be equal to one. In this more general case, we find that the relationship (55) holds according as

$$(1 - \zeta_{AC\cdot B})(1 - \zeta_{BC\cdot A})(\zeta_{C\cdot AB}\zeta_{ABC})^{1/2} + (1 - \zeta_{C\cdot AB})(1 - \zeta_{ABC})(\zeta_{AC\cdot B}\zeta_{BC\cdot A})^{1/2} \gtreqless 0, \quad (59)$$

where

$$\zeta_{C\cdot AB} = (\tau^{C})^4, \zeta_{ABC} = (\tau^{ABC})^4. \quad (60)$$

Note that (59) is a generalization of (56).

Consider now the four-way table, where variables A, B, C, D denote the four dichotomous variables in the table. Formula (12) applies here for the saturated model, with the letters P, C, N, S in (12) replaced by A, B, C, D, respectively. We shall use the symbol $\zeta_{AB\cdot kl}$ to denote the conditional expected odds-ratio between variables A and B when the level of the C and D variables is set at k and l, respectively (for $k = 1,2$; $l = 1,2$). As in (7), we find that in the special case where the three-factor and four-factor τ's are equal to one, then

$$\zeta_{AB\cdot kl} = (\tau^{AB})^4, \quad (61)$$

for $k = 1,2$ and $l = 1,2$. In the more general case where the three-factor and four-factor τ's need not be equal to one, then

$$(\zeta_{AB\cdot 11}\,\zeta_{AB\cdot 22}\,\zeta_{AB\cdot 12}\,\zeta_{AB\cdot 21})^{1/4} = (\tau^{AB})^4. \quad (62)$$

We shall use the symbol $\zeta_{AB\cdot CD}$ to denote the geometric mean of the four expected odds-ratios on the left side of (62). Thus, from (62) we see that

$$\zeta_{AB\cdot CD} = (\tau^{AB})^4. \quad (63)$$

Letting F_{ijk}^{ABC} denote the expected frequency in cell (i,j,k) in the marginal $2 \times 2 \times 2$ table $\{ABC\}$, we shall use the symbol $\upsilon_{AB\cdot k}$ to denote the expected odds-ratio between variables A and B when variable C is at level k (for $k = 1,2$) in the $\{ABC\}$ table. In the special case where all three-factor and four-factor τ's in (12) are equal to one, we then find that

$$\upsilon_{AB\cdot k} \gtreqless \zeta_{AB\cdot CD} \text{ (for } k = 1,2) \quad (64)$$

according as

$$(1 - \zeta_{AD\cdot BC})(1 - \zeta_{BD\cdot AC}) \gtreqless 0, \quad (65)$$

where

$$\zeta_{AD\cdot BC} = (\tau^{AD})^4, \zeta_{BD\cdot AC} = (\tau^{BD})^4. \tag{66}$$

In addition, we note that

$$v_{AB\cdot 2} \gtreqless v_{AB\cdot 1} \tag{67}$$

according as

$$(1 - \zeta_{AD\cdot BC})(1 - \zeta_{BD\cdot AC})(1 - \zeta_{CD\cdot AB})(1 - \zeta_{D\cdot ABC}) \gtreqless 0, \tag{68}$$

where

$$\zeta_{D\cdot ABC} = (\tau^{D})^4, \zeta_{CD\cdot AB} = (\tau^{CD})^4. \tag{69}$$

We shall now consider the general case where the three-factor and four-factor τ's need not be equal to one. We shall use the symbol $\zeta_{AB\cdot kD}$ to denote the geometric mean of $\zeta_{AB\cdot k1}$ and $\zeta_{AB\cdot k2}$, that is,

$$\zeta_{AB\cdot kD} = \sqrt{\zeta_{AB\cdot k1}\, \zeta_{AB\cdot k2}}. \tag{70}$$

From (12) and (70), we obtain the following:

$$\zeta_{AB\cdot 1D} = (\tau^{AB}\tau^{ABC})^4,$$
$$\zeta_{AB\cdot 2D} = (\tau^{AB}/\tau^{ABC})^4. \tag{71}$$

We also find that

$$v_{AB\cdot k} \gtreqless \zeta_{AB\cdot kD} \tag{72}$$

for $k = 1$ according as the following quantity is greater than, equal to, or less than zero:

$$(1 - \zeta_{AD\cdot BC}\zeta_{ACD\cdot B})(1 - \zeta_{BD\cdot AC}\zeta_{BCD\cdot A})\,(\zeta_{D\cdot ABC}\zeta_{CD\cdot AB}\zeta_{ABD\cdot C}\zeta_{ABCD})^{1/2} \tag{73a}$$
$$+ (1 - \zeta_{D\cdot ABC}\zeta_{CD\cdot AB})(1 - \zeta_{ABD\cdot C}\zeta_{ABCD})\,(\zeta_{AD\cdot BC}\zeta_{BD\cdot AC}\zeta_{ACD\cdot B}\zeta_{BCD\cdot A})^{1/2}$$

and for $k = 2$ according as the following quantity is greater than, equal to, or less than zero:

$$(\zeta_{ACD\cdot B} - \zeta_{AD\cdot BC})(\zeta_{BCD\cdot A} - \zeta_{BD\cdot AC})(\zeta_{D\cdot ABC}\zeta_{CD\cdot AB}\zeta_{ABD\cdot C}\zeta_{ABCD})^{1/2} \tag{73b}$$
$$+ (\zeta_{CD\cdot AB} - \zeta_{D\cdot ABC})(\zeta_{ABCD} - \zeta_{ABD\cdot C})\,(\zeta_{AD\cdot BC}\zeta_{BD\cdot AC}\zeta_{ACD\cdot B}\zeta_{BCD\cdot A})^{1/2}$$

where

$$\zeta_{ACD\cdot B} = (\tau^{ACD})^4, \ldots, \zeta_{ABCD} = (\tau^{ABCD})^4. \tag{74}$$

Note that (73) is a generalization of (65). In addition, we note that relationship (67) holds true according as the quantity

$$\frac{\{\zeta^{1/2}_{*}[(\zeta_{AD\cdot BC}\zeta_{ACD\cdot B}) + (\zeta_{BD\cdot AC}\zeta_{BCD\cdot A})] + \zeta^{1/2}_{**}[(\zeta_{D\cdot ABC}\zeta_{CD\cdot AB}) + (\zeta_{ABD\cdot C}\zeta_{ABCD}}{\{\zeta^{1/2}_{*}[1 + \zeta_{**}] + \zeta^{1/2}_{**}[1 + \zeta_{*}]\}\zeta^{2}_{ABC\cdot D}}$$

is greater than, equal to, or less than the quantity (75)

$$\frac{\{\zeta^{1/2}_{*}[(\zeta_{AD\cdot BC}\zeta_{BCD\cdot A}) + (\zeta_{BD\cdot AC}\zeta_{ACD\cdot B})] + \zeta^{1/2}_{**}[(\zeta_{D\cdot ABC}\zeta_{ABCD}) + (\zeta_{ABD\cdot C}\zeta_{CD\cdot AB}}{\{\zeta^{1/2}_{*}[(\zeta_{AD\cdot BC}\zeta_{BD\cdot AC}) + (\zeta_{ACD\cdot B}\zeta_{BCD\cdot A})] + \zeta^{1/2}_{**}[(\zeta_{D\cdot ABC}\zeta_{ABD\cdot C}) + (\zeta_{CD\cdot AB}\zeta_{ABCD}}$$

where

$$\zeta_{*} = \zeta_{D\cdot ABC}\zeta_{CD\cdot AB}\zeta_{ABD\cdot C}\zeta_{ABCD},$$

$$\zeta_{**} = \zeta_{AD\cdot BC}\zeta_{BD\cdot AC}\zeta_{ACD\cdot B}\zeta_{BCD\cdot A}. \qquad (76)$$

The relationship (75) is a generalization of (68). In the special case where the three-factor and four-factor τ's are equal to one, we find that (75) simplifies to (68). It is also possible to obtain other simplifications of (75) by assuming that other specified sets of τ's are set equal to one.

The results presented in the preceding paragraph generalize the relationships (65) and (68) to the case where the three-factor and four-factor τ's need not be equal to one. A simpler kind of generalization can be obtained in the following way.

Consider now the five-way table, where variables A, B, C, D, E denote the five dichotomous variables in the table. We shall use the symbol $\zeta_{AB\cdot klm}$ to denote the conditional expected odds-ratio between variables A and B when the level of the variables C, D, E is set at k, l, m, respectively (for $k = 1,2$; $l = 1,2$; $m = 1,2$). As in (61), we find that in the special case where the three-factor, four-factor, and five-factor τ's are equal to one, then

$$\zeta_{AB\cdot klm} = (\tau^{AB})^{4}, \qquad (77)$$

for $k = 1,2$; $l = 1,2$; $m = 1,2$. We shall use the symbol $\zeta_{AB\cdot CDE}$ to denote the geometric mean of the six expected odds-ratios corresponding to the six joint levels of variables C, D, E. As earlier, we see that

$$\zeta_{AB\cdot CDE} = (\tau^{AB})^{4}. \qquad (78)$$

Letting F^{ABCD}_{ijkl} denote the expected frequency in cell (i,j,k,l) in the marginal $2 \times 2 \times 2 \times 2$ table $\{ABCD\}$, we shall use the symbol $\upsilon_{AB\cdot kl}$ to denote the expected odds-ratio between variables A and B when the level of variables C and D is set at k and l, respectively, in the $\{ABCD\}$ table. In the special case where all three-factor, four-factor, and five-factor τ's are equal to one, we then find that

$$v_{AB\cdot kl} \gtreqless \zeta_{AB\cdot CDE} \tag{79}$$

(for $k = 1,2$; $l = 1,2$) according as

$$(1 - \zeta_{AE\cdot BCD})(1 - \zeta_{BE\cdot ACD}) \gtreqless 0, \tag{80}$$

where

$$\zeta_{AE\cdot BCD} = (\tau^{AE})^4, \zeta_{BE\cdot ACD} = (\tau^{BE})^4. \tag{81}$$

In addition, we note that

$$v_{AB\cdot 2l} \gtreqless v_{AB\cdot 1l} \tag{82}$$

according as

$$(1 - \zeta_{AE\cdot BCD})(1 - \zeta_{BE\cdot ACD})(1 - \zeta_{CE\cdot ABD})(1 - \zeta_{E\cdot ABCD}) \gtreqless 0, \tag{83}$$

where

$$\zeta_{E\cdot ABCD} = (\tau^{E})^4, \zeta_{CE\cdot ABD} = (\tau^{CE})^4. \tag{84}$$

Similarly, we find that

$$v_{AB\cdot k2} \gtreqless v_{AB\cdot k1} \tag{85}$$

according as

$$(1 - \zeta_{AE\cdot BCD})(1 - \zeta_{BE\cdot ACD})(1 - \zeta_{DE\cdot ABC})(1 - \zeta_{E\cdot ABCD}) \gtreqless 0. \tag{86}$$

The relationships (65) and (68) have now been extended by (80), (83), (86) to the case where an additional variable is taken into account. By comparing (56), (65), and (80), we find that the corresponding extension to the case where any number of additional variables are taken into account is straightforward. A similar remark applies also to the extension of (68), (83), and (86).

For various applications of the results presented in the above section see Goodman (1972*a*).

APPENDIX

How to Calculate the Estimate of the Expected Frequencies under a Given Unsaturated Model

We shall now describe an iterative procedure that can be used to calculate the maximum-likelihood estimate $\hat{F}_{ijkl}$ of the expected frequency F_{ijkl} under any given unsaturated model of the kind considered in the present article. For illustrative purposes, we shall again consider model (5) (i.e., H_1 of table 6), and shall show how the $\hat{F}_{ijkl}$ entries in table 2 were calculated. As we noted earlier (see n. 24 above), the iterative procedure for calculating $\hat{F}_{ijkl}$ under model (5) insures that the entries in the five marginal tables given in table 3, which were obtained from the data in table 1, will be equal to the corresponding entries in the five marginal tables calculated from the $\hat{F}_{ijkl}$ in table 2.

The entries in the five two-way marginal tables given in table 3 were calculated from the data in table 1 by applying formula (3) and similar formulae. We shall denote these entries as f^{PC}_{ij}, f^{PN}_{ik}, f^{CN}_{jk}, f^{CS}_{jl}, f^{NS}_{kl} (for $i = 1,2$; $j = 1,2$; $k = 1,2$; $l = 1,2$), corresponding to the two-way marginal tables $\{PC\}$, $\{PN\}$, $\{CN\}$, $\{CS\}$, $\{NS\}$, respectively. These five two-way marginal tables were listed in table 6 as the marginal tables that are fitted under H_1.

In order to calculate the $\hat{F}_{ijkl}$ under H_1, the iterative procedure starts with an initial value denoted by $\hat{F}^{(0)}_{ijkl}$; the first step changes $\hat{F}^{(0)}_{ijkl}$ to $\hat{F}^{(1)}_{ijkl}$; the second step change. $\hat{F}^{(1)}_{ijkl}$ to $\hat{F}^{(2)}_{ijkl}$; the third step changes $\hat{F}^{(2)}_{ijkl}$ to $\hat{F}^{(3)}_{ijkl}$ etc. Table A1 gives the numerical values of the $\hat{F}^{(0)}_{ijkl}$, $\hat{F}^{(1)}_{ijkl}$, $\hat{F}^{(2)}_{ijkl}$, . . . that are obtained with the iterative procedure. This procedure can be described as follows:

We begin the procedure simply by setting

$$\hat{F}^{(0)}{}_{ijkl} = 1 \tag{A0}$$

(for $i = 1,2$; $j = 1,2$; $k = 1,2$; $l = 1,2$). (See the column for step 0 in table A1.) For the first step, calculate

$$\hat{F}^{(1)}{}_{ijkl} = \hat{F}^{(0)}{}_{ijkl}\, f^{PC}{}_{ij} / [\hat{F}^{(0)}]^{PC}{}_{ij}, \tag{A1}$$

where

$$[\hat{F}^{(0)}]^{PC}{}_{ij} = \sum_{k=1}^{2} \sum_{l=1}^{2} \hat{F}^{(0)}{}_{ijkl}. \tag{A1a}$$

(Comparing [A1a] with formula [3], we see that $[\hat{F}^{(0)}]^{PC}_{ij}$ is defined in the same way as f^{PC}_{ij}, but with the f_{ijkl} in formula [3] replaced in [A1a] by the $\hat{F}^{(0)}_{ijkl}$.) For the second step, calculate

TABLE A1

VALUES OF $\hat{F}^{(0)}_{ijkl}$, $\hat{F}^{(1)}_{ijkl}$, $\hat{F}^{(2)}_{ijkl}$, . . . THAT ARE OBTAINED BY THE ITERATIVE PROCEDURE USED TO CALCULATE THE ESTIMATE OF THE EXPECTED FREQUENCIES PRESENTED IN TABLE 2

	Iterative Steps										
P C N S	0	1	2	3	4	5	6	7	8	24	25
+ + + + ...	1.00	43.75	49.70	58.05	69.43	83.29	80.31	75.94	74.40	74.80	74.80
+ + + −	1.00	43.75	49.70	58.05	46.66	37.32	35.99	34.03	33.34	34.47	34.48
+ + − + ...	1.00	43.75	37.80	32.09	38.38	31.48	30.35	32.83	33.64	32.28	32.28
+ + − −	1.00	43.75	37.80	32.09	25.80	29.41	28.36	30.68	31.43	33.45	33.45
+ − + + ...	1.00	18.75	21.30	16.94	11.06	13.27	14.26	13.49	14.01	14.75	14.75
+ − + −	1.00	18.75	21.30	16.94	22.81	18.24	19.61	18.54	19.26	17.98	17.98
+ − − + ...	1.00	18.75	16.20	18.47	12.07	9.89	10.64	11.51	11.32	11.30	11.30
+ − − −	1.00	18.75	16.20	18.47	24.88	28.36	30.49	32.98	32.45	30.97	30.97
− + + + ...	1.00	34.00	23.93	27.95	33.44	40.11	42.26	45.30	44.38	42.94	42.94
− + + −	1.00	34.00	23.93	27.95	22.47	17.97	18.94	20.30	19.89	19.79	19.79
− + − + ...	1.00	34.00	44.07	37.41	44.75	36.69	38.67	37.31	38.22	35.99	35.99
− + − −	1.00	34.00	44.07	37.41	30.07	34.28	36.13	34.86	35.71	37.29	37.29
− − + + ...	1.00	55.50	39.07	31.06	20.29	24.34	23.73	25.43	26.42	28.51	28.51
− − + −	1.00	55.50	39.07	31.06	41.84	33.46	32.63	34.97	36.32	34.76	34.76
− − − + ...	1.00	55.50	71.93	82.03	53.58	43.94	42.84	41.34	40.66	42.43	42.44
− − − −	1.00	55.50	71.93	82.03	110.47	125.94	122.80	118.49	116.57	116.29	116.29

$$\hat{F}^{(2)}_{ijkl} = \hat{F}^{(1)}_{ijkl} f^{PN}_{ik}/[\hat{F}^{(1)}]^{PN}_{ik}, \tag{A2}$$

where

$$[\hat{F}^{(1)}]^{PN}_{ik} = \sum_{j=1}^{2} \sum_{l=1}^{2} \hat{F}^{(1)}_{ijkl}. \tag{A2a}$$

For the third step, calculate

$$\hat{F}^{(3)}_{ijkl} = \hat{F}^{(2)}_{ijkl} f^{CN}_{jk}/[\hat{F}^{(2)}]^{CN}_{jk}, \tag{A3}$$

where

$$[\hat{F}^{(2)}]^{CN}_{jk} = \sum_{i=1}^{2} \sum_{l=1}^{2} \hat{F}^{(2)}_{ijkl}. \tag{A3a}$$

The fourth and fifth steps are similarly calculated using the f^{CS}_{jl} and f^{NS}_{kl}, respectively, in the calculations (with the $[\hat{F}^{(3)}]^{CS}_{jl}$ and $[\hat{F}^{(4)}]^{NS}_{kl}$ defined in a similar way to the corresponding quantities given by [A1a], [A2a], [A3a]).

After the fifth step is calculated, we say that the "first cycle of iterations" has been completed. (Recall that there were five marginal tables to be fitted—the five steps described above corresponded to the fitting of these five marginal tables.) We then begin the "second cycle of iterations" with the sixth step. For the sixth step, calculate

$$\hat{F}^{(6)}_{ijkl} = \hat{F}^{(5)}_{ijkl} f^{PC}_{ij}/[\hat{F}^{(5)}]^{PC}_{ij}, \tag{A6}$$

where

$$[\hat{F}^{(5)}]^{PC}_{ij} = \sum_{k=1}^{2} \sum_{l=1}^{2} \hat{F}^{(5)}_{ijkl}. \tag{A6a}$$

Note that the sixth step (i.e., the first step of the second cycle of iterations) repeats step 1 with the $\hat{F}^{(0)}$ in step 1 replaced now by $\hat{F}^{(5)}$ (compare [A1] with [A6]). Similarly, the seventh step repeats step 2 with the $\hat{F}^{(1)}$ in step 2 replaced now by $\hat{F}^{(6)}$; and the eighth step repeats step 3 with the $\hat{F}^{(2)}$ in step 3 replaced now by $\hat{F}^{(7)}$; etc.

As we noted earlier, table A1 gives the numerical values of the $\hat{F}^{(0)}_{ijkl}$ $\hat{F}^{(1)}_{ijkl}$, $\hat{F}^{(2)}_{ijkl}$, . . . that are obtained by carrying out the calculations described in the preceding two paragraphs. (All calculations have been carried out to more decimal places than are reported in table A1.) The reader can use the numerical values in table A1 to check both his understanding of the iterative procedure described above and the results to be presented next.

From (A1) we find that

$$\sum_{k=1}^{2} \sum_{l=1}^{2} \hat{F}^{(1)}{}_{ijkl} = f^{PC}{}_{ij}. \tag{A7}$$

In other words, the $\hat{F}^{(1)}_{ijk}$ fit the marginal table $\{PC\}$. From (A2) we find that

$$\sum_{j=1}^{2} \sum_{l=1}^{2} \hat{F}^{(2)}{}_{ijkl} = f^{PN}{}_{ik}, \tag{A8}$$

i.e., the $F^{(2)}_{ijkl}$ fit the marginal table $\{PN\}$. From (A3) we find that

$$\sum_{i=1}^{2} \sum_{l=1}^{2} \hat{F}^{(3)}{}_{ijkl} = f^{PN}{}_{jk}, \tag{A9}$$

i.e., the $\hat{F}^{(3)}_{ijkl}$ fit the marginal table $\{CN\}$. Similarly we find that the $\hat{F}^{(4)}_{ijkl}$ and $\hat{F}^{(5)}_{ijkl}$ fit the marginal tables $\{CS\}$ and $\{NS\}$, respectively. Thus, at the end of the first cycle of iterations, we have in turn fitted each of the five two-way marginal tables that were listed in table 6 as the marginal tables fitted under H_1 (see marginal tables in table 3). However, although the $F^{(5)}_{ijkl}$ fit the marginal table $\{NS\}$, they will usually not fit the other four marginal tables ($\{PC\}$, $\{PN\}$, $\{CN\}$, $\{CS\}$), and so we proceed to the second cycle of iterations. (From [A6] we see that the $\hat{F}^{(6)}_{ijkl}$ fit the marginal table $\{PC\}$; but they will usually not fit the other four marginal tables, $\{PN\}$, $\{CN\}$, $\{CS\}$, $\{NS\}$.) The iterative procedure is continued until we reach, say, the *u*th step, where the $\hat{F}^{(u)}_{iikl}$ fit all five marginal tables. Note that the $\hat{F}^{(25)}_{ijkl}$ in table A1 actually fit these five marginal tables.

For expository purposes, the present Appendix focused upon the iterative procedure for the case where model (5) (i.e., H_1 of table 6) is the particular unsaturated model under consideration. This iterative procedure can be applied more generally to a wide range of unsaturated models. Using notation that is similar to the notation appearing in the present Appendix, a description of the iterative procedure for this wide range of models was given by formula (3.10) in Goodman (1970). For related material, the reader is also referred to Ku and Kullback (1968), Bishop (1969), Bishop and Mosteller (1969), Goodman (1969*b*, 1970), Haberman (1970).

The iterative procedure referred to above is not the only way to calculate the $\hat{F}_{ijkl}$ in some cases. For certain kinds of unsaturated models, explicit formulae for the $\hat{F}_{ijkl}$ are available (see, e.g., formulae [3.5]–[3.7] and [4.1]–[4.2] in Goodman 1970). In addition, note should also be taken of the fact that, for various kinds of unsaturated models, the $\hat{F}_{ijkl}$ can be calculated by a step-by-step method that is quite different from the iterative procedure considered in the present Appendix (see, e.g., Goodman 1971*b*). For the models to which they are applicable, both the explicit formulae and the step-by-step method are easier to apply than the iterative procedure considered here; but the iterative procedure has the advantage that it is applicable to a wider range of unsaturated models—a range that includes all of the models to which the explicit formulae and the step-by-step method are applicable, and other models as well. As we noted earlier, a single computer program (which applies the iterative procedure) can be used to calculate the $\hat{F}_{ijkl}$, and other statistics of interest, for any set of models in this wide range of unsaturated models.

The computer program referred to above can be used with any unsaturated model that is "hierarchical" (see, e.g., Goodman 1970). Model (5) and the equivalent model (8) are examples of hierarchical models since, for each interaction effect (e.g., λ^{PC}) in model (8), there is also included in the model the corresponding main effects (λ^{P} and λ^{C}). More generally, an unsaturated model is hierarchical if, corresponding to the superscript of each λ parameter in the model, all subsets of the letters in that superscript appear themselves as superscripts of λ parameters that are also included in the model. (Corresponding to, say, the superscript *PC*, we have the two subsets *P* and *C*.) All of the models considered in the present article were examples of hierarchical models.

Before closing this Appendix, let us briefly consider "nonhierarchical" models. As an example of such a model, consider, say, the model obtained from (8) by setting the λ^{S} parameter in (8) equal to zero. This model states, among other things, that the main effect λ^{S} is nil, but the corresponding interaction effects λ^{CS} and λ^{NS} need not be nil. (It is nonhierarchical because λ^{S} is not included in the model despite the fact the λ^{CS} and λ^{NS} are.) This model differs from model (8) only in that λ^{S} is included now with the λ parameters that are set equal to zero. The maximum-likelihood estimates of the parameters under this model would be equal to those obtained under model (8) (see table 4) if the estimated value of λ^{S} under model (8) actually turned out to be zero. To the extent that the estimated value of λ^{S} under model (8) is close to zero (see table 4), the maximum-likelihood estimates of the parameters under model (8) will be close to those of the nonhierarchical model introduced in the present paragraph.[53] In cases where this approximation may not be

[53] The maximum likelihood estimates under model (8) can serve as consistent estimates of the corresponding parameters under the nonhierarchical model; but they are not the maximum-likelihood estimates under the latter model, except when the estimated value of λ^{S} under model (8) actually turns out to be zero. In estimating the parameters in the nonhierarchical model, the maximum-likelihood estimates under model (8) will not be as efficient as the maximum-likelihood estimates under the nonhierarchical model.

adequate,[54] the maximum-likelihood estimates under the nonhierarchical model can be calculated by methods that are more complicated than those presented herein (see Haberman 1970).

The remarks in the preceding paragraph pertaining to a particular nonhierarchical model (and the hierarchical model that is related to it) can be extended in a straightforward way to apply to other nonhierarchical models as well.

REFERENCES

Bahadur, R. R. 1961. "A Representation of the Joint Distribution of Responses to *N* Dichotomous Items." In *Studies in Item Analysis and Prediction*, edited by H. Solomon. Stanford, Calif.: Stanford University Press.

Birch, M. W. 1963. "Maximum Likelihood in the Three-Way Contingency Tables." *Journal of the Royal Statistical Society* (ser. B) 25:220–33.

Bishop, Y. M. M. 1969. "Full Contingency Tables, Logits, and Split Contingency Tables." *Biometrics* 25:383–400.

Bishop, Y. M. M., and F. Mosteller. 1969. "Smoothed Contingency Tables." In *The National Halothane Study*. Washington, D.C.: U.S. Government Printing Office.

Blalock, H. J., Jr. 1964. *Causal Inferences in Nonexperimental Research*. Chapel Hill: University of North Carolina Press.

Davis, J. A. 1971. *Elementary Survey Analysis*. Englewood Cliffs, N.J.: Prentice-Hall.

Duncan, O. D. 1970. "Partials, Partitions, and Paths." In *Sociological Methodology 1970*, edited by E. F. Borgatta and G. W. Bohrnstedt. San Francisco: Jossey-Bass.

Goldberger, A. S. 1964. *Econometric Theory*. New York: Wiley.

Goodman, L. A. 1964. "Simple Methods for Analyzing Three-Factor Interactions in Contingency Tables. *Journal of the American Statistical Association* 59:319–52.

———. 1965. "On the Multivariate Analysis of Three Dichotomous Variables." *American Journal of Sociology* 71:290–301.

———. 1968. "The Analysis of Cross-classified Data: Independence Quasi-Independence, and Interactions in Contingency Tables with or without Missing Entries." *Journal of the American Statistical Association* 63:1091–1131.

———. 1969*a*. "How to Ransack Social Mobility Tables and Other Kinds of Cross-Classification Tables." *American Journal of Sociology* 75:1–40.

———. 1969*b*. "On Partitioning χ^2 and Detecting Partial Association in Three-Way Contingency Tables." *Journal of the Royal Statistical Society* (ser. B) 31:486–98.

———. 1970. "The Multivariate Analysis of Qualitative Data: Interactions among Multiple Classifications." *Journal of the American Statistical Association* 65:226–56.

———. 1971*a*. "The Analysis of Multidimensional Contingency Tables: Stepwise Procedures and Direct Estimation Methods for Building Models for Multiple Classifications." *Technometrics* 13:33–61.

———. 1971*b*. "Partitioning of Chi-Square, Analysis of Marginal Contingency Tables, and Estimation of Expected Frequencies in Multidimensional Contingency Tables." *Journal of the American Statistical Association* 66:339–44.

———. 1972*a*. "Causal Analysis of Data from Panel Studies and Other Kinds of Surveys." *American Journal of Sociology* (forthcoming).

———. 1972*b*. "A Modified Multiple Regression Approach to the Analysis of Dichotomous Variables." *American Sociological Review* 37:28–46.

Haberman, S. J. 1970. The General Log-linear Model. Ph.D. thesis. University of Chicago.

Kendall, P. L., and P. F. Lazarsfeld. 1950. "Problems of Survey Analysis." In *Continuities in Social Research*, edited by R. K. Merton and P. F. Lazarsfeld. New York: Free Press.

Ku, H. H., and S. Kullback. 1968. "Interaction in Multidimensional Contingency Tables: An Information Theoretic Approach." *Journal of Research of the National Bureau of Standards* 72B:159–99.

Lazarsfeld, P. F. 1961. "The Algebra of Dichotomous Systems." In *Studies in Item Analysis and Prediction*, edited by H. Solomon. Stanford, Calif.: Stanford University Press.

———. 1968. "Survey Analysis II. The Analysis of Attribute Data." In *International Encyclopedia of the Social Sciences*, edited by D. Sills. Vol. 15. New York: Macmillan.

[54] If the estimated value of λ^S is far from zero, the nonhierarchical model may not provide an adequate description of the data.

Wilner, D. M., R. P. Walkley, and S. W. Cook. 1955. *Human Relations in Interracial Housing: A Study of the Contact Hypothesis*. Minneapolis: University of Minnesota Press.

Wiorkowski, J. J. 1970. "Estimation of the Proportion of the Variance Explained by Regression, When the Number of Parameters in the Model May Depend on the Sample Size." *Technometrics* 12:915–19.

Yule, G. U., and M. G. Kendall. 1950. *An Introduction to the Theory of Statistics*. 14th ed. rev. New York: Hafner.

Leo A. Goodman

Reprinted from: THE AMERICAN JOURNAL OF SOCIOLOGY,
Vol. 77, No. 6, May 1972.

Chapter 4

The Multivariate Analysis of Qualitative Data: Interactions among Multiple Classifications

For the m-way contingency table, we discuss both the direct estimation of the multiplicative interactions among the m variables, and the indirect testing of hypotheses pertaining to these interactions. We consider, among other things, hierarchical hypotheses pertaining to the interactions among the m variables, including hypotheses that can be expressed in terms of one or more of the following kinds of concepts: (a) the usual concepts of independence and equiprobability; (b) concepts describing conditional properties (e.g., conditional independence) pertaining to a subset of the m variables, given the level of some of the remaining variables; (c) concepts related to the usual logit-analysis or to a generalized form of logit-analysis; and (d) concepts related to a more general log-linear model. Methods of partitioning these hypotheses are introduced which provide, among other things, insight into the relationship between tests applied to the m-way table and tests applied to marginal tables formed from the m-way table. We also show, by example, how the combined use of direct estimation and indirect testing can lead to the discovery of hypotheses (models) that fit the data in the m-way table better than the hypotheses that have been fitted in the earlier literature.

1. INTRODUCTION

Different authors have proposed different methods for the multivariate analysis of m qualitative variables ($m=2, 3, 4, \cdots$); i.e., for the analysis of m-way contingency tables. In the present article, we shall show that some of these methods, which might appear at first sight to be hardly comparable, actually do complement each other, and that a systematic (but flexible) combined application of these methods to the data in m-way tables can lead to a more complete analysis of the data than has heretofore been possible.

We shall discuss here both the direct estimation of the multiplicative interactions among the m variables, and the indirect testing of hypotheses pertaining to these interactions. Earlier work on estimation problems that are related (more or less) to the problems considered here appears in [13, 22, 23, and 37]; earlier related work on indirect testing appears in [4, 7, 26, 33, and 40]. In the present article, we shall make use of some of the methods developed earlier; we shall also introduce new methods of estimation and of testing; and we shall see how the estimation and testing methods can be used to complement each other.

The classical chi-square test applied to the 2-way table tests the hypothesis that the table's two qualitative variables are independent of each other; and,

---

Leo A. Goodman is professor of statistics and sociology, University of Chicago, and research associate at the Population Research Center at the University. His publications include approximately 80 articles in journals and books published in the United States, England, Russia, Japan and India. The author is indebted to Y. M. Bishop, W. G. Cochran, J. Fennessey, S. E. Fienberg, S. Haberman, H. H. Ku, S. Kullback, N. Mantel, J. W. Pratt, and T. Pullum for helpful comments. This research was supported in part by Research Contract No. NSF GS 1905 from the Division of the Social Sciences of the National Science Foundation.

similarly, a generalized form of this test applied to the m-way table tests certain kinds of hypotheses about the table's m variables. The classical chi-square approach has been criticized because (1) it does not provide estimates of the "effects" of the table's m variables on each other, and (2) its application to the m-way table is complicated when $m>2$ (see, e.g. [46]). We shall provide extensions and simplifications of this approach that will meet these criticisms.

One of the alternatives to the classical approach has been an analysis based on "logits" (see, e.g. [7 and 16]). This analysis is applicable when (a) the m variables can be viewed in an asymmetric way, with one of the variables considered the "dependent" variable and the other $m-1$ variables considered "independent" variables, and (b) the dependent variable is dichotomous. With logit-analysis, the original, say, $I \times J \times 2$ contingency table (when $m=3$) is viewed as an $I \times J$ factorial arrangement whose (i, j) cell contains the "logit" pertaining to the dependent variable at level (i, j) of the independent variables. (The "logit" is one-half the logarithm of the ratio of the observed frequencies in the two classes of the dependent variable in cell (i, j).) We shall show here how to extend a part of this analysis to the case where the dependent variable is not dichotomous but is polytomous. (For a different extension, see [10].) We shall consider both the case where the m-way table can be viewed asymmetrically, and the case where it should be viewed in a symmetric way (when the mutual relationships among all m variables are of interest).

For illustrative purposes, we shall re-examine a single set of data that has already been analyzed in six publications: [7, 14, 16, 38, 44 and 46]. We shall see that the application of the methods and points of view developed in the present article further enrich our understanding of these data. The data (Table 1) describe a cross-classification of five dichotomized variables. The methods which we shall present here can be applied to analyze the relationship among m dichotomies ($m=2, 3, \cdots$), and more generally the relationship among m polytomies. Before discussing Table 1, we first consider the relationship among m polytomies for $m=2$, 3, and 4.

Table 1. CROSS-CLASSIFICATION OF INDIVIDUALS ACCORDING TO FIVE DICHOTOMIZED VARIABLES[a]

	Radio				*No radio*			
	Solid reading		*No solid reading*		*Solid reading*		*No solid reading*	
	Newspapers							
Knowledge:	Good	Poor	Good	Poor	Good	Poor	Good	Poor
Lectures	23	8	8	4	27	18	7	6
No lectures	102	67	35	59	201	177	75	156
	No newspapers							
Knowledge:	Good	Poor	Good	Poor	Good	Poor	Good	Poor
Lectures	1	3	4	3	3	8	2	10
No lectures	16	16	13	50	67	83	84	393

[a] The five variables classify individuals according to whether they (1) read newspapers, (2) listen to radio, (3) read books and magazines (solid reading), (4) attend lectures, and (5) whether their knowledge of cancer is good or poor.

2. THE 2-WAY CONTINGENCY TABLE

2.1 The Direct Estimation of Interactions Among Variables

Consider the $I \times J$ population table, with the row and column variables denoted by A and B, respectively. Let P_{ij} denote the probability that an observation will fall in cell (i, j) $(i=1, 2, \cdots, I; j=1, 2, \cdots, J)$. Thus

$$\sum_{i,j} P_{ij} = 1. \tag{2.1}$$

Let $\nu_{ij} = \log P_{ij}$, where log refers to the natural logarithm throughout. (For simplicity, we assume that all $P_{ij} > 0$.) As in the analysis of variance (following, e.g. [4 and 8]) we can decompose ν_{ij} as follows:

$$\nu_{ij} = \mu + \lambda_i{}^A + \lambda_j{}^B + \lambda_{ij}{}^{AB}, \tag{2.2}$$

where

$$\sum_i \lambda_i{}^A = \sum_j \lambda_j{}^B = \sum_i \lambda_{ij}{}^{AB} = \sum_j \lambda_{ij}{}^{AB} = 0. \tag{2.3}$$

The λ's in (2.2) represent the possible "effects" of the two variables on ν_{ij}: The main effects are $\lambda_i{}^A$ and $\lambda_j{}^B$, the AB interaction effects are the $\lambda_{ij}{}^{AB}$. Because of (2.1) and (2.2), μ satisfies the condition that

$$\{\exp \mu\} \left\{ \sum_{i,j} \exp(\lambda_i{}^A + \lambda_j{}^B + \lambda_{ij}{}^{AB}) \right\} = 1. \tag{2.4}$$

We shall use a dot subscript to denote an average over the corresponding index: $\nu_{i.} = \sum_j \nu_{ij}/J$, $\nu_{.j} = \sum_i \nu_{ij}/I$, $\nu_{..} = \sum_{i,j} \nu_{ij}/IJ$. As in the analysis of variance, from (2.2) we obtain

$$\lambda_i{}^A = \nu_{i.} - \nu_{..}, \qquad \lambda_j{}^B = \nu_{.j} - \nu_{..}, \qquad \lambda_{ij}{}^{AB} = \nu_{ij} - \nu_{i.} - \nu_{.j} + \nu_{..}. \tag{2.5}$$

Further understanding of the meaning of the λ's can be obtained from (2.5) in the special case where $I = J = 2$. Then,

$$\begin{aligned} \lambda_1{}^A &= (\nu_{11} + \nu_{12} - \nu_{21} - \nu_{22})/4 = (1/4) \sum_j \log(P_{1j}/P_{2j}) \\ \lambda_1{}^B &= (\nu_{11} - \nu_{12} + \nu_{21} - \nu_{22})/4 = (1/4) \sum_i \log(P_{i1}/P_{i2}) \\ \lambda_{11}{}^{AB} &= (\nu_{11} - \nu_{12} - \nu_{21} + \nu_{22})/4 = (1/4)\{\log[(P_{11}P_{22})/(P_{12}P_{21})]\}. \end{aligned} \tag{2.6}$$

Thus, $\lambda_{11}{}^{AB}$ is proportional to the logarithm of the usual cross-product ratio, which can also be written as the odds-ratio $(P_{11}/P_{21})/(P_{12}/P_{22})$ or $(P_{11}/P_{12})/(P_{21}/P_{22})$. The importance of this ratio in the analysis of the 2×2 contingency table has long been recognized (see, e.g., [21 and 40] and literature cited there).

From (2.6) we see that $\lambda_{11}{}^{AB}$ is proportional to the difference between the two log-odds $\nu_{11} - \nu_{21}$ and $\nu_{12} - \nu_{22}$; or, equivalently, to the difference between the two log-odds $\nu_{11} - \nu_{12}$ and $\nu_{21} - \nu_{22}$. Also, $\lambda_1{}^A$ is proportional to the sum of the two log-odds $\nu_{11} - \nu_{21}$ and $\nu_{12} - \nu_{22}$; and similarly for $\lambda_1{}^B$. We find that $\lambda_{11}{}^{AB} = 0$ if and only if variables A and B are independent of each other. Furthermore, if it is assumed that $\lambda_{11}{}^{AB} = 0$, then $\lambda_1{}^A = 0$ if and only if the two classes of variable A are equiprobable. If no assumption is made about the value of $\lambda_{11}{}^{AB}$, then

$\lambda_1^A = 0$ if and only if the odds for Class 1 of variable A, given that variable B is at Level 1, "balance" the odds for Class 1 of variable A, given that variable B is at Level 2 (i.e., the sum of the corresponding log-odds is zero). λ_{11}^{AB} serves as a measure of the inequality between these two odds values (i.e., a measure of the association between A and B); and λ_1^A serves as a measure of the "imbalance" between the same two odds values (i.e., how different from zero is the sum of the corresponding log-odds). Similar remarks apply to λ_1^B.

For a sample of n observations from the general $I \times J$ table, let f_{ij} denote the number of observations in cell (i, j), and let

$$y_{ij} = \log f_{ij}. \tag{2.7}$$

(For simplicity, we assume that $f_{ij} > 0$.) Note that $\sum_{i,j} f_{ij} = n$. Maximum-likelihood estimates of the λ's can be obtained by replacing the ν's in (2.5) by the corresponding y's. We denote these estimates as $\hat{\lambda}_i^A$, $\hat{\lambda}_j^B$, $\hat{\lambda}_{ij}^{AB}$. With these estimates inserted in (2.2), the observed data are fitted perfectly.

Each of the $\hat{\lambda}$'s can be expressed in the form

$$\hat{\lambda} = \sum_{i,j} a_{ij} y_{ij}, \tag{2.8}$$

where the a_{ij} are constants that depend upon which $\hat{\lambda}$ is being calculated, and where $\sum_{i,j} a_{ij} = 0$. (When $I = J = 2$, we note from (2.6) that a_{ij} is $+1/4$ or $-1/4$.) Writing $\hat{\lambda}$ as (2.8), the variance of $\hat{\lambda}$ can be estimated by

$$S_{\hat{\lambda}}^2 = \sum_{i,j} a_{ij}^2 / f_{ij} \tag{2.9}$$

(see, e.g. [21, 37, 42, and 45]). The standardized value of $\hat{\lambda}$ (viz., $\hat{\lambda}/S_{\hat{\lambda}}$) can be used to test the hypothesis that $\lambda = 0$. For simultaneous confidence intervals and multiple-tests pertaining to the λ's, see [21, 23]. (For a table of critical values for making simultaneous inferences, see [25].)

Because of (2.3) we see that

$$\lambda_I^A = -\sum_{i=1}^{I-1} \lambda_i^A, \quad \lambda_J^B = -\sum_{j=1}^{J-1} \lambda_j^B, \quad \lambda_{iJ}^{AB} = -\sum_{j=1}^{J-1} \lambda_{ij}^{AB}, \quad \lambda_{Ij}^{AB} = -\sum_{i=1}^{I-1} \lambda_{ij}^{AB}.$$

Thus, λ_I^A, λ_J^B, and λ_{iJ}^{AB}, λ_{Ij}^{AB} are linear functions of the remaining $I-1$, $J-1$, and $(I-1)(J-1)$ parameters λ_i^A, λ_j^B, and λ_{ij}^{AB} (for $i < I$, $j < J$). We can therefore view (2.2) in terms of the basic sets of $I-1$, $J-1$, and $(I-1)(J-1)$ parameters. Other linear functions of these parameters may also be of interest. (For example, when $I = 3$, we can consider two linear functions of the λ^A's that represent linear and quadratic components associated with variable A, assuming the levels of A are equally spaced; see, e.g. [13 and 28].) Since the estimates of these linear functions can be expressed in the form (2.8), the methods noted in the preceding paragraph can also be applied to these estimates.

Before calculating the $\hat{\lambda}$ for a particular sample, we recommend replacing f_{ij} in (2.7)–(2.9) by $f_{ij} + (1/2)$. This adjustment of the f's reduces both the asymptotic bias and mean-squared-error of $\hat{\lambda}$, and it also has some merit when applied in (2.9). (For example, when samples of size n_i are drawn from the ith row $(i = 1, 2, \cdots, I)$ of a $I \times J$ population table in which $J = 2$, and the parameters

λ_{ij}^{AB} and λ_j^B (or linear functions of them) are of interest, then the adjustment applied to (2.9) will reduce both the asymptotic bias and mean-squared-error of the estimate of the variance of the $\hat{\lambda}$'s.) This follows from results in [27]; see also [18]. For an alternative adjustment that can be applied in (2.9), see [23].

As we have noted, (2.2) and (2.5) are analogous to similar formulae in the analysis of variance; and a part of (2.6) has its counterpart in the 2×2 factorial experiment (see, e.g. [12 and 32]). Of course, the present situation differs from the analysis of variance because of (2.4), and because we are dealing with a sample from a cross-classification rather than with independent observations from populations that are normally distributed and homoscedastic. The estimates of the λ's defined by (2.5) form a subset of the $\hat{\lambda}$'s considered earlier by Goodman [23]. We recommend that the $\hat{\lambda}$'s (with their estimated variances) be used to make simultaneous inferences as described in [23], but we shall use them here primarily as a guide in selecting the hypotheses that are to be fitted to the data from among the hypotheses to be described in the next section. For an example in which the $\hat{\lambda}$'s are used both to make simultaneous inferences and also to guide the selection of the hypotheses that are fitted to the data, see [28].

2.2 Hierarchical Hypotheses and Tests Based on Fitted Marginals

The methods referred to above for making simultaneous inferences can be used to obtain confidence intervals for any set of λ's as well as to test hypotheses about these λ's. In the present section, we shall consider only hypotheses pertaining to certain sets of λ's—hypotheses that can be tested by the more conventional chi-square methods. We shall re-examine conventional tests (and related tests) in a way that should facilitate the presentation of generalizations later.

Table 2 presents three different kinds of hypotheses that can be tested by conventional chi-square methods, expressing each hypothesis in terms of the basic sets of λ's that are assumed to be zero under the hypothesis. The basic sets of $I-1$, $J-1$, and $(I-1)(J-1)$ parameters λ_i^A, λ_j^B, and λ_{ij}^{AB} (for $i<I$, $j<J$) will be denoted here by λ^A, λ^B, λ^{AB}, respectively; and, for short, by the symbols (A), (B), and (AB) in Table 2. Expressed in more conventional terms, Hypothesis 1 of Table 2 states that variables A and B are independent of each other; Hypothesis 2 states that the classes of variable B are equiprobable, given the level of variable A; and Hypothesis 3 states that the classes of the joint variable AB are equiprobable. Introducing some notation for the above expressions, Hypotheses 1, 2, and 3 will be denoted by $[\overline{A}\otimes\overline{B}]$, $[\overline{B}=\Phi|\overline{A}]$, and $[\overline{AB}=\Phi]$, respectively. By interchanging the letters A and B in Hypothesis 2, we obtain a different hypothesis, but one of the same kind; see Column 4 of Table 2.

To obtain distributions other than the equiprobable (i.e., uniform) distribution in the expressions for Hypotheses 2 and 3 of Table 2, we can replace Hypothesis 2 (expressed in terms of the λ's that are assumed zero under the hypothesis) by the hypothesis that $\lambda^{AB}=0$, that the λ^B are specified values with some $\lambda^B\neq0$, and that the λ^A are unspecified; and we can replace Hypothesis 3 by the hypothesis that all the λ's are specified values with some $\lambda\neq0$. Discussion

Table 2. **EACH HIERARCHICAL HYPOTHESIS PERTAINING TO THE $I \times J$ CONTINGENCY TABLE**

Kind of hypothesis	*Parameters set at zero*	*Degrees of freedom*	*Number of hypotheses of this kind*	*Fitted marginals*	*Description in conventional terms*
1	(AB)	$(I-1)(J-1)$	1	$\{A\}$, $\{B\}$	$[\overline{A} \otimes \overline{B}]$
2	(AB), (B)	$I(J-1)$	2	$\{A\}$	$[\overline{B} = \Phi \mid \overline{A}]$
3	(AB), (B), (A)	$IJ-1$	1	n	$[\overline{AB} = \Phi]$

here is limited to hypotheses of the kind described in Table 2 (and in Tables 3 and 4), although extensions to the more general kind of hypothesis are possible.

The following formulae provide maximum-likelihood estimates of the expected frequency F_{ij} in cell (i, j) of the $I \times J$ table, under Hypotheses 1, 2, 3, respectively, in Table 2:

$$\text{(a) } \hat{F}_{ij} = f_i^A f_j^B / n; \qquad \text{(b) } \hat{F}_{ij} = f_i^A / J; \qquad \text{(c) } \hat{F}_{ij} = n/(IJ). \tag{2.10}$$

Here f_i^A and f_j^B denote the observed row and column marginals; i.e., the marginals pertaining to variables A and B, respectively. Note that (a) of (2.10) is such that $\sum_j \hat{F}_{ij} = f_i^A$ and $\sum_i \hat{F}_{ij} = f_j^B$; (b) of (2.10) is such that $\sum_j \hat{F}_{ij} = f_i^A$; (c) of (2.10) is such that $\sum_{i,j} \hat{F}_{ij} = n$. Thus, the $\hat{F}$'s fit the following marginals: f_i^A and f_j^B for Hypothesis 1; f_i^A for Hypothesis 2; and for Hypothesis 3 only n has been fitted. This is expressed in Column 5 of Table 2 using $\{A\}$ and $\{B\}$ to denote the sets of frequencies f_i^A and f_j^B, respectively. For each hypothesis, the marginals that are fitted correspond to the λ's that are not assumed to be zero under the hypothesis.

The degrees of freedom of the appropriate chi-square statistic used to test these hypotheses can be calculated by the usual considerations. For Hypotheses 1 and 3, we obtain $(I-1)(J-1)$ and $IJ-1$, respectively; and for Hypothesis 2 we obtain $I(J-1)$ since this chi-square statistic is the sum of the I different chi-square statistics obtained in testing that the J classes in the ith row are equiprobable (for $i = 1, 2, \cdots, I$). For each hypothesis, the degrees of freedom are equal to the number of λ's in the basic sets that are assumed zero under the hypothesis.

Let Y and Z denote subsets of the letters $A, B, \cdots$. For example, Y might be AB, and Z might be A, in which case we say that Y contains Z. Each hypothesis in Table 2 is hierarchical in the sense that, if λ^Z is any of the λ's that are assumed zero under the given hypothesis (say, H_0), then λ^Y is also assumed zero under H_0 for all Y that contain Z. (The symbol λ^Z refers to the λ's in the basic set with Z in the superscript; e.g., when Z is AB, then λ^Z is the basic set of λ_{ij}^{AB} for $i < I, j < J$.) An equivalent condition for H_0 to be hierarchical is that, if λ^Y is any of the λ's that are not assumed to be zero under H_0, then λ^Z is not assumed to be zero under H_0 for all Z contained in Y. For each hypothesis in Table 2, the maximum-likelihood estimate $\hat{F}_{ij}$ of F_{ij} could be calculated as a product of a set of "effects" pertaining to the non-zero λ's (under the hypothesis), fitting the $\hat{F}_{ij}$ to the marginals pertaining to the non-zero λ's (see (2.10) and comments following it). This is true in general for hierarchical hypotheses [4].

For a given hypothesis, let $\{Y_1\}, \{Y_2\}, \cdots, \{Y_T\}$ be the set of marginals that are fitted. For Hypotheses 1 and 2 in Table 2, T is 2 and 1, respectively. The usual chi-square test of Hypothesis 1 is "valid" when sampling from the $I \times J$ multinomial (with n given), when sampling independently in each row (with $\{A\}$ given) or independently in each column (with $\{B\}$ given), and when sampling with both $\{A\}$ and $\{B\}$ given. (By "valid" we mean that the test statistic has the appropriate asymptotic chi-square distribution under the hypothesis when the specified sampling procedure is applied.) More generally, the chi-square tests discussed in the present article will be valid when sampling (a) with no marginals given (with n given), (b) with the set of fitted marginals $\{Y_1\}, \{Y_2\}, \cdots, \{Y_T\}$ given, or (c) with a subset of this set given.[1]

From Columns 1, 2, and 4 of Table 2, we see that there are three different *kinds* of hierarchical hypotheses about the 2-way table, and four hierarchical hypotheses that are actually different. (We shall see later that the number of different *kinds* of hierarchical hypotheses about the 3-way and 4-way tables is 8 and 27, respectively; and the number of hierarchical hypotheses that are actually different is 18 and 166, respectively.) In analyzing a given contingency table, for guidance in selecting from among these hypotheses the ones that should be fitted to the data, the standardized values of the $\hat{\lambda}$'s (presented in Section 2.1 and 3.1) can be used, together with prior information (if any) concerning the magnitude of the λ's in the particular context under investigation.

3. THE 3-WAY CONTINGENCY TABLE

3.1 The Direct Estimation of Interactions Among Variables

Consider the $I \times J \times K$ population table, with the row, column, and layer variables denoted by A, B, and C, respectively. Let P_{ijk} denote the probability that an observation will fall in cell (i, j, k) $(i=1, 2, \cdots, I; j=1, 2, \cdots, J; k=1, 2, \cdots, K)$, and let $\nu_{ijk}=\log P_{ijk}$. (For simplicity, we assume that $P_{ijk}>0$.) As in (2.2), the ν_{ijk} can be decomposed, with the sum $\lambda_i^A+\lambda_j^B+\lambda_k^C+\lambda_{ij}^{AB}+\lambda_{ik}^{AC}+\lambda_{jk}^{BC}+\lambda_{ijk}^{ABC}$ replacing the corresponding sum in (2.2). The conditions (2.3) and (2.4) can also be extended in a straightforward fashion. The λ's represent the possible effects of the variables A, B, and C on ν_{ijk}: The main effects are λ_i^A, λ_j^B, λ_k^C; the interaction effects are λ_{ij}^{AB}, λ_{ik}^{AC}, λ_{jk}^{BC} for 2-factor effects, and λ_{ijk}^{ABC} for 3-factor effects.

As in (2.5), we obtain

$$
\begin{aligned}
\lambda_i^A &= \nu_{i..} - \nu_{...}, \qquad \lambda_{ij}^{AB} = \nu_{ij.} - \nu_{i..} - \nu_{.j.} + \nu_{...}, \\
\lambda_{ijk}^{ABC} &= \nu_{ijk} - \nu_{ij.} - \nu_{i.k} - \nu_{.jk} + \nu_{i..} + \nu_{.j.} + \nu_{..k} - \nu_{...},
\end{aligned} \tag{3.1}
$$

with similar formulae for λ_j^B, λ_k^C, λ_{ik}^{AC}, λ_{jk}^{BC}. As in (2.6), the λ's can be expressed in a particularly simple form when $I=J=K=2$; e.g.,

$$
\begin{aligned}
\lambda_{111}^{ABC} &= (1/8)(\nu_{111} - \nu_{121} - \nu_{211} + \nu_{221} - \nu_{112} + \nu_{122} + \nu_{212} - \nu_{222}) \\
&= (1/8)\{\log[(P_{111}P_{221})/(P_{121}P_{211})] - \log[(P_{112}P_{222})/(P_{122}P_{212})]\}.
\end{aligned} \tag{3.2}
$$

Thus, λ_{111}^{ABC} is proportional to the logarithm of the ratio of the odds-ratio

[1] This statement is a generalization of Birch's result [4] for the special case of sampling with only one of the fitted marginals (say $\{Y_1\}$) given, or with no marginals given. For a proof of the more general result, see [29].

pertaining to the A and B variables when variable C is at Level 1 and the corresponding odds-ratio when variable C is at Level 2; and it serves as a measure of the inequality of these two odds-ratios. (Note that λ_{111}^{ABC} is invariant when variables A, B, and C are interchanged.) Also, λ_{11}^{AB} serves as a measure of the "imbalance" between these two odds-ratios (see related comment in Section 2.1); and λ_1^A serves as a measure of the "imbalance" between the odds for Class 1 of variable A (i.e., P_{1jk}/P_{2jk}) at the different levels (j, k) of the joint variable BC. (Note that (3.1) is analogous to a similar formula in the analysis of variance, and a part of (3.2) has its counterpart in the $2\times2\times2$ factorial experiment; see, e.g. [12 and 32].)

For a sample of n observations from the general $I\times J\times K$ table, let f_{ijk} denote the number of observations in cell (i, j, k), and let $y_{ijk}=\log f_{ijk}$. (For simplicity, we assume that $f_{ijk}>0$.) As in Section 2.1, the maximum-likelihood estimate $\hat{\lambda}$ of a particular λ can be obtained by replacing the ν's in (3.1) by the corresponding y's. With these estimates, the observed data are fitted perfectly, as in Section 2.1.

Each of the $\hat{\lambda}$'s can be expressed in the form

$$\hat{\lambda} = \sum_{i,j,k} a_{ijk}y_{ijk}, \tag{3.3}$$

where the a_{ijk} are constants that depend upon which $\hat{\lambda}$ is being calculated, and where $\sum_{i,j,k} a_{ijk}=0$. (When $I=J=K=2$, a_{ijk} is $+1/8$ or $-1/8$; see (3.2).) As in (2.9), with $\hat{\lambda}$ written as (3.3), the variance of $\hat{\lambda}$ can be estimated by

$$S_{\hat{\lambda}}^2 = \sum_{i,j,k} a_{ijk}^2/f_{ijk}. \tag{3.4}$$

The standardized value of $\hat{\lambda}$ (viz., $\hat{\lambda}/S_{\hat{\lambda}}$) can be used to test the null hypothesis that $\lambda=0$. Since the remarks in Section 2.1 after (2.9) can be extended directly to the 3-way table, we shall not cover this ground here.

3.2 Hierarchical Hypotheses and Tests Based on Fitted Marginals

As noted in Section 2.2, methods of simultaneous inference based upon the $\hat{\lambda}$'s can be used to obtain confidence intervals for any set of λ's as well as to test hypotheses about these λ's (see [22 and 23]). In the more conventional chi-square analysis of the 3-way table, only hypothesis testing is discussed, and only hypotheses pertaining to certain sets of λ's are considered. We shall now reexamine conventional and related tests.

Table 3 presents eight different kinds of hypotheses that can be tested by conventional chi-square methods, expressing each hypothesis in terms of the basic sets of λ's that are assumed zero under the hypothesis. The basic sets of $I-1$, $J-1$, $K-1$, $(I-1)(J-1)$, $\cdots$, $(I-1)(J-1)(K-1)$ parameters λ_i^A, λ_j^B, λ_k^C, λ_{ij}^{AB}, $\cdots$, λ_{ijk}^{ABC} (for $i<I$, $j<J$, $k<K$) will be denoted here by λ^A, λ^B, λ^C, λ^{AB}, $\cdots$, λ^{ABC}, respectively; and, for short, by the symbols (A), (B), (C), (AB), $\cdots$, (ABC) in Table 3. Hypothesis 1 in Table 3 states that the 3-factor interactions λ^{ABC} are zero; a hypothesis that cannot be expressed in more conventional terms (using the concepts of independence, equiprobability, conditional independence, and/or conditional equiprobability). (For a more complete discussion of the meaning of Hypothesis 1, see, e.g. [22].)

Hypotheses 2 to 8 in Table 3 can be expressed in these more conventional terms as indicated in Column 6 of the table. For example, Hypothesis 2 states that variables B and C are conditionally independent, given the level of variable $A(\lambda^{BC}=0$ and $\lambda^{Y}=0$ for all Y that include BC); Hypothesis 5 states that variables A, B, and C are mutually independent ($\lambda^{Y}=0$ for all Y containing two or more of the letters A, B, C); Hypothesis 6 states that (a) variables A and B are independent of each other, and (b) the classes of variable C are equiprobable, given the level of the joint variable $AB(\lambda^{Y}=0$ for all Y, except for λ^{A} and λ^{B}). (The equivalence between the description of Hypotheses 2 to 8 in terms of the λ's that are assumed zero and their description in more conventional terms, as in Column 6 of Table 3, is an extension of Birch's result [4] for the special case of Hypotheses 2, 3, and 5. Various generalizations of this result will be presented later herein.) By interchanging the letters A, B, and C in Hypothesis 2, we obtain two other hypotheses, but each of these hypotheses is of the same kind; see Column 4 of Table 3, for Hypotheses 2, 3, 4, 6, 7.

The following formulae provide maximum-likelihood estimates of the expected frequency F_{ijk} in cell (i, j, k) of the $I\times J\times K$ table for Hypotheses 2 to 8, respectively, in Table 3:

$$\text{(a)}\ \hat{F}_{ijk}=f_{ij}^{AB}f_{ik}^{AC}/f_{i}^{A};\quad \text{(b)}\ \hat{F}_{ijk}=f_{ij}^{AB}f_{k}^{C}/n;\quad \text{(c)}\ \hat{F}_{ijk}=f_{ij}^{AB}/K;$$
$$\text{(d)}\ \hat{F}_{ijk}=f_{i}^{A}f_{j}^{B}f_{k}^{C}/n^{2};\quad \text{(e)}\ \hat{F}_{ijk}=f_{i}^{A}f_{j}^{B}/(Kn);\quad \text{(f)}\ \hat{F}_{ijk}=f_{i}^{A}/(JK);\quad (3.5)$$
$$\text{(g)}\ \hat{F}_{ijk}=n/(IJK).$$

Here f_{ij}^{AB} denotes the observed row-column marginal pertaining to the joint variable AB, and the other 2-way and 1-way marginals are similarly denoted. (For simplicity, we assume in (a) of (3.5) that $f_{i}^{A}>0$.) As with (2.10), the $\hat{F}$'s of (3.5) fit the marginals that correspond to the λ's that are not assumed to be zero under the corresponding hypothesis. The fitted marginals are listed in Column 5 of Table 3 using $\{AB\}$ to denote the set of marginal frequencies f_{ij}^{AB}; and similar notation for the other 2-way and 1-way sets of marginal frequencies. (For short, we actually list in Column 5 the minimal set of fitted marginals, as defined later in this section.)

Table 3. EACH HIERARCHICAL HYPOTHESIS PERTAINING TO THE $I\times J\times K$ CONTINGENCY TABLE

Kind of hypothesis	*Parameters set at zero*	*Degrees of freedom*	*Number of hypotheses of this kind*	*Fitted marginals*	*Description in conventional terms*
1	(ABC)	$(I-1)(J-1)(K-1)$	1	$\{AB\}, \{AC\}, \{BC\}$	—
2	$(ABC), (BC)$	$I(J-1)(K-1)$	3	$\{AB\}, \{AC\}$	$[\overline{B}\otimes\overline{C}\mid\overline{A}]$
3	$(ABC), (BC), (AC)$	$(IJ-1)(K-1)$	3	$\{AB\}, \{C\}$	$[\overline{AB}\otimes\overline{C}]$
4	$(ABC), (BC), (AC), (C)$	$IJ(K-1)$	3	$\{AB\}$	$[\overline{C}=\Phi\mid\overline{AB}]$
5	$(ABC), (BC), (AC), (AB)$	$IJK-I-J-K+2$	1	$\{A\}, \{B\}, \{C\}$	$[\overline{A}\otimes\overline{B}\otimes\overline{C}]$
6	$(ABC), (BC), (AC), (AB), (C)$	$IJK-I-J+1$	3	$\{A\}, \{B\}$	$[\overline{A}\otimes\overline{B}]\cap[\overline{C}=\Phi\mid\overline{AB}]$
7	$(ABC), (BC), (AC), (AB), (C), (B)$	$I(JK-1)$	3	$\{A\}$	$[\overline{BC}=\Phi\mid\overline{A}]$
8	$(ABC), (BC), (AC), (AB), (C), (B), (A)$	$IJK-1$	1	n	$[\overline{ABC}=\Phi]$

The calculation of $\hat{F}_{ijk}$ under Hypothesis 1 of Table 3 will be discussed later in this section; but we note here that, for this hypothesis too, the $\hat{F}_{ijk}$ fit the marginals that correspond to the λ's that are not assumed to be zero under the hypothesis (viz., f_{ij}^{AB}, f_{ik}^{AC}, f_{jk}^{BC}, and the one-way marginals—see [15]).

The degrees of freedom of the appropriate chi-square statistic used to test each hypothesis in Table 3 can be calculated by the usual considerations; e.g., for Hypothesis 2 we obtain $I(J-1)(K-1)$ since this chi-square statistic is the sum of the I different chi-square statistics obtained in testing that the column and layer variables are independent of each other in the ith $J \times K$ table (for $i=1, 2, \cdots, I$). The degrees of freedom can also be calculated as the number of λ's in the basic sets that are assumed zero under the hypothesis, or as the difference between $(IJK-1)$ and the number of λ's in the basic sets that are not assumed to be zero under the hypothesis (e.g., for Hypothesis 6 we obtain $IJK-1-(I-1)-(J-1)=IJK-I-J+1$).

Each hypothesis in Table 3 is hierarchical in the sense defined in Section 2.2. We have noted here that, for these hypotheses, the marginals that are fitted in calculating the $\hat{F}$'s correspond to the λ's that are not assumed to be zero under the hypothesis, as in Section 2.2. (If the marginal $\{Y\}$ is fitted, then the marginal $\{Z\}$ is fitted for any Z contained in Y.) The degrees of freedom are also calculated as in Section 2.2; and we can apply here the comment in that section about the validity of the test when sampling with no marginals given, with the set of fitted marginals given, or with a particular subset of this set given.

For a given hierarchical hypothesis, let $\{Y_1\}, \{Y_2\}, \cdots, \{Y_T\}$ denote the minimal set of marginals that are fitted. The minimal set is obtained by excluding from the set of fitted $\{Y\}$'s any marginal $\{Z\}$ if Z is contained in at least one of the Y's. (The marginal $\{Z\}$ is automatically fitted in this case.) From Table 3 we see that for the 3-way table there are at most three marginals that are included in the minimal set (i.e., $T \leq 3$).

If there is a set Z of letters (the set Z could contain one or more of the letters representing the dimensions of the contingency table) none of which appear in any of the Y's in the set of fitted $\{Y\}$'s, then the corresponding hypothesis states (among other things) that the classes of the (joint) variable Z are equiprobable, given the level of the remaining variables. Hypotheses 4, 6, 7, and 8 in Table 3, and Hypotheses 2 and 3 in Table 2, are examples of this. (See also Hypotheses 4, 9, 11, 13–17, 27 in Table 4, and Hypothesis 5 in Table 6.)

If there is a set Z of letters that is contained in each of the Y's in the minimal set of fitted $\{Y\}$'s, then the corresponding hypothesis can be described by conditional properties (e.g., conditional equiprobability, conditional independence) that apply conditionally on the level of the (joint) variable Z. Hypotheses 2, 4, 7 in Table 3, and Hypothesis 2 in Table 2, are examples of this. (See also Hypotheses 1, 2, 4, 5, 9, 13, 16, 19 in Table 4, and Hypotheses 18, 20, 24, 32, 34, 35 in Table 6.) At each level of the (joint) variable Z, the conditional properties pertain to a $(m-m')$-way table, where m' is the number of letters in Z, and m is the number of dimensions in the original contingency table. For example, if $m'=m-2$, then the conditional properties describe a hypothesis about a 2-way table, at each level of the (joint) variable Z; see Hypotheses 2 and 7 in Table 3, Hypotheses 1 and 13 in Table 4, Hypotheses 24 and 35 in Table 6.

If there is a set Z of two (or more) letters such that no letter in Z is included in any particular Y in the minimal set of fitted $\{Y\}$'s unless the other letters in Z are also included in the same Y, then the corresponding hypothesis describes (among other things) a property of the joint variable Z, rather than a property of the separate variables represented by the separate letters in Z. (If the set Z described in the preceding paragraph, or in the paragraph which preceded that one, contains two or more letters, then the remarks in the present paragraph will also apply to the corresponding hypotheses considered in the above two paragraphs.) For the case described at the beginning of the present paragraph, the corresponding hypothesis actually pertains to a contingency table of fewer dimensions than the original one, since the joint variable Z (viewed as a single dimension) can replace the several dimensions represented by the separate letters in Z. Hypotheses 3, 4, 7, 8 in Table 3, and Hypothesis 3 in Table 2, are examples of this. (See also Hypotheses 1–4, 8, 10, 11, 13, 15, 16, 17 in Table 4, and Hypotheses 5, 24, 25, 27, 35, 36, 38 in Table 6.)

Each of the kinds of hypotheses considered in the preceding three paragraphs can also be described in terms of the λ's that are assumed zero under the hypothesis. We remark here on only one point. If Z^* is a specified subset of m^* letters from among the m letters ($m^* < m$) pertaining to the dimensions in the m-way table, let us consider the hypothesis that the $\lambda^{Z^*}=0$ and that the $\lambda^{Y}=0$ for all Y that contain Z^*. This hypothesis can be expressed as a hypothesis describing the conditional property that $\lambda^{Z^*}=0$ in the m^*-way table, given the level of the joint variable represented by the letters not included in Z^*. For example, with $m=3$, Hypothesis 4 in Table 3 describes conditional equiprobability ($m^*=1$), and Hypothesis 2 describes conditional independence ($m^*=2$). See also Hypothesis 2 in Table 2, Hypotheses 1, 4, 19 in Table 4, and Hypotheses 24 and 35 in Table 6.

We shall now express the $\hat{F}$'s in (3.5) in a more general form that will facilitate later generalization. For a given hypothesis, if the Y's in the minimal set $\{Y_1\}, \{Y_2\}, \cdots, \{Y_T\}$ of fitted marginals are mutually exclusive (i.e., if there are no letters that are contained in two (or more) of the Y's), then the $\hat{F}$'s can be expressed as

$$\hat{F} = \left(\prod_{t=1}^{T} f^{Y_t}\right) \Big/ (Gn^{T-1}), \tag{3.6}$$

where G is the product of the number of classes pertaining to each of the variables with letters that were not contained in any of the Y's. (If the set of such variables is empty, then $G=1$.) Formulas (b) to (g) of (3.5) are particular examples of (3.6). If there is only one set of letters (say, Z) that is included in two (or more) of the Y's in the minimal set (i.e., if all non-empty intersections of the Y's are equal to Z), then

$$\hat{F} = \left(\prod_{t=1}^{T} f^{Y_t}\right) \Big/ [Gn^{h}(f^{Z})^{h'-1}], \tag{3.7}$$

where h is the number of Y's that do not contain Z, and h' is the number of Y's that contain Z. Formula (a) of (3.5) is a particular example of (3.7). (For simplicity, we assume in (3.7) that $f^Z > 0$; and in the rest of the article we shall assume that the observed f's are greater than zero, though weaker assumptions are possible at each point in the article.)

For each hypothesis in Tables 2 and 3, except Hypothesis 1 of Table 3, the $\hat{F}$'s can be calculated using (3.6) or (3.7). For Hypothesis 1 of Table 3, $\{Y_1\} = \{AB\}$, $\{Y_2\} = \{AC\}$, $\{Y_3\} = \{BC\}$, and the $\hat{F}$'s can be calculated by the following iterative scaling method (see, e.g. [7, 26, 33, and 40]): For the initial step, set $\hat{F}_{ijk}{}^{(0)} = 1$. For the first cycle, calculate

$$\hat{F}_{ijk}{}^{(1)} = \hat{F}_{ijk}{}^{(0)} f_{ij}{}^{AB}/\hat{F}_{ij+}{}^{(0)}, \qquad \hat{F}_{ijk}{}^{(2)} = \hat{F}_{ijk}{}^{(1)} f_{ik}{}^{AC}/\hat{F}_{i+k}{}^{(1)},$$
$$\hat{F}_{ijk}{}^{(3)} = \hat{F}_{ijk}{}^{(2)} f_{jk}{}^{BC}/\hat{F}_{+jk}{}^{(2)}, \tag{3.8}$$

where the + subscript denotes summation over the corresponding index. For the second cycle, add 3 to each superscript in parentheses in (3.8), and then apply formulae obtained thereby. For the third cycle, add 6, etc. This procedure is continued until the following set of equations are satisfied to the desired degree of accuracy, for some integer value of u:

$$\hat{F}_{ij+}{}^{(u)} = f_{ij}{}^{AB}, \qquad \hat{F}_{i+k}{}^{(u)} = f_{ik}{}^{AC}, \qquad \hat{F}_{+jk}{}^{(u)} = f_{jk}{}^{BC}. \tag{3.9}$$

The procedure converges rapidly (at a geometric rate) to the correct $\hat{F}_{ijk}$ (see, e.g. [31]).

To facilitate later generalization, we describe iterative scaling in the following more general terms, considering $\{Y_1\}, \{Y_2\}, \cdots, \{Y_T\}$ as the minimal set of fitted marginals. In this case, for the initial step set $\hat{F}^{(0)} = 1$. For the first cycle, calculate

$$\hat{F}^{(t)} = \hat{F}^{(t-1)} f^{Y_t}/(\hat{F}^{(t-1)})^{Y_t}, \qquad \text{for } t = 1, 2, \cdots, T, \tag{3.10}$$

where f^{Y_t} pertains to the marginal $\{Y_t\}$ calculated from the contingency table describing the f's, and $(\hat{F}^{(t-1)})^{Y_t}$ pertains to the marginal corresponding to $\{Y_t\}$ but calculated from the table describing the $\hat{F}^{(t-1)}$'s. For the second cycle, calculate (3.10) for $t = T+1, T+2, \cdots, 2T$, etc.

If this iterative scaling method is applied for any hypothesis in Table 2 or 3, except Hypothesis 1 of Table 3, then $\hat{F}^{(T+1)} = \hat{F}^{(T)}$, and the correct $\hat{F}$'s are obtained at the end of the first cycle. (These $\hat{F}$'s will be equal to the corresponding quantities obtained by explicit formulae (3.6) – (3.7).) Hypotheses of this kind we shall call elementary. For Hypothesis 1 of Table 3, the $\hat{F}^{(T+1)}$ will usually differ from $\hat{F}^{(T)}$, and additional cycles will be required. Hypotheses of this kind we shall call non-elementary. The differences between elementary and non-elementary hypotheses will be discussed further later.

We shall now comment on the relationship between some of the hypotheses in Table 3 and the usual hypotheses considered in the analysis of the logits pertaining to variable C, when C is dichotomous; i.e., when $K = 2$. For the $I \times J \times 2$ population table, the logit ψ_{ij} pertaining to variable C in cell (i, j) is one-half the corresponding log-odds; i.e., $\psi_{ij} = (\nu_{ij1} - \nu_{ij2})/2$ (see, e.g., [17]). From the decomposition of ν_{ijk} described in Section 3.1 (see also (2.2)–(2.3)), we find that

$$\psi_{ij} = \lambda_1{}^{C} + \lambda_{i1}{}^{AC} + \lambda_{j1}{}^{BC} + \lambda_{ij1}{}^{ABC}. \tag{3.11}$$

Hypothesis 1 in Table 3 (viz., the hypothesis that $\lambda^{ABC} = 0$) is equivalent to the hypothesis that the logit ψ_{ij} can be expressed as a sum of the main effects of the general mean and of the variables A and B on C (i.e., $\lambda_1{}^{C}$, $\lambda_{i1}{}^{AC}$, $\lambda_{j1}{}^{BC}$, respectively). (From (3.11) we see that the parameters λ^{A}, λ^{B}, λ^{AB} are irrelevant here; but since the marginal $\{AB\}$ is fitted (implicitly) in the logit-analysis pertaining to variable C, this would be equivalent to the exclusion of these parameters from the set of λ's that are assumed zero under the hypothesis.) In general, the usual hypotheses considered in logit-analysis pertaining to a dichotomous variable C will be equivalent to the corresponding hypotheses about the λ's pertaining to the contingency table [7]. The above expression of Hypothesis 1 in terms of the logits, will be denoted by $\psi(C|AB)=\lambda^C+\lambda^{AC}+\lambda^{BC}$. Similarly, the expression of Hypothesis 2, 3, and 4 in term of logits can be denoted by $\psi(C|AB)=\lambda^C+\lambda^{AC}$, $\psi(C|AB)=\lambda^C$, and $\psi(C|AB)=0$, respectively.

The remarks in the preceding paragraph pertain to the case where variable C is dichotomous. But the corresponding hypotheses described in Table 3 pertain, more generally, to the case where variable C has K classes ($K\geqq 2$). In this case, (3.11) is replaced by

$$\Psi_{ij} = \Lambda^C + \Lambda_i^{AC} + \Lambda_j^{BC} + \Lambda_{ij}^{ABC}, \tag{3.12}$$

where Ψ_{ij} and the Λ's are vectors with $K-1$ entries:

$$\begin{aligned} \Psi_{ij} &= \{(\nu_{ij1} - \nu_{ijk})/2, \text{ for } k = 2, 3, \cdots, K\}; \\ \Lambda^C &= \{(\lambda_1^C - \lambda_k^C)/2, \text{ for } k = 2, 3, \cdots, K\}; \\ \Lambda_i^{AC} &= \{(\lambda_{i1}^{AC} - \lambda_{ik}^{AC})/2, \text{ for } k = 2, 3, \cdots, K\}; \text{ etc.} \end{aligned} \tag{3.13}$$

Each "hierarchical" hypothesis about the logit-vector Ψ_{ij} pertaining to variable C is equivalent to a corresponding hypothesis about the λ's pertaining to the contingency table, a hypothesis for which the set of fitted marginals includes the joint marginal pertaining to all of the remaining variables (viz., $\{AB\}$ in the 3-way table). The expression of Hypothesis 1 of Table 3 in terms of logit-vectors, can be denoted by $\Psi(C|AB)=\Lambda^C+\Lambda^{AC}+\Lambda^{BC}$; and similarly for Hypotheses 2 to 4. In testing each of these hypotheses, the degrees of freedom can be calculated by the methods described earlier herein, or by the following generalization of the calculation made in the usual logit-analysis: viz., $(IJ-V)\cdot(K-1)$, where V is the number of vector parameters (the Λ's) in the basic sets that are fitted to the logit-vectors. (For Hypotheses 1 to 4 in Table 3, V is $I+J-1$, I, 1, and 0, respectively.)

In the preceding generalization of logit-analysis, the logit-vector Ψ_{ij} was defined by comparing ν_{ij1} with ν_{ijk}, for all $k\neq 1$. In other words, Class 1 of variable C was used as the point of reference. The remarks in the preceding paragraph would continue to be true even if the logit-vector Ψ_{ij} is defined using some other class of variable C as the point of reference, and a corresponding change is made in the definition of the Λ's. If, say, $\Psi(C|AB)=\Lambda^C+\Lambda^{AC}+\Lambda^{BC}$ when Class 1 of variable C is used as the point of reference, then the equation will also be correct when some other class of variable C is used instead.

4. THE 4-WAY CONTINGENCY TABLE

The methods presented in Sections 2.1 and 3.1 can be directly generalized to apply to the 4-way table. We shall now extend the methods in Sections 2.2 and 3.2.

Table 4 lists the hierarchical hypotheses pertaining to the $I\times J\times K\times L$ table. To save space, we have not listed the set of λ's that are assumed zero under each hypothesis. For a given hypothesis, these parameters correspond to the marginals that are not fitted. (Recall that if the marginal $\{Y\}$ is fitted, then the marginal $\{Z\}$ is automatically fitted for all Z contained in Y.) Also to save space, we give the degrees of freedom in the case where $I=J=K=L=2$; the extension to the more general case is straightforward.

Table 4. EACH HIERARCHICAL HYPOTHESIS PERTAINING TO THE 4-WAY CONTINGENCY TABLE

Kind of hypothesis	*Fitted marginals and description in conventional terms*		*Degrees of freedom*[a]	*Number of hypotheses of this kind*
	I. Elementary hypotheses			
1	$\{ABC\}$, $\{ABD\}$	$[\bar{C}\otimes\bar{D}\mid\bar{A}\bar{B}]$	4	6
2	$\{ABC\}$, $\{AD\}$	$[\bar{B}\bar{C}\otimes\bar{D}\mid\bar{A}]$	6	12
3	$\{ABC\}$, $\{D\}$	$[\bar{A}\bar{B}\bar{C}\otimes\bar{D}]$	7	4
4	$\{ABC\}$	$[\bar{D}=\Phi\mid\bar{A}\bar{B}\bar{C}]$	8	4
5	$\{AB\}$, $\{AC\}$, $\{AD\}$	$[\bar{B}\otimes\bar{C}\otimes\bar{D}\mid\bar{A}]$	8	4
6	$\{AB\}$, $\{AC\}$, $\{BD\}$	$[\bar{B}\otimes\bar{C}\mid\bar{A}]\cap[\bar{A}\bar{C}\otimes\bar{D}\mid\bar{B}]$	8	12
7	$\{AB\}$, $\{AC\}$, $\{D\}$	$[\bar{B}\otimes\bar{C}\mid\bar{A}]\cap[\bar{A}\bar{B}\bar{C}\otimes\bar{D}]$	9	12
8	$\{AB\}$, $\{CD\}$	$[\bar{A}\bar{B}\otimes\bar{C}\bar{D}]$	9	3
9	$\{AB\}$, $\{AC\}$	$[\bar{B}\otimes\bar{C}\mid\bar{A}]\cap[\bar{D}=\Phi\mid\bar{A}\bar{B}\bar{C}]$	10	12
10	$\{AB\}$, $\{C\}$, $\{D\}$	$[\bar{A}\bar{B}\otimes\bar{C}\otimes\bar{D}]$	10	6
11	$\{AB\}$, $\{C\}$	$[\bar{A}\bar{B}\otimes\bar{C}]\cap[\bar{D}=\Phi\mid\bar{A}\bar{B}\bar{C}]$	11	12
12	$\{A\}$, $\{B\}$, $\{C\}$, $\{D\}$	$[\bar{A}\otimes\bar{B}\otimes\bar{C}\otimes\bar{D}]$	11	1
13	$\{AB\}$	$[\bar{C}\bar{D}=\Phi\mid\bar{A}\bar{B}]$	12	6
14	$\{A\}$, $\{B\}$, $\{C\}$	$[\bar{A}\otimes\bar{B}\otimes\bar{C}]\cap[\bar{D}=\Phi\mid\bar{A}\bar{B}\bar{C}]$	12	4
15	$\{A\}$, $\{B\}$	$[\bar{A}\otimes\bar{B}]\cap[\bar{C}\bar{D}=\Phi\mid\bar{A}\bar{B}]$	13	6
16	$\{A\}$	$[\bar{B}\bar{C}\bar{D}=\Phi\mid\bar{A}]$	14	4
17	n	$[\bar{A}\bar{B}\bar{C}\bar{D}=\Phi]$	15	1
	II. Non-elementary hypotheses			
18	$\{ABC\}$, $\{ABD\}$, $\{ACD\}$, $\{BCD\}$		1	1
19	$\{ABC\}$, $\{ABD\}$, $\{ACD\}$		2	4
20	$\{ABC\}$, $\{ABD\}$, $\{CD\}$		3	6
21	$\{ABC\}$, $\{AD\}$, $\{BD\}$, $\{CD\}$		4	4
22	$\{ABC\}$, $\{AD\}$, $\{BD\}$		5	12
23	$\{AB\}$, $\{AC\}$, $\{AD\}$, $\{BC\}$, $\{BD\}$, $\{CD\}$		5	1
24	$\{AB\}$, $\{AC\}$, $\{AD\}$, $\{BC\}$, $\{BD\}$		6	6
25	$\{AB\}$, $\{AC\}$, $\{AD\}$, $\{BC\}$		7	15
26	$\{AB\}$, $\{AC\}$, $\{BC\}$, $\{D\}$		8	4
27	$\{AB\}$, $\{AC\}$, $\{BC\}$		9	4

[a] Degrees of freedom for the 2^4 table.

Formula (3.6) can be applied to obtain the $\hat{F}$'s for Hypotheses 3, 4, 8, 10–17 in Table 4; (3.7) can be applied for Hypotheses 1, 2, 5, 7, 9. Formula (3.10) can be applied for any of the hypotheses in this table, yielding correct values at the end of the first cycle for Hypotheses 1–17 (the elementary hypotheses), and requiring additional cycles for the remaining hypotheses. Note that Hypothesis 6 is an elementary hypothesis, but neither (3.6) nor (3.7) can be applied to it. For this particular hypothesis,

$$\hat{F}_{ijkl} = f_{ij}^{AB} f_{ik}^{AC} f_{jl}^{BD} / f_i^{A} f_j^{B}. \tag{4.1}$$

Formula (4.1) is a special case of the more general formula (4.2) given next.

For a given hypothesis, if the minimal set $\{Y_1\}$, $\{Y_2\}$, $\cdots$, $\{Y_T\}$ of fitted marginals is such that all non-empty intersections of the Y's are equal to Z' or Z'', where Z' and Z'' are mutually exclusive sets of letters, then $\hat{F}$ can be expressed as

$$\hat{F} = \left(\prod_{t=1}^{T} f^{Y_t} \right) \Big/ \left[G n^{h-h^*+1} (f^{Z'})^{h'-1} (f^{Z''})^{h''-1} \right], \tag{4.2}$$

where G is defined as in (3.6); h is the number of Y's that contain neither Z' nor Z'', h^* is the number of Y's that contain both Z' and Z'' ($h^*=0$ or 1); h' and h'' are the number of Y's that contain Z' and Z'', respectively. Formula (4.1) is a particular example of (4.2), and are so (3.6) and (3.7) (with $f^Z=n$ when Z is the empty set). If all non-empty intersections of the Y's in the minimal set are equal to Z' or Z'', where Z' and Z'' are distinct but not mutually exclusive sets of letters, then one of these sets is contained in the other (say, $Z'' \subset Z'$), and (4.2) can be applied with the denominator replaced by

$$\left[G n^{h} (f^{Z'})^{h'-1} (f^{Z''})^{h''} \right],$$

where h'' is now the number of Y's that contain Z'' but not Z'. Still more general formulae could be presented, but we shall not do so here.

The elementary hypotheses in Table 4 are described in more conventional terms in Column 3 (Part I) of the table. Any hypothesis for which (3.6) is appropriate can be described in terms of the concepts of independence, equiprobability, and/or conditional equiprobability. (For example, Hypothesis 4 in Table 4 states that the classes of variable D are conditionally equiprobable, given the level of the joint variable ABC.) Any hypotheses for which (3.7) or (4.2) is appropriate can be described in terms of the concepts of independence, conditional independence, and/or conditional equiprobability. (For example, Hypothesis 6 in Table 4 states that (a) variables B and C are conditionally independent, given the level of variable A, and (b) variable D and the joint variable AC are conditionally independent, given the level of variable B.) When (3.7) is appropriate, the hypothesis describes (among other things) conditional independence among certain variables, given the level of the (joint) variable Z (see (3.7)). When (4.2) is appropriate, a similar (but more complicated) type or remark applies.

Each elementary hypothesis in Table 4 (see Column 3, Part I) can be partitioned into four (or fewer) subhypotheses using a method which we now describe in more general terms. For the m-way table, let $W_1, W_2, \cdots, W_m$ denote the table's m variables in some specified order (e.g., for $m=4$, the $W_1, \cdots, W_4$ could denote variables A, B, C, D, respectively). Let H denote the hypothesis obtained by the intersection of m subhypotheses, where the sth subhypothesis $H_s (s=1, 2, \cdots, m)$ is one of the following hypotheses pertaining to the conditional probability $P_s(x)$ that class x of variable W_s will be observed, given the level of the joint variable $W_1W_1 \cdots W_{s-1}$: (a) $P_s(x)$ is independent of x for the range of possible x values (i.e., the classes of variable W_s are conditionally equi-probable—$[\overline{W}_s = \Phi | \overline{W_1W_2 \cdots W_{s-1}}]$); (b) $P_s(x)$ is independent of the level of the joint variable $W_1W_2 \cdots W_{s-1}$ (i.e., $[\overline{W}_s \otimes \overline{W_1W_2 \cdots W_{s-1}}]$); (c) $P_s(x)$ is independent of the level of the joint variable $V_1V_2 \cdots V_t$ (for some specified value of $t<s-1$), given the level of the joint variable $V_{t+1} \cdots V_{s-1}$, where $V_1, V_2, \cdots, V_{s-1}$ denotes some specified permutation of the variables

$W_1, W_2, \cdots, W_{s-1}$ (i.e., $[\overline{W_s} \otimes \overline{V_1V_2 \cdots V_t} | \overline{V_{t+1} \cdots V_{s-1}}]$); (d) $P_s(x)$ is unspecified. (Hypothesis (d) is vacuous; and so are (b) and (c) when $s=1$.)

Let H^* denote any hypothesis in Tables 2–4. If H^* is an elementary hypothesis, then it is equivalent to a hypothesis of the form H just described, since the corresponding explicit formula for $\hat{F}/n$ is actually a product of the usual estimated conditional probabilities $\hat{P}_s$ for a subhypothesis H_s, for $s=1, 2, \cdots, m$ (for some ordering of the contingency table's m variables). (Note that $H \equiv H^*$ if and only if (a) the minimal set of marginals used in the calculation of $\hat{P}_s$ (for $s=1, 2, \cdots, m$) is equal to the minimal set of fitted marginals under H^*, and (b) the sum of the degrees of freedom for testing H_s (for $s=1, 2, \cdots, m$) equals the degrees of freedom for testing H^*.) Conversely, if H^* is equivalent to a hypothesis of the form H, then H^* is elementary and $\hat{F}/n$ under H^* is the product of the $\hat{P}_s$ under H_s, for $s=1, 2, \cdots, m$. The preceding remark pertaining to the m-way table with $m=2, 3, 4$ (see Tables 2–4) also holds for $m=5, 6$; and for $m>6$ it will hold if a somewhat more general definition of an "elementary" hypothesis is used (see [29]).

Consider now the non-elementary hypotheses in Table 4. Hypothesis 18 states that the 4-factor interactions λ^{ABCD} are zero. Hypothesis 19 states that the 3-factor λ's are zero in the conditional 3-way table that is obtained when the level of variable A is given. We shall comment later on Hypotheses 20–22. Hypothesis 23 states that the 3-factor and 4-factor λ's are zero, etc.

Let us now examine some of these hypotheses from the point of view of the usual logit-analysis pertaining to variable D, when D is dichotomous. Hypothesis 21 states that the logit pertaining to variable D can be expressed as a sum of the main effects of the general mean and of the variables A, B, and C on D; i.e., $\psi(D|ABC) = \lambda^D + \lambda^{AD} + \lambda^{BD} + \lambda^{CD}$. Hypothesis 22 states that $\psi(D|ABC) = \lambda^D + \lambda^{AD} + \lambda^{BD}$, and similarly for Hypotheses 2, 3, and 4. Hypothesis 1 states that $\psi(D|ABC) = \lambda^D + \lambda^{AD} + \lambda^{BD} + \lambda^{ABD}$; and similarly for Hypotheses 18, 19, and 20. The remarks at the end of Section 3.2 concerning the analysis of logit-vectors can be directly extended here to the case where variable D is polytomous rather than dichotomous.

Various results presented in Sections 2.2 and 3.2 can be directly extended to the 4-way table and, more generally, to the m-way table. Note, for example, the conditions that indicate that a hypothesis (a) states (among other things) that the classes of a (joint) variable Z are conditionally equiprobable, or (b) describes conditional properties that apply conditionally on the level of a (joint) variable Z, or (c) describes the conditional property that the $\lambda^{Z*}=0$ in the m^*-way table ($m^*<m$) given the level of the remaining variables, or (d) describes (among other things) a property of a joint variable Z rather than a property of the separate variables that form the joint variable, or (e) describes a generalized form of logit-analysis. Note also the alternative ways of calculating the degrees of freedom. We can also apply here the earlier remarks on the validity of the chi-square tests when sampling is done with certain sets of marginals given.

5. THE 5-WAY CONTINGENCY TABLE: AN EXAMPLE

Consider now the 5-way table (Table 1), a cross-classification of 1,729 individuals with respect to five dichotomized variables pertaining to the presence or absence of newspapers (N), radio (R), solid reading (S), lectures (L), and whether knowledge (K) of cancer is good or poor. We begin with the direct estimation of the main effects and interaction effects among the five variables. Before applying the direct extension of (2.6), (3.2), and (3.4) to calculate the $\hat{\lambda}$'s and their estimated variance, we replace each f by $f+(1/2)$ as recommended earlier herein. Table 5 gives the value of each $\hat{\lambda}$ and its standardized value. The estimate of the standard deviation of each $\hat{\lambda}$ is $S=.059$. To facilitate the use of the standardized values of the $\hat{\lambda}$'s as a guide, we have listed them in Table 5 in decreasing order of their absolute values, first considering the one-factor $\hat{\lambda}$'s, then the 2-factor $\hat{\lambda}$'s, etc.

Table 5. ESTIMATE OF MAIN EFFECTS AND INTERACTION EFFECTS AMONG FIVE VARIABLES IN THE 5-WAY CONTINGENCY TABLE (TABLE 1)[a]

Variable	*Effect*	*Standardized value*	*Variable*	*Effect*	*Standardized value*
L	−1.17	−19.73	KLS	−.14	−2.30
R	− .43	− 7.23	LRS	−.09	−1.50
N	.41	6.94	KLN	.09	1.45
K	− .14	− 2.37	LNR	−.08	−1.31
S	.08	1.39	KLR	.06	1.00
NS	.31	5.25	KRS	−.05	− .76
KN	.23	3.95	KNR	−.02	− .29
LR	.21	3.62	KNS	.01	.17
NR	.14	2.44	NRS	−.01	− .13
KS	.14	2.38	LNS	−.00	− .01
KR	.12	1.99	$KLNS$	.06	1.08
LS	.11	1.81	$KNRS$	.06	.98
KL	.10	1.61	$KLRS$	−.05	− .84
LN	.09	1.48	$LNRS$	.04	.69
RS	− .03	− .52	$KLNR$	−.02	− .36
			$KLNRS$	.05	.89

[a] The five variables pertain to the presence or absence of newspaper (N), radio (R), solid reading (S), lectures (L), and whether knowledge (K) of cancer is good or poor.

Denoting the five variables of Table 1 by K, L, N, R, S, we also denote $\hat{\lambda}_{11111}^{KLNRS}$ by $\hat{\lambda}^{KLNRS}$, $\hat{\lambda}_{1111}^{KLNS}$ by $\hat{\lambda}^{KLNS}$, etc. Considering their absolute values, we see (Table 5) that the 5-factor $\hat{\lambda}^{KLNRS}$ is larger than three out of the five 4-factor $\hat{\lambda}$'s; that the 4-factor $\hat{\lambda}^{KLNS}$ is larger than six out of the ten 3-factor $\hat{\lambda}$'s; that the 3-factor $\hat{\lambda}^{KLS}$ is larger than five out of the ten 2-factor $\hat{\lambda}$'s. (We are commenting here on the ranking of the absolute values of the $\hat{\lambda}$'s, not on the statistical significance of the differences among the $\hat{\lambda}$'s. The statistical significance of the difference between two particular $\hat{\lambda}$'s can be determined by using as a gauge the estimated standard deviation of the difference, which can be calculated from a formula analogous to (3.4) after the difference has been written in a form analogous to (3.3).)

In some of the earlier literature, methods were developed for fitting the data in the m-way table under the assumption that all interactions of a specified order or higher order (e.g., all 3-factor, 4-factor, · · · , m-factor interactions) are zero. (See [20, 43]; and also [1], which utilized a different definition of interaction from the one used here.) Direct examination of the $\hat{\lambda}$'s in Table 5 suggests that this assumption (which was made in the earlier literature) can be improved upon. We shall now illustrate this point in a different way.

Applying the testing methods described earlier herein to Table 1, we present in Table 6 the chi-square goodness-of-fit values obtained in testing various hierarchical hypotheses concerning the λ's. (A single computer program can be used to calculate these values, and other statistics of interest, for any set of hierarchical hypotheses.) Letting H_u denote the u-th hypothesis in Table 6, we find, for example, that the test of H_3, that all sixteen 3-factor and higher-order λ's are zero, yields a chi-square of 21.2 (with $31-5-10=16$ df). The test of H_6, that the 3-factor and higher-order λ's, except for λ^{KLS}, are zero, yields a chi-square of 15.2 (with $16-1=15$ df); and the test of H_7, that $\lambda^{RS}=0$ and the 3-factor and higher-order λ's, except for λ^{KLS}, are zero, yields a chi-square of 15.6 (with $15+1=16$ df). Thus, we could improve upon the fit of H_3 (which is a hypothesis of the kind referred to in the preceding paragraph): the goodness-of-fit value of 21.2 (with 16 df) could be reduced to 15.6 (with 16 df) or to 15.2 (with 15 df). We shall discuss methods for studying the difference between these kinds of chi-square values in Section 6.

Of course, if we reduce the set of interactions that are assumed zero (i.e., if some of the interactions in the original set that were assumed zero are then not assumed to be zero), then a better fit to the data can be expected. Except in the special case where the $\hat{F}$'s are exactly the same when the original set of interactions are assumed zero as when the reduced set of interactions are assumed zero, the goodness-of-fit statistic will be reduced for the reduced set. (Compare H_3 and H_6, or H_7 and H_6, or H_5, H_4, H_3, H_2, H_1.) If the set of 3-factor and higher-order λ's, which were assumed zero in H_3, is reduced by not assuming that $\lambda^{KLS}=0$, or that $\lambda^{KLS}=\lambda^{LRS}=0$, or that $\lambda^{KLS}=\lambda^{LRS}=\lambda^{KLN}=0$, or that $\lambda^{KLS}=\lambda^{LRS}=\lambda^{KLN}=\lambda^{LNR}=0$, or that $\lambda^{KLS}=\lambda^{LRS}=\lambda^{KLN}=\lambda^{LNR}=\lambda^{KLR}=0$, then we obtain in turn H_6, H_8, H_9, H_{10}, H_{11}, with chi-square values 15.2, 12.8, 10.6, 9.1, 7.7, and with degrees of freedom 15, 14, 13, 12, 11, respectively. The advantages of improving the fit should be weighed against the disadvantages of introducing additional parameters (reducing the set of λ's that are assumed zero).

To gain a better understanding as to why the particular hypotheses reported above were studied (e.g., why λ^{KLS}, rather than, say λ^{LNS}, was not assumed to be zero), and why the chi-square values came out as they did, see the $\hat{\lambda}$'s in Table 5. Some caution is required in the use of these $\hat{\lambda}$'s as a simple guide in selecting which hypotheses to fit the to data, since such guidance is not always foolproof. For example, although $\hat{\lambda}^{KNS}$ (and its standardized value) is smaller than $\hat{\lambda}^{KLR}$ (and its standardized value), H_{12} fits the data somewhat better than does H_{11}, despite the fact that the only difference between these hypotheses is that λ^{KLR} is assumed to be zero under H_{12} but not under H_{11}, whereas λ^{KNS} is assumed to be zero under H_{11} but not under H_{12}. (Ten other λ's are assumed zero under both H_{11} and H_{12}.) Using the $\hat{\lambda}$'s as a more complex guide, we see that, although $\hat{\lambda}^{KNS}$ is small, $\hat{\lambda}^{KNSL}$ and $\hat{\lambda}^{KNSR}$ are of the same order of magnitude as $\hat{\lambda}^{KLR}$; which can help to explain the fits obtained with H_{11} and H_{12}. Further guidance can be obtained by replacing the $\hat{\lambda}$'s (and their standardized values) by the corresponding maximum-likelihood estimates (and their

Table 6. CHI-SQUARE VALUES FOR SOME HYPOTHESES PERTAINING TO THE 5-WAY CONTINGENCY TABLE (TABLE 1)

Hypothesis	*Fitted marginals*	*Degrees of freedom*	*Goodness-of-fit chi-square*	*Likelihood-ratio chi-square*
1	All 4-factor marginals	1	1.0	1.0
2	All 3-factor marginals	6	3.3	3.2
3	All 2-factor marginals	16	21.2	19.6
4	All 1-factor marginals	26	751.3	596.8
5	n	31	3811.8	2666.2
6	{KLS} and 2-factor marginals	15	15.2	15.5
7	{KLS} and 2-factor marginals except {RS}	16	15.6	16.0
8	{KLS}, {LRS} and 2-factor marginals	14	12.8	13.1
9	{KLS}, {LRS}, {KLN} and 2-factor marginals	13	10.6	10.1
10	{KLS}, {LRS}, {KLN}, {LNR} and 2-factor marginals	12	9.1	9.1
11	{KLS}, {LRS}, {KLN}, {LNR}, {KLR} and 2-factor marginals	11	7.7	7.6
12	{KLS}, {LRS}, {KLN}, {LNR}, {KNS} and 2-factor marginals	11	6.3	6.0
13	{LNRS}, {KN}, {KS}, }KR}, {KL}	11	13.6	13.6
14	{LNRS}, {KN}, {KLS}	11	15.9	16.3
15	{LNRS}, {KN}, {KS}, {KL}	12	20.2	20.0
16	{LNRS}, {KN}, {KS}, {KR}	12	19.3	18.5
17	{LNRS}, {KN}, {KR}, {KLS}	10	9.9	10.3
18	{LNRS}, {KLS}, {KLN}	10	13.4	13.2
19	{LNRS}, {KLS}, {KLN}, {KR}	9	7.1	7.0
20	{LNRS}, {KLS}, {KLN}, {KLR}	8	5.6	5.6
21	{LNRS}, {KLS}, {KLN}, {KNS}, {KR}	8	4.5	4.3
22	{LNRS}, {KLS}, {KLR}, {KNS}	8	5.9	6.0
23	{LNRS}, {KLN}, {KLR}, {KNS}	8	7.1	6.7
24	{LNRS}, {KLNS}	8	9.2	9.2
25	{LNRS}, {KLNS}, {KR}	7	3.2	3.3
26	{LNRS}, {KLS}, {KLN}, {KLR}, {KNS}	7	2.8	2.8
27	{LNRS}, {K}	15	207.5	216.3
28	{KLNS}, {RL}, {RN}, {RK}, {RS}	11	9.1	8.4
29	{KLNS}, {RK}, {RLN}	11	6.9	6.8
30	{KLNS}, {RL}, {RN}, {RK}	12	9.5	8.9
31	{KLNS}, {RN}, {RK}, {RLS}	10	6.2	6.0
32	{KLNS}, {RLN}, {RKL}	10	5.7	5.6
33	{KLNS}, {RLN}, {RKL}, {RS}	9	5.1	5.0
34	{KLNS}, {RLN}, {RKL}, {RLS}	8	3.5	3.5
35	{KLNS}, {KLNR}	8	5.3	5.3
36	{KLNS}, {KLNR}, {RS}	7	4.7	4.7
37	{KLNS}, {RLS}, {RLN}, {RKL}, {RNS}	7	2.1	2.1
38	{KLNS}, {R}	15	100.7	102.5

standardized values) obtained when certain sets of λ's are assumed zero (see Section 7 below). (The $\hat{\lambda}$'s in Table 5 are the maximum-likelihood estimates when none of the λ's are assumed zero.)

We shall now comment on H_{13} to H_{38}, which can be expressed as hypotheses about the logits of K (H_{13} to H_{27}) or about the logits of R (H_{28} to H_{38}). To gain

Table. 7. STANDARDIZED VALUE OF MAIN EFFECTS AND INTERACTION EFFECTS OF REMAINING VARIABLES ON KNOWLEDGE VARIABLE (K) AND ON RADIO VARIABLE (R)

1. Effect on the K variable		*2. Effect on the R variable*	
Variable	*Standardized value*	*Variable*	*Standardized value*
K	−2.37	*R*	−7.23
KN	3.95	*RL*	3.62
KS	2.38	*RN*	2.44
KR	1.99	*RK*	1.99
KL	1.61	*RS*	− .52
KLS	−2.30	*RLS*	−1.50
KLN	1.45	*RLN*	−1.31
KLR	1.00	*RKL*	1.00
KRS	− .76	*RKS*	− .76
KNR	− .29	*RKN*	− .29
KNS	.17	*RNS*	− .13
KLNS	1.08	*RKNS*	.98
KNRS	.98	*RKLS*	− .84
KLRS	− .84	*RLNS*	.69
KLNR	− .36	*RKLN*	− .36
KLNRS	.89	*RKLNS*	.89

a better understanding of the results reported next, see the standardized values of the λ's in Tables 5 and 7.

Considering the usual hypothesis of logit-analysis (H_{13}) pertaining to variable K, with each logit expressed as a sum of the main effects of the general mean and of the variables L, N, R, and S on K (i.e., $\psi(K|LNRS) = \lambda^K + \lambda^{KL} + \lambda^{KN} + \lambda^{KR} + \lambda^{KS}$), we obtain a chi-square of 13.6 with $16-5=11$ df. (This chi-square agrees with the corresponding value in Bishop [7], but it differs somewhat from the value 13.85 reported by Dyke and Patterson [16] who applied two cycles of the usual iterative operations of logit-analysis (calculating working logits, inverting matrices, etc.). More accurate methods were used in the present article; the appropriate marginals for the $\hat{F}$'s calculated here and for the observed f's agree to four decimal places in this case.) Although Dyke and Patterson state that their logit-analysis "provides a satisfactory fit," by comparing H_{13} with H_{12} we see that the chi-square goodness-of-fit value can be reduced from 13.6 to 6.3, with 11 df in each case.

In view of the decrease in the chi-square value noted above, we comment further on the difference between the usual hypothesis of logit-analysis (H_{13}) and the hypothesis which we introduced here (H_{12}) for comparative purposes. Hypotheses H_{13} and H_{12} differ with respect to the set of marginal tables used to explain the data in the 5-way table: For H_{13} the set is $\{LNRS\}$, $\{KN\}$, $\{KS\}$, $\{KR\}$, $\{KL\}$; for H_{12} the set is $\{KLS\}$, $\{LRS\}$, $\{KLN\}$, $\{LNR\}$, $\{KNS\}$, $\{KR\}$. The improvement in the fit noted above is obtained by replacing an explanation (H_{13}) based upon the association present in a 4-way table and in four 2-way tables by an explanation (H_{12}) based upon the association present in five 3-way tables and in one 2-way table. The number of λ's that are assumed zero is the same for each hypothesis. The parameters λ^{KLS}, λ^{KLN}, λ^{KNS} are assumed zero under H_{13} but not under H_{12}; the parameters λ^{LNRS}, λ^{LNS}, λ^{NRS} are assumed zero under H_{12} but not under H_{13}. (Eight other λ's are assumed zero under both H_{13} and H_{12}.)

We shall consider now the other hypotheses about the logits of K. Considering H_{19} (i.e., $\psi(K|LNRS)=\lambda^K+\lambda^{KL}+\lambda^{KN}+\lambda^{KR}+\lambda^{KS}+\lambda^{KLN}+\lambda^{KLS}$), we obtain a chi-square of 7.1 with 9 df. Hypotheses H_{14} to H_{18} can be described in similar terms, and their chi-squares can be compared. For example, compare H_{19} with H_{18} and H_{17}; H_{18} with H_{14} and H_{13}; H_{17} with H_{14} and H_{13}; H_{16} with H_{15} and H_{13}; H_{15} with H_{14} and H_{13}.

Hypotheses H_{20} to H_{23} state that $\psi(K|LNRS)=\lambda^K+\lambda^{KL}+\lambda^{KN}+\lambda^{KR}+\lambda^{KS}+\lambda^{KZ'}+\lambda^{KZ''}+\lambda^{KZ'''}$, where Z', Z'', Z''', are three out of the following four sets of letters: (1) LN, (2) LR, (3) LS, (4) NS. The smallest chi-square is obtained here for H_{21}, with the sets (1), (3), (4). This result is consistent with a finding made by Cox and Snell [14] using quite different methods from those used there. The remarks made earlier herein, in comparing H_{11} and H_{12}, apply here too.

Hypotheses H_{24} to H_{26} can also be described in similar terms, and their chi-squares can be compared. For example, compare H_{26} with H_{25} and H_{20}; and H_{25} with H_{24}. Note that H_{24}, a particular hypothesis about the logits pertaining to variable K, can also be expressed as $[\overline{K}\otimes\overline{R}|\overline{LNS}]$.

Considering H_{27} (i.e., $\psi(K|LNRS)=\lambda^K$, which can also be expressed as $[\overline{K}\otimes\overline{LNRS}]$), we obtain a chi-square of 207.5, with 15 df. In assessing the magnitudes of the chi-square values obtained for the logit-analyses pertaining to H_{13} to H_{26}, the chi-square for H_{27} can serve as a base line for comparisons. (Other possible base lines might be the chi-squares values for H_5 or H_4, which could serve also as base lines for the chi-square values for H_1 to H_3 and H_6 to H_{12}.) The relative difference between the chi-square values for H_{27} and H_u (for u from 13 to 26) can be interpreted as a kind of (squared) multiple correlation coefficient in the sense described next. Since the chi-square can be expressed as a measure of the "weighted unexplained variation" when the observed logits pertaining to variable K are compared with their estimated expected values under a null hypothesis (see, e.g. [16]), the relative difference between the chi-square values noted above can be interpreted as the relative decrease in the "weighted unexplained variation" obtained in the case where the additive effects on K, which are included in H_u, are taken into account in contrast to the case (H_{27}) where only the main effect of a general mean on K (i.e., λ^K) is taken into account.

Hypotheses H_{28} to H_{38} are about the logits pertaining to variable R. The chi-square values for these hypotheses are in general less than the corresponding values obtained in the logit-analysis pertaining to variable K. (H_{36} is an exception.) Generally speaking, a better fit to the data is obtained with the logit-analysis of variable R than with the earlier logit-analysis.

For each chi-square statistic considered here, Table 6 gives the degrees of freedom of its asymptotic distribution under the null hypothesis. This is the appropriate number of degrees of freedom to use in testing the hypothesis *if* the hypothesis were decided upon before the data were studied. On the other hand, if a set of hypotheses were tested simultaneously (or if the particular hypothesis that was tested was contained within a larger set of hypotheses that were studied), the degrees of freedom could be adjusted in a similar way to the adjustment made in calculating simultaneous confidence intervals (see, e.g. [21, 28a]). This adjustment will limit the risks of rejecting hypotheses that are true, even when the hypotheses are suggested by the data. (Of course, the risks of accepting false hypotheses are also affected if the hypotheses are suggested by the data.)

6. PARTITIONING CHI-SQUARE AND TABLES FOR THE ANALYSIS OF ASSOCIATION

The usual chi-square goodness-of-fit statistic, which we used above to measure the agreement between the observed frequency f in the contingency table and the corresponding estimate $\hat{F}$ of the expected frequency F under a given hypothesis, can be written as

$$\Sigma(f - \hat{F})^2/\hat{F}, \tag{6.1}$$

where the summation is over all cells in the table. The corresponding chi-square statistic based upon the likelihood-ratio criterion can be written as

$$X^2 = 2\Sigma f \log(f/\hat{F}). \tag{6.2}$$

Because (6.1) is more familiar to many readers, we used it above; but, in certain contexts, (6.2) has some advantages (see [2, 19, 24, 28, and 30]). In the remaining sections, we shall use only (6.2). (Columns 4 and 5 of Table 6 give the values of (6.1) and (6.2) for hypotheses pertaining to Table 1.)

6.1 The Partitioning of Conditional Independence

We first return to the $I \times J \times K$ table whose variables are A, B, and C; and we denote the uth hypothesis in Table 3 by $H_u{}^+$. Hypothesis $H_1{}^+$ states that $\lambda^{ABC} = 0$, and $H_2{}^+$ states that $\lambda^{ABC} = \lambda^{BC} = 0$. Under the assumption that $H_1{}^+$ is true, the λ^{BC} measure the partial association between variables B and C (see [5, 6 and 26]). Let $H_2{}^{*+}$ denote the hypothesis that this partial association is zero (i.e., $\lambda^{BC} = 0$, *assuming* $H_1{}^+$ is true). Thus, $H_2{}^+$ states that both $H_1{}^+$ and $H_{2*}{}^+$ are true.

To test $H_1{}^+$ we calculate $X^2(H_1{}^+)$ from (6.2), with $\hat{F}$ obtained by iterative scaling (see (3.8)). To test $H_2{}^+$ we calculate $X^2(H_2{}^+)$ from (6.2) with $\hat{F}$ obtained from (a) of (3.5). These X^2's can be used in calculating the chi-square statistic $X^2(H_{2*}{}^+)$ based upon the likelihood-ratio criterion for testing $H_{2*}{}^+$, since

$$\begin{aligned} X^2(H_{2*}{}^+) &= X^2(H_2{}^+) - X^2(H_1{}^+) = 2\Sigma f \log[\hat{F}_{(1)}/\hat{F}_{(2)}] \\ &= 2\Sigma \hat{F}_{(1)} \log[\hat{F}_{(1)}/\hat{F}_{(2)}], \end{aligned} \tag{6.3}$$

where $\hat{F}_{(u)}$ is the $\hat{F}$ used in testing $H_u{}^+$ (for $u = 1, 2$); see [26]. (The final equality in (6.3) holds because $\hat{F}_{(1)}$ fits the 2-way marginals of f.) The statistic $X^2(H_{2*}{}^+)$ will have an asymptotic chi-square distribution under $H_{2*}{}^+$, with $(J-1)(K-1)$ degrees of freedom.

The test of $H_{2*}{}^+$ based upon $X^2(H_{2*}{}^+)$ provides a more powerful method for detecting partial association (if present) than the more usual test of conditional independence (i.e., the test based upon $X^2(H_2{}^+)$) applied for this purpose. This follows from the fact that $X^2(H_2{}^+)$ can be partitioned into $X^2(H_{2*}{}^+)$ and an asymptotically independent statistic (viz., $X^2(H_1^+)$) whose distribution does not depend upon whether or not there is partial association (assuming, of course, that H_1^+ is true).

The method of partitioning $X^2(H_2^+)$ just presented can be displayed in tabular form as Table 8, which is somewhat analogous to the usual analysis of variance table. Here, however, the component in the second row of Table 8 has been defined assuming that the component in the first row of this table is zero for the population 3-way table.

Table 8. ANALYSIS OF CONDITIONAL DEPENDENCE IN A 3-WAY CONTINGENCY TABLE

Component due to	*Likelihood-ratio chi-square*	*Degrees of freedom*
Three-factor interaction	$2\Sigma f \log[f/\hat{F}_{(1)}]$	$(I-1)(J-1)(K-1)$
Partial association between B and C	$2\Sigma f \log[\hat{F}_{(1)}/\hat{F}_{(2)}]$	$(J-1)(K-1)$
Conditional dependence between B and C, given the level of A	$2\Sigma f \log[f/\hat{F}_{(2)}]$	$I(J-1)(K-1)$

Table 8 is related to tables presented in work by Kullback and his coauthors. However, the methods for calculating the quantity corresponding to $\hat{F}_{(1)}$ in $X^2(H_1^+)$ and $X^2(H_{2*}^+)$, which were used in Tables 7.1 and 8.1 of Chapter 8 in [34], and in Tables 4.3 and 4.4 in [35], need to be amended to obtain the correct quantity presented herein. (This difficulty was appreciated in [33], as can be surmised by a comparison of the numerical values in Table 4.1 there.) Also, contrary to Ku and Kullback [33] (see their Tables 3.3 and 3.5), the quantity corresponding to the $X^2(H_{2*}^+)$ component is not due to the "two-way marginals interaction with the row variable A"; but rather to the partial association between variables B and C (i.e., the λ^{BC}) assuming H_1^+ is true, as we just stated. (The expression "two-way marginals interaction with the row variable A" is a translation to the 3-way table of the expression used in Tables 3.3 and 3.5 of [33] for the corresponding components in the 4-way table.) From (6.3) herein, we see that $X^2(H_{2*}^+)$ is proportional to a weighted average of the $\log[\hat{F}_{(1)}/\hat{F}_{(2)}]$, where $\hat{F}_{(1)}$ is obtained when the fitted marginals are $\{AB\}$, $\{AC\}$, and $\{BC\}$, and $\hat{F}_{(2)}$ is obtained when the fitted marginals are $\{AB\}$ and $\{AC\}$, assuming that the "interactions," in a certain sense, between the two-way marginals are zero. Furthermore, while Ku and Kullback [33] note that the formulae in their tables corresponding to $X^2(H_1^+)$ are the quantities obtained after a first iteration (further iterations are necessary), similar remarks should be made about the formulae they give in Tables 3.3 and 3.5 corresponding to $X^2(H_{2*}^+)$ and to the X^2 statistic for testing the hypothesis that the 4-factor λ's are zero in the 4-way table. (The remark about $X^2(H_{2*}^+)$ is a consequence of the corresponding remark about $X^2(H_1^+)$.)

6.2 The Partitioning of Mutual Independence

Hypothesis H_3^+ of Table 3 can be partitioned as follows:

$$[\overline{AB} \otimes \overline{C}] \equiv [\overline{A} \otimes \overline{C}] \cap [\overline{B} \otimes \overline{C} \mid \overline{A}]. \tag{6.4}$$

Because of (6.4), we find that $X^2(H_3^+)$ can also be partitioned:

$$X^2(H_3^+) = X^2(H_{AC}^*) + X^2(H_2^+), \tag{6.5}$$

where H_{AC}^* denotes the hypothesis that $[\bar{A} \otimes \bar{C}]$, and $X^2(H_{AC}^*)$ is the usual chi-square statistic based upon the likelihood-ratio criterion for testing H_{AC}^* in the 2-way marginal table for variables A and C. (Formula (6.5) can be derived directly, using (a) and (b) of (3.5) for the $\hat{F}$'s in $X^2(H_2^+)$ and $X^2(H_3^+)$, respectively.)

Hypothesis H_5^+ of Table 3 can be partitioned as follows:

$$[\bar{A} \otimes \bar{B} \otimes \bar{C}] \equiv [\bar{A} \otimes \bar{B}] \cap [\overline{AB} \otimes \bar{C}]; \tag{6.6}$$

and from (6.4) and (6.6) we obtain

$$[\bar{A} \otimes \bar{B} \otimes \bar{C}] \equiv [\bar{A} \otimes \bar{B}] \cap [\bar{A} \otimes \bar{C}] \cap [\bar{B} \otimes \bar{C} \mid \bar{A}]. \tag{6.7}$$

Combining this partitioning with the partitioning in Section 6.1, we find that

$$X^2(H_5^+) = X^2(H_{AB}^*) + X^2(H_{AC}^*) + X^2(H_{2*}^+) + X^2(H_1^+), \tag{6.8}$$

where $X^2(H_{AB}^*)$ is the likelihood-ratio chi-square for testing the hypothesis $[\bar{A} \otimes \bar{B}]$.

The partitioning of $X^2(H_3^+)$ using (6.5), or $X^2(H_5^+)$ using (6.8), can be displayed in tabular forms similar to Table 8. Applying (6.8), we obtain a table for the analysis of mutual dependence between the variables in the 3-way contingency table (see [26]).

Although H_5^+ states that $\lambda^{AB} = \lambda^{AC} = \lambda^{BC} = \lambda^{ABC} = 0$, we see from (6.8) that $X^2(H_5^+)$ need not equal $X^2(H_{AB}^*) + X^2(H_{AC}^*) + X^2(H_{BC}^*) + X^2(H_1^+)$. Hypothesis H_{BC}^* is different from H_{2*}^+; and the statistics $X^2(H_{BC}^*)$ and $X^2(H_{2*}^+)$ can differ greatly. This point seems to have been overlooked in the partitioning method proposed by Lancaster [36]. (For other comments on Lancaster's method, see [15, 22, 26, and 42].)

The remark in the preceding paragraph can be extended to the 4-way table. In this context, consider the hypothesis (H_{12} in Table 4) that the 2-factor λ's are zero and that the 3-factor and 4-factor λ's are also zero (this is the hypothesis $[\bar{A} \otimes \bar{B} \otimes \bar{C} \otimes \bar{D}]$); and the hypothesis ($H_{23}$ in Table 4) that the 3-factor and 4-factor λ's are zero. Contrary to the impression the reader might get from Table 4.3 in [33], we find that, in general, the X^2 statistic for testing H_{12} need not equal, or approximate closely, the sum of the X^2's for testing independence in each of the 2-way marginal tables and the X^2 for testing H_{23}, although for certain kinds of contingency tables they can be approximately equal. (Ku and Kullback [33] stated that they "should [be] equal," and that their difference "represents the effect of the marginal restraints.") For the 3-way table, we noted that one out of the three X^2's calculated from the 2-way marginal tables can give different results from those obtained with the partitioning (6.8) (compare H_{2*}^+ with H_{BC}^*); for the 4-way table, three or four out of the six X^2's calculated from the 2-way marginal tables can give different results.

The X^2 statistic for testing the hypothesis of mutual independence in the 4-way table (H_{12} in Table 4) can be partitioned by a direct extension of the method used to obtain (6.6) and (6.7) (see (6.10)–(6.11) below); and, if desired, this partitioning can be partitioned further by the methods of Section 6.1 to obtain an extension of (6.8). The X^2 statistic can also be partitioned into a different set of components by the general method of Section 6.4 below. The research worker's choice of which method of partitioning to apply should depend upon which set of component hypotheses are of particular interest to him. (The method of partitioning presented in Table 4.3 by Ku and Kullback [33], to which we referred in the preceding paragraph, does not conform to any of the methods proposed herein, and it is defective in the sense that (*a*) the sum of the X^2 statistics for their so-called "components" can differ greatly from the X^2 statistic for testing the over-all hypothesis, and (*b*) the interpretation of the magnitudes of some of these "components" can lead to incorrect conclusions; for further details, see [28].) The above remarks can be directly extended to the *m*-way table.

6.3 The Analysis of Marginal Tables

For a given 3-way table, consider the hypothesis H_{AC}^* that variables A and C are marginally independent. Although Birch [4] has stated that H_{AC}^* cannot be expressed conveniently in terms of λ parameters of the kind discussed herein for the 3-way table, we shall now show that some justification for testing H_{AC}^* can be expressed directly in terms of these λ's.

From (6.5) we obtain

$$X^2(H_{AC}^*) = X^2(H_3^+) - X^2(H_2^+). \tag{6.9}$$

Thus, $X^2(H_{AC}^*)$ is equal to the chi-square statistic based upon the likelihood-ratio criterion for testing H_3^+ assuming that H_2^+ is true; i.e., for testing that $\lambda^{AC}=0$, assuming that $\lambda^{BC}=\lambda^{ABC}=0$. Thus, in testing H_3^+ assuming that H_2^+ is true, we obtain the same X^2 that would be obtained in testing H_{AC}^*.

The justification just presented for the test of H_{AC}^* was based upon the partitioning of H_3^+ into H_{AC}^* and H_2^+ (see (6.4)–(6.5)). For the *m*-way table, tests of hypotheses about certain kinds of marginal tables can be justified similarly, using the partitioning methods described by the following general rules: Letting $Z_0, Z_1, \cdots, Z_t$ denote mutually exclusive sets of the m letters representing the variables in the *m*-way table, we obtain

$$\begin{aligned}[\overline{Z_1 Z_2 \cdots Z_{t-1}} \otimes \overline{Z}_t \,|\, \overline{Z}_0] \equiv {} & [\overline{Z}_1 \otimes \overline{Z}_t \,|\, \overline{Z}_0] \cap [\overline{Z}_2 \otimes \overline{Z}_t \,|\, \overline{Z_0 Z_1}] \cap \cdots \\ & \cap [\overline{Z}_{t-1} \otimes \overline{Z}_t \,|\, \overline{Z_0 Z_1 \cdots Z_{t-2}}]\end{aligned} \tag{6.10}$$

$$\begin{aligned}[\overline{Z}_1 \otimes \overline{Z}_2 \otimes \cdots \otimes \overline{Z}_t \,|\, \overline{Z}_0] \equiv {} & [\overline{Z}_1 \otimes \overline{Z}_2 \,|\, \overline{Z}_0] \cap [\overline{Z_1 Z_2} \otimes \overline{Z}_3 \,|\, \overline{Z}_0] \cap \cdots \\ & \cap [\overline{Z_1 Z_2 \cdots Z_{t-1}} \otimes \overline{Z}_t \,|\, \overline{Z}_0].\end{aligned} \tag{6.11}$$

Formulae (6.4) and (6.6) are special cases of (6.10) and (6.11), respectively. Using (6.10) and (6.11), tests of certain kinds of hypotheses about independence (or conditional independence) in the *m*-way table can be partitioned into tests of hypotheses about marginal independence (or conditional marginal independence) in tables of fewer dimensions and other tests of independence or conditional independence in the *m*-way table.

We also obtain

$$[\overline{Z}_2 = \Phi \mid \overline{Z_0 Z_1}] \equiv [\overline{Z}_2 = \Phi \mid \overline{Z}_0] \cap [\overline{Z}_1 \otimes \overline{Z}_2 \mid \overline{Z}_0] \tag{6.12}$$

$$[\overline{Z_1 Z_2 \cdots Z_t} = \Phi \mid \overline{Z}_0] \equiv [\overline{Z}_1 = \Phi \mid \overline{Z}_0] \cap [\overline{Z}_2 = \Phi \mid \overline{Z_0 Z_1}] \cap \cdots \cap [\overline{Z}_t = \Phi \mid \overline{Z_0 Z_1 \cdots Z_{t-1}}]. \tag{6.13}$$

Formulae (6.12)–(6.13) can be used to partition, e.g., Hypotheses 2 and 3 of Table 2, and Hypotheses 4, 7, 8 of Table 3. Using (6.12) and (6.13), tests of certain kinds of hypotheses about equiprobability (or conditional equiprobability) in the m-way table can be partitioned into tests of hypotheses about marginal equiprobability (or conditional marginal equiprobability) in tables of fewer dimensions and other tests of conditional equiprobability in the m-way table. For particular applications of some of the general rules (6.10)–(6.13), see, e.g. [35].

The rules (6.10)–(6.13) do not exhaust the ways of partitioning hypotheses about the m-way table into hypotheses about marginal tables of fewer dimensions and other hypotheses about the m-way table. For example, the hypothesis (H_7 in Table 4) that $[\overline{B} \otimes \overline{C} \mid \overline{A}]$ and $[\overline{ABC} \otimes \overline{D}]$ can be partitioned into the hypothesis $[\overline{B} \otimes \overline{D}]$ and the hypothesis (H_6 in Table 4) that $[\overline{B} \otimes \overline{C} \mid \overline{A}]$ and $[\overline{AC} \otimes \overline{D} \mid \overline{B}]$. This follows from the fact that

$$[\overline{ABC} \otimes \overline{D}] \equiv [\overline{B} \otimes \overline{D}] \cap [\overline{AC} \otimes \overline{D} \mid \overline{B}]. \tag{6.14}$$

Similarly, Hypothesis H_7 in Table 4 can also be partitioned into the hypothesis $[\overline{A} \otimes \overline{D}]$ and the hypothesis (H_5 in Table 4) that $[\overline{B} \otimes \overline{C} \otimes \overline{D} \mid \overline{A}]$. This follows from the fact that

$$[\overline{ABC} \otimes \overline{D}] \equiv [\overline{A} \otimes \overline{D}] \cap [\overline{BC} \otimes \overline{D} \mid \overline{A}] \tag{6.15}$$

and

$$[\overline{B} \otimes \overline{C} \mid \overline{A}] \cap [\overline{BC} \otimes \overline{D} \mid \overline{A}] \equiv [\overline{B} \otimes \overline{C} \otimes \overline{D} \mid \overline{A}]. \tag{6.16}$$

The examples in the preceding paragraph illustrate a general rule that can be applied in order to partition some of the other elementary hypotheses about the m-way table. Since an elementary hypothesis can be described in terms of the concepts of independence, equiprobability, conditional independence, and /or conditional equiprobability, the partitioning rules (6.10)–(6.13) can be applied (when appropriate) to one or more of the subhypotheses that form the elementary hypothesis. (Note how (6.7) was obtained from (6.6).) Some of the parts obtained by the partitioning of the subhypotheses can then be recombined to form a different set of subhypotheses. (Note how we used (6.14), or (6.15)–(6.16).)

The partitioning of hypotheses about the m-way table need not be limited to elementary hypotheses. As an example of the partitioning of non-elementary hypotheses, consider the hypothesis (H_{22} in Table 4) that $\Psi(D \mid ABC) = \Lambda^D + \Lambda^{AD} + \Lambda^{BD}$ in the 4-way table $\{ABCD\}$. This hypothesis can be partitioned into (a) the hypothesis that $\Psi(D \mid AB) = \Lambda^D + \Lambda^{AD} + \Lambda^{BD}$ in the 3-way marginal table $\{ABD\}$ (which is equivalent to the hypothesis of zero 3-factor interaction in $\{ABD\}$), and (b) the hypothesis that $[\overline{C} \otimes \overline{D} \mid \overline{AB}]$ (H_1 in Table 4). More generally, letting $Z_1, Z_2, \cdots, Z_t$ denote mutually exclusive sets of the m letters representing variables in the m-way table, consider any hierarchical hypothesis about the logit (or logit-vector) pertaining to the (joint) variable Z_t in the m-way table that can be expressed as a sum of the main effects (if any) of the general mean, of the variables represented by at most $t-2$ of the remaining Z's (say, $Z_2, \cdots, Z_{t-1}$), and of some of their interactions. This hypothesis can be partitioned into (a) the same hypothesis about the logit (or logit-vector) pertaining to the (joint) variable Z_t applied to the marginal table obtained when Z_1 is ignored, and (b) the hypothesis that $[\overline{Z_1 \otimes Z_t} | \overline{Z_2 Z_3 \cdots Z_{t-1}}]$. A similar remark also applies if the hypothesis about the logit (or logit-vector) was a conditional hypothesis, given the level of variable Z_0 (where $Z_0, Z_1, \cdots, Z_t$ are mutually exclusive). The partioning (6.12) is a special case of this.

6.4 A More General Method of Partitioning

Let $\mathfrak{S}'$ and $\mathfrak{S}^*$ denote mutually exclusive basic sets of λ's, and let $\mathfrak{S}$ denote the union of $\mathfrak{S}'$ and $\mathfrak{S}^*$. (For example, with $m=3$, if $\mathfrak{S}'$ is the basic set of λ^{ABC}, and $\mathfrak{S}^*$ is the basic set of λ^{BC}, then $\mathfrak{S}$ is the union of the λ^{ABC} and λ^{BC}.) Let H, H', and H^* denote the hypotheses that the λ's in $\mathfrak{S}$, $\mathfrak{S}'$, and $\mathfrak{S}^*$, respectively, are zero; and let $[H^* | H']$ denote the hypothesis that H^* is true, assuming that H' is true. (For example, if H and H' are Hypotheses 2 and 1 respectively in Table 3, then $[H^* | H']$ is the hypothesis H_{2*}^+ of Section 6.1.) Note that $[H^* | H']$ is equivalent to the hypothesis that H is true, assuming that H' is true.

The X^2 statistics can be calculated from (6.2) to test the hypotheses H and H'. (Here we assume that H and H' are hierarchical.) These statistics can also be used in calculating the chi-square statistic $X^2(H^* | H')$ based upon the likelihood-ratio criterion for testing H^* assuming that H' is true, since

$$X^2(H^* \mid H') = X^2(H) - X^2(H') = 2\Sigma f \log[\hat{F}'/\hat{F}] = 2\Sigma \hat{F}' \log[\hat{F}'/\hat{F}], \quad (6.17)$$

where $\hat{F}$ and $\hat{F}'$ are the estimated F under H and H', respectively (see, e.g. [41]). (The final equality in (6.17) holds because $\hat{F}'$ fits the marginals of f that are fitted under H'; see, e.g. (6.3), a special case of (6.17).) The statistic $X^2(H^* | H')$ will have an asymptotic chi-square distribution under $[H^* | H']$, with the degrees of freedom equal to the number of λ's in $\mathfrak{S}^*$ (i.e., to the difference between the degrees of freedom for $X^2(H)$ and $X^2(H')$).

For a given hypothesis H^*, the statistic (6.17) yields, generally speaking, different chi-squares for different H'; but, in some cases, different H' can yield the same chi-square. For example, if H^* states that $\lambda^{BC}=0$, then $X^2(H^* | H')$ is equal to $X^2(H_{2*}^+)$ of (6.3) when H' is H_1^+ of Table 3; but it is equal to $X^2(H_{BC}^*)$ (where H_{BC}^* is defined as in Section 6.2) when the set $\mathfrak{S}'$ corresponding to H' is any one of the following four sets of λ's: (1) λ^{ABC} and λ^{AB}; (2) λ^{ABC} and λ^{AC}; (3) λ^{ABC}, λ^{AB}, λ^{AC}; (4) λ^{ABC}, λ^{AB}, λ^{AC}, λ^{A} (see Sections 6.2 and 6.3 for related results).

As in Table 8, we can partition $X^2(H)$ into $X^2(H')$ and $X^2(H^* | H')$. The partitioning of $X^2(H_2^+)$ in Table 8, and of $X^2(H_3^+)$ in (6.5), are examples of this. In (6.5), the statistic corresponding to $X^2(H^* | H')$ could be expressed more simply as $X^2(H_{AC}^*)$; and it was not necessary to assume that H' (i.e.,

H_2^+) was true in order to test the component hypothesis H_{AC}^* using $X^2(H_{AC}^*)$. With this simplification, the statistic $X^2(H^*|H')$ (i.e., $X^2(H_{AC}^*)$) will have an asymptotic chi-square distribution under H_{AC}^*, even when H' is not true. A similar simplification occurs in the partitioning based upon (6.6) and (6.7), and the methods of Section 6.3.

The methods just described for partitioning $X^2(H)$ into two components can be applied successively to partition it into more than two components (e.g., the partitioning (6.8) can be obtained thus); and the components can be displayed in tabular form similar to Table 8.

6.5 The Partitioning of Conditional Properties

Instead of partitioning $X^2(H_2^+)$ in Table 8, with its $I(J-1)(K-1)$ degrees of freedom, into two components, with $(I-1)(J-1)(K-1)$ and $(J-1)(K-1)$ degrees of freedom, respectively, it can be partitioned into I components, each with $(J-1)(K-1)$ degrees of freedom. The ith component ($i=1, 2, \cdots, I$) is the X^2 obtained in testing for independence between the columns and layers of the $J \times K$ table at the ith level of the row variable. This partitioning is useful when the relationship between variables B and C, at the ith level of variable A, is of interest.

More generally, consider any hypothesis H that can be described by conditional properties (e.g., conditional independence, conditional equiprobability) that apply conditionally on the level of the (joint) variable Z (see related comments in Section 3.2). The X^2 for H can be partitioned into the asymptotically independent components obtained by testing the property at each level of variable Z; and the components can be displayed in tabular form. This partitioning provides an alternative to the partitioning of H in Section 6.4. (It will often prove worthwhile to carry out an even more detailed separate analysis at each level of variable Z, using, e.g., the appropriately signed square-root of the goodness-of-fit chi-square when the chi-square has one degree of freedom, or the appropriately signed quantities corresponding to single-degree-of-freedom components when the chi-square has more than one degree of freedom—see, e.g. [11].)

7. ESTIMATING PARAMETERS IN UNSATURATED MODELS AND RELATED TESTS

We noted in Sections 2.1 and 3.1 that the $\hat{\lambda}$'s were maximum-likelihood estimates of the corresponding λ's in the case where none of the λ's are assumed zero on *a priori* grounds (i.e., in the "saturated" model). We now provide maximum-likelihood estimates of the λ's in "unsaturated" models in which some of the λ's are assumed zero.

Consider first the $I \times J \times K$ table, assuming that $\lambda^{ABC}=0$. Under this assumption, the maximum-likelihood estimate $\tilde{\lambda}$ of λ can be obtained by first calculating the $\hat{F}$'s by iterative scaling (see (3.8)), and then calculating $\tilde{\lambda}$ from (3.1) with the ν's in that formula replaced by the corresponding $\log \hat{F}$'s. (In this case, $\tilde{\lambda}^{ABC}=0$.) As noted in Section 3.2, the assumption that $\lambda^{ABC}=0$ is equivalent to an assumption about the logit (or logit-vector) pertaining to variable C (viz., $\Psi_{ij}=\Lambda^C+\Lambda_i^{AC}+\Lambda_j^{BC}$); and the maximum-likelihood estimates of the Λ's can be obtained from (3.13) with the λ's in that formula replaced by the corresponding $\tilde{\lambda}$'s. (The calculation of $\tilde{\lambda}^C$, $\tilde{\lambda}^{AC}$, $\tilde{\lambda}^{BC}$ just described could replace the usual logit-analysis when variable C is dichotomous.)

More generally, assume that a given hierarchical hypothesis H' about the m-way table is true. Under this assumption, the $\hat{F}$'s can be calculated by iterative scaling (see (3.10)) (or from (3.6), (3.7), (4.2) and generalizations thereof when H' is elementary); and the maximum-likelihood estimate $\tilde{\lambda}$ of λ can be calculated from (2.5), (3.1), and extensions thereof, replacing the ν's by the corresponding $\log\hat{F}$'s. When H' can be expressed as a hypothesis about the logit (or logit-vector) pertaining to a variable Z, the maximum-likelihood estimates of the effects (the Λ's) can be obtained as in the preceding paragraph. The preceding calculation of the $\tilde{\lambda}$'s could replace the usual logit-analysis when variable Z is dichotomous, since the λ's are calculated without the iterative inversion of matrices required in logit-analysis. Although this method has the disadvantage that it does not calculate automatically the estimated variance-covariance matrix of the $\tilde{\lambda}$'s (as in logit-analysis), this disadvantage can be mitigated somewhat, since a test of the hypothesis $[H^*|H']$ that a given set of $\tilde{\lambda}$'s do not differ significantly from zero (assuming that H' is true) can be based upon (6.17). (Recall that H and H' in (6.17) are limited to hierarchical hypotheses.)

To illustrate the preceding point, we assume now that H' is true, where H' is H_{26} in Table 6 (viz., $\psi(K|LNRS) = \lambda^K + \lambda^{KL} + \lambda^{KN} + \lambda^{KR} + \lambda^{KS} + \lambda^{KLN} + \lambda^{KLR} + \lambda^{KLS} + \lambda^{KNS}$); and we consider the hypothesis that $\tilde{\lambda}^{KLS}$ does not differ significantly from zero. To test this hypothesis using (6.17), the appropriate H is H_{23} of Table 6, and thus a chi-square value of $6.7 - 2.8 = 3.9$ is obtained, with one degree of freedom.

For some further discussion of the methods introduced in this section, and for various examples of the application of these methods, we refer the reader to [28].

REFERENCES

[1] Bahadur, R. R., "A Representation of the Joint Distribution of Responses to N Dichotomous Items," in H. Solomon, ed., *Studies in Item Analysis and Prediction*, Stanford, Calif.: Stanford University Press, 1961.

[2] ———, "Rates of Convergence of Estimates and Test Statistics," *Annals of Mathematical Statistics*, 38 (1967), 303–24.

[3] Bartlett, M. S., "Contingency Table Interactions," *Journal of the Royal Statistical Society Supplement*, 2 (1935), 248–52.

[4] Birch, M. W., "Maximum Likelihood in Three-Way Contingency Tables," *Journal of the Royal Statistical Society*, Series B, 25 (1963), 220–33.

[5] ———, "The Detection of Partial Association, I: The 2×2 Case," *Journal of the Royal Statistical Society*, Series B, 26 (1964), 313–24.

[6] ———, "The Detection of Partial Association, II: The General Case," *Journal of the Royal Statistical Society*, Series B, 27 (1965), 111–23.

[7] Bishop, Y. M. M., "Full Contingency Tables, Logits, and Split Contingency Tables," *Biometrics*, 25 (1969), 383–400.

[8] ———, "Calculating Smoothed Contingency Tables," Appendix to Chapter IV-3, *The National Halothane Study*, Washington, D. C.: U. S. Government Printing Office, 1969.

[9] ——— and Mosteller, F., "Smoothed Contingency Table Analysis," Chapter IV-3,

The National Halothane Study, Washington, D. C.: U. S. Government Printing Office, 1969.

[10] Bock, R. D., "Estimating Multinomial Response Relations," in R. C. Bose, ed., *Contributions to Statistics and Probability: Essays in Memory of S. N. Roy*, Chapel Hill, N. C.: University of North Carolina Press, 1969.

[11] Cochran, W. C., "Some Methods for Strengthening the Common χ^2 Test," *Biometrics*, 10 (1954), 417–52.

[12] ——— and Cox, G. M., *Experimental Designs*, New York: John Wiley and Sons, Inc., 1950.

[13] Cox, D. R. and Lauh, E., "A Note on the Graphical Analysis of Multidimensional Contingency Tables," *Technometrics*, 9 (1967), 481–8.

[14] ——— and Snell, E. J., "A General Definition of Residuals," *Journal of the Royal Statistical Society*, Series B, 30 (1968), 248–65.

[15] Darroch, J. N., "Interactions in Multifactor Contingency Tables," *Journal of the Royal Statistical Society*, Series B, 24 (1962), 251–63.

[16] Dyke, G. V. and Patterson, H. D., "Analysis of Factorial Arrangements When the Data Are Proportions," *Biometrics*, 8 (1952), 1–12.

[17] Fisher, R. A. and Yates, F., *Statistical Tables for Biological, Agricultural and Medical Research*, 6th Ed., New York: Hafner Publishing Co, Inc., 1963.

[18] Gart, J. J. and Zweifel, J. R., "On the Bias of Various Estimators of the Logit and Its Variance," *Biometrika*, 54 (1967), 181–7.

[19] Good, I. J., "Saddle-Point Methods for the Multinomial Distribution," *Annals of Mathematical Statistics*, 28 (1957), 861–81.

[20] ———, "Maximum Entropy for Hypothesis Formulation, Especially for Multidimensional Contingency Tables," *Annals of Mathematical Statistics*, 34 (1963), 911–34.

[21] Goodman, L. A., "Simultaneous Confidence Limits for Cross-Product Ratios in Contingency Tables," *Journal of the Royal Statistical Society*, Series B, 26 (1964), 86–102.

[22] ———, "Simple Methods for Analyzing Three-Factor Interactions in Contingency Tables," *Journal of the American Statistical Association*, 59 (1964), 319–52.

[23] ———, "Interactions in Multidimensional Contingency Tables," *Annals of Mathematical Statistics*, 35 (1964), 632–46.

[24] ———, "The Analysis of Cross-Classified Data: Independence, Quasi-Independence, and Interactions in Contingency Tables with or without Missing Entries," *Journal of the American Statistical Association*, 63 (1968), 1091–131.

[25] ———, "How to Ransack Social Mobility Tables and Other Kinds of Cross Classification Tables," *American Journal of Sociology*, 75 (1969), 1–40.

[26] ———, "On Partitioning χ^2 and Detecting Partial Association in Three-Way Contingency Tables," *Journal of the Royal Statistical Society*, Series B, 31 (1969).

[27] ———, "On Estimating the Logit and Its Variance in Binomial and Multinomial Populations and in Contingency Tables," Unpublished manuscript.

[28] ———, "The Analysis of Multidimensional Contingency Tables: Stepwise Procedures and Direct Estimation Methods for Building Models for Multiple Classifications," *Techometrics*, 12 (1970).

[28a] ———, "A Simple Simultaneous Test Procedure for Quasi-Independence in Contingency Tables," (Manuscript submitted for publication).

[29] Haberman, S., Unpublished report, Department of Statistics, University of Chicago.

[30] Hoeffding, W., "Asymptotically Optimal Tests for Multinomial Distributions," *Annals of Mathematical Statistics*, 36 (1965), 369–408.

[31] Ireland, C. T. and Kullback, S., "Contingency Tables with Given Marginals," *Biometrika*, 55 (1968), 179–88.

[32] Kempthorne, O., *The Design and Analysis of Experiments*, New York: John Wiley and Sons, Inc., 1952.

[33] Ku, H. H. and Kullback, S., "Interaction in Multidimensional Contingency Tables:

An Information Theoretic Approach," *Journal of Research of the National Bureau of Standards*, 72B (1968), 159–99.

[34] Kullback, S., *Information Theory and Statistics*, New York: John Wiley and Sons, Inc., 1959.

[35] ———, Kupperman, M., and Ku, H. H., "An Application of Information Theory to the Analysis of Contingency Tables, with a Table of 2n log n, n=1 (1) 10,000," *Journal of Research of the National Bureau of Standards*, 66B (1962), 217–43.

[36] Lancaster, H. C., "Complex Contingency Tables Treated by the Partition of χ^2," *Journal of the Royal Statistical Society*, Series B, 13 (1951), 242–9.

[37] Lindley, D. V., "The Bayesian Analysis of Contingency Tables," *Annals of Mathematical Statistics*, 35 (1964), 1622–43.

[38] Lombard, H. L. and Doering, C. R., "Treatment of the Four-Fold Table by Partial Correlation as It Relates to Public Health Problems," *Biometrics*, 3 (1947), 123–8.

[39] Mantel, N., "Models for Complex Contingency Tables and Polychotomous Dosage Response Curves," *Biometrics*, 22 (1966), 83–95.

[40] Mosteller, F., "Association and Estimation in Contingency Tables," *Journal of the American Statistical Association*, 63 (1968), 1–28.

[41] Neyman, J., "Contributions to the Theory of the χ^2 Test," *Proceedings of the Berkeley Symposium on Mathematical Statistics and Probability*, University of California Press, 1949, 239–73.

[42] Plackett, R. L., "A Note on Interactions in Contingency Tables," *Journal of the Royal Statistical Society*, Series B, 24 (1962), 162–6.

[43] Tsao, R. F., "A Second Order Exponential Model for Multidimensional Dichotomous Contingency Tables, with Applications in Medical Diagnosis," Cambridge Scientific Center, International Business Machine Corporation, 1967.

[44] Winsor, C. P., "Factorial Analysis of a Multiple Dichotomy," *Human Biology*, 20 (1948), 159–204.

[45] Woolf, B., "On Estimating the Relation between Blood Group and Disease," *Annals of Human Genetics*, (19), 251–3.

[46] Yates, F., *Sampling Methods for Censuses and Surveys*, Third Edition, London: Charles Griffin and Co. Ltd., 1960.

Leo A. Goodman

Reprinted from: the JOURNAL OF THE AMERICAN STATISTICAL ASSOCIATION, March 1970, Vol. 65, pp. 226-256. Used with permission.

Chapter 5

The Analysis of Multidimensional Contingency Tables: Stepwise Procedures and Direct Estimation Methods for Building Models for Multiple Classifications

For the m-way contingency table, we consider models that describe the possible multiplicative interactions among the m variables of the table, and we show how to select models that fit the data in the table, using methods that are, in part, somewhat analogous to the usual stepwise procedures in regression analysis. (The m variables here are dichotomous or, more generally, polytomous variables.) These methods can be applied to build models for any of the following situations: (a) the m variables are response variables and the mutual relationships among the variables are of interest; (b) one of the variables is a response variable and the other $m - 1$ variables are factors that may affect the response; (c) m' of the variables are response variables ($1 < m' < m$) and the other m-m' variables are factors that may affect the m' variables and the mutual relationships among the m' variables. For illustrative purposes, we analyze a 4-way table (actually, a 3×2^3 table), considering both linear and quadratic interaction effects.

The Problem

The merits of building models to describe the relationship between a "dependent" variable and m-1 other "independent" variables have been discussed in the literature, and the utility of stepwise regression procedures in selecting the models that fit the data has been noted, in the case where both the dependent and independent variables are quantitative variables (or where some, or all, of the independent variables are qualitative (i.e., dummy) variables); see, e.g., Draper and Smith (1966). However, these regression procedures are not directly applicable to the case where the dependent variable is a qualitative variable (e.g., a dichotomous or, more generally, a polytomous variable). In the present article, we shall provide various methods (stepwise procedures and direct estimation techniques) that are suited to the case where all of the variables (including the dependent variable) are qualitative.

This research was supported in part by Research Contract No. NSF GS 1905 from the Division of the Social Sciences of the National Science Foundation. For helpful comments, the author is indebted to C. Bingham, Y. M. Bishop, S. E. Fienberg, S. J. Haberman, G. G. Koch, S. Kullback, G. Sande, and H. Smith, Jr.

Received Jan. 1969; revised Aug. 1969.

The m-way contingency table describes the joint-distribution of m qualitative (polytomous) variables. In addition to considering here the situation where one of the m variables is a "dependent" variable and the other m-1 variables are "independent" variables (i.e., the situation which is directly analogous to that considered in the usual regression analysis), we shall also consider the situation where m' of the variables are the "dependent" variables and the other m-m' variables are "independent" variables ($1 \leq m' \leq m$) that may affect the m' dependent variables and the mutual relationships among them. When $m' = m$, we have the more usual contingency table situation in which the mutual relationships among the m variables are of interest; when $m' = 1$, we have an m-1 dimensional factorial arrangement in which the data in each cell consist of an observed proportion (if the response variable is dichotomous) or an observed frequency distribution (if the response variable is polytomous); when $1 < m' < m$, we have a m-m' dimensional factorial arrangement in which the data in each cell consist of an observed m'-way contingency table.

To illustrate the methods proposed here, we shall analyze a 4-way contingency table (Table 1), which was studied earlier by Ries and Smith (1963), Cox and Lauh (1967), and Ku and Kullback (1968). Applying the methods introduced herein to Table 1, we shall see how the results obtained with these methods further enrich our understanding of the data in the table. With this 3×2^3 contingency table, we study the case where one of the variables is trichotomous and the other three variables are dichotomous, and we consider both linear and quadratic interaction effects pertaining to the trichotomous variable (assuming that the levels of the trichotomous variable are equally spaced). Although an assumption is made about the spacing of the trichotomous variable in this particular example, the general methods proposed here can be applied in cases where no assumption about spacing are made, as well as in cases where such assumptions can be made. Although only trichotomous and dichotomous variables are considered in the particular example, the methods can be applied more generally to polytomous variables.

We shall study Table 1 from several different points of view: (a) considering one of the variables to be the "dependent" variable and the other three variables to be "independent variables"; (b) considering the four variables to be response variables, and examining the mutual relationships among them; (c) rearranging the data in Table 1 so that the variables of the rearranged table reflect more directly the "concordance" between certain variables in Table 1, and then analyzing the relationships among the variables of the rearranged table; (d) rearranging the data in Table 1 in ways other than (c), and then performing the corresponding analysis.

In conjunction with the stepwise procedures presented herein for the analysis of multiple classifications, we shall also introduce several different ways of assessing the relative merits of the various models that are fitted to the data, including the use of a measure that is somewhat analogous to the usual multiple correlation coefficient, but is applicable to qualitative variables. The relationship between some of the methods proposed in the present article and the methods appearing in some of the earlier literature will be discussed in the final section herein.

TABLE 1

Cross-Classification of Sample of 1008 Consumers According to (1) the Softness of the Laundry Water Used, (2) the Previous Use of Detergent Brand M, (3) the Temperature of the Laundry Water Used, (4) the Preference for Detergent Brand X over Brand M in a Consumer Blind Trial

		Previous User of M		Previous Non-User of M	
Water Softness	Brand Preference	High Temperature	Low Temperature	High Temperature	Low Temperature
Soft	X	19	57	29	63
	M	29	49	27	53
Medium	X	23	47	33	66
	M	47	55	23	50
Hard	X	24	37	42	68
	M	43	52	30	42

THE MODELS

Consider the 4-way $I \times J \times K \times L$ contingency table, where the 4 dimensions pertain to variables A, B, C, D, respectively. Let $f_{ijk\ell}$ denote the observed frequency in cell (i, j, k, ℓ) of the table $(i = 1, 2, \cdots, I; j = 1, 2, \cdots, J; k = 1, 2, \cdots, K; \ell = 1, 2, \cdots, L)$ when a sample of n observations is drawn from the corresponding population table, and let $F_{ijk\ell} = E\{f_{ijk\ell}\}$. Since $\sum_{i,j,k,\ell} f_{ijk\ell} = n$, we see that

$$\sum_{i,j,k,\ell} F_{ijk\ell} = n. \tag{1}$$

Let $\xi_{ijk\ell} = \log F_{ijk\ell}$, where log refers to the natural logarithm throughout. (For simplicity, we assume that $F_{ijk\ell} > 0$.) As in the analysis of variance (following, e.g., Birch, 1963; Bishop, 1969a), we can decompose $\xi_{ijk\ell}$ as follows:

$$\xi_{ijk\ell} = \theta + \lambda_i^A + \lambda_j^B + \lambda_k^C + \lambda_\ell^D + \lambda_{ij}^{AB} + \lambda_{ik}^{AC} + \lambda_{i\ell}^{AD} + \lambda_{jk}^{BC} + \lambda_{j\ell}^{BD} + \lambda_{k\ell}^{CD} + \lambda_{ijk}^{ABC} + \lambda_{ij\ell}^{ABD} + \lambda_{ik\ell}^{ACD} + \lambda_{jk\ell}^{BCD} + \lambda_{ijk\ell}^{ABCD}, \tag{2}$$

where the λ's satisfy the usual kinds of conditions:

$$\sum_i \lambda_i^A = 0, \cdots, \sum_i \lambda_{ij}^{AB} = \sum_j \lambda_{ij}^{AB} = 0, \cdots, \sum_i \lambda_{ijk}^{ABC} = \sum_j \lambda_{ijk}^{ABC} = \sum_k \lambda_{ijk}^{ABC} = 0, \cdots, \sum_i \lambda_{ijk\ell}^{ABCD} = \sum_j \lambda_{ijk\ell}^{ABCD} = \sum_k \lambda_{ijk\ell}^{ABCD} = \sum_\ell \lambda_{ijk\ell}^{ABCD} = 0. \tag{3}$$

The λ's represent the possible "effects" of the 4 variables on $\xi_{ijk\ell}$: The main effects are λ_i^A, λ_j^B, λ_k^C, λ_ℓ^D; the interaction effects are λ_{ij}^{AB}, λ_{ik}^{AC}, $\cdots$, $\lambda_{ijk\ell}^{ABCD}$. (For further insight into the meaning of the λ's, see Goodman, 1970.) From (1) and (2), we see that θ satisfies the condition that

$$\{\exp \theta\}\{\sum_{i,j,k,\ell} \exp(\lambda_i^A + \lambda_j^B + \cdots + \lambda_{ijk\ell}^{ABCD})\} = n. \tag{4}$$

Formula (2) describes the "saturated" model in which all possible "effects" are included. We shall provide methods here for selecting "unsaturated" models (i.e., models in which some of the λ's in (2) are zero) that fit the data in the table. Model (2) and the corresponding unsaturated models are suitable for the case where the 4 variables (A, B, C, D) are response variables. The following modification of model (2) will be suitable for the case where m' of the variables are response variables and the other 4-m' variables are factors that may affect the m' variables and the mutual relationships among the m' variables ($1 \leq m' < 4$).

Consider first the case where $m' = 1$, and where variable D is the response variable. Let $\Psi_{ijk}^{\bar{D}}$ denote the L-1 dimensional vector

$$\Psi_{ijk}^{\bar{D}} = \{(\xi_{ijk1} - \xi_{ijk\ell})/2, \quad \text{for} \quad \ell = 2, 3, \cdots, L\}. \tag{5}$$

Thus, $\Psi_{ijk}^{\bar{D}}$ denotes the L-1 logits (viz., $\frac{1}{2}\log[F_{ijk1}/F_{ijk\ell}\}$, for $\ell = 2, 3, \cdots, L$) at the level (i, j, k) of variables A, B, C. From (2) we obtain

$$\Psi_{ijk}^{\bar{D}} = \Lambda^{\bar{D}} + \Lambda_i^{A\bar{D}} + \Lambda_j^{B\bar{D}} + \Lambda_k^{C\bar{D}} + \Lambda_{ij}^{AB\bar{D}} + \Lambda_{ik}^{AC\bar{D}} + \Lambda_{jk}^{BC\bar{D}} + \Lambda_{ijk}^{ABC\bar{D}}, \tag{6}$$

where the Λ's are the L-1 dimensional vectors

$$\begin{aligned} \Lambda^{\bar{D}} &= \{(\lambda_1^D - \lambda_\ell^D)/2, \quad \text{for} \quad \ell = 2, 3, \cdots, L\}, \\ \Lambda_i^{A\bar{D}} &= (\{\lambda_{i1}^{AD} - \lambda_{i\ell}^{AD})/2, \quad \text{for} \quad \ell = 2, 3, \cdots, L\}, \cdots, \\ \Lambda_{ij}^{AB\bar{D}} &= \{(\lambda_{ij1}^{ABD} - \lambda_{ij\ell}^{ABD})/2, \quad \text{for} \quad \ell = 2, 3, \cdots, L\}, \cdots. \end{aligned} \tag{7}$$

Note that the Λ vectors satisfy conditions similar to (3). The Λ's represent the possible "effects" of the variables A, B, C on the logit-vector $\Psi_{ijk}^{\bar{D}}$: The effect of the general mean is $\Lambda^{\bar{D}}$, the main effects are $\Lambda_i^{A\bar{D}}$, $\Lambda_j^{B\bar{D}}$, $\Lambda_k^{C\bar{D}}$; the interaction effects are $\Lambda_{ij}^{AB\bar{D}}, \cdots, \Lambda_{ijk}^{ABC\bar{D}}$. In the special case when $L = 2$ (i.e., when D is dichotomous), formula (6) includes the usual models of logit-analysis (see, e.g., Dyke and Patterson, 1952; Bishop, 1969a), and when $L > 2$ we have a generalization thereof (see Goodman, 1970).

Formula (6) describes the "saturated" model in which all possible "effects" are included. We shall provide methods for selecting "unsaturated" models (i.e., models in which some of the Λ's in (6) are zero) that fit the data. These methods can be applied in any of the following situations: (a) a sample of n observations is drawn from the 4-way population table; (b) a sample of $f_{ijk.}$ observations is drawn from the population of responses on variable D at level (i, j, k) of variables A, B, C (for $i = 1, 2, \cdots, I$; $j = 1, 2, \cdots, J$; $k = 1, 2, \cdots, K$) (i.e., the marginal table $\{f_{ijk.}\}$—pertaining to variables A, B, C—is fixed); (c) some other marginal table, pertaining to a subset of the variables A, B, C, is fixed. (For related comments, see Bhapkar and Koch, 1968a, b).

Consider now the case where variables C and D are response variables ($m' = 2$), and variables A and B are factors that may affect variables C and D

and the relationship between C and D. Let $\Psi_{ij}^{\bar{C}\bar{D}}$ denote the KL-1 dimensional vector

$$\Psi_{ij}^{\bar{C}\bar{D}} = \{(\xi_{ij11} - \xi_{ijk\ell})/2, \quad \text{for all} \quad (k, \ell) \quad \text{except} \quad (1, 1)\}. \tag{8}$$

Thus, $\Psi_{ij}^{\bar{C}\bar{D}}$ denotes the KL-1 logits (viz., $\frac{1}{2}$ log $[F_{ij11}/F_{ijk\ell}]$, for all (k, ℓ) except (1, 1)) at level (i, j) of variables A and B. From (2) we obtain

$$\Psi_{ij}^{\bar{C}\bar{D}} = \Lambda^{\bar{C}} + \Lambda_i^{A\bar{C}} + \Lambda_j^{B\bar{C}} + \Lambda_{ij}^{AB\bar{C}} + \Lambda^{\bar{D}} + \Lambda_i^{A\bar{D}} + \Lambda_j^{B\bar{D}} + \Lambda_{ij}^{AB\bar{D}}$$
$$+ \Lambda^{\bar{C}\bar{D}} + \Lambda_i^{A\bar{C}\bar{D}} + \Lambda_j^{B\bar{C}\bar{D}} + \Lambda_{ij}^{AB\bar{C}\bar{D}}, \tag{9}$$

where the Λ's are the KL-1 dimensional vectors

$$\Lambda^{\bar{C}} = \{(\lambda_1^C - \lambda_\ell^C)/2, \quad \text{for all } (k, \ell) \text{ except } (1, 1)\}$$
$$\Lambda_i^{A\bar{C}} = \{(\lambda_{i1}^{AC} - \lambda_{i\ell}^{AC})/2, \quad \text{for all } (k, \ell) \text{ except } (1, 1)\} \tag{10}$$
$$\vdots$$
$$\Lambda_{ij}^{AB\bar{C}\bar{D}} = \{(\lambda_{ij11}^{ABCD} - \lambda_{ijk\ell}^{ABCD})/2, \quad \text{for all } (k, \ell) \text{ except } (1, 1)\}.$$

(To avoid a more complicated notation, the symbols used to denote the Λ's in (10) are similar to the symbols used in (7), but now the vectors are KL-1 dimensional rather than L-1 dimensional.) The Λ's in (10) represent the possible "effects" of the variables A and B on the logit-vector $\Psi_{ij}^{\bar{C}\bar{D}}$. Remarks similar to those made following (7), and in the preceding paragraph, apply here too (when appropriately modified).

Consider now the case where variables B, C, D, are response variables ($m' = 3$), and variable A is a factor that may affect variables B, C, and D, and the mutual relationship among B, C, D. The appropriate modification of (2) for this case ($m' = 3$) is evident from the modifications (6) and (9) presented above for $m' = 1$ and 2.

When D is the response variable ($m' = 1$), then the unsaturated models pertain only to λ's in (2) with superscripts that include D (see (6)–(7)), and the other λ's in (2) are irrelevant in a certain sense (i.e., the other λ's in (2) do not affect $\Psi_{ijk}^{\bar{D}}$—see (6)) and are left unspecified. When C and D are the response variables ($m' = 2$), then the unsaturated models pertain only to λ's in (2) with superscripts that include C or D, or CD (see (9)–(10)), and the other λ's in (2) are irrelevant (in a certain sense) and are left unspecified. When there are m' response variables, then the unsaturated models pertain only to λ's in (2) with superscripts that include at least one of the m' letters corresponding to the response variables, and the other λ's in (2) are irrelevant (in a certain sense) and are left unspecified. Thus, model (2) can be applied in the case where $m' = 1, 2, 3,$ or 4, as long as the unsaturated models that are considered pertain to the appropriate set of λ's in (2). (The λ's in (2) that are set at zero in the unsaturated model cannot include any of the λ's that are irrelevant (in a certain sense) and are left unspecified. The latter λ's will therefore be included among the λ's in (2) that are not set at zero in the unsaturated model).

In the case where $m' = 1, 2, 3,$ or 4, the methods which we shall provide here can be applied in any of the following situations: (a) a sample of n observations is drawn from the 4-way population table; (b) a sample is drawn from the m'-way population table of responses on the m' response variables at each joint level of the 4-m' factor variables (i.e., the (4-m')-way marginal table of f's, pertaining to the 4-m' factor variables, is fixed); (c) some other marginal table, pertaining to a subset of the 4-m' factor variables, is fixed. In situation (a), only n is fixed; and in situations (b) or (c), one marginal table is fixed. Applying Birch's result (1963) on sampling procedures, we find that the maximum-likelihood estimate $\hat{F}_{ijk\ell}$ of $F_{ijk\ell}$ (under an appropriate model in situations (b) or (c)) will be equal to the corresponding estimate of $F_{ijk\ell}$ in situation (a). Thus, it will be sufficient to consider only situation (a) using models based upon (2) which are appropriate for the case at hand (i.e., m' = 1, 2, 3, or 4).

Estimates of the λ's in the Saturated Model

The estimation method presented in this section is not an essential part of the stepwise procedures that will be proposed in later sections, but the estimates obtained by this method can provide (a) partial guidance in determining the direction of the steps to be taken in the stepwise procedures (thus avoiding some of the steps that would turn out in the end to have been unnecessary), and (b) further insight into the results obtained with the stepwise procedures. In later sections, we shall show how this estimation method can be used in conjunction with the stepwise procedures. In the present section, we shall see that the estimates obtained by the estimation method proposed here can provide some guidance in the selection of models that could fit the data in the contingency table even in the case where stepwise procedures (of the kind described in later sections) are not utilized.

We shall use a dot subscript to denote an average over the corresponding index: $\xi_{ijk.} = \sum_\ell \xi_{ijk\ell}/L$, $\xi_{ij..} = \sum_{k,\ell} \xi_{ijk\ell}/KL$, etc. As in the analysis of variance, from (2) we obtain

$$\begin{aligned}\lambda_i^A &= \xi_{i...} - \xi_{....}\,, \quad \lambda_{ij}^{AB} = \xi_{ij..} - \xi_{i...} - \xi_{.j..} + \xi_{....}\,,\\ \lambda_{ijk}^{ABC} &= \xi_{ijk.} - \xi_{ij..} - \xi_{i.k.} - \xi_{.jk.} + \xi_{i...} + \xi_{.j..} + \xi_{..k.} - \xi_{....}\,,\\ \lambda_{ijk\ell}^{ABCD} &= \xi_{ijk\ell} - \xi_{ijk.} - \xi_{ij.\ell} - \xi_{i.k\ell} - \xi_{.jk\ell} + \xi_{ij..} + \xi_{i.k.} + \xi_{i..\ell} + \xi_{.jk.}\\ &\quad + \xi_{.j.\ell} + \xi_{..k\ell} - \xi_{i...} - \xi_{.j..} - \xi_{..k.} - \xi_{...\ell} + \xi_{....},\end{aligned} \tag{11}$$

with similar formulae for the other λ's. Let

$$y_{ijk\ell} = \log f_{ijk\ell}\,. \tag{12}$$

(For simplicity, we assume that $f_{ijk\ell} > 0$.) The maximum-likelihood estimate $\hat{\lambda}$ of a particular λ (for the saturated model) can be obtained by replacing the ξ's in (11) by the corresponding y's. With these estimates inserted in (2), the observed data are fitted perfectly.

Let $\lambda^A, \lambda^B, \lambda^C, \lambda^D, \lambda^{AB}, \cdots, \lambda^{ABCD}$ denote the basic sets of $I - 1$, $J - 1$, $K - 1$, $L - 1$, $(I - 1)(J - 1)$, $\cdots$, $(I - 1)(J - 1)(K - 1)(L - 1)$ parameters $\lambda_i^A, \lambda_j^B, \lambda_k^C, \lambda_\ell^D, \lambda_{ij}^{AB}, \cdots, \lambda_{ijk\ell}^{ABCD}$(for $i < I, j < J, k < K, \ell < L$). Because of (3), it is sufficient to consider these basic sets of parameters. For the 2^4 case ($I = J = K = L = 2$), we see that $\lambda^A = \lambda_1^A, \cdots, \lambda^{AB} = \lambda_{11}^{AB}, \cdots,$ $\lambda^{ABCD} = \lambda_{1111}^{ABCD}$, and we can estimate these parameters by

$$\hat{\lambda}^A = \sum_{j,k,l} (y_{1jkl} - y_{2jkl})/16, \cdots,$$
$$\hat{\lambda}^{AB} = \sum_{k,l} (y_{11kl} - y_{21kl} - y_{12kl} + y_{22kl})/16, \tag{13}$$

etc. For the 3×2^3 case ($I = 3$, $J = K = L = 2$), the basic sets consist of the λ's with subscript 1 or 2 corresponding to an A superscript, and subscript 1 corresponding to superscripts B, C, and D. In this case, we can also consider linear and quadratic components associated with variable A (as in the analysis of variance), assuming the levels of variable A to be equally spaced. Denoting these components as λ_L^A, λ_Q^A, λ_L^{AB}, λ_Q^{AB}, etc., we find that

$$\hat{\lambda}_L^A = (\hat{\lambda}_1^A - \hat{\lambda}_3^A)/2 = \sum_{j,k,l} (y_{1jkl} - y_{3jkl})/16,$$
$$\hat{\lambda}_Q^A = (\hat{\lambda}_1^A + \hat{\lambda}_3^A - 2\hat{\lambda}_2^A)/4 = \sum_{j,k,l} (y_{1jkl} + y_{3jkl} - 2y_{2jkl})/32, \tag{14}$$

etc. (Cox and Lauh, 1967, considered some of these $\hat{\lambda}$'s for the 3×2^3 case—they considered only the $\hat{\lambda}$'s that include D among the letters in the superscript.)

Note that each $\hat{\lambda}$ can be expressed in the form

$$\hat{\lambda} = \sum_{i,j,k,l} a_{ijkl} y_{ijkl}, \tag{15}$$

where the a_{ijkl} are constants that depend upon which $\hat{\lambda}$ is being calculated, and where $\sum_{i,j,k,l} a_{ijkl} = 0$. The variance of $\hat{\lambda}$ can be estimated by

$$S_{\hat{\lambda}}^2 = \sum_{i,j,k,l} a_{ijkl}^2 / f_{ijkl} \tag{16}$$

(see, e.g., Goodman, 1964a; Lindley, 1964). The standardized value of $\hat{\lambda}$ can then be calculated as $\hat{\lambda}/S_{\hat{\lambda}}$. Before calculating this quantity for a particular sample, we recommend replacing f_{ijkl} by $f_{ijkl} + \frac{1}{2}$ in (12), (15), and (16) (see Goodman, 1970).

We now calculate the $\hat{\lambda}$'s and their standardized values for the 4 variables of Table 1: water softness (S), previous use (U), water temperature (T), and brand preference (P). Water softness (S) has three levels in the 3×2^3 table, and we calculate $\hat{\lambda}$'s associated with its linear and quadratic components, S_L and S_Q. To facilitate the use of the standardized values (of the $\hat{\lambda}$'s) as a guide, we have listed them in Table 2 in decreasing order of their absolute values, first considering the one-factor $\hat{\lambda}$'s, then the 2-factor $\hat{\lambda}$'s, etc. (The linear and quadratic components associated with a given effect pertaining to variable S have been listed together, using the standardized value with the larger absolute value (in the pair) to determine their position in Table 2.)

If a given effect (i.e., a given λ) is nil, then the asymptotic distribution of the corresponding standardized value (i.e., $\hat{\lambda}/S_{\hat{\lambda}}$) will be the normal distribution with zero mean and unit variance. Thus, to judge the magnitude of a standardized value in Table 2, we can use the usual normal distribution to begin with. From Table 2, we see that the $\hat{\lambda}$'s pertaining to the main effect T, and to the

TABLE 2

Estimate of Main Effects and Interaction Effects of the Four Variables in Table 1: The Four Variables Pertain to (1) Water Softness (S), (2) Previous Use (U), (3) Water Temperature (T), and (4) Brand Preference (P)

Variable	Estimate of Effect	Standardized Value
T	-.282	-8.45
U	-.042	-1.25
S_L	-.035	- .86
S_Q	-.016	- .47
P	-.015	- .44
PU	-.154	-4.64
S_LT	-.101	-2.45
S_QT	.005	.15
PT	-.063	-1.89
S_LU	.004	.10
S_QU	-.041	-1.17
TU	.026	.77
S_LP	.005	.12
S_QP	.010	.30
S_LPU	.076	1.85
S_QPU	.024	.67
PTU	-.049	-1.48
S_LTU	-.023	- .55
S_QTU	-.036	-1.02
S_LPT	-.018	- .43
S_QPT	-.003	- .10
S_LPTU	-.023	- .56
S_QPTU	.022	.62

interaction effects PU and ST (linear component of S), are statistically significant at the .05 level; the $\hat{\lambda}$ pertaining to the interaction effect PT is significant at the .06 level; and the $\hat{\lambda}$ pertaining to the interaction effect SPU (linear component of S) is significant at the .07 level. Since we are concerned here with a set of λ's, rather than with a single given λ, the multiple-test procedures described, e.g., by Goodman (1964a, b) would be appropriate. Applying these procedures, we find, for example, that (a) the $\hat{\lambda}$'s pertaining to the main effect T and to the interaction effect PU are statistically significant at the .05 level when all 23 $\hat{\lambda}$'s in Table 2 are considered; and (b) if we consider only the eighteen 2-factor, 3-factor, and 4-factor interactions, then the $\hat{\lambda}$ pertaining to the interaction effect PU is significant at the .05 level. (See Goodman, 1969b, for a table of critical constants for these multiple-tests.) Use of the multiple-test procedures will limit the risks of including any effects (i.e., λ's) that are actually nil in the set of effects for which the corresponding null hypotheses are rejected (i.e., for which the corresponding $\hat{\lambda}$'s are statistically significant), even when the hypotheses are suggested by the data. (Of course, if the hypotheses are suggested by the data, this will affect the risks of excluding λ's that are actually non-zero from the set of effects for which the corresponding null hypotheses are rejected.)

In order to fit the data in the contingency table, we shall build an unsaturated model (i.e., a model in which some of the λ's in (2) are assumed zero, and the remaining λ's in (2) are the parameters of the model) by starting with the set of λ's for which the corresponding null hypotheses are rejected by the above multiple-test procedure. As parameters in the unsaturated models we include, with this set of λ's, the λ's that are needed to make the model hierarchical in the sense defined at the beginning of the next section. (If the set consists of λ^T and λ^{PU}, then λ^P and λ^U are needed.) If the model thus obtained fits the data well when the method of fitting described in the next section is used, our task is completed. If a better fit is desired, the magnitudes of the standardized values of the $\hat{\lambda}$'s in Table 2 can serve as a guide in determing which additional λ's should be included in the unsaturated model. (For some cautionary remarks on the use of Table 2 as a simple guide, see Goodman, 1970.) By including additional λ's in the model, the fit can be improved; and so the researcher must weigh in each particular case the advantages of the improved fit against the disadvantages of having introduced additional parameters in the model. Different researchers will weigh these advantages and disadvantages differently.

At this point in the exposition, we draw the reader's attention to three different unsaturated models that are obtained using Table 2 as a guide. (These are not the only models obtained using Table 2, but they seem to be particularly worthy of attention.) First, the model M_1 obtained using the multiple-test procedure, with the λ's consisting of λ^T, λ^{PU} and the other λ's that are needed to make this model hierarchical. Second, the model M_2 with the λ's consisting of λ^T, λ^{PU}, λ^{ST}, λ^{PT}, and the other λ's that are needed to make this model hierarchical. Third, the model M_3 with the λ's consisting of λ^T, λ^{PU}, λ^{ST}, λ^{PT}, λ^{SPU}, and the other λ's that are needed to make this model hierarchical. The number of parameters in M_1 , M_2 and M_3 , is 4, 9, and 15, respectively. (Here we count the λ's in the basic sets of parameters that are included in the model.) We shall discuss the fit of these three models in the next section.

Tests for Unsaturated Models

Consider again the 4-way $I \times J \times K \times L$ contingency table, where the 4 dimensions pertain to variables A, B, C, D, respectively. Let Y and Z denote subsets of the letters A, B, C, D. For example, Y might be ABC, and Z might be BC, in which case we say that Y contained Z. We shall now consider any unsaturated model H' (i.e., a model in which some of the λ's in (2) are assumed zero) that is hierarchical in the sense that, if λ^Z is any of the λ's that are assumed zero under H', then λ^Y is also assumed zero (under H') for all Y that contain Z. (The symbol λ^Z refers to the λ's in the basic set with Z as the superscript.) An equivalent condition for H' to be hierarchical is that, if λ^Y is any of the λ's that are not assumed to be zero under H', then λ^Z is also not assumed to be zero (under H') for all Z contained in Y.

The maximum-likelihood estimate $\hat{F}'_{ijkl}$ of F'_{ijkl} , under a given hierarchical model H', can be calculated by explicit formulae for certain kinds of "elementary" models (see Goodman, 1970), or by iterative scaling otherwise (Ku and Kullback, 1968; Mosteller, 1968; Bishop, 1969a; Goodman, 1969a, 1970). To compare $\hat{F}'_{ijkl}$ with the observed f_{ijkl}, we calculate

$$X^2(H') = 2 \sum_{i,j,k,l} f_{ijkl} \log [f_{ijkl}/\hat{F}'_{ijkl}], \tag{17}$$

which is the χ^2 statistic based upon the likelihood-ratio criterion for testing H'. (Although the more usual goodness-of-fit χ^2 statistic is somewhat less preferable in the present context (see, e.g., Goodman, 1970), we include its numerical value together with that of (17) in Table 3, which we shall discuss later.) The asymptotic distribution of (17) is the χ^2 distribution under H', with degrees of freedom equal to the number of λ's in the basic sets that are assumed zero under H'.

For the given model H', let $\mathfrak{S}'$ denote the basic sets of λ's that are assumed zero under H'. For a different model, say, H, let $\mathfrak{S}$ denote the basic sets of λ's that are assumed zero under H. Consider the case where $\mathfrak{S}$ includes the basic sets in $\mathfrak{S}'$ (in addition to other basic sets), and let $\mathfrak{S}^*$ denote the basic sets of λ's that are included in $\mathfrak{S}$ but not in $\mathfrak{S}'$. Let H^* denote the hypothesis that the λ's in $\mathfrak{S}^*$ are zero, and let $[H^* \mid H']$ denote the hypothesis that H^* is true *assuming* that H' is true. (Note that $[H^* \mid H']$ is equivalent to the hypothesis that H is true assuming that H' is true.) We consider the case where H and H' are hierarchical. To test H, we calculate $X^2(H)$ by (17), with H' and $\hat{F}'$ replaced in (17) by H and $\hat{F}$, respectively. To test $[H^* \mid H']$, we calculate

$$X^2[H^* \mid H'] = X^2(H) - X^2(H') = 2 \sum_{i,j,k,l} f_{ijkl} \log [\hat{F}'_{ijkl}/\hat{F}_{ijkl}]$$

$$= 2 \sum_{i,j,k,l} \hat{F}'_{ijkl} \log [\hat{F}'_{ijkl}/\hat{F}_{ijkl}], \tag{18}$$

which is the χ^2 statistic based upon the likelihood-ratio criterion for testing H assuming that H' is true. The asymptotic distribution of (18) is the χ^2 distribution under $[H^* \mid H']$, with degrees of freedom equal to the number of λ's in the basic sets in $\mathfrak{S}^*$. For a more complete discussion of the tests based upon (17) and (18), see, e.g., Goodman (1970).

As was just noted, we consider the case where the models H and H' are hierarchical. The methods proposed herein can be directly extended to the case where one or both of these models are not hierarchical, but the calculation of the maximum-likelihood estimate $\hat{F}_{ijkl}$ of F_{ijkl}, under a given non-hierarchical model, is more complicated than the corresponding calculation for hierarchical models (see Haberman, 1970). Throughout this article we shall confine our attention to hierarchical models.

We shall now apply the tests based upon (17) and (18) to the models M_1, M_2, and M_3 introduced at the end of the preceding section. Calculating the statistic (17) for the data in Table 1, we obtain $X^2(M_1) = 22.85$, $X^2(M_2) = 11.89$, and $X^2(M_3) = 5.25$, with degrees of freedom equal to 19, 14, and 8, respectively. Calculating the statistic (18), we obtain $X^2[M_1 \mid M_2] = 10.96$, $X^2[M_1 \mid M_3] = 17.60$, and $X^2[M_2 \mid M_3] = 6.64$, with degrees of freedom equal to 5, 11, and 6, respectively. Note that the value of $X^2[M_1 \mid M_2]$ is nearly statistically significant at the .05 level, and the value of $X^2[M_1 \mid M_3]$ is statistically significant at the .10 level; but the value of $X^2[M_2 \mid M_3]$ is not statistically significant. Thus, the addition of the parameters λ^{ST}, λ^{PT}, and λ^{S} to the parameters of M_1 (to form M_2) improved the fit significantly over M_1 at nearly the .05 level, and the further addition of the parameters λ^{SPU}, λ^{SP}, and λ^{SU} to the parameters of M_2 (to form M_3) improved the fit significantly over M_1 at the .10 level; but the additional parameters in M_3 did not improve the fit significantly over M_2. In the comparison of these models, note should also be taken of the fact that the relative decrease in the X^2 value obtained with (17) was .5 when M_1 was compared with M_2, it was .8 when M_1 was compared with M_3, and it was .6 when M_2 was compared with M_3.

In the next section, we shall apply the tests based upon (17) and (18) using stepwise procedures that do not depend upon the earlier analysis of the $\hat{\lambda}$'s in Table 2. (Recall that the models M_1, M_2, and M_3, which we considered above, were obtained with the earlier analysis of the $\hat{\lambda}$'s.) As we shall see, the stepwise procedures can be carried out without the earlier analysis of the $\hat{\lambda}$'s, but when the standardized values of the $\hat{\lambda}$'s are available, they can provide partial guidance and further insight. (See related comment at the beginning of the preceding section.)

Some Stepwise Procedures

For a given hypothesis H, let $\bar{\mathfrak{S}}$ denote the basic sets of λ's that are not assumed to be zero under H. (Thus each basic set of λ's is either in $\bar{\mathfrak{S}}$ or in $\mathfrak{S}$ as defined in the preceding section.) Let $\bar{\mathfrak{S}}_m$ denote a modification of $\bar{\mathfrak{S}}$ that is obtained by removing from $\bar{\mathfrak{S}}$ any λ^Z if Z is contained in any other Y for which λ^Y is also in $\bar{\mathfrak{S}}$. Let T denote the number of basic sets in $\bar{\mathfrak{S}}_m$, and let $Y_1, Y_2, \cdots, Y_T$ denote the Y superscripts of the λ's in the T sets of $\bar{\mathfrak{S}}_m$. If H is hierarchical, then $\bar{\mathfrak{S}}$ consists of $\bar{\mathfrak{S}}_m$ and any λ^Z for which Z is contained in some Y_i, for $i = 1, 2, \cdots, T$. Letting $\{Y_i\}$ denote the corresponding λ^{Y_i}, we shall call $\{Y_1\}, \{Y_2\}, \cdots, \{Y_T\}$ the abbreviated list of estimated parameters under H. (The full list of estimated parameters under H consists of the λ's in $\bar{\mathfrak{S}}$.)

Various hierarchical hypotheses pertaining to the 4-way table (Table 1), with variables S, U, T, P, are considered in Table 3. Each hypothesis H is described by the abbreviated list of estimated parameters under H. Let H_i denote the i-th hypothesis in Table 3. Hypothesis H_1 states that there are no estimated parameters in $\bar{\mathfrak{S}}_m$ (or in $\bar{\mathfrak{S}}$) and that all λ's are zero (i.e., that all cells in the 4-way table are equiprobable). Hypothesis H_2 states that the estimated parameters are the 1-factor λ's (viz., λ^S, λ^T, λ^U, λ^P), and that all 2-factor and higher-order λ's are zero (i.e., that the 4 variables are mutually independent). Hypothesis H_3 states that the estimated parameters are the 2-factor λ's and also the 1-factor λ's, and that the 3-factor and 4-factor λ's are zero. Hypothesis H_4 states that the estimated parameters are the 3-factor λ's and also the 1-factor and 2-factor λ's, and that the 4-factor λ's are zero. (The models M_1, M_2, and M_3 considered in the preceding two sections correspond to hypotheses H_7, H_{10}, and H_{18}, respectively.) For each H_i, Table 3 gives the chi-square values

TABLE 3

Chi-Square Values for Some Hypotheses Pertaining to Table 1

Hypothesis	Abbreviated List of Estimated Parameters	Degrees of Freedom	Likelihood Ratio Chi-Square	Goodness-of-Fit Chi-Square
1	None	23	118.63	115.71
2	{S}, {P}, {T}, {U}	18	42.93	43.90
3	{SP} and all other pairs	9	9.85	9.87
4	{SPT} and all other triplets	2	0.74	0.74
5	{PU}, {S}, {T}	17	22.35	23.13
6	{PU}, {S}	18	95.56	94.78
7	{PU}, {T}	19	22.85	23.54
8	{PU}, {PT}	18	18.49	18.74
9	{PT}, {U}	19	39.07	39.34
10	{PU}, {PT}, {ST}	14	11.89	11.92
11	{PU}, {PT}, {S}	16	17.99	18.33
12	{PU}, {ST}	15	16.25	16.73
13	{PU}, {ST}, {PT}, {TU}, {SU}	11	10.06	10.15
14	{PU}, {ST}, {PT}, {TU}	13	11.19	11.28
15	{SPU}, {PTU}, {STU}	4	0.90	0.90
16	{SPU}, {PTU}, {ST}	6	2.25	2.24
17	{SPU}, {ST}, {PT}, {TU}	7	4.52	4.47
18	{SPU}, {ST}, {PT}	8	5.25	5.20
19	{SPU}, {ST}	9	9.56	9.59
20	{SPU}, {PT}	10	11.30	11.17
21	{SP}, {SU}, {PU}, {ST}, {PT}	10	10.59	10.55
22	{STU}, {P}	11	32.83	32.47
23	{STU}, {PU}	10	12.24	12.20
24	{STU}, {PU}, {PT}	9	8.44	8.43
25	{STU}, {PT}	10	28.46	28.25
26	{STU}, {PU}, {PT}, {PS}	7	8.23	8.22
27	{STU}, {PTU}, {PS}	6	5.50	5.49
28	{STU}, {PTU}	8	5.66	5.65

obtained with (17) and with the goodness-of-fit statistic, and the corresponding degrees of freedom.

From Table 3, we see that H_1 and H_2 do not fit the data, but H_3 and H_4 do. (The percentiles of the corresponding χ^2 distributions are used for purposes of comparison.) We now determine which of the six 2-factor λ's should be added to the $\bar{\mathfrak{S}}$ of H_2 in order to improve the fit obtained with H_2. Calculating (17) for each of the corresponding six hypotheses, we find that H_5 led to the best fit. (Note also the magnitudes of the standardized values in Table 2 that pertain to each of these hypotheses.)

The addition of $\{PU\}$ to the $\bar{\mathfrak{S}}$ of H_2 improved the fit significantly, as can be seen by applying (18) with $H = H_2$ and $H' = H_5$. (Here we calculate the difference 42.93 − 22.35 (comparing H_2 with H_5), and we judge the magnitude obtained using the χ^2 distribution with 18 − 17 = 1 degrees of freedom.) The inclusion of $\{T\}$ in H_5 contributed significantly to the fit (compare H_6 with H_5); but the inclusion of $\{S\}$ did not (compare H_7 with H_5). The reader can gain further insight into why these calculations turned out as they did by examining the magnitudes of the standardized values in Table 2 that pertain to the various hypotheses considered above.

We now determine which of the other five 2-factor λ's should be added to the $\mathfrak{S}$ of H_7 in order to improve the fit. (Since we consider only hierarchical hypotheses here, if a 2-factor λ that includes the letter S in the superscript is added to the $\mathfrak{S}$ of H_7, then we also add λ^S to it.) Calculating (17) for each of the corresponding five hypotheses, we find that H_8 led to the best fit. The addition of $\{PT\}$ to the $\mathfrak{S}$ of H_7 improved the fit significantly (compare H_7 with H_8); and the inclusion of $\{PU\}$ in H_8 contributed significantly to the fit of H_8 (compare H_9 with H_8).

We now determine which of the other four 2-factor λ's should be added to the $\mathfrak{S}$ of H_8 in order to improve the fit. Calculating (17) for each of the corresponding four hypotheses, we find that H_{10} led to the best fit. The addition of $\{ST\}$ and $\{S\}$ to the $\mathfrak{S}$ of H_8 did not improve the fit significantly over H_8 (compare H_8 with H_{10}); but $\{ST\}$ is significantly different from zero within the model H_{10} (compare H_{11} with H_{10})—it is significant at just the .05 level. The other two 2-factor λ's in H_{10} are also significantly different from zero within H_{10} (compare, e.g., H_{12} with H_{10}).

The procedure described above was stepwise in the usual sense, since it re-examined each of the λ's incorporated in the model (except for the λ's that corresponded to non-hierarchical hypotheses), for each new model that was considered. This procedure is usually preferable to the corresponding forward-selection procedure, though the latter is simpler to apply (see, e.g., Draper and Smith, 1966). Applied to Table 1, forward-selection proceeds from H_2 to H_5 to H_{12} to H_{10}. We have summarized these steps in rows 1b.1 to 1b.4 of Table 4. The χ^2 values in Table 4 are $X^2(H_2)$, $X^2(H_3)$, $X^2(H_2 \mid H_3)$, for rows 1, 1a, 1b; and they are $X^2(H_2 \mid H_5)$, $X^2(H_5 \mid H_{12})$, $X^2(H_{12} \mid H_{10})$, $X^2(H_{10} \mid H_3)$, for rows 1b.1 to 1b.4, respectively. Note that $X^2(H_2)$ has been partitioned in rows 1a and 1b; and, in more detail, in rows 1a and 1b.1 to 1b.4.

Let us now consider the corresponding backward-elimination procedure (see, e.g., Draper and Smith, 1966). Applied to Table 1, backward-elimination proceeds from H_3 to H_{13} to H_{14} to H_{10}. We have summarized these steps in rows 1b.1′ to 1b.4′ of Table 4. The χ^2 values here are $X^2(H_{13} \mid H_3)$, $X^2(H_{14} \mid H_{13})$, $X^2(H_{10} \mid H_{14})$, $X^2(H_2 \mid H_{10})$, for rows 1b.1′ to 1b.4′, respectively. The χ^2 values in rows 1b.1′ to 1b.3′ partition $X^2(H_{10} \mid H_3)$, which appeared in row 1b.4. The χ^2 value in row 1b.4′ was partitioned in rows 1b.1 to 1b.3.

The three χ^2 values in rows 1b.1 to 1b.3 (which are used to test the 2-factor interactions $\{PU\}$, $\{ST\}$, and $\{PT\}$ within certain unsaturated models) are actually equal to the corresponding χ^2 values obtained in testing the usual hypothesis of independence between the 2 variables of the 2-way marginal table (formed from the 4-way table of f's), for the marginal tables corresponding to the three 2-factor interactions; but this is not the case for the three χ^2 values

TABLE 4

Analysis of Association in Table 1 by Forward Selection, and by Backward Elimination

	Source of Variation	df	χ^2
1.	Total mutual dependence among S, P, T, U	18	42.93*
1a.	Due to residual 3-factor and 4-factor interactions	9	9.85
1b.	Due to 2-factor interactions (assuming that 3-factor or 4-factor interactions are nil)	9	33.08*
	Partition of (1b) by Forward Selection		
	1b.1. Due to {PU}	1	20.58*
	1b.2. Due to {ST}|{PU}	2	6.10*
	1b.3. Due to {PT}|{PU}, {ST}	1	4.36*
	1b.4. Due to {SP}, {SU}, {TU}|{PU}. {ST}, {PT}	5	2.04
	Partition of (1b) by Backward Elimination		
	1b.1'. Due to {SP}|{PU}, {ST}, {PT}, {TU}, {SU}	2	0.22
	1b.2'. Due to {SU}|{PU}, {ST}, {PT}, {TU}	2	1.13
	1b.3'. Due to {TU}|{PU}, {ST}, {PT}	1	0.69
	1b.4'. Due to {PU}, {ST}, {PT}	4	31.04*

* Denotes significance at the .05 level.

in rows 1b.1′ to 1b.3′. Also, although the χ^2 value in row 1b.4′ is equal to the sum of the χ^2 values calculated for the 2-way marginal tables corresponding to the three 2-factor interactions, this is not the case for the χ^2 value in row 1b.4. (For further details, see Goodman, 1970.)

The stepwise methods described above need not be limited to an analysis of 2-factor interactions. To illustrate this point, we shall analyze the 3-factor and 4-factor interactions by the backward-elimination procedure. Applied to Table 1, backward-elimination proceeds from H_4 to H_{15} to H_{16} to H_{17} ; and from H_{17} we could proceed to H_3 . These steps have been summarized in Table 5. The χ^2 values in Table 5 are $X^2(H_3)$, $X^2(H_4)$, $X^2(H_3 \mid H_4)$, for rows 1, 1a, 1b; and they are $X^2(H_{15} \mid H_4)$, $X^2(H_{16} \mid H_{15})$, $X^2(H_{17} \mid H_{16})$, $X^2(H_3 \mid H_{17})$, for rows 1b.1 to 1b.4 in this table. Note that $X^2(H_3)$, which appeared also in row 1a of Table 4, has been partitioned in rows 1a and 1b of Table 5; and, in more detail, in rows 1a and 1b.1 to 1b.4 of this table.

When considering H_{17} in the preceding paragraph, we investigated whether the single 3-factor interaction $\{SPU\}$ in H_{17} contributed significantly to the fit. (The interaction $\{SPU\}$ was actually significant at the .10 level.) If we had instead investigated whether the 2-factor interactions in H_{17} contributed significantly, we would have proceeded from H_{17} to H_{18} (compare H_{18} with H_{17}). Note that H_{18} fits better than does H_3 . We could stop now at H_{18} , or we could proceed further by backward elimination. (Both the presence of $\{PT\}$ and that

TABLE 5

Analysis of 3-Factor and 4-Factor Interactions in Table 1 by Backward Elimination

	Source of Variation	df	χ^2
1.	Total due to residual 3-factor and 4-factor interactions	9	9.85
1a.	Due to residual 4-factor interactions	2	0.74
1b.	Due to 3-factor interactions (assuming that 4-factor interactions are nil)	7	9.11
	Partition of (1b) by Backward Selection		
	1b.1. Due to {SPT}|{SPU}, {PTU}, {STU}	2	0.16
	1b.2. Due to {STU}|{SPU}, {PTU}	2	1.35
	1b.3. Due to {PTU}|{SPU}	1	2.27
	1b.4. Due to {SPU}	2	5.33+

+ Denotes significance at the .10 level.

of $\{ST\}$ contributed significantly to the fit of H_{18}, at just the .05 level (compare H_{19} with H_{18}; H_{20} with H_{18}); but the contribution of $\{SPU\}$ was not quite significant at that level (compare H_{21} with H_{18}).)

If we proceed from H_{17} to H_3 as noted in the paragraph before the preceding one, then we would proceed further (as earlier in this section) from H_3 to H_{13} to H_{14} to H_{10} by the backward-elimination procedure (see rows 1b.1′ to 1b.4′ of Table 4). If we proceed from H_{17} to H_{18} as noted in the preceding paragraph, then we could proceed further from H_{18} to H_{21} (the contribution of $\{SPU\}$ was not quite significant at the .05 level) and eventually to H_{10}. Thus, all of the procedures presented in this section will lead toward H_{10} when the .05 level is used at each step. (It should be recalled, however, that if the step from H_8 to H_{10} is taken in the first procedure discussed in this section, justification for this step rests, in part, upon the fact that each of the 2-factor interaction effects in H_{10} contributed significantly to the fit of H_{10} (at the .05 level); despite the fact that, when considering the set consisting of the interaction effect $\{ST\}$ and the main effect $\{S\}$, the inclusion of this set in H_{10} did not improve the fit significantly over H_8.) On the other hand, the procedure presented in the preceding paragraph will stop at H_{18} when the .10 level is used. Recall that H_{10} and H_{18} correspond to the models M_2 and M_3, respectively, which we discussed and compared in the preceding two sections. As we noted earlier, the choice between these two models will depend upon the weight given by the researcher to the advantages of the improved fit (as measured by $X^2(M_2 \mid M_3)$ and by the relative index $X^2(M_2 \mid M_3)/X^2(M_2)$) and the disadvantages of having introduced additional parameters (from 9 parameters for M_2 to 15 parameters for M_3). Note that the introduction of the additional parameters corresponds to a reduction in the degrees of freedom (from 14 df for M_2 to 8 df for M_3).

At various points in the stepwise procedures described in this section, we could have avoided unnecessary calculations by using the standardized values in Table 2 for partial guidance. For examples of this, the reader should note the magnitudes of the standardized values in Table 2 that pertain to the various hypotheses of interest, as suggested earlier in this section. (Considering the standardized values in Table 2, the reader could anticipate, for example, that (a) the addition of $\{PU\}$ to H_2 would lead to a better fit than would be obtained with the addition of any other 2-factor interaction to H_2 ; (b) the inclusion of $\{T\}$ in H_5 would contribute significantly to the fit; (c) the inclusion of $\{S\}$ in H_5 would not contribute significantly to the fit; etcetera.) Table 2 could also be used in the selection of the starting point for the stepwise procedures. For example, using the multiple-test procedure described in an earlier section herein, we found that (a) the standardized values for $\{T\}$ and $\{PU\}$ were significantly different from zero when all 23 standardized values in Table 2 are considered, and (b) the standardized value for $\{PU\}$ was significantly different from zero when the eighteen 2-factor, 3-factor, and 4-factor interactions are considered; and so we could consider H_7 or H_5 (rather than H_2 or H_1) in Table 3 as a starting point for our stepwise procedures. (Recall that H_7 corresponds to the model M_1 which we discussed in the preceding two sections.)

The utilization of Table 2, as suggested in the preceding paragraph, would lead to the elimination of some of the calculations that were made at the start and during the stepwise procedures; but the final outcome would in this particular case remain unchanged. Caution is required in the use of the standardized λ's as a simple guide, since such guidance is not always foolproof (see Goodman, 1970). We shall use the term "abbreviated stepwise procedure" to refer to the modified form of a stepwise procedure that is obtained when certain kinds of calculations are omitted; viz., the calculations that have outcomes that can be anticipated (with some degree of assurance) by the direct examination of the standardized λ's. Although the (unabbreviated) stepwise procedures introduced in this section can be applied without calculating the standardized λ's, it will sometimes be easier to use abbreviated stepwise procedures, of the kind described above, when the standardized λ's have been calculated.

When the selection of a model to fit the data was based mainly upon the analysis of the standardized λ's (see the discussion of models M_2 and M_3—i.e., H_{10} and H_{18}—in the preceding two sections), the final outcome was somewhat similar to that obtained in the present section. The similarity of outcome was obtained in the particular case considered here, but this need not happen always. In cases where the selection based mainly upon the analysis of the standardized λ's differs from that obtained with the stepwise procedures, it need not be assumed that the latter selection is always preferable. Even when all the stepwise procedures considered in the present section lead to the same model, say M, it may be possible to find a different model, say M', (with the same number of degrees of freedom) that fits the data still better than M. (See the related remarks in, e.g., Beale, Kendall, and Mann , 1967, concerning the various regression procedures when quantitative variables are considered.) The analysis of the standardized λ's might provide M', or it might suggest a variety of models (some of which might not be obtained with the stepwise procedures), and these models should be compared to determine which of them fit the data particularly well. We do not recommend that the selections based mainly on the analysis of the standardized $\hat{\lambda}$'s should be relied upon solely; but suggest rather a combined form of analysis (when the standardized $\hat{\lambda}$'s have been calculated) in which the stepwise procedures, or the abbreviated stepwise procedures, can be used to check whether it is possible to find models that fit the data better than the models that are obtained when selection is based mainly upon the analysis of the standardized $\hat{\lambda}$'s. Search procedures that are more extensive than the stepwise procedures could also supplement the analysis.

All of the chi-square values given in Table 3 were calculated by a single general computer program. (See the remarks earlier herein pertaining to the calculation of the maximum-likelihood estimate $\hat{F}'_{ijkl}$ of F'_{ijkl}, under a given hierarchical model H', and the corresponding calculation of the chi-square values.) For any hierarchical hypothesis that may be of interest to the researcher, he need only provide the abbreviated list of the parameters that are to be estimated under the hypothesis, and the general computer program will readily calculate the chi-square values given in Table 3 and other relevant statistics. Since this is so easy, the researcher could apply the various stepwise procedures suggested herein, and also possibly other more extensive search procedures, whenever a thorough analysis of the data is desired. (The fact that the various stepwise procedures led to similar results in this particular case does not mean that they will do so always.) The researcher who wishes to choose among the various stepwise procedures (and/or among other more extensive search procedures) should examine the related remarks made by Draper and Smith (1966) concerning the relative merits of the various regression procedures when quantitative variables are analyzed.

Before closing this section, we note that the levels of significance of the various χ^2 tests that can be made with the stepwise procedures discussed in this section should be interpreted with caution, as is the case with the usual stepwise regression analysis. We have used the levels of significance here as a simple way of taking into account the degrees of freedom in judging the magnitude of the corresponding chi-square values, but this simple method could be supplemented (or replaced) by other (perhaps better) methods that would meet this purpose. It should be noted that the relative differences between chi-square values obtained for different hypotheses, as well as the actual differences (between these chi-square values) which we used here, are of interest.

Stepwise Logit Analysis

As we noted in an earlier section, to analyze the logits pertaining to, say, variable P in Table 1, it will be sufficient to consider models based upon (2) (with the symbols A, B, C, D replaced for convenience by S, T, U, P, respectively), limiting consideration to the unsaturated models that pertain only to λ's with superscripts that include P. These unsaturated models are obtained whenever the abbreviated list $\overline{\mathfrak{S}}_m$ of estimated parameters includes $\{STU\}$. For example, hypothesis H_{26} of Table 3 describes the unsaturated model in which the logit $\Psi^{\bar{P}}_{ijk}$ pertaining to variable P is the sum of the effects $\Lambda^{\bar{P}}$, $\Lambda^{S\bar{P}}_i$, $\Lambda^{T\bar{P}}_j$, $\Lambda^{U\bar{P}}_k$; i.e., the "main effects" of the general mean and of each of the three variables S, T, U on P (see equation (6) and related comments). Hypotheses H_{22} to H_{28} (and also H_4 and H_{15}) of Table 3 describe various unsaturated models about the $\Psi_{ijk}^{\bar{P}}$.

The methods described in the preceding sections (the stepwise procedures and the direct estimation techniques) can now be directly applied to models in which $\{STU\}$ is included in $\mathfrak{S}_m$. With the stepwise procedure, we move from H_{22} to H_{23} to H_{24}. (The addition of $\{PU\}$ to the $\mathfrak{S}$ of H_{22} improved the fit significantly (compare H_{22} with H_{23}); the addition of $\{PT\}$ to the $\mathfrak{S}$ of H_{23} improved the fit significantly (compare H_{23} with H_{24}) at the .10 level; the inclusion of $\{PU\}$ in H_{24} contributed significantly to the fit of H_{24} (compare H_{25} with H_{24}); but the addition of $\{PS\}$ to the $\mathfrak{S}$ of H_{24} did not improve the fit significantly (compare H_{24} with H_{26}).) From H_{24} we could proceed to H_{28}. (The addition of $\{PTU\}$ to the $\mathfrak{S}$ of H_{24} improved the fit significantly at just the .10 level; compare H_{24} with H_{28}.)

We have summarized the above steps in rows 1, 1a, 1b, and 2b.1 of Table 6. The other rows in Table 6 give additional χ^2 values that are obtained when

TABLE 6

Analysis of the Logits of Variable P in Table 1 by Forward Selection and Backward Elimination

	Source of Variation	df	χ^2
1.	Total variation of logits of P	11	32.83*
1a.	Due to {PU}	1	20.58*
1b.	Due to {PT}|{PU}	1	3.80+
1c.	Due to {PS}|{PU}, {PT}	2	0.22
1d.	Due to residual interaction effects of 2 or more factors on P	7	8.23
	Partition of (1d) by Forward Selection		
1d.1.	Due to {PTU}|{PU}, {PT}, {PS}	1	2.73+
1d.2.	Due to other residual interaction effects on P	6	5.50
2.	Total variation of logits due to {PU}, {PT}, {PS}, {PTU} (assuming that other interaction effects on P are nil)	5	27.33*
2a.	Due to {PS}|{PU}, {PT}, {PTU}	2	0.16
2b.	Due to {PU}, {PT}, {PTU}	3	27.17*
	Partition of (2b) by Backward Elimination		
2b.1.	Due to {PTU}|{PU}, {PT}	1	2.79+
2b.2.	Due to {PU}, {PT}	2	24.38*

* Denotes significance at the .05 level.
+ Denotes significance at the .10 level.

we consider the "main effects" on $\Psi_{ijk}^{\bar{P}}$ of (a) variable U, (b) variable T (in addition to U), (c) variable S (in addition to U and T), and (d.1) the "interaction effects" of variables T and U (in addition to the effects of (a), (b), (c)). The χ^2 values in Table 6 are $X^2(H_{22})$, $X^2(H_{22} \mid H_{23})$, $X^2(H_{23} \mid H_{24})$, $X^2(H_{24} \mid H_{26})$, $X^2(H_{26})$ for rows 1 and 1a to 1d, respectively; $X^2(H_{26} \mid H_{27})$ and $X^2(H_{27})$ for rows 1d.1 and 1d.2; $X^2(H_{22} \mid H_{27})$, $X^2(H_{28} \mid H_{27})$, $X^2(H_{22} \mid H_{28})$ for rows 2, 2a, 2b; $X^2(H_{24} \mid H_{28})$ and $X^2(H_{22} \mid H_{24})$ for rows 2b.1 and 2b.2. Note that $X^2(H_{22} \mid H_{27})$ has been partitioned by the χ^2 values in rows 2a and 2b; and, in more detail, in rows 2a, 2b.1, and 2b.2; and also in row 1a, 1b, 1c, 1d.1. Similarly, $X^2(H_{22} \mid H_{28})$ has been partitioned by the χ^2 values in rows 2b.1 and 2b.2; and, in more detail, in rows 1a, 1b, 2b.1.

The methods considered above can be used to test any given hierarchical model for the logit $\Psi_{ijk}^{\bar{P}}$ pertaining to variable P. In addition, for a given model, say H', these methods can be used to test the null hypothesis H^* that a specific subset of the parameters in H' are nil. (We can consider any specific subset of the parameters that yields a hierarchical model when the corresponding null hypothesis H^* is true and H' is true.) For example, to test the model H_{28} in Table 3, we calculate the χ^2 value $X^2(H_{28}) = 5.66$ with 8 degrees of freedom (see Table 3), and to test the null hypotheses that, say, the interaction effect $\lambda_{jk}^{TU\bar{P}}$ in model H_{28} is zero (assuming that H_{28} is true) we calculate the χ^2 value $X^2(H_{24} \mid H_{28}) = 2.79$ with one degree of freedom (see row 2b.1 in Table 6). In the next section, we shall show how to estimate the parameters in model H'.

Before closing this section, it should also be noted that the methods described in this section for the analysis of the logits $\Psi_{ijk}^{\bar{P}}$ (see equation (6)) can be directly extended to the analysis of the $\Psi_{ij}^{\bar{U}\bar{P}}$ (see equation (9)), or the analysis of the $\Psi_{i}^{\bar{T}\bar{U}\bar{P}}$. For the sake of brevity, we shall not spell this out here.

Estimates of the λ's in Unsaturated Models

Table 2 provided estimates of the λ's for the saturated model (2). For the unsaturated model H', the maximum-likelihood estimate $\hat{\lambda}$ of a particular λ in H' can be obtained by replacing the ξ values in (11) by the corresponding $\log \hat{F}'$ values, where $\hat{F}'$ is the maximum-likelihood estimate of the corresponding F (under H'), which we considered earlier herein. Table 7 gives the $\hat{\lambda}$ for models H_4, H_{10}, H_{18}, H_{28} of Table 3. (We selected these models for illustrative purposes; other models in Table 3 may also be of interest.) These four models fit the data well, and so the differences in the corresponding estimates under the different models are not great. Had we chosen some models that did not fit the data well, greater differences would have been noted.

Since the $\hat{\lambda}$ values (under H') are calculated using the $\hat{F}'$ values (as just described) which were calculated, in turn, from the marginal tables (formed from the observed 4-way table of f's) corresponding to the λ's in the abbreviated list of the parameters in H' (see, e.g., Birch, 1963; Goodman, 1970), these marginal tables will also serve as a summary picture of the relationship among the variables. For example, under H_{10} of Table 3, a summary picture of the relationships among the 4 variables in the observed 4-way table can be obtained simply from the three 2-way marginal tables (formed from the observed 4-way

TABLE 7
Estimates of Main Effects and Interaction Effects Among Four Variables in Table 1 Under Hypotheses 4, 10, 18, 28 of Table 3

Variable	Hypotheses 4	10	18	28
T	-.286	-.278	-.278	-.287
U	-.042	-.043	-.045	-.042
S_L	-.034	-.045	-.033	-.045
S_Q	-.019	-.021	-.020	-.020
P	-.016	-.017	-.016	-.015
PU	-.155	-.144	-.143	-.157
S_LT	-.097	-.100	-.100	-.102
S_QT	.000	-.003	-.001	-.001
PT	-.065	-.068	-.068	-.067
S_LU	.005	.000	.012	.005
S_QU	-.043	.000	-.032	-.045
TU	.026	.000	.000	.024
S_LP	.005	.000	.010	.000
S_QP	.011	.000	.015	.000
S_LPU	.082	.000	.088	.000
S_QPU	.018	.000	.021	.000
PTU	-.050	.000	.000	-.056
S_LTU	-.022	.000	.000	-.025
S_QTU	-.037	.000	.000	-.039
S_LPT	-.016	.000	.000	.000
S_QPT	-.004	.000	.000	.000
S_LPTU	.000	.000	.000	.000
S_QPTU	.000	.000	.000	.000

table) corresponding to $\{PU\}$, $\{PT\}$, and $\{ST\}$. Similarly, if we consider, say, the model H_{28} of Table 3, a summary picture under this model can be obtained from the two 3-way marginal tables corresponding to $\{STU\}$ and $\{PTU\}$.

Since the $\hat{\lambda}$'s calculated above provide the maximum-likelihood estimates of the parameters in any hierarchical unsaturated model for the ξ_{ijk} based upon (2), they can also provide maximum-likelihood estimates of the parameters in any hierarchical unsaturated model for the logit $\Psi_{ijk}^{\bar{D}}$ (or the corresponding logit-vector) based upon (6), and similarly for the $\Psi_{ij}^{\bar{C}\bar{D}}$ (see (9)), or for the $\Psi_i^{\bar{B}\bar{C}\bar{D}}$. For example, for the logit-vector $\Psi_{ijk}^{\bar{D}}$ under a given hierarchical unsaturated model based upon (6), we first consider the corresponding hierarchical unsaturated model H' based upon (2) with the parameters λ_{ijk}^{ABC} included in the abbreviated list of parameters in H', and we calculate the $\hat{\lambda}$ under H' as described in the first paragraph of this section. Then the maximum-likelihood estimate $\tilde{\Lambda}$ for the corresponding parameter Λ in the model based upon (6) can be calculated simply by replacing the λ's in (7) by the corresponding $\tilde{\lambda}$'s.

To illustrate the application of the above procedure to Table 1, consider, say, the following model based upon (6):

$$\Psi_{ijk}^{\bar{P}} = \Lambda^{\bar{P}} + \Lambda_j^{T\bar{P}} + \Lambda_k^{U\bar{P}} + \Lambda_{jk}^{TU\bar{P}}. \tag{19}$$

The corresponding model based upon (2) is H_{28} of Table 3, and from the final column of Table 7 we see that

$$\tilde{\Lambda}^{\bar{P}} = -.015, \qquad \tilde{\Lambda}_1^{T\bar{P}} = -.067, \qquad \tilde{\Lambda}_1^{U\bar{P}} = -.157, \qquad \tilde{\Lambda}_{11}^{TU\bar{P}} = -.056. \tag{20}$$

Because of (3) we find that the other terms in (19) are as follows:

$$\tilde{\Lambda}_2^{T\bar{P}} = .067, \quad \tilde{\Lambda}_2^{U\bar{P}} = .157, \quad \tilde{\Lambda}_{22}^{TU\bar{P}} = -.056, \quad \tilde{\Lambda}_{12}^{TU\bar{P}} = \tilde{\Lambda}_{21}^{TU\bar{P}} = .056. \tag{21}$$

In the preceding section we noted that $X^2(H_{24} \mid H_{28})$ can be used to test the statistical significance of the $\tilde{\Lambda}^{TU\bar{P}}$. Similarly, the statistical significance of the $\tilde{\Lambda}^{T\bar{P}}$ and $\tilde{\Lambda}^{TU\bar{P}}$ can be tested with $X^2(H_{23} \mid H_{28})$; the statistical significance of the $\tilde{\Lambda}^{U\bar{P}}$ and $\tilde{\Lambda}^{TU\bar{P}}$ can be tested with $X^2(H_{25} \mid H_{28})$; etc.

It should be noted that the method of estimating the λ's, which we described in the first paragraph of this section, would provide all the $\tilde{\lambda}$'s, including the $\tilde{\lambda}_i^S$ (for $i = 1, 2, 3$), $\tilde{\lambda}_{i1}^{SP}$ (for $i = 1, 2, 3$), etc. These $\tilde{\lambda}$'s can be used to calculate the corresponding $\tilde{\Lambda}$'s as noted above. However, Table 7 actually gives the estimated linear and quadratic components rather than the corresponding $\tilde{\lambda}_i^S$ (for $i = 1, 2, 3$), $\tilde{\lambda}_{i1}^{SP}$ (for $i = 1, 2, 3$), etc. From (14) we find that

$$\tilde{\lambda}_L^S = (\tilde{\lambda}_1^S - \tilde{\lambda}_3^S)/2, \qquad \tilde{\lambda}_Q^S = (\tilde{\lambda}_1^S + \tilde{\lambda}_3^S - 2\tilde{\lambda}_2^S)/4, \tag{22}$$

and thus we also see that

$$\tilde{\lambda}_1^S = \tilde{\lambda}_L^S + \tfrac{2}{3}\tilde{\lambda}_Q^S, \qquad \tilde{\lambda}_2^S = -\tfrac{4}{3}\tilde{\lambda}_Q^S, \qquad \tilde{\lambda}_3^S = -\tilde{\lambda}_L^S + \tfrac{2}{3}\tilde{\lambda}_Q^S. \tag{23}$$

Similar formulae can be obtained for the other $\tilde{\lambda}$'s that include S among the letters in their superscripts. Thus, the numerical results given in Table 7 can be translated into the corresponding $\tilde{\lambda}$'s using (23) and similar formulae.

In the preceding paragraph, we showed how to express the $\tilde{\lambda}_i^S$ (for $i = 1, 2, 3$), and some of the other $\tilde{\lambda}$'s, in terms of the corresponding linear and quadratic components, and vice versa. It should also be noted that the models considered herein (e.g., (2), (6), (9)) can also be written more explicitly in terms of linear and quadratic components (and possibly higher order components), when some of the variables in the contingency table have three (or more) levels, and when these levels are assumed to be equally spaced. For the sake of brevity, we shall not go into these matters here.

A Multiple Correlation Coefficient for Multiple Classifications, and Some Related Measures

For simplicity, let us consider again the model (19), which corresponds to H_{28} of Table 3. The statistic $X^2(H_{28})$ is related to a measure of the "weighted unexplained variation" when the observed logits $\Psi_{ijk}^{\bar{P}}$ pertaining to variable P are compared with their estimated expected values under the model (H_{28}) that takes into account the main effect of the general mean on variable P (viz., $\Lambda^{\bar{P}}$), the main effects of variables T and U on P (viz., $\Lambda_j^{T\bar{P}}$ and $\Lambda_k^{U\bar{P}}$), and the interaction effect of variables T and U on P (viz., $\Lambda_{jk}^{TU\bar{P}}$). (We shall use the term "weighted unexplained variation" below to refer to the statistic (17), which actually compares the observed frequency f with the corresponding $\hat{F}$ under the model, and which is asymptotically equivalent (under the model) to the goodness-of-fit statistic $\sum (f - \hat{F})^2/\hat{F}$; see columns 4 and 5 of Table 3.) Thus, the statistic $X^2(H_{22} \mid H_{28})/X^2(H_{22})$ can be interpreted as the relative decrease in the "weighted unexplained variation" obtained in the case where the additive effects on variable P, which are included in H_{28} (viz., $\Lambda^{\bar{P}}$, $\Lambda_j^{T\bar{P}}$, $\Lambda_k^{U\bar{P}}$, $\Lambda_{jk}^{TU\bar{P}}$), are taken into account in contrast to the case (H_{22}) where only the main effect of the general mean on variable P (viz., $\Lambda^{\bar{P}}$) is taken into account. The statistic $X^2(H_{22} \mid H_{28})/X^2(H_{22})$ is thus somewhat analogous to the usual R^2, the square of the multiple correlation coefficient, which is sometimes referred to as the coefficient of multiple determination.

More generally, consider any model H_i pertaining to the logit $\Psi_{ijk}^{\bar{P}}$ of variable P; e.g., the models H_{22} to H_{28}, and H_4 and H_{15} of Table 3. For a given H_i, we would use $X^2(H_{22} \mid H_i)/X^2(H_{22})$ as a coefficient of multiple determination. Even more generally, consider now any model H_i based upon (2). In this case, the statistic $X^2(H_1 \mid H_i)/X^2(H_1)$ could be interpreted as a kind of coefficient of multiple determination. In some situations, the use of H_1 in the above index can be replaced by some other relevant model; e.g., H_2 in some cases where all the λ parameters in H_2 are included among the parameters of H_i.

In addition to the indices just introduced (which were analogous to the coefficient of multiple determination), it is also worth noting that, in judging the merits of model H_i, an index that is somewhat analogous to the residual (weighted) mean square is obtained by dividing $X^2(H_i)$ by the degrees of freedom d_i pertaining to H_i. This index too can be used to assess the merits of H_i.

Return now to the case where the model H_i pertains to the logits of variable P (see (6)). The statistic $X^2(H_i)$ can be used to test the hypothesis H_i (with d_i degrees of freedom); and, when H_i can be assumed to be true, the statistic $X^2(H_{22} \mid H_i)$ can be used to test the hypothesis that, except for $\Lambda^{\bar{P}}$ (the main effect of the general mean on the logits of variable P), all of the other additive effects on the logits of variable P in the model H_i are nil. The number of degrees of freedom associated with $X^2(H_{22} \mid H_i)$ is $d_{22} - d_i$ (i.e., $11 - d_i$). The percentile corresponding to the calculated value of $X^2(H_{22} \mid H_i)$ can also be used to assess the merits of H_i.

More generally, consider now any model H_i based upon (2). In this case, the statistic $X^2(H_1 \mid H_i)$ will be of interest, and the percentile corresponding to it (calculated using the χ^2 distribution with $23\text{-}d_i$ degrees of freedom) can serve as an index. Also, as earlier in this section, the use of H_1 in this index can in some cases be replaced by some other relevant model (e.g., H_2 in some cases where all the λ parameters in H_2 are included among the parameters of H_i), and the degrees of freedom should be changed accordingly.

Consider now the case where the models H_i and H_j both pertain to the logits of variable P (see (6)), and where the λ parameters in H_j are included among the parameters of H_i. The statistic $X^2(H_j \mid H_i)/X^2(H_j)$ can then be interpreted as the relative decrease in the "weighted unexplained variation" obtained in the case where the additive effects on variable P, which are included in H_i , are taken into account in contrast to the case where only the additive effects on variable P, which are included in H_j , are taken into account. The statistic $X^2(H_j \mid H_i)/X^2(H_j)$ is thus somewhat analogous to the square of the multiple-partial correlation coefficient (a coefficient of multiple-partial determination) that pertains to the multiple correlation of variable P with the additive effects that are in H_i but not in H_j , holding constant (in a certain sense) the additive effects that are in H_j .

With respect to the λ parameters that are in H_i , consider now the special case where only one of these parameters (say, λ^*) is not in H_j (i.e., where $d_i = d_j - 1$). In this case, the square-root of $X^2(H_j \mid H_i)/X^2(H_j)$ can be used as a kind of measure of the partial correlation between variable P and the additive effect λ^*, holding constant (in a certain sense) the additive effects that are in H_j . The appropriate sign of the square-root should be taken to be the sign of the maximum-likelihood estimate $\tilde{\lambda}^*$ of λ^* under H_i , which we discussed in an earlier section.

The methods and measures described in the preceding two paragraphs can be directly extended to the more general case where H_i and H_j are models that are based upon (2) rather than (6).

In closing the present section, we note that the various measures introduced in it can be applied to the m-way table for $m > 2$. In the case where $m = 2$, some of these measures are applicable (e.g., measures that use the model corresponding to H_1 as a base for comparisons), but others are not.

Some Additional Ways to View the Multidimensional Contingency Table

In the analysis of Table 1 presented in the section on stepwise logit analysis herein, we found that all of the models that fit the data well included the estimated parameter $\{PU\}$. (Recall also that the standardized value corresponding to $\{PU\}$ in Table 2 was significantly large.) Because of this, a separate analysis of the logits $\Psi^{\bar{P}}_{ijk}$ at each of the two levels of variable U might also be of interest (i.e., an analysis of each of the corresponding 3×2^2 tables). The various methods presented herein can be applied directly to each of these subtables. Application of these methods to the two subtables of Table 1 does, in fact, lead to interesting findings about Table 1. However, since these findings were obtained by direct application of the methods already described herein (applied now to the 3×2^2 subtables), to save space we shall not report them here. Although we shall not pursue the matter here, it nevertheless should be emphasized that, at each level of a particular variable (or a particular set of m^* variables) in the m-way table, we have a $(m - 1)$ way subtable (or a $(m - m^*)$-way subtable), and each of these subtables can be analyzed separately by the methods presented herein.

Consider now the 3×2^3 table that can be formed from Table 1 (with variables S, T, U, P) by replacing variables P and U by the following variables P^* and U^*. The variable P^* classifies individuals according to whether their brand preference (X or M) was "concordant" or "discordant" with their previous use or non-use of M; i.e., brand preference M was concordant with previous use of M and discordant with previous non-use of M; and brand preference X was concordant with previous non-use of M and discordant with previous use of M. The variable U^* classifies individuals according to whether their previous use or non-use of M was "concordant" or "discordant" with the temperature of the laundry water used; i.e., previous use of M was concordant with high temperature and discordant with low temperatures; and previous non-use of M was concordant with low temperature and discordant with high temperature. Thus, the 3×2^3 table pertaining to variables S, T, U^*, and P^* can be formed from Table 1 by a rearrangement of the cell frequencies in that table. For example, the observed f of 19 corresponds to soft water usage, high temperature, concordance of previous usage with temperature, and discordance of brand preference with previous usage. Similarly, the observed f of 52 corresponds to hard water usage, low temperature, discordance of previous usage with temperature, and concordance of brand preference with previous usage. This rearrangement of Table 1 also leads to a reordering of the interpretation attributed to the standardized values calculated in Table 2, which we describe next.

Let us first consider the table obtained when variable P is changed to P^* as defined above, but the other variables (S, T, U) remain unchanged. The change from P in Table 1 to P^* in the modified table will change λ^{PU} to λ^{P^*} and λ^{P} to λ^{P^*U}, it will change λ^{PTU} to λ^{P^*T} and λ^{PT} to λ^{P^*TU}, etc. Let us next consider the table obtained when variable U in Table 1 is changed to U^* as defined above, but the other variables (S, T, P) remain unchanged. The change from U in Table 1 to U^* in the modified table will change λ^{TU} to λ^{U^*} and λ^{U} to λ^{TU^*}, it will change λ^{STU} to λ^{SU^*} and λ^{SU} to λ^{STU^*}, etc. In order to determine the change in the λ's obtained when variable P is changed to P^* and variable U is changed to U^*, first make the changes associated with the change from P to P^*; and then, to the new set of λ's thus obtained, make the changes associated with the change from U to U^*. The final outcome is given in Table 8. (Each standardized $\hat{\lambda}$ with a P^* among the superscripts also had its sign changed, because it was brand preference M (rather than X) that was concordant with previous usage of M.)

From Table 8 we see that, generally speaking, the change from P to P^* and U to U^* decreased the magnitude of the absolute value of the higher order standardized $\hat{\lambda}$'s. Indeed, if we calculate $X^2(H_i)$ for the modified table, we find that $X^2(H_1) = 118.63$ (as in Table 1 since the modified table was no more than a rearrangement of the f's in Table 1); but $X^2(H_2) = 22.17$ (in contrast to a value of 42.93 for Table 1). Thus, the mutual dependence among the four variables (S, T, U^*, P^*) in the modified table was less than the corresponding mutual dependence in Table 1. We also find that $X^2(H_3) = 3.02$ and $X^2(H_4) = .10$ for the modified table, in contrast to values of 9.85 and .74, respectively, for Table 1. Furthermore, if we consider, for example, the model in which the abbreviated list of estimated parameters consists of $\{SP^*\}$, $\{ST\}$, and $\{P^*U^*\}$, we find that the corresponding χ^2 value is 7.25 with 13 df for the modified table; in contrast with, say, a χ^2 value of 11.19 with 13 df for H_{14} of Table 3 (applied to Table 1). Similarly, if we consider, say, the logit model in which

TABLE 8

Comparison of the Standardized Value of the Estimated Main Effects and Interaction Effects of the Four Variables in Table 1 and in the Modified Table Formed by the Following Four Variables: (1) Water Softness (S); (2) Concordance of Previous Use with Water Temperature (U); (3) Water Temperature (T); (4) Concordance of Brand Preference with Previous Use (P*).*

Variable	Standardized Value for Table 1	Standardized Value for Modified Table 1
T	-8.45	-8.45
U	-1.25	.77
S_L	- .86	- .86
S_Q	- .47	- .47
P	- .44	4.64
PU	-4.64	1.89
$S_L T$	-2.45	-2.45
$S_Q T$	.15	.15
PT	-1.89	1.48
$S_L U$	.10	- .55
$S_Q U$	-1.17	-1.02
TU	.77	-1.25
$S_L P$	.12	-1.85
$S_Q P$	.30	- .67
$S_L PU$	1.85	.43
$S_Q PU$	.67	.10
PTU	-1.48	.44
$S_L TU$	- .55	.10
$S_Q TU$	-1.02	-1.17
$S_L PT$	- .43	.56
$S_Q PT$	- .10	- .62
$S_L PTU$	- .56	- .12
$S_Q PTU$	.62	- .30

the abbreviated list of estimated parameters consists of $\{STU^*\}$, $\{P^*S\}$, and $\{P^*U^*\}$, we find that the corresponding χ^2 value is 3.60 with 8 df for the modified table; in contrast with, say, a χ^2 value of 5.66 with 8 df for H_{28} of Table 3 (applied to Table 1).

Comments on Some Related Work

We shall comment briefly on the relationship between some of the methods proposed in the present article and the methods appearing in some of the earlier literature.

In their study of Table 1, Cox and Lauh (1967) analyzed the $\hat{\lambda}$'s that include P among the letters in the superscript. (These $\hat{\lambda}$'s are included among the $\hat{\lambda}$'s of Table 2 herein.) We have noted herein that, if variable P is not the only response variable (e.g., variables S, T, or U may also be response variables), then other $\hat{\lambda}$'s (i.e., $\hat{\lambda}$'s that do not include P in the superscript) in Table 2 will also be of interest. We have also noted that the analysis of the $\hat{\lambda}$'s suggested herein, using constants that are appropriate for multiple-tests, will limit the risks of including any effects (i.e., λ's) that are actually nil in the set of effects for which the corresponding null hypotheses are rejected; and we have shown that the analysis of the $\hat{\lambda}$'s (using the methods presented in the Cox–Lauh article or the direct estimation methods proposed herein) can be supplemented by a variety of modes of analysis, including the use of indirect testing methods (and related stepwise procedures) pertaining to the study of unsaturated models in order to select the models that fit the data well.

In their study of Table 1, Ries and Smith (1963) analyzed a variety of hypotheses about the 4-way table and about marginal tables formed from the 4-way table (by ignoring one or more of the 4 variables). Their hypotheses can be described in conventional terms using the usual concepts of independence and conditional independence. Each of their hypotheses about the 4-way table is equivalent to an unsaturated model based on equation (2). (For example, the hypothesis that variables P and S are conditionally independent, given the level of variables T and U, is equivalent to H_{28} of Table 3.) The χ^2 values for their hypotheses about the 4-way table would be equal to the corresponding χ^2 values in Table 3; and the χ^2 values for their hypotheses about the marginal tables formed from the 4-way table can be calculated as differences between χ^2 values of the kind appearing in Table 3. (For further details, see Goodman, 1970.) In the present article, we have presented a set of unsaturated models based on a single equation (viz., equation (2)), and a set of methods for analyzing these models, that can be used to provide both a unified way of viewing the 4-way table, and a systematic way of analyzing a wide range of hypotheses about the table—a range that includes all the hypotheses considered by Ries and Smith (1963), and other hypotheses as well.

Some of the tables presented herein are related to tables presented in work by Kullback and his co-authors. For example, considering the χ^2 value in row 1b in Table 4 herein (i.e., $X^2(H_2 \mid H_3)$), the partitioning of this χ^2 presented in rows 1b.1 to 1b.4 (and in rows 1b.1′ to 1b.4′) of that table is related to the two different methods of partitioning presented in Tables 4.3 and 4.7 of Ku and Kullback (1968). However, each of their methods needs to be emended to obtain the correct quantities presented herein. With their first method (in which they consider each of the six 2-way marginal tables formed from the 4-way table, and use as a "component" of $X^2(H_2 \mid H_3)$ the usual χ^2 value obtained in testing the hypothesis of independence between the 2 variables of the 2-way marginal table), some of their so-called "components" will differ from the correct values obtained by the methods proposed herein. (Considering the six χ^2 values presented in rows 1b.1 to 1b.3 and rows 1b.1′ to 1b.3′ of Table 4 herein, three of these values will differ from the corresponding "components" in Table 4.3 of Ku and Kullback.) The "components" obtained by the first method presented in the Ku–Kullback article are defective in the sense that (a) the sum of the six "components" of $X^2(H_2 \mid H_3)$ can differ greatly from $X^2(H_2 \mid H_3)$, and (b) the interpretation of the magnitudes of some of these "components" can lead to incorrect conclusions (see Goodman, 1970, for further details). The second method of partitioning $X^2(H_2 \mid H_3)$ presented by Ku and Kullback has the defect that, even when the researcher has specified the particular order in which the 2-factor interactions are to be considered, the "components" obtained by this method are still not uniquely defined. If a constant c is added to one "component" and subtracted from another, the quantities thus obtained would also satisfy the definition of the "components" as presented in the Ku–Kullback article; additional equations that would norm the "components" need to be introduced explicitly in order to uniquely define these quantities. (The second method of partitioning presented in the Ku–Kullback article has been modified recently by Ku, Varner, and Kullback, 1969.)

In addition to the defects noted above pertaining to two of the partitioning methods in the Ku–Kullback article, there are also other defects in some of the other partitioning methods presented in that article and in some of the earlier articles by Kullback and his co-authors. For example, in Table 3.2 of the Ku–Kullback article, the formula for the component attributed to the "addition of the marginal restraint $p_{.ik.}$" was based only upon the first iteration of the iterative scaling method, despite the fact that additional iterations are necessary in this case in order to obtain results that would be correct. A similar remark applies to the components in Table 7.1 of Chap. 8 in Kullback (1959), Tables 4.3, 4.11, and 4.12 of Kullback, Kupperman and Ku (1962), and Tables 3.3, 3.4, and 3.5 of Ku and Kullback (1968). (The latter article states that the formula for the component attributed to "second-order interaction" in Tables 3.3-3.5 was based only upon the first iteration, but a similar cautionary remark should be made about (a) the other components in those tables, (b) the component referred to above from Table 3.2 of their article, and (c) the components in the tables cited above from their earlier publications.) Also, an incorrect formula was used for the component attributed to 3-factor interaction in the "algebraic analysis" presented in Table 8.1 of Chap. 8 in Kullback (1959), and Tables 4.4, 4.9, and 4.10 of Kullback, Kupperman, and Ku (1962). Furthermore, in Tables 3.3-3.5, 4.4, and 4.5 of the Ku–Kullback article, incorrect expressions were used to describe what some of the components in the tables can be attributed to. (For further details on these matters, see Goodman, 1969, 1970, 1971.)

The methods for selecting models, which were proposed by Bishop (1969b), are also related to (but different from) those presented herein. We shall now comment briefly upon the following selection method used by Bishop. Let V_1, V_2, V_3, V_4 denote variables A, B, C, D, respectively, of the 4-way table. For a given pair (V_i, V_j) of variables $(i < j)$, let H_{ij} denote the model in which $\lambda^{V_iV_j} = 0$ and $\lambda^Y = 0$ for all Y that contain V_iV_j. (For discussion of the interpretation of H_{ij}, see Bishop, 1969b, and Goodman, 1970.) For the 4-way table, there are six different H_{ij}, for $i < j$. Fit each model H_{ij} (for $i < j$) to the data in the 4-way table, and let $\mathcal{B}$ denote the set of V_iV_j for which the corresponding H_{ij} fit the data well. Then let H denote the model in which (a) $\lambda^Z = 0$ for all Z in $\mathcal{B}$, and (b) $\lambda^Y = 0$ for all Y that contain three or more letters (i.e., three or more of the V's). Fit model H to the data, and select this model if it fits the data well. If it does not fit well, then consider in turn models H_3 and H_4 of Table 3 herein.

The models selected by the above method will usually not fit the data as well as the models obtained by the stepwise procedures introduced herein. (The method described in the preceding paragraph was presented in Bishop (1969b); a somewhat different method presented in Bishop's unpublished thesis (1967) will not be discussed here.) It should be noted that, even in cases where, say, H_{12} fit the data perfectly (and the other H_{ij} did not fit well), the model H obtained by the above method (with $\mathcal{B}$ taken as V_1V_2) need not fit the data as well as the model H' obtained when $\mathcal{B}$ is taken as, say, V_3V_4. Indeed, if we consider each of the six models H'_{ij} (for $i < j$) in which (a) $\lambda^{V_iV_j} = 0$ and (b) $\lambda^Y = 0$ for all Y that contain three or more V's, then the model H obtained by the method described in the preceding paragraph (with H taken as H'_{12} in the situation considered above where the model H_{12} fit the data perfectly and the other H_{ij} did not fit well) could provide a less adequate fit of the data than any of the other five models H'_{ij}, for $i < j$. (This could happen, for example, in some cases where $\lambda^{V_1V_2} = 0$ and $\lambda^Y = 0$ for all Y that contain V_1V_2—thus H_{12} fits perfectly—but $\lambda^{V_1V_3V_4}$ and $\lambda^{V_2V_3V_4}$ are not equal to zero.) Caution must be exercised in the use of Bishop's method described (1969b) in the preceding paragraph.

In closing, we comment briefly upon the method suggested by Grizzle, Starmer, and Koch (1969) which is also somewhat related to the methods proposed herein. We shall consider their approach in the present context, where hierarchical log-linear models (see (2)) or logit models (see (6) and (9)) are of interest. (Their approach can also be applied in other contexts.) They have suggested that the usual methods for analyzing linear models (in which certain quadratic forms are minimized) can be applied to contingency tables. They note that the estimates obtained by these methods (minimizing quadratic forms) differ from the maximum-likelihood estimates in the present context, and they cite results by Rao (1965) to indicate that their estimates have a somewhat larger variance than the maximum-likelihood estimates; but they nevertheless advocate the use of their estimates, suggesting that the somewhat larger variance "is not a high price to pay for the simplicities that result." But we find that the application of their methods to study the kinds of hypotheses considered in the present article for the 4-way contingency table (or in Goodman, 1970, for the 5-way table) is not as simple as the application of the methods proposed herein, which are (as we noted earlier) based upon maximum-likelihood estimation. (The Grizzle–Starmer–Koch article notes that, for each hypothesis of interest, the application of their method requires the use of a different "C matrix" and/or a "K matrix," and also that the necessary "design matrix X" is different from the usual design matrix used in the corresponding multiple regression analysis, if the dependent variable in the logit-analysis is polytomous (rather than dichotomous); additional complications arise when considering hypotheses pertaining to the general log-linear model (2).) For hierarchical hypotheses of the kind considered herein (e.g., the 28 hypotheses considered in Table 3 for the 4-way contingency table, and the various hypotheses for the m-way table (for $m = 2, 3, 4, 5$) considered in Goodman, 1970), we would recommend the use of the various techniques developed in the present article.

REFERENCES

[1] BEALE, E. M. L., KENDALL, M. G., and MANN, D. W. (1967). The discarding of variables in multivariate analysis. *Biometrika 54*, 357–366.

[2] BHAPKAR, V. P., and KOCH, G. G. (1968a). On the hypothesis of 'no interaction' in contingency tables. *Biometrics 24*, 567–594.

[3] BHAPKAR, V. P., and KOCH, G. G. (1968b). Hypothesis of 'no interaction' in multidimensional contingency tables. *Technometrics 10*, 107–124.

[4] BIRCH, M. W. (1963). Maximum likelihood in three-way contingency tables. *J. Roy. Statist. Soc. Ser. B 25*, 220–233.

[5] BISHOP, Y. M. M. (1969a). Full contingency tables, logits, and split contingency tables. *Biometrics 25*, 383–400.

[6] BISHOP, Y. M. M. (1969b). Calculating smoothed contingency tables. Appendix to Chap. IV-3 of *The National Halothane Study*, U. S. Govt. Printing Office, Washington, D. C.

[7] BISHOP, Y. M. M., and FIENBERG, S. E. (1969). Incomplete two-dimensional contingency tables. *Biometrics 25* 119–128.

[8] BISHOP, Y. M. M., and MOSTELLER, F. (1969). Smoothed contingency tables. Chap. IV-3 of *The National Halothane Study*, U. S. Govt. Printing Office, Washington, D. C.

[9] COX, D. R., and LAUH, E. (1967). A note on the graphical analysis of multidimensional contingency tables. *Technometrics 9*, 481–488.

[10] DRAPER, N., and SMITH, H. (1966). *Applied Regression Analysis*. Wiley, New York.

[11] DYKE, G. V., and PATTERSON, H. D. (1952). Analysis of factorial arrangements when the data are proportions. *Biometrics 8*, 1–12.

[12] GOODMAN, L. A. (1964a). Interactions in multidimensional contingency tables. *Ann. Math. Statist. 35*, 632–646.

[13] GOODMAN, L. A. (1964b). Simultaneous confidence limits for cross-product ratios in contingency tables. *J. Roy. Statist. Soc. Ser. B 26*, 86–102.

[14] GOODMAN, L. A. (1969a). On partitioning χ^2 and detecting partial association in three-way contingency tables. *J. Roy. Statist. Soc. Ser. B 31*, 486–98.

[15] GOODMAN, L. A. (1969b). How to ransack social mobility tables and other kinds of cross-classification tables. *Amer. J. Sociology 75*, 1–40.

[16] GOODMAN, L. A. (1970). The multivariate analysis of qualitative data: Interactions among multiple classifications. *J. Amer. Statist. Assoc. 65*, 226–256.

[17] GOODMAN, L. A. (1971). The partitioning of chi-square, the analysis of marginal contingency tables, and the estimation of expected frequencies in multidimensional contingency tables. *J. Amer. Statist. Assoc. 66*. In press.

[18] GRIZZLE, J. E., STARMER, C. F., and KOCH, G. G. (1969). Analysis of categorical data by linear models. *Biometrics 25*, 489–504.

[19] HABERMAN, S. J. (1970). The general log-linear model. Ph.D. Thesis, Department of Statistics, University of Chicago.

[20] KU, H. H., and KULLBACK, S. (1968). Interaction in multidimensional contingency tables: An information theoretic approach. *J. Res. National Bureau of Standards 72B*, 159–199.

[21] KU, H. H., Varner, R., and KULLBACK, S. (1969). On the analysis of multidimensional contingency tables. Unpublished manuscript.

[22] KULLBACK, S. (1959). *Information Theory and Statistics*. Wiley, New York.

[23] KULLBACK, S., KUPPERMAN, M., and KU, H. H. (1962). An application of information theory to the analysis of contingency tables, with a table of 2n log n, $n = 1(1)$ 10,000. *J. Res. National Bureau of Standards 66B*, 217–243.

[24] LINDLEY, D. V. (1964). The Bayesian analysis of contingency tables. *Ann. Math. Statist. 35*, 1622–43.

[25] MOSTELLER, F. (1968). Association and estimation in contingency tables. *J. Amer. Statist. Assoc. 63*, 1–28.

[26] RAO, C. R. (1965). Criteria of estimation in large samples. *Contributions to Statistics* pp. 345–62, Pergamon Press, New York.

[27] RIES, P. N. and SMITH, H. (1963). The use of chi-square for preference testing in multidimensional problems. *Chemical Engineering Progress 59*, 39–43.

Leo A. Goodman

Reprinted from: TECHNOMETRICS, Vol. 13, No. 1, February 1971, pp. 33-61. Used with permission.

Chapter 6

Causal Analysis of Data from Panel Studies and Other Kinds of Surveys[1]

This article presents new models and methods that are particularly appropriate for the causal analysis of data obtained in panel studies and in other kinds of surveys. For a given path diagram, we provide methods for estimating the magnitude of the various effects represented in the diagram and also methods for determining whether the path-diagram model (described by a given system of equations) is congruent with the data. We present an overall test of the entire system of equations in the model, as well as separate tests of each equation in the system, and we also provide methods for comparing the merits and demerits of a variety of path diagrams pertaining to the data. To illustrate the various advantages of the techniques proposed, we shall reanalyze three different sets of data: (*a*) The famous two-attribute turnover table (the 16-fold table) on voting intention analyzed earlier by Lazarsfeld (1948), Lipset et al. (1954), Campbell (1963), Lazarsfeld, Berelson, and Gaudet (1968), Boudon (1968), and Lazarsfeld (1970); (*b*) a two-attribute turnover table on student attitudes analyzed earlier by Coleman (1964); and (*c*) the four-way cross-classification pertaining to the "contact hypothesis" analyzed earlier by Wilner, Walkley, and Cook (1955), Davis (1971), and Goodman (1972*b*). New insights into these data will be obtained.

1. INTRODUCTION

Let us begin by describing the data that will be reanalyzed here for illustrative purposes. The first example is the two-attribute turnover table in Lazarsfeld (1948), Lipset et al. (1954), and Lazarsfeld, Berelson, and Gaudet (1968), reproduced below as table 1. This table cross-classifies the responses of 266 people, each interviewed at two successive points in time, with respect to their vote intention (Republican versus Democrat) and their opinion of a particular candidate (for or against the Republican candidate). The second example is a two-attribute turnover table in Coleman (1964), reproduced below as table 2. This table cross-classifies the responses of 3,398 schoolboys, each interviewed at two successive points in time, with respect to questions about their (self-perceived) membership in the "leading crowd" (in it or out of it) and their attitude concerning it

[1] This research was supported in part by research contract NSF GS 2818 from the Division of the Social Sciences of the National Science Foundation. For helpful comments, the author is indebted to O. D. Duncan, A. Goldberger, J. Kasarda, D. McFarland, I. R. Savage, S. Schooler, A. Stinchcombe.

TABLE 1

OBSERVED CROSS-CLASSIFICATION OF 266 PEOPLE, IN INTERVIEWS AT TWO SUCCESSIVE POINTS IN TIME, WITH RESPECT TO TWO DICHOTOMOUS VARIABLES*

First Interview: Vote Intention	First Interview: Candidate Opinion	Second Interview: Vote Intention + / Candidate Opinion +	Vote Intention + / Candidate Opinion −	Vote Intention − / Candidate Opinion +	Vote Intention − / Candidate Opinion −
+	+	129	3	1	2
+	−	11	23	0	1
−	+	1	0	12	11
−	−	1	1	2	68

SOURCES.—Lazarsfeld (1948), Lipset et al. (1954), and Lazarsfeld et al. (1968), among others.
NOTE.—With respect to vote intention, + and − denote Republican and Democrat, respectively; with respect to opinion of the candidate, a favorable opinion is denoted + and an unfavorable opinion is denoted −. Vote intention and candidate opinion will be denoted by the letters *A* and *B*, respectively, in the first interview; and by the letters *C* and *D*, respectively, in the second interview.
* Variables = (1) vote intention (Republican vs. Democrat); (2) favorableness of the respondent's opinion of the Republican candidate.

(whether they think that membership in it does not require going against one's principles sometimes or whether they think that it does). The third example is the four-way cross-classification in Wilner, Walkley, and Cook (1955), reproduced below as table 3. This table cross-classifies 608 white women in public housing with respect to proximity to a Negro family, favorableness of local norms toward Negroes; frequency of contacts with Negroes, and favorableness of respondent's attitudes (sentiments) toward Negroes in general.

TABLE 2

OBSERVED CROSS-CLASSIFICATION OF 3,398 SCHOOLBOYS, IN INTERVIEWS AT TWO SUCCESSIVE POINTS IN TIME, WITH RESPECT TO TWO DICHOTOMOUS VARIABLES*

First Interview: Membership	First Interview: Attitude	Second Interview: Membership + / Attitude +	Membership + / Attitude −	Membership − / Attitude +	Membership − / Attitude −
+	+	458	140	110	49
+	−	171	182	56	87
−	+	184	75	531	281
−	−	85	97	338	554

SOURCE.—Coleman (1964).
NOTE.—With respect to (self-perceived) membership in the "leading crowd," being in it is denoted + and being out of it is denoted −; with respect to attitude concerning the "leading crowd," a favorable attitude is denoted + and an unfavorable attitude is denoted −. Membership and attitude will be denoted by the letters *A* and *B*, respectively, in the first interview; and by the letters *C* and *D*, respectivly, in the second interview.
* Variables = (1) self-perceived membership in "leading crowd"; (2) favorableness of attitude concerning the "leading crowd."

TABLE 3

OBSERVED CROSS-CLASSIFICATION OF 608 WHITE WOMEN LIVING IN PUBLIC HOUSING PROJECTS, WITH RESPECT TO FOUR DICHOTOMIZED VARIABLES*

			Sentiment	
Proximity	Norms	Contact	+	−
+	+	+	77	32
+	+	−	14	19
+	−	+	30	36
+	−	−	15	27
−	+	+	43	20
−	+	−	27	36
−	−	+	36	37
−	−	−	41	118

SOURCE.—The observed frequencies in the above table were recalculated from the percentage table in Wilner et al. (1955).

NOTE.—Proximity, norms, contact, and sentiment will be denoted by the letters *P, N, C,* and *S,* respectively.

* Variables = (1) proximity to a Negro family; (2) favorableness of local norms toward Negroes; (3) frequency of contact with Negroes; (4) favorableness of respondent's attitudes (sentiments) toward Negroes in general.

Each of the three tables described above can be viewed as a four-way cross-classification table in which four dichotomous variables are cross-classified. In each of tables 1 and 2, two of the four variables (the two variables at the first interview) were observed prior in time to the remaining two variables (the two variables at the second interview), whereas in table 3 the four variables were observed at the same time (i.e., in a single interview). The methods we shall propose can be applied both in the case where there is an obvious temporal ordering of the variables (as in tables 1 and 2) and also in the case where that kind of temporal ordering is not immediately evident (as in table 3).

Tables 1 and 2 describe data obtained in panel studies in which each respondent was interviewed twice with respect to the same two dichotomous attributes, while table 3 describes data obtained in a survey in which each respondent was interviewed once with respect to four dichotomous attributes. The methods presented in the present article can be applied more generally to the case where each respondent is interviewed T times ($T = 1, 2, \ldots$) with respect to a set of m polytomous attributes ($m = 1, 2, \ldots$).[2] (For tables 1 and 2, we have $T = 2$ and $m = 2$; whereas for table 3, we have $T = 1$ and $m = 4$.) The methods proposed here can be applied even more generally to the case where each respondent is interviewed T times, where the first interview is with respect to m_1 polytomous attributes ($m_1 =$

[2] Of course, some surveys and panel studies collect information pertaining to a set of quantitative (continuous) variables in addition to (or instead of) information pertaining to a set of dichotomous (or polytomous) variables. The methods presented here are limited to the analysis of polytomous or polytomized variables, which would of course include the analysis of dichotomous or dichotomized variables.

1, 2, . . .), where the second interview is with respect to m_2 polytomous attributes ($m_2 = 1, 2, \ldots$), . . . , where the Tth interview is with respect to m_T polytomous attributes ($m_T = 1, 2, \ldots$); and where the set of attributes in different interviews need not be the same. (For tables 1 and 2, we have $m_1 = m_2 = 2$, but in general we could have $m_1 \neq m_2$.)

In the present article, we shall present a broad class of models that can be used to study data of the kind described above, and we shall also provide methods for determining which of the models in this class are congruent with a given set of data.

The usual models and methods of causal analysis (see, e.g., Wright 1954, Duncan 1966) were designed for the study of quantitative variables. These models and methods make various assumptions (either implicitly or explicitly) that are violated when the variables under investigation are dichotomous (or polytomous) rather than quantitative (continuous). The models and methods presented in the present article were designed specifically for the analysis of dichotomous or polytomous variables. We shall comment in our final section on the relation between our methods and the usual methods of causal analysis, and also on the relation between our analysis of tables 1, 2, and 3 and earlier analyses of these data.

The format used for tables 1 and 2 displays the data by cross-classifying the joint response on the first set of two variables (the two variables at the first interview) with the joint response on the second set of two variables (the two variables at the second interview). The format used here for table 3 displays the data by cross-classifying the joint response on the first three variables (the variables pertaining to proximity, norms, and contact) with the response on the fourth variable (the variable pertaining to sentiment). Despite these different formats, we have already noted that each of these three tables is actually a four-way cross-classification in which four dichotomous variables (say, variables A, B, C, D) are cross-classified. (For example, with respect to table 1, we use the letters A, B, C, and D to denote vote intention at the first interview, opinion of the particular candidate at the first interview, vote intention at the second interview, and opinion of the particular candidate at the second interview, respectively.)

From the four-way table pertaining to variables A, B, C, D, we can obtain a two-way table pertaining to variables A and B (ignoring variables C and D), a two-way table pertaining to variables A and C (ignoring variables B and D), etc. There will be six two-way tables obtained from the four-way table. (For example, from table 1 we obtain the six two-way tables presented in table 4.) Similarly, from this four-way table, we can obtain a three-way table pertaining to variables A, B, C (ignoring variable D), a three-way table pertaining to variables A, B, D (ignoring variable

TABLE 4

TWO-WAY MARGINAL TABLE FOR SIX PAIRS OF VARIABLES, CALCULATED FROM DATA IN TABLE 1*

Table I: {AB}

		B +	B −
A	+	135	35
A	−	24	72

Table II: {AC}

		C +	C −
A	+	166	4
A	−	3	93

Table III: {AD}

		D +	D −
A	+	141	29
A	−	16	80

Table IV: {BC}

		C +	C −
B	+	133	26
B	−	36	71

Table V: {BD}

		D +	D −
B	+	143	16
B	−	14	93

Table VI: {CD}

		D +	D −
C	+	142	27
C	−	15	82

* The six pairs of variables are as follows: (I) vote intention at interview 1 and candidate opinion at interview 1 ({*AB*}); (II) vote intention at interview 1 and vote intention at interview 2 ({*AC*}); (III) vote intention at interview 1 and candidate opinion at interview 2 ({*AD*}); (IV) candidate opinion at interview 1 and vote intention at interview 2 ({*BC*}); (V) candidate opinion at interview 1 and candidate opinion at interview 2 ({*BD*}); (VI) vote intention at interview 2 and candidate opinion at interview 2 ({*CD*}).

C), etc. There will be four such three-way tables obtained from the four-way table. In general, information about the joint relationships among the four variables *A*, *B*, *C*, *D* may be lost when the four-way table is replaced by the three-way tables described above, and even more information about these relationships may be lost when the four-way table is replaced by the two-way tables. In the present article, we shall present methods for determining whether we can ignore the information that may be lost by replacing the four-way table by tables of smaller dimension. For each of the four-way tables considered here (tables 1, 2, 3), we shall find that the four-way table can be replaced by the corresponding two-way tables (e.g., table 1 can be replaced by table 4) without the loss of much information. Indeed, we shall find that we can even ignore some of the two-way tables without losing much information about the joint relationships among the four variables.

In our analysis of table 1, we shall find that much of the information about the joint relationships among the four variables in the four-way table is contained in only *three* two-way tables (table $\{AC\}$, $\{BD\}$, and $\{CD\}$ in table 4). We shall present a model that uses these three two-way tables to estimate the joint relationships among the four variables in table 1, and we shall summarize the results obtained with the model in a simple path diagram (see fig. 4 below). This model fits the observed data quite well. In addition to this model, we shall provide several other models (see, e.g., figs. 2 and 5), which also fit the observed data well—even better than the first model. These models, and the corresponding path diagrams (figs. 2, 4, and 5), provide new interpretations of the data in table 1.

We shall analyze tables 2 and 3 using the same general methods used to analyze table 1, but our conclusions about these tables will differ from those obtained in table 1. Compare, for example, figures 4 and 9; figures 2, 7, 11, and 12; figures 5, 10, 13, and 14 later in this article.

Before closing this Introduction (sec. 1), we note that the rest of the article is divided into five main parts (secs. 2–6). The methods and models proposed in the article are applied to tables 1, 2, and 3 in sections 2, 3, and 4, respectively; section 5 includes a more general discussion of these methods and models; and section 6 discusses the relation between our results and earlier work. Section 2 is divided into the following subsections: 2.1 presents and compares some models that fit the data in table 1; 2.2 shows how these models can be expressed as systems of simultaneous equations that describe the "effects" (if any) of the variables (A, B, C, D) upon each other; 2.3 presents path diagrams corresponding to these models; 2.4 shows how to test whether each of the effects in the path diagram is statistically significant; 2.5 introduces additional kinds of models that fit the data; 2.6 shows how the apparent time reversal in the models can be interpreted in terms of latent variables and indicator variables; 2.7 describes various consequences that can be derived from the models; 2.8 introduces still another kind of model that will, in some cases, have certain advantages over the models discussed in the earlier sections; 2.9 presents a general technique (based upon the "saturated" model) that can be used by the research worker to assist him in his search for models that fit a given set of data, and it also includes a method for determining whether certain models that appear (at first sight) to be different from each other are or are not actually equivalent (in a certain sense) to each other. Section 6 is also divided into subsections; 6.1 compares the analyses of tables 1, 2, and 3 in the earlier literature with the analysis presented here; 6.2 compares our methods with those obtained using the usual models of causal analysis and multiple regression; 6.3 compares our multiplicative-effects models with various kinds of additive-effects models.

2. THE ANALYSIS OF TABLE 1

2.1. Some Models That Fit the Data

As we have already noted, the letters A, B, C, D will be used in our analysis of table 1 to denote the four variables in the table (vote intention at the first interview, opinion of the particular candidate at the first interview, vote intention at the second interview, and opinion of the particular candidate at the second interview, respectively). A positive (+) response on a given variable will be denoted by the number 1, and a negative (—) response by the number 2. Let (i, j, k, l) denote the cell in the four-way table (table 1) corresponding to the case where the respondent is classified as i on variable A $(i = 1 \text{ or } 2)$, j on variable B $(j = 1 \text{ or } 2)$, k on variable C $(k = 1 \text{ or } 2)$, and l on variable D $(l = 1 \text{ or } 2)$. Let f_{ijkl} denote the observed frequency in cell (i, j, k, l). From table 1, we see, for example, that $f_{1111} = 129$, $f_{1112} = 3$, $f_{1121} = 1$, etc. For some specified hypothesis about the system of variables in the four-way table, we shall use the symbol F_{ijkl} to denote the expected frequency in cell (i, j, k, l), calculated under the assumption that the specified hypothesis is true.[3]

Let n denote the total sample size (e.g., $n = 266$ in table 1). From the definition of the f_{ijkl}, we see that

$$\sum_{i=1}^{2} \sum_{j=1}^{2} \sum_{k=1}^{2} \sum_{l=1}^{2} f_{ijkl} = n. \tag{1}$$

Because equation (1) is satisfied for the observed frequencies f_{ijkl}, we find that the expected frequencies F_{ijkl} will satisfy the same kind of condition. Thus, we obtain equation (2):

$$\sum_{i=1}^{2} \sum_{j=1}^{2} \sum_{k=1}^{2} \sum_{l=1}^{2} F_{ijkl} = n. \tag{2}$$

From the four-way table $\{ABCD\}$ (pertaining to variables A, B, C, and D), we can obtain the two-way marginal table $\{AB\}$ (pertaining to variables A and B, ignoring variables C and D) in the following way. Letting $f^{AB}{}_{ij}$ denote the observed frequency in cell (i, j) of the two-way table $\{AB\}$ (where $i = 1, 2$; $j = 1, 2$), we can calculate $f^{AB}{}_{ij}$ by the formula

$$f^{AB}{}_{ij} = \sum_{k=1}^{2} \sum_{l=1}^{2} f_{ijkl}. \tag{3}$$

[3] In other words, the symbol F_{ijkl} will denote the ***expected value*** of the observed frequency in cell (i, j, k, l) of the table, under the specified hypothesis. We shall use the term "expected" to refer to the expected value F_{ijkl} under some specified hypothesis and also to certain quantities that are based upon the F_{ijkl}.

Similar formulae can be applied in order to calculate the other two-way marginal tables. Calculated from the data in the four-way table (table 1), we presented in table 4 the two-way marginal table $\{AB\}$ and also the other five two-way marginal tables, $\{AC\}$, $\{AD\}$, $\{BC\}$, $\{BD\}$, $\{CD\}$.

We shall now introduce a set of 10 parameters (τ^A, τ^B, τ^C, τ^D, τ^{AB}, τ^{AC}, τ^{AD}, τ^{BC}, τ^{BD}, τ^{CD}), and shall consider the hypothesis H that states that the expected frequencies F_{ijkl} can be expressed in terms of these parameters in the following way:

$$F_{ijkl} = \eta\tau^A{}_i\tau^B{}_j\tau^C{}_k\tau^D{}_l\tau^{AB}{}_{ij}\tau^{AC}{}_{ik}\tau^{AD}{}_{il}\tau^{BC}{}_{jk}\tau^{BD}{}_{jl}\tau^{CD}{}_{kl}, \tag{4}$$

where

$$\tau^A{}_1 = \tau^A, \tau^A{}_2 = 1/\tau^A, \tau^{AB}{}_{11} = \tau^{AB}{}_{22} = \tau^{AB}, \tau^{AB}{}_{12} = \tau^{AB}{}_{21} = 1/\tau^{AB}, \text{etc.}, \tag{5}$$

and where η is a constant that is introduced to insure that the F_{ijkl} satisfy condition (2).[4] From formulae (4)–(5), we see that the effect of, say, the τ^A parameter upon the F_{ijkl} is to introduce the multiplicative factor τ^A when variable A is at level 1, and the multiplicative factor $1/\tau^A$ when variable A is at level 2. (The parameter τ^A is the "main effect" of variable A on the F_{ijkl}.) Similarly, we see that the effect of, say, the τ^{AB} parameter is to introduce the multiplicative factor τ^{AB} when both variables A and B are at the same level, and the multiplicative factor $1/\tau^{AB}$ when variables A and B are at different levels. (The parameter τ^{AB} is the "interaction effect" of variables A and B on the F_{ijkl}.) The other τ parameters can be interpreted in a similar way.

From formulae (4)–(5), we also find that

$$[(F_{11kl}F_{22kl})/(F_{12kl}F_{21kl})] = (\tau^{AB})^4. \tag{6}$$

The quantity on the left of the equality sign in (6) is called the "cross-product ratio" or the "odds-ratio" in the 2×2 table that describes the "expected relationship" between variables A and B when the levels of the remaining variables C and D are set at k and l, respectively.[5] Formula (6) states that, when the remaining variables (C, D) are "held constant" at level (k, l), then the "expected odds-ratio" pertaining to the relationship between variables A and B will have the same value, $(\tau^{AB})^4$, at each of

[4] From (5) we see that the following conditions are satisfied:

$$\prod_{i=1}^{2} \tau^A{}_i = 1, \prod_{i=1}^{2} \tau^{AB}{}_{ij} = \prod_{j=1}^{2} \tau^{AB}{}_{ij} = 1, \text{ etc.}$$

[5] Since the quantity on the left side of the equality sign in (6) is based upon the expected frequencies F_{ijkl} (under model [4]–[5]), we shall call this quantity the "expected odds-ratio." This terminology is consistent with usage as described at the very end of n. 3.

the four possible levels (for $k = 1, 2$; $l = 1, 2$) under model (4)–(5). This provides a clearer meaning for the parameter τ^{AB} introduced in the preceding paragraph. A similar kind of interpretation can also be obtained for the other τ parameters introduced in the preceding paragraph.

Using only the two-way marginal tables (see table 4), we can estimate the τ parameters in (4), and can also estimate the F_{ijkl} under model (4)–(5) (i.e., under hypothesis H). (Some comments on the estimation procedures are included later in this section and in secs. 2.2 and 6.1.) We can then compare the observed f_{ijkl} in table 1 with the corresponding estimate of the F_{ijkl} under H, by calculating either the usual goodness-of-fit χ^2 statistic

$$\sum_{i=1}^{2} \sum_{j=1}^{2} \sum_{k=1}^{2} \sum_{l=1}^{2} (f_{ijkl} - F_{ijkl})^2 / F_{ijkl}, \tag{7}$$

or the corresponding χ^2 based upon the likelihood-ratio statistic:[6]

$$2 \sum_{i=1}^{2} \sum_{j=1}^{2} \sum_{k=1}^{2} \sum_{l=1}^{2} f_{ijkl} \log [f_{ijkl}/F_{ijkl}]. \tag{8}$$

The χ^2 value obtained from (7) or (8) can be assessed by comparing its numerical value with the percentiles of the tabulated χ^2 distribution. The degrees of freedom for testing H will be $16 - 1 - 10 = 5$ (since [*a*] there are 16 cells in table 1, [*b*] there is the restriction of the form [2] to which the F_{ijkl} are subject, and [*c*] there are 10 τ parameters that are estimated in model [4]).[7]

Using (7), we obtain a goodness-of-fit χ^2 value of 0.74; and using (8), we obtain a likelihood-ratio χ^2 value of 0.71. Since there were 5 df under H, the model fits the data very well indeed.

Model (4) is a particular example of the kind of model listed by Goodman (1970) in his table 4 for the four-way table. (There are 166 different models listed in his table 4.) For each of these models, table 4 in Goodman (1970) gave the corresponding degrees of freedom when the four-way table is a 2^4 table, and that article also described various ways of calculating the degrees of freedom when some (or all) of the variables in the four-way table are not necessarily dichotomous. A single computer program can be used to calculate the estimate of the F_{ijkl}, and the corresponding χ^2 values (7) or (8), for any set of models listed in table 4 of Goodman

[6] Throughout this paper, "log" denotes the natural logarithm. The term "0 log 0" is taken to be zero.

[7] More generally, the degrees of freedom can be calculated by subtracting the number of τ parameters estimated in the model from the number of cells in the table minus one.

(1970). For related material, see, for example, Birch (1963), Ku and Kullback (1968), Bishop (1969), Goodman (1969, 1970, 1971*a*).

We shall now consider a number of different models pertaining to table 1 and shall list in table 5 the χ^2 values obtained in testing these models. We include both the χ^2 values (7) and (8) in table 5; but, in the present

TABLE 5

χ^2 VALUES FOR SOME MODELS PERTAINING TO TABLE 1

Model	Fitted Marginals	df	Likelihood-Ratio χ^2	Goodness-of-Fit χ^2
H_1	{AB}, {AC}, {AD}, {BC}, {BD}, {CD}	5	0.71	0.74
H_2	{AB}, {AC}, {AD}, {BC}, {BD}	6	16.98	46.33
H_3	{AB}, {AC}, {AD}, {BC}, {CD}	6	103.23	121.63
H_4	{AB}, {AC}, {AD}, {BD}, {CD}	6	6.55	7.98
H_5	{AB}, {AC}, {BC}, {BD}, {CD}	6	1.46	1.49
H_6	{AB}, {AD}, {BC}, {BD}, {CD}	6	173.50	220.10
H_7	{AC}, {AD}, {BC}, {BD}, {CD}	6	7.83	11.84
H_8	{AB}, {AC}, {BC}, {BD}	7	58.81	75.98
H_9	{AB}, {AC}, {BC}, {CD}	7	104.00	121.17
H_{10}	{AB}, {AC}, {BD}, {CD}	7	6.74	7.55
H_{11}	{AB}, {BC}, {BD}, {CD}	7	215.33	225.91
H_{12}	{AC}, {BC}, {BD}, {CD}	7	8.60	11.37
H_{13}	{AC}, {BD}, {CD}, {AD}	7	7.85	11.93
H_{14}	{AC}, {BD}, {CD}	8	8.62	11.45
H_{15}	{AC}, {BD}, {AB}	8	58.96	74.88
H_{16}	{AC}, {BD}	9	136.64	137.59
H_{17}	{AC}, {CD}, {B}	9	181.82	164.40
H_{18}	{BD}, {CD}, {A}	9	293.04	239.25
H_{19}	{ABC}, {BCD}	4	0.74	0.69
H_{20}	{ABC}, {BD}, {CD}	5	1.45	1.47
H_{21}	{ABD}, {AC}, {BC}, {CD}	4	0.31	0.28
H_{22}	{ACD}, {AB}, {BC}, {BD}	4	0.70	0.73
H_{23}	{BCD}, {AB}, {AC}	5	0.75	0.70
H_{24}	{ABD}, {ACD}	4	5.89	7.44
H_{25}	{ACD}, {BCD}	4	7.07	10.85
H_{26}	{BCD}, {AC}	6	7.88	10.48
H_{27}	{ACD}, {BD}	6	7.81	11.83
H_{28}	{ABC}, {AD}, {BD}	5	16.97	46.19
H_{29}	{ABC}, {AD}, {CD}	5	103.22	121.56
H_{30}	{ABC}, {AD}, {BD}, {CD}	4	0.71	0.74

context, (8) has some advantages (see, e.g., Goodman 1968, 1970). In the remaining discussion, we shall use only the χ^2 value based upon (8).

The hypothesis H described by model (4) is listed as H_1 in table 5. In this model, there are six two-factor τ's (τ^{AB}, τ^{AC}, τ^{AD}, τ^{BC}, τ^{BD}, τ^{CD}); and corresponding to these six two-factor τ's, we used the six two-way marginal tables in table 4 ($\{AB\}$, $\{AC\}$, $\{AD\}$, $\{BC\}$, $\{BD\}$, $\{CD\}$) in order to estimate the τ's and the F_{ijkl}. The six two-way tables are listed in table 5 as the "fitted marginals" under H_1.

We next consider the models H_2–H_7 in table 5. In each of these models, there are five two-factor τ's. One of the six two-factor τ's in H_1 (i.e., in model [4]) has been deleted in each of the models H_2–H_7. Accordingly, to estimate the τ's and the F_{ijkl} under the given model, we used the five two-way marginal tables corresponding to the five two-factor τ's in the given model. For example, model H_5 differs from model H_1 in that it uses the five two-way tables ($\{AB\}$, $\{AC\}$, $\{BC\}$, $\{BD\}$, $\{CD\}$), but not the two-way table $\{AD\}$, to estimate the τ's and the F_{ijkl} under the model. Model H_5 includes the corresponding five two-factor τ's (τ^{AB}, τ^{AC}, τ^{BC}, τ^{BD}, τ^{CD}), but not the two-factor τ^{AD}. Under model H_5, the F_{ijkl} can be expressed as follows:

$$F_{ijkl} = \eta \tau^A{}_i \tau^B{}_j \tau^C{}_k \tau^D{}_l \tau^{AB}{}_{ij} \tau^{AC}{}_{ik} \tau^{BC}{}_{jk} \tau^{BD}{}_{jl} \tau^{CD}{}_{kl}. \tag{9}$$

(Compare [9] with [4].)

From table 5 we see that, under model H_5, the χ^2 value based on (8) is 1.46, with 6 df.[8] Thus, model H_5 also fits the data very well.

Model H_1 included the τ^{AD} parameter (see [4]), but model H_5 did not (see [9]). (Except for the τ^{AD} parameter, models H_1 and H_5 include the same set of parameters.) Thus, the χ^2 value for model H_5 ($X^2[H_5] = 1.46$, with 6 df) could be compared with the χ^2 value for model H_1 ($X^2[H_1] =$ 0.71, with 5 df), in order to see how much improvement in the fit is obtained by including the τ^{AD} parameter in model H_1. The difference $X^2(H_5) - X^2(H_1)$ can be used to test the null hypothesis that $\tau^{AD} = 1$ in model H_1. This difference can be assessed by comparing its numerical value with the percentiles of the χ^2 distribution. Since $X^2(H_5)$ had 6 df and $X^2(H_1)$ had 5 df, their difference will have 1 df (see Goodman 1970). Since $X^2(H_5) - X^2(H_1) = 0.75$, with 1 df, we conclude that the τ^{AD} parameter can be deleted from model H_1, thus obtaining model H_5.

For each of the models discussed, we could present the corresponding four-way table of the estimated expected frequencies F_{ijkl} under the model. To save space, we present only the corresponding χ^2 values (in table 5); but for some of the models that deserve special attention in the present article, we shall also present the corresponding table of the estimated expected frequencies. The estimated expected frequencies under model H_5 are presented in table 6. Comparing tables 1 and 6, we see how well model H_5 fits the data.

[8] Comparing models (9) and (4) (i.e., H_5 and H_1), we see that the former model has 1 df more than the latter model, since H_5 has one less parameter to estimate. There are $16 - 1 - 9 = 6$ df under model (9), since there are nine τ parameters that are estimated in model (9).

TABLE 6

ESTIMATE OF EXPECTED FREQUENCIES IN FOUR-WAY CONTINGENCY TABLE (TABLE 1), UNDER MODEL H_5 (IN TABLE 5) DESCRIBED BY FORMULA (9), USING THE TWO-WAY MARGINAL TABLES (IN TABLE 4), EXCEPT FOR TABLE $\{AD\}$

		Second Interview			
	Vote Intention	+	+	−	−
	Candidate Opinion	+	−	+	−
First Interview					
Vote Intention	Candidate Opinion				
+	+	128.33	3.73	1.56	1.39
+	−	12.03	21.92	0.02	1.04
−	+	0.92	0.03	12.20	10.86
−	−	0.73	1.33	1.23	68.71

2.2. A Different View of the Models

We shall now express model H_5 in terms that will provide further understanding of its meaning. To do so we must first introduce some additional notation. We shall use the symbol $\omega^{\bar{D}}_{ijk\cdot}$ to denote the observed odds in favor of a + response on variable D (i.e., the observed odds that variable D will take on the value 1), when the joint level of the remaining variables (A, B, C) is (i, j, k). Thus, $\omega^{\bar{D}}_{ijk\cdot}$ is defined as follows:

$$\omega^{\bar{D}}_{ijk\cdot} = f_{ijk1}/f_{ijk2} \, . \tag{10}$$

(For example, from table 1 we see that $\omega^{\bar{D}}_{111\cdot} = 129/3 = 43.00$; i.e., the observed odds are 43 to 1 that there will be a + response on variable D, when the joint level of the remaining variables $[A, B, C]$ is $[1, 1, 1]$.) Similarly, we shall use the symbol $\omega^{\bar{C}}_{ij\cdot l}$ to denote the observed odds in favor of a + response on variable C (i.e., the observed odds that variable C will take on the value 1), when the joint level of the remaining variables (A, B, D) is (i, j, l). Thus, $\omega^{\bar{C}}_{ij\cdot l}$ is defined as follows:

$$\omega^{\bar{C}}_{ij\cdot l} = f_{ij1l}/f_{ij2l} \, . \tag{11}$$

In addition, we also define the following observed odds:

$$\omega^{\bar{B}}_{i\cdot kl} = f_{i1kl}/f_{i2kl} \, , \tag{12}$$

$$\omega^{\bar{A}}_{\cdot jkl} = f_{1jkl}/f_{2jkl} \, . \tag{13}$$

Corresponding to these observed odds, we now define the following "expected odds" pertaining to variables D, C, B, and A, respectively:[9]

[9] The quantities on the right side of the equality sign in (14)–(17) were obtained by replacing the f_{ijkl} in the formulae for the observed odds (i.e., formulae [10]–[13])

$$\Omega^{\bar{D}}{}_{ijk\cdot} = F_{ijk1}/F_{ijk2}\ , \tag{14}$$

$$\Omega^{\bar{C}}{}_{ij\cdot l} = F_{ij1l}/F_{ij2l}\ , \tag{15}$$

$$\Omega^{\bar{B}}{}_{i\cdot kl} = F_{i1kl}/F_{i2kl}\ , \tag{16}$$

$$\Omega^{\bar{A}}{}_{\cdot jkl} = F_{1jkl}/F_{2jkl}\ . \tag{17}$$

In section 2.1 we introduced the τ parameters, and we expressed the expected frequencies F_{ijkl} in terms of them (see, e.g., [4] and [9]). We next define a new set of parameters (the γ parameters) in terms of the τ parameters as follows:

$$\gamma^{\bar{D}} = (\tau^{D}{}_{1})^2, \gamma^{B\bar{D}}{}_{j} = (\tau^{BD}{}_{j1})^2, \gamma^{C\bar{D}}{}_{k} = (\tau^{CD}{}_{k1})^2. \tag{18}$$

Using (5), (14), and (18), we see from model (9) that

$$\Omega^{\bar{D}}{}_{ijk\cdot} = \gamma^{\bar{D}}\gamma^{B\bar{D}}{}_{j}\gamma^{C\bar{D}}{}_{k}\ . \tag{19}$$

Formula (19) states that the expected odds $\Omega^{\bar{D}}{}_{ijk\cdot}$ pertaining to variable D are "dependent" (in a certain sense) upon variables B and C, but not upon variable A.[10] Similarly, from model (9) we obtain the following additional formulae:

$$\Omega^{\bar{C}}{}_{ij\cdot l} = \gamma^{\bar{C}}\gamma^{A\bar{C}}{}_{i}\gamma^{B\bar{C}}{}_{j}\gamma^{D\bar{C}}{}_{l}\ , \tag{20}$$

$$\Omega^{\bar{B}}{}_{i\cdot kl} = \gamma^{\bar{B}}\gamma^{A\bar{B}}{}_{i}\gamma^{C\bar{B}}{}_{k}\gamma^{D\bar{B}}{}_{l}\ , \tag{21}$$

$$\Omega^{\bar{A}}{}_{\cdot jkl} = \gamma^{\bar{A}}\gamma^{B\bar{A}}{}_{j}\gamma^{C\bar{A}}{}_{k}, \tag{22}$$

where

$$\gamma^{\bar{C}} = (\tau^{C}{}_{1})^2, \gamma^{\bar{B}} = (\tau^{B}{}_{1})^2, \gamma^{\bar{A}} = (\tau^{A}{}_{1})^2,$$
$$\gamma^{A\bar{C}}{}_{i} = (\tau^{AC}{}_{i1})^2, \gamma^{B\bar{C}}{}_{j} = (\tau^{BC}{}_{j1})^2, \text{etc.} \tag{23}$$

Note should be taken of the following relationships:

$$\gamma^{A\bar{B}}{}_{1} = \gamma^{B\bar{A}}{}_{1}, \gamma^{A\bar{C}}{}_{1} = \gamma^{C\bar{A}}{}_{1}, \gamma^{B\bar{C}}{}_{1} = \gamma^{C\bar{B}}{}_{1}, \text{etc.} \tag{24}$$

Formula (9) expressed the expected frequencies F_{ijkl} in terms of the τ parameters, while formulae (19)–(22) expressed the expected odds ($\Omega^{\bar{D}}{}_{ijk\cdot}$,

---

by the corresponding F_{ijkl}. Because the quantities on the right side of the equality sign in (14)–(17) were based upon the expected frequencies F_{ijkl}, we call these quantities the "expected odds." As in n. 5, the terminology here, too, is consistent with usage as described at the very end of n. 3.

10 The bar over the superscript D in $\Omega^{\bar{D}}{}_{ijk\cdot}$ and in the γ's in formula (19) signifies that the expected odds pertain to variable D and that these expected odds are the "dependent" variable (i.e., the "predicted" variable) in this formula. The absence of a bar over the superscripts B and C in the γ's in formula (19) signifies that variables B and C are used as "independent" variables (i.e., "predictor" or "explanatory" variables) in this formula. Further insight into the meaning of the γ's in (19) will be obtained later when formulae (25a)–(25d) are introduced (for related matters, see Goodman 1972*a*).

$\Omega^{\bar{C}}_{ij\cdot l}$, $\Omega^{\bar{B}}_{i\cdot kl}$, $\Omega^{\bar{A}}_{\cdot jkl}$) in terms of the γ parameters. Having already presented estimates of the F_{ijkl} (see table 6), we can insert these estimates in formula (6) (and in related formulae) to obtain the corresponding estimates of the τ parameters,[11] and we can insert the estimated τ's in (18) and (23) to obtain the corresponding estimates of the γ parameters.[12] The estimated γ's are presented in table 7. We shall now use the estimated γ's in table 7 to obtain further insight into the meaning of the system of equations (19)–(22).

TABLE 7

ESTIMATE OF γ PARAMETERS IN MODEL (19)–(22) AND CORRESPONDING β PARAMETERS IN MODEL (30)–(33), FOR FOUR-WAY CROSS-CLASSIFICATION TABLE (TABLE 1)

Variable	γ Parameters	β Parameters
A	1.45	0.37
B	1.30	0.26
C	1.26	0.23
D	0.78	−0.24
AB	2.91	1.07
AC	33.06	3.50
AD	1.00	0.00
BC	0.36	−1.03
BD	7.92	2.07
CD	5.54	1.71

Let us first focus our attention on equation (19). By applying the estimated values of the γ's in table 7 to equation (19), we see that the estimate of $\Omega^{\bar{D}}_{ijk\cdot}$ can be expressed as follows:[13]

$$\Omega^{\bar{D}}_{111\cdot} = (0.78)(7.92)(5.54) = 34.44; \tag{25a}$$

$$\Omega^{\bar{D}}_{121\cdot} = (0.78)\left(\frac{1}{7.92}\right)(5.54) = 0.55; \tag{25b}$$

$$\Omega^{\bar{D}}_{112\cdot} = (0.78)(7.92)\left(\frac{1}{5.54}\right) = 1.12; \tag{25c}$$

$$\Omega^{\bar{D}}_{122\cdot} = (0.78)\left(\frac{1}{7.92}\right)\left(\frac{1}{5.54}\right) = 0.02. \tag{25d}$$

[11] A general set of formulae that expresses the τ parameters in terms of the F_{ijkl} can be obtained, e.g., from Goodman (1970, 1972*b*).

[12] The γ parameters can also be estimated directly from the estimated Ω's using a general set of formulae that expresses the γ parameters in terms of the Ω's (see Goodman 1972*a*). The Ω's are estimated by inserting the estimated F_{ijkl} (see table 6) into formulae (14)–(17).

[13] All calculations in this paper were carried out to more significant digits than are reported here.

Comparing (25a) with (25b), we see the effect of $\gamma^{B\bar{D}}{}_1$ (i.e., the effect of variable B on the estimate of the expected odds pertaining to variable D). Comparing (25a) with (25c), we see the effect of $\gamma^{C\bar{D}}{}_1$ (i.e., the effect of variable C on the estimate of the expected odds pertaining to variable D). Comparing (25a) with (25d), we see the joint effect of both $\gamma^{B\bar{D}}{}_1$ and $\gamma^{C\bar{D}}{}_1$. Since $\gamma^{A\bar{D}}{}_1 = 1$ in the model considered here, the effect of $\gamma^{A\bar{D}}{}_1$ is nil (i.e., variable A does not effect the estimate of the expected odds pertaining to variable D); and thus the $\Omega^{\bar{D}}{}_{1jk\cdot}$ in (25a)–(25d) are equal to the corresponding $\Omega^{\bar{D}}{}_{2jk\cdot}$

The remarks in the preceding paragraph pertain to formula (19). Similar remarks can be made about formulae (20), (21), and (22).

When model (9) holds true, then the set of four equations (19)–(22) will also hold true. Conversely, if the set of four equations (19)–(22) holds true, then model (9) will also hold true. As we shall see later (when table 10 is introduced in sec. 2.9), the set of two equations (19)–(20) is actually equivalent to model (9),[14] and so is any other pair of equations from among the four equations (19)–(22), except for the pair (20)–(21).[15] (Any set of three equations from among the four equations [19]–[22] will also be equivalent to model [9].)

Model (9) expresses the F_{ijkl} as a product of certain "main-effect" and "interaction-effect" parameters (the τ's). The corresponding equations (19)–(22) express the Ω's as a product of certain "main-effect" parameters (the γ's). These formulae can also be expressed in an additive form via logarithms. Letting G_{ijkl} denote the natural logarithm of F_{ijkl} (i.e., $G_{ijkl} = \log F_{ijkl}$, where "log" denotes the natural logarithm), we see from (9) that the G_{ijkl} can be expressed as follows:

$$G_{ijkl} = \theta + \lambda^{A}{}_{i} + \lambda^{B}{}_{j} + \lambda^{C}{}_{k} + \lambda^{D}{}_{l} + \lambda^{AB}{}_{ij} + \lambda^{AC}{}_{ik} + \lambda^{BC}{}_{jk} + \lambda^{BD}{}_{jl} + \lambda^{CD}{}_{kl}\,, \qquad (26)$$

[14] The set of τ parameters that are taken equal to one when (19) and (20) are true will be equal to the set of τ parameters that are taken equal to one when (9) is true, as we shall see later from table 10. Thus, the estimated expected frequencies F_{ijkl} under model (19)–(20) will be equal to the corresponding quantities under model (9).

[15] I define two models (say H and H′) as equivalent here if the estimated expected frequencies under H are equal to the corresponding quantities under H′, for all sets of observed frequencies f_{ijkl}. In other words, H and H′ are equivalent if they provide the same model for the F_{ijkl}. Analysis of the f_{ijkl} is of no help in distinguishing between models that are equivalent in this sense. (Although the concept of "equivalence" here is different from what it would be in the analysis of quantitative variables, a similar point was made in that context by Duncan [1969].) However, it is important to note that two models that are equivalent in the sense defined above (e.g., the model described by [19]–[22] and the model described by [19]–[20]; see figs. 1 and 2 in sec. 2.3) need not be identical. The order of priority of the variables in each model can differ; the set of consequences implied by each model can differ; etc. These differences can be used to distinguish between "equivalent" models; but additional kinds of data and/or additional substantive considerations would have to be introduced in order to do this.

where

$$\theta = \log \eta,\ \lambda^A{}_i = \log \tau^A{}_i,\ \lambda^B{}_j = \log \tau^B{}_j, \text{ etc.} \tag{27}$$

Corresponding to the τ parameters discussed earlier in this section, the λ parameters are defined as

$$\lambda^A = \log \tau^A,\ \lambda^B = \log \tau^B, \text{ etc.} \tag{28}$$

From (5), (27), and (28), we obtain the following relationship:[16]

$$\lambda^A = \lambda^A{}_1 = -\lambda^A{}_2,\ \lambda^{AB} = \lambda^{AB}{}_{11} = \lambda^{AB}{}_{22} = -\lambda^{AB}{}_{12} = -\lambda^{AB}{}_{21}, \text{ etc.} \tag{29}$$

Letting Φ denote the natural logarithm of the corresponding Ω, we see from (19)–(22) that the Φ's can be expressed as follows:

$$\Phi^{\bar{D}}{}_{ijk\cdot} = \beta^{\bar{D}} + \beta^{B\bar{D}}{}_j + \beta^{C\bar{D}}{}_k, \tag{30}$$

$$\Phi^{\bar{C}}{}_{ij\cdot l} = \beta^{\bar{C}} + \beta^{A\bar{C}}{}_i + \beta^{B\bar{C}}{}_j + \beta^{D\bar{C}}{}_l, \tag{31}$$

$$\Phi^{\bar{B}}{}_{i\cdot kl} = \beta^{\bar{B}} + \beta^{A\bar{B}}{}_i + \beta^{C\bar{B}}{}_k + \beta^{D\bar{B}}{}_l, \tag{32}$$

$$\Phi^{\bar{A}}{}_{\cdot jkl} = \beta^{\bar{A}} + \beta^{B\bar{A}}{}_j + \beta^{C\bar{A}}{}_k, \tag{33}$$

where

$$\beta^{\bar{D}} = \log \gamma^{\bar{D}},\ \beta^{B\bar{D}}{}_j = \log \gamma^{B\bar{D}}{}_j, \text{ etc.} \tag{34}$$

From (18), (23), (28), and (34), we see that

$$\beta^{\bar{D}} = 2\lambda^D{}_1,\ \beta^{B\bar{D}}{}_j = 2\lambda^{BD}{}_{j1},\ \beta^{C\bar{D}}{}_k = 2\lambda^{CD}{}_{k1}, \text{ etc.}; \tag{35a}$$

and from (29) and (35*a*) we obtain

$$\beta^{B\bar{D}} = \beta^{B\bar{D}}{}_1 = -\beta^{B\bar{D}}{}_2,\ \beta^{C\bar{D}} = \beta^{C\bar{D}}{}_1 = -\beta^{C\bar{D}}{}_2, \text{ etc.} \tag{35b}$$

Equation (30) states that $\Phi^{\bar{D}}{}_{ijk\cdot}$, which is the logarithm of the expected odds $\Omega^{\bar{D}}{}_{ijk\cdot}$, can be expressed as a sum of a "general mean" ($\beta^{\bar{D}}$) and the "main effects" of variable B ($\beta^{B\bar{D}}{}_j$) and of variable C ($\beta^{C\bar{D}}{}_k$). From formulae (30) and (35b) we see that the effect of, say, the $\beta^{B\bar{D}}$ parameter upon the $\Phi^{\bar{D}}{}_{ijk\cdot}$ is to introduce the additive term $\beta^{B\bar{D}}$ when variable B is at level 1, and the additive term $-\beta^{B\bar{D}}$ when variable B is at level 2. (In other words, $\Phi^{\bar{D}}{}_{ijk\cdot}$ is increased by $\beta^{B\bar{D}}$ when variable B is at level 1 and decreased by $\beta^{B\bar{D}}$ when variable B is at level 2.) The other β parameters in (30)–(33) can be interpreted in a similar way.

From (35b) we see that the parameter $\beta^{\bar{D}}$ in (30) is equal to the aver-

[16] From (29) we see that the following conditions are satisfied:

$$\sum_{i=1}^{2} \lambda^A{}_i = 0,\ \sum_{i=1}^{2} \lambda^{AB}{}_{ij} = \sum_{j=1}^{2} \lambda^{AB}{}_{ij} = 0, \text{ etc.}$$

age of the eight values of the $\Phi^{\bar{D}}_{ijk\cdot}$ (for $i=1, 2$; $j=1, 2$; $k=1, 2$), and the other β parameters in (30) can also be expressed in terms of the $\Phi^{\bar{D}}_{ijk\cdot}$ (see Goodman 1972*a*). These expressions for the β parameters in terms of the $\Phi^{\bar{D}}_{ijk\cdot}$ may provide further insight into the meaning of these parameters. Some additional insight may also be obtained by expressing (30) in an equivalent multiple-regression form as we do later in formula (56), using dummy variables. In a similar way, the β parameters in formulae (31)–(33) can be expressed in terms of the Φ's in these formulae, and each of these formulae also can be expressed in an equivalent multiple-regression form.

Having already presented estimates of the γ parameters (see table 7), we can insert these estimates in formula (34) to obtain the corresponding estimates of the β parameters.[17] The estimated β's are also included in table 7.

We use the estimated γ's in table 7, in formulae (25a)–(25d), in order to gain further insight into the meaning of equation (19) and thus into the meaning of the system of equations (19)–(22). In a similar way we can use the estimated β's in table 7, in formulae directly analogous to (25a)–(25d), in order to gain further insight into the meaning of equation (30) and thus into the meaning of the system of equations (30)–(33). The estimated β's can also be used, as we shall see in section 2.3, in order to obtain path diagrams that are somewhat analogous to those used in the causal analysis of quantitative variables.

The system of simultaneous equations (30)–(33) is somewhat analogous to the kinds of systems of simultaneous linear equations that are used in the causal analysis of quantitative variables.[18] We shall now show how to use equations like (30)–(33) to obtain appropriate methods for the causal analysis of dichotomous variables.

2.3. Path Diagrams Corresponding to the Models

We noted in section 2.2 that model (9), which is equivalent to model (26), could also be expressed in the equivalent forms (19)–(22) and (30)–(33). This model was also included in table 5 as model H_5, and the expected

[17] In actual practice, we first estimate G_{ijkl} as the logarithm of the estimated F_{ijkl} (see table 6), and we then estimate the λ parameters (and the β parameters—see [35a]) from the estimated G_{ijkl}, using a general set of formulae that expresses the λ parameters (and the β parameters) in terms of the G_{ijkl} (see Goodman 1970). The γ parameters are then estimated from the estimated β's using (34). (From [34] we see that $\gamma^{\bar{D}} = \exp \beta^{\bar{D}}$, etc., where "exp" denotes the exponential function. The γ parameters can be estimated from the corresponding estimated β's using a table of natural logarithms in, so to speak, inverted order (see [34]), or equivalently using a table of the exponential function.)

[18] This analogy will be discussed further in sec. 6.2.

frequencies estimated under the model were presented in table 6. Expressing the model in the form (30)–(33), we provide in figure 1 the corresponding path diagram, which summarizes some of the information obtained when the estimated β's in table 7 are inserted in (30)–(33).[19]

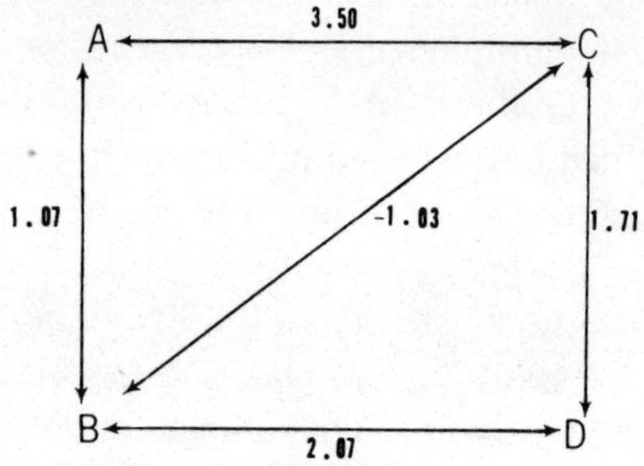

FIG. 1.—Path diagram for the system of eqq. (30)–(33), for the four-way cross-classification table (table 1). See model H_5 in table 5.

In section 2.2 we also noted that the system of equations (19)–(20) and the equivalent system of equations (30)–(31) were also equivalent to the model referred to in the preceding paragraph. (For further comments on this point, see nn.14 and 15 and the discussion of table 10 in sec. 2.9.) The path diagram corresponding to the system of equations (30)–(31) is presented in figure 2.[20]

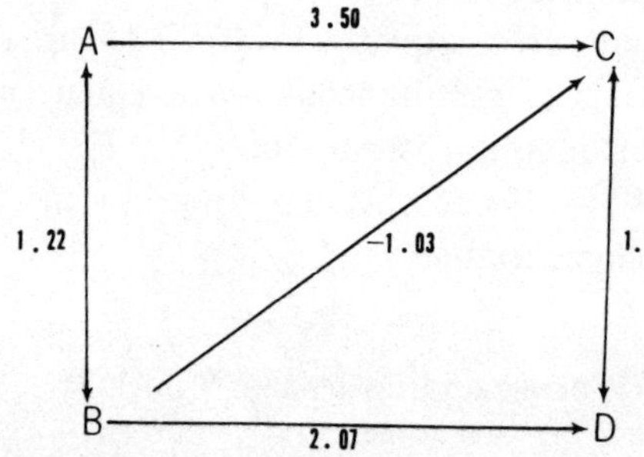

FIG. 2.—Path diagram for the system of eqq. (30)–(31), for the four-way cross-classification table (table 1). See model H_5 in table 5.

19 In fig. 1, the two arrows pointing to D correspond to $\beta^{B\bar{D}}$ and $\beta^{C\bar{D}}$ in (30); the three arrows pointing to C correspond to $\beta^{A\bar{C}}$, $\beta^{B\bar{C}}$, $\beta^{D\bar{C}}$ in (31); the three arrows pointing to B correspond to $\beta^{A\bar{B}}$, $\beta^{C\bar{B}}$, $\beta^{D\bar{B}}$ in (32); the two arrows pointing to A correspond to $\beta^{B\bar{A}}$ and $\beta^{C\bar{A}}$ in (33). A double-headed arrow is used (rather than two separate single-headed arrows) to take note of the fact that $\beta^{A\bar{C}} = \beta^{C\bar{A}}$, $\beta^{B\bar{C}} = \beta^{C\bar{B}}$, etc., which is a consequence of (24).

20 In fig. 2, the two arrows pointing to D correspond to $\beta^{B\bar{D}}$ and $\beta^{C\bar{D}}$ in (30); the three arrows pointing to C correspond to $\beta^{A\bar{C}}$, $\beta^{B\bar{C}}$, $\beta^{D\bar{C}}$ in (31). With respect to the double-headed arrow between C and D, see the final sentence in n. 19. With respect to the double-headed arrow between A and B in fig. 2, see the comments in the next paragraph.

As in figure 1, the numerical values presented in figure 2 were obtained from table 7, except for the estimated β^{AB} parameter in figure 2. Since there are no arrows pointing to A or B from C or D in figure 2, the appropriate estimate of β^{AB} was obtained by an analysis of the two-way marginal table $\{AB\}$ (see table 4), as would also be the case in the usual causal analysis (for quantitative variables) corresponding to figure 2. For the two-way table $\{AB\}$, using the same kind of notation as in (9) and (26), we can write the "saturated" model in either the following multiplicative form or the equivalent additive form:

$$F^{AB}_{ij} = \eta\tau^{A}_{i}\tau^{B}_{j}\tau^{AB}_{ij} \, ,$$

$$G^{AB}_{ij} = \theta + \lambda^{A}_{i} + \lambda^{B}_{j} + \lambda^{AB}_{ij} \, . \tag{36}$$

Using the same kind of notation as in (30)–(33), we find that the above saturated model is equivalent to

$$\Phi^{\bar{B}}_{i\cdot} = \beta^{\bar{B}} + \beta^{A\bar{B}}_{i} \, , \tag{37}$$

or to

$$\Phi^{\bar{A}}_{\cdot j} = \beta^{\bar{A}} + \beta^{B\bar{A}}_{j} \tag{38}$$

(see Goodman 1970, 1972*b*). Model (36) is equivalent to (37) or (38) or to the system of equations (37)–(38).[21] The path diagram corresponding to (37)–(38) is represented by the A-B double-headed arrow in figure 2.[22] The β^{AB} parameter in (37)–(38), and the other parameters in this model and in (36), were estimated using the methods already presented in Goodman (1970, 1972*b*) for the analysis of "saturated" models.[23]

From figure 2, we see that the largest effect is that of A on C, and next comes the effect of B on D. Then comes the C-D effect, and after that the A-B effect.[24] In this model, the effect of A on D is nil, and we find that the effect of B on C turned out to be a negative one. (Some of the above

21 See n. 15 above.

22 The arrow pointing to A corresponds to $\beta^{B\bar{A}}$ in (38), and the arrow pointing to B corresponds to $\beta^{A\bar{B}}$ in (37). Note that $\beta^{B\bar{A}} = \beta^{A\bar{B}}$, as in the final sentence of n. 19.

23 In estimating the parameters of the saturated models, Goodman (1970, 1972*b*) recommended that each f_{ijkl} be changed to $f_{ijkl} + \frac{1}{2}$ before the formulae presented there are applied; whereas, in estimating the parameters of a given "unsaturated" model, a corresponding change was not recommended in the estimated F_{ijkl}. Since all the β parameters in fig. 2 (except for the β^{AB} parameter) were estimated for an "unsaturated" model, we have not made the "½ correction" in estimating those β's; and for the sake of consistency we have also not done so in estimating the β^{AB} parameter. (If the "½ correction" had been made in estimating the β^{AB} parameter, the 1.22 value in fig. 2 would have been 1.21.)

24 In comparing these effects, it should be kept in mind that the A-B effect in fig. 2 was estimated from the two-way table $\{AB\}$, whereas all the other effects in this figure were estimated from the four-way table $\{ABCD\}$.

results may seem surprising at first sight, but we shall provide a simple explanation for them in section 2.8.)

Both figures 1 and 2 describe nonrecursive "causal" models. Note, for example, that variable C affects the log-odds pertaining to variable D (i.e., $\Phi^{\bar{D}}_{ijk\cdot}$) in (30), and variable D in turn affects the log-odds pertaining to variable C (i.e., $\Phi^{\bar{C}}_{ij\cdot l}$) in (31). The A-B double-headed arrow in figure 2 can represent either the nonrecursive "causal" model (37)–(38), or the "noncausal" model (36), pertaining to the two-way table $\{AB\}$. Although model (37)–(38) is equivalent to (36) in the sense defined in footnote 15, the interpretation of these two models is somewhat different.[25]

The techniques required for the analysis of nonrecursive systems when the variables are quantitative are usually more complicated than are those required for a corresponding analysis of recursive systems.[26] Recursive systems are, of course, easier to comprehend. On the other hand, the techniques presented in the present article for the analysis of nonrecursive systems (e.g., figs. 1 and 2 above) when the variables are dichotomous (or polytomous) are not more difficult to apply than are those required for the analysis of recursive systems.[27] We shall discuss the analysis of recursive systems when figure 5 is introduced in section 2.8 and in section 5 where some more general models will also be discussed.

2.4. The Statistical Significance of the Effects

We noted earlier that model (9) (i.e., H_5 in table 5) fits the data very well indeed. Since the path diagrams in both figures 1 and 2 are equivalent to model (9), these path diagrams are congruent with the observed data. We shall now show how to test whether a given effect in these diagrams is statistically significant.

Consider first the C-D effect in figures 1 and 2. Model H_5 included a C-D effect (as represented by the τ^{CD} parameter), but model H_8 in table 5 does not. (Except for the τ^{CD} parameter, models H_5 and H_8 include the same set of parameters.) Thus, the χ^2 value for model H_8 ($X^2[H_8] =$ 58.81) could be compared with the χ^2 value for model H_5 ($X^2[H_5] =$

[25] For the sake of simplicity, we use the A-B double-headed arrow to represent both model (37)–(38) and model (36), although it might be preferable to replace the straight double-headed arrow by, say, a curved double-headed arrow when representing the latter model. The interpretation of these two models will be discussed further in sec. 2.6.

[26] See, e.g., Duncan and Featherman 1972.

[27] This statement applies when we compare the broad class of models that we shall introduce for the analysis of nonrecursive systems with the class of models that we shall introduce for the analysis of recursive systems (see secs. 2.8 and 5). This statement does not apply to nonrecursive systems that are not included in this broad class of models. Although many kinds of nonrecursive systems are included in this broad class (see sec. 5), there are kinds of nonrecursive systems that are not included.

1.46) in order to see how much improvement in the fit is obtained by including the τ^{CD} parameter in model H_5. The difference $X^2(H_8) - X^2(H_5)$ can be used to test the null hypothesis that $\tau^{CD} = 1$ in model H_5. (From [18], [26], [33], and [34], we see that when $\tau^{CD} = 1$, then $\gamma^{C\bar{D}}{}_1 = \gamma^{D\bar{C}}{}_1 = 1$, $\lambda^{CD} = 0$, and $\beta^{C\bar{D}}{}_1 = \beta^{D\bar{C}}{}_1 = 0$.) Since $X^2(H_8) - X^2(H_5) = 57.35$, using the χ^2 distribution with 1 df, we see that the null hypothesis is rejected. (For further discussion of this method of testing whether a given effect is statistically significant, see Goodman 1970, 1971*a*.)

The procedure described above can be applied to obtain a test of the null hypothesis pertaining to each of the effects in figures 1 and 2 (except for the *A-B* effect in fig. 2). By comparing the χ^2 values for H_8–H_{12} in table 5 with H_5, we find that each of these effects is statistically significant.

To obtain a test of the null hypothesis pertaining to the *A-B* effect in figure 2, we use the usual test of the null hypothesis that variables *A* and *B* are statistically independent in the two-way marginal table $\{AB\}$ (see table 4).[28] Calculating the usual likelihood-ratio χ^2 for this test, we obtain a χ^2 value of 77.68, with 1 df.[29] Thus, the *A-B* effect in figure 2 is also statistically significant.

The methods presented above for testing whether a given effect in the path diagram is statistically significant were all based upon the calculation of χ^2 statistics. An alternative procedure, based upon the estimation of the standard deviation of the estimated effects, will be presented in section 2.9.

2.5. Some Additional Models That Fit the Data

Section 2.4 focused attention on model H_5 of table 5 (expressed in various forms). On the other hand, comparison of the χ^2 values of $X^2(H_2)$ to $X^2(H_7)$ in table 5 might suggest that models H_4 and H_7 should not be completely ignored. Models H_5, H_4, and H_7 were obtained from model H_1

28 If $\beta^{A\bar{B}}{}_1 = \beta^{B\bar{A}}{}_1 = 0$ in (37)–(38), then $\lambda^{AB}{}_{11} = 0$ in (36), and variables *A* and *B* are statistically independent in the two-way table $\{AB\}$.

29 The likelihood-ratio χ^2 can be calculated either from the two-way table $\{AB\}$, or as the difference $X^2(H_{16}) - X^2(H_{15})$ in table 5, or as the difference between the X^2 values obtained for any two other "Markov-type" models for the $\{ABCD\}$ table which differ only to the extent that one model includes the $\{AB\}$ table among the set of fitted marginals and the other model does not (see Goodman 1970, 1971*b*). (Aside from the $\{AB\}$ table, both models need to include the letter *A* and the letter *B* among the letters in the remaining set of fitted marginals [e.g., in $\{AC\}$ and $\{BD\}$].) The corresponding goodness-of-fit statistic can be calculated from the two-way table $\{AB\}$, but the other method described above for calculating χ^2s (as differences between the corresponding χ^2s for certain models for table $\{ABCD\}$) will not give the same value as obtained for the goodness-of-fit statistic calculated from table $\{AB\}$. The numerical value of the goodness-of-fit statistic calculated from table $\{AB\}$ is 75.55.

by deleting from it the parameters τ^{AD}, τ^{BC}, and τ^{AB}, respectively.[30] We next consider the model in which all three parameters (τ^{AD}, τ^{BC}, τ^{AB}) have been deleted, thus obtaining model H_{14} of table 5. Since $X^2(H_{14}) = 8.62$, with 8 df, this model provides a rather good fit.

Model H_{14} expresses the expected frequencies F_{ijkl} in terms of the τ parameters in the following way:

$$F_{ijkl} = \eta\, \tau^{A}{}_{i}\, \tau^{B}{}_{j}\, \tau^{C}{}_{k}\, \tau^{D}{}_{l}\, \tau^{AC}{}_{ik}\, \tau^{BD}{}_{jl}\, \tau^{CD}{}_{kl}\,. \tag{39}$$

Formula (39) was obtained by deleting the three parameters (τ^{AD}, τ^{BC}, τ^{AB}) from (4). Could the fit of model H_{14} be improved, in a statistically significant way, by the insertion of τ^{AD}, or τ^{BC}, or τ^{AB} into the model (i.e., into [39])? To test whether a statistically significant improvement would be obtained by the insertion of τ^{AD} into (39), we compare $X^2(H_{14})$ with $X^2(H_{13})$, calculating $X^2(H_{14}) - X^2(H_{13}) = 0.77$, with 1 df. Similarly, by comparing $X^2(H_{14})$ with $X^2(H_{12})$ or $X^2(H_{14})$ with $X^2(H_{10})$, we can test whether a statistically significant improvement would be obtained by the insertion of τ^{BC} or τ^{AB}, respectively, into (39). These comparisons indicate that the fit of model H_{14} cannot be improved, in a statistically significant way, by inserting either τ^{AD}, or τ^{BC}, or τ^{AB} into it.[31] (On the other hand, comparison of $X^2[H_{14}]$ with $X^2[H_5]$ indicates that the insertion of *both* τ^{BC} and τ^{AB} in model H_{14} would lead to a statistically significant improvement; but the model thus obtained would be H_5, which we have already considered.) We shall now focus our attention on model H_{14}.

Since model H_{14} includes fewer two-factor τ parameters than did model H_5, the equations corresponding to (19)–(22) and (30)–(33) will be somewhat simpler in the present case. From (39), we obtain for model H_{14} the following equations corresponding to (30)–(33):

$$\Phi^{\bar{D}}{}_{ijk\cdot} = \beta^{\bar{D}} + \beta^{B\bar{D}}{}_{j} + \beta^{C\bar{D}}{}_{k}\,, \tag{40}$$

$$\Phi^{\bar{C}}{}_{ij\cdot l} = \beta^{\bar{C}} + \beta^{A\bar{C}}{}_{i} + \beta^{D\bar{C}}{}_{l}, \tag{41}$$

$$\Phi^{\bar{B}}{}_{i\cdot kl} = \beta^{\bar{B}} + \beta^{D\bar{B}}{}_{l}\,, \tag{42}$$

$$\Phi^{\bar{A}}{}_{\cdot jkl} = \beta^{\bar{A}} + \beta^{C\bar{A}}{}_{k}\,. \tag{43}$$

[30] Although the inclusion of the τ^{AD} parameter in model H_1 did not improve the fit of the model in a statistically significant way, the inclusion of the other τ parameters did.

[31] The above statement is true when the χ^2 statistic (8) is used, as we do throughout this article. However, if the statistic (7) is used instead, then the numerical values obtained would suggest that the insertion of the τ^{AB} parameter into model H_{14} would be statistically significant at the 5% level.

When the observed frequencies in the cross-classification are large enough (see, e.g., tables 2 and 3), the conclusions obtained using statistic (8) will usually agree with those obtained using (7); but when some of the observed frequencies are very small (as in table 1), this need not be the case. When different conclusions are obtained, it would be prudent to suspend judgment until additional data can be obtained.

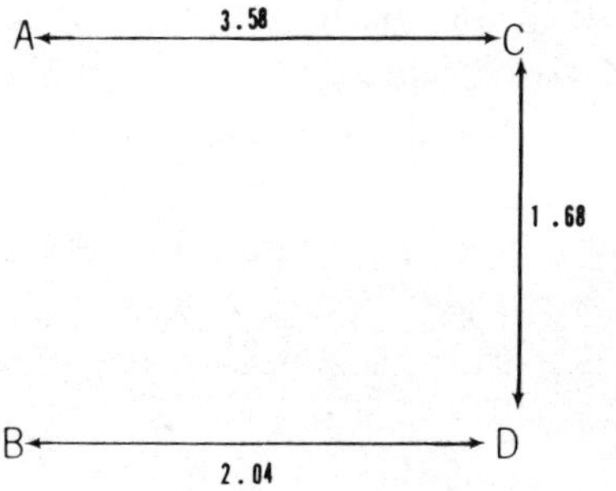

FIG. 3.—Path diagram for the system of eqq. (40)–(43), for the four-way cross-classification table (table 1). See model H_{14} in table 5.

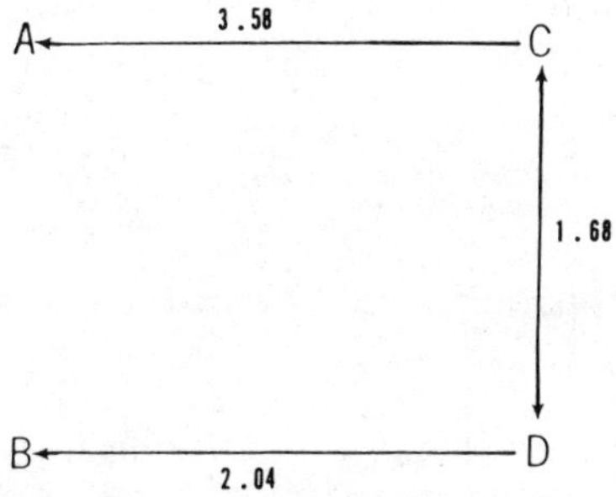

FIG. 4.—Path diagram for the system of eqq. (42)–(43), for the four-way cross-classification table (table 1). See model H_{14} in table 5.

The β's in (40)–(43) can be estimated using the same procedure that was used in section 2.2 to estimate the β's in (30)–(33). Figures 3 and 4 provide path diagrams corresponding to the system of equations (40)–(43) and the system of equations (42)–(43), respectively;[32] and the numerical values presented in these diagrams were obtained from the estimated β parameters in these equations.[33] As we shall see later (when table 10 is

[32] In fig. 3, the two arrows pointing to D correspond to $\beta^{B\bar{D}}$ and $\beta^{C\bar{D}}$ in (40); the two arrows pointing to C correspond to $\beta^{A\bar{C}}$ and $\beta^{D\bar{C}}$ in (41); the arrow pointing to B corresponds to $\beta^{D\bar{B}}$ in (42); the arrow pointing to A corresponds to $\beta^{C\bar{A}}$ in (43). In fig. 4, the arrow pointing to B and the one pointing to A correspond to $\beta^{D\bar{B}}$ in (42) and $\beta^{C\bar{A}}$ in (43), respectively; and, since there are no arrows pointing to C or D from A or B in this figure, the C-D double-headed arrow here corresponds to the β^{CD} parameter obtained in an analysis of the two-way marginal table $\{CD\}$. For related comments, see the earlier discussion of the β^{AB} parameter in fig. 2; if the letters A and B in the earlier discussion are replaced by C and D, the comments there will apply directly to the β^{CD} parameter in fig. 4.

[33] The C-D effect in fig. 4, which can be estimated from the saturated model for the two-way table $\{CD\}$, is equal to the corresponding value obtained in fig. 3 using equations (40)–(43). This equality is due to the fact that the A-B effect, the A-D effect, and the B-C effect are nil under the model; and we see from (39) that the τ^{CD} parameter (and thus the β^{CD} parameter) is not affected when variables A and B are collapsed to obtain the two-way table $\{CD\}$ (see, e.g., Goodman 1971*b*, 1972*b*). Similarly, we also find that the β^{AC} and β^{BD} parameters in figs. 3 and 4 can be estimated from the "saturated" models for tables $\{AC\}$ and $\{BD\}$, respectively.

introduced in sec. 2.9), the path diagrams in both figures 3 and 4 are equivalent to model H_{14}.[34] Since model H_{14} fits the data rather well, so do these path diagrams. (The meaning of these path diagrams will be discussed in sec. 2.6.)

The same procedures used in section 2.4 to test whether a given effect in figures 1 and 2 is statistically significant can be used now with the effects in figures 3 and 4. For example, compare H_{16}–H_{18} in table 5 with H_{14}.[35] Each of the effects in these figures actually is statistically significant.

Note should be taken of the fact that under model H_{14} the "expected odds-ratio" between variables A and B will be equal to one, when the levels of variables C and D are set at levels k and l, respectively.[36] This means that, under model H_{14}, variable A will be statistically independent of variable B, when the joint level of the remaining variables (C, D) is held constant.[37] It does *not* mean that variables A and B will be statistically independent in the two-way marginal table $\{AB\}$.

2.6. Comments on Apparent Time Reversal, Latent Variables, and Double-headed Arrows

In figures 1, 3, and 4, there are arrows pointing from C to A and from D to B, whereas it is usually assumed in causal analysis that arrows should point from temporally prior to temporally posterior variables. To better understand the meaning of this apparent time reversal, let us first focus our attention on figure 4.

Figure 4 expresses, in part, the notion that an individual's "true colors" will show up "in the end" (i.e., at the second interview rather than the

[34] In other words, both model (40)–(43) and model (42)–(43) are equivalent to model H_{14} as described by (39). For example, the set of τ parameters that are taken equal to one when (42) and (43) are true will be equal to the set of τ parameters that are taken equal to one when (39) is true, as we shall see later from table 10. Thus, the estimated expected frequencies F_{ijkl} under model (42)–(43) will be equal to the corresponding quantities under model (39) (see related comments in nn. 14–15).

[35] When $\{CD\}$ is deleted from the set of fitted marginals in model H_{14}, the letter C and the letter D still appear among the letters in the remaining set of fitted marginals (see model H_{16}). However, when $\{BD\}$ is deleted, the letter B no longer appears among the letters in that set. Because of this, the one-way marginal array $\{B\}$ needed to be included among the set of fitted marginals in model H_{17}. A similar remark applies also to model H_{18}.

[36] See formula (6) above. Since the τ^{AB} parameter is equal to one under model H_{14}, the corresponding "expected odds-ratio" will also be equal to one.

[37] Various consequences can be derived from model H_{14}, and they will be discussed more fully in sec. 2.7. The consequence noted above (that variables A and B are statistically independent, when the level of $[C, D]$ is held constant) means that, under model H_{14}, the observed association between variables A and B in the two-way marginal table $\{AB\}$ becomes "spurious" when the level of (C, D) is taken into account.

first). Here variables C and D, which were observed at the second interview, are viewed as indicators for some more fundamental variables (say, the "latent variables" C^* and D^*, respectively) which affect variables A and B observed at the first interview. The latent variables (C^*, D^*) are prior to the observed variables (A, B) in the sense that the (C^*, D^*) affect the (A, B); but the (C^*, D^*) remain unobserved until the second interview when their indicator variables (C, D) are observed. Figure 4 *assumes* that variables (C, D) are *accurate* indicators for (C^*, D^*),[38] and so it does not distinguish between (C, D) and (C^*, D^*).[39]

The remark in the final paragraph of section 2.5 can now be expressed as follows: One of the consequences of the model described by figure 4 is that variable A will be statistically independent of variable B, when the level of (C^*, D^*) is taken into account. It is also possible to show that, under the model described by figure 4, the changes from the level of variable C^* (or C) to the level of variable A are independent of the changes from the level of variable D^* (or D) to the level of variable B (see Goodman 1973*a*).

The introduction of the latent variables (C^*, D^*) helps us to better understand the meaning of the apparent time reversal in figure 4. While figure 4 displays this time reversal, the corresponding path diagram in figure 4*A* does not. The latter path diagram corresponds to model H_{15} in table 5, while figure 4 corresponded to model H_{14}, as we noted earlier. Since $X^2(H_{15}) = 58.96$ and $X^2(H_{14}) = 8.62$ in table 5, each with 8 df,

[38] Since the model described by fig. 4 fits the data rather well, there does not seem to be a real need here to modify and complicate the model under consideration by viewing variables (C, D) as *fallible* indicators of the latent variables (C^*, D^*). The model described by fig. 4 is more parsimonious than the corresponding model obtained when variables (C, D) are viewed as fallible indicators. The former model is a special case of the latter model in which fallibility is nil. Generally speaking, if we wish to compare two models, say H and H′ where H is a special case of H′, we would usually investigate whether the fit obtained with H could be improved, in a statistically significant way, by replacing H by H′. Even if the improvement is not statistically significant, we might still wish to use the more general model if it is particularly meaningful in the substantive context under consideration. (Of course, in analyzing any data, the assessment of *substantive* significance is important, and it *cannot* be replaced by an assessment of *statistical* significance.) In the present context, we did not use the more general model (i.e., the model in which variables $[C, D]$ are viewed as fallible indicators) because (*a*) the model described by fig. 4 fit the data rather well, and (*b*) the analysis of the model in which variables (C, D) are fallible indicators requires statistical techniques that are different from those presented here. By applying techniques that are appropriate for the analysis of the more general model (see Goodman 1973*c*), we would find that the more general model did not improve the fit in a statistically significant way.

[39] The letters (C, D) in fig. 4 denote the latent variables (C^*, D^*) and their (accurate) indicator variables (C, D). In this figure, the letters (C, D) are placed to the right of the letters (A, B) (since they denote the variables $[C, D]$), but they could have been placed to the left of (A, B) (since they also denote the variables $[C^*, D^*]$).

FIG. 4*A*.—Path diagram corresponding to model H_{15} in table 5. This path diagram does not fit the data in table 1.

we see that the path diagram in figure 4*A* does not fit the data, but the one in figure 4 fits the data rather well.

We noted earlier that the *C-D* double-headed arrow in figure 4 and the *A-B* double-headed arrow in figure 2 could be interpreted in similar ways. The *A-B* double-headed arrow in figure 2 suggests that variables *A* and *B* are not statistically independent (see model [36]); and it also can suggest that variable *A* affects *B* and vice versa (see model [37]–[38]). Similarly, the *C-D* double-headed arrow in figure 4 suggests that variables *C* and *D* are not statistically independent; and it also can suggest that variable *C* affects *D* and vice versa (see nn. 25 and 32). (As we noted earlier, the *A-B* double-headed arrow in fig. 2 and the *C-D* double-headed arrow in fig. 4 pertain to the analysis of the two-way marginal tables $\{AB\}$ and $\{CD\}$, respectively, since there are no arrows pointing to *A* or *B* from *C* or *D* in fig. 2 and no arrows pointing to *C* or *D* from *A* or *B* in fig. 4.)

The absence of any arrows between *A* and *B* in figure 4 is not inconsistent with the presence of the double-headed arrow between *A* and *B* in figure 2. The interpretation of this absence in figure 4 was presented in the third paragraph preceding this one (and in the final paragraph of sec. 2.5), and the interpretation of its presence in figure 2 was presented in the preceding paragraph.

The meaning of the other double-headed arrows in figures 1–3 is different from that described above for the *A-B* double-headed arrow in figure 2 and the *C-D* double-headed arrow in figure 4. Consider, for example, the *C-D* double-headed arrow in figures 1 and 2. Since there are arrows pointing to *C* and *D* from *A* and *B* in figures 1 and 2, the double-headed arrow in these figures does *not* pertain to the two-way marginal table $\{CD\}$. Instead, it pertains to the effect of variable *C* on *D*, when the effects of the other variables upon variable *D* are taken into account (see [30]); and it also pertains to the effect of variable *D* on *C*, when the effects of the other variables upon variable *C* are taken into account (see [31]). A similar kind of interpretation can be given for the other double-headed arrows in figures 1–3 (aside from the *A-B* double-headed arrow in fig. 2); see footnotes 19, 20, and 32.

The above remarks should shed some light on the meaning of the path diagrams in figures 1–4. Further light will be shed on this matter in sections 2.7 and 2.8.

We shall use the symbol $[\bar{A} \otimes \bar{D} \mid \bar{B}, \bar{C}]$ to denote the hypothesis that variable A will be statistically independent of variable D when the joint level of the remaining variables (B, C) is held constant. In other words, $[\bar{A} \otimes \bar{D} \mid \bar{B}, \bar{C}]$ is the hypothesis of *conditional* independence between variables A and D, *given* the level of the remaining variables (B, C). If model H_5 in table 5 is true (see fig. 2), then hypothesis $[\bar{A} \otimes \bar{D} \mid \bar{B}, \bar{C}]$ will also be true.[40] Hypothesis $[\bar{A} \otimes \bar{D} \mid \bar{B}, \bar{C}]$ is actually equivalent to model H_{19} in table 5 (see Goodman 1970); and we can use this model to test the hypothesis. Since $X^2(H_{19}) = 0.74$, with 4 df, we see that this hypothesis is congruent with the data.

Model H_5 states that $[\bar{A} \otimes \bar{D} \mid \bar{B}, \bar{C}]$ is true (i.e., that H_{19} is true); and, in addition, it states that there is no three-factor interaction among the variables A, B, C in the three-way marginal table $\{ABC\}$, and that there is no three-factor interaction among the variables B, C, D in the three-way marginal table $\{BCD\}$.[41] Model H_5 implies model H_{19}, but model H_{19} does *not* in general imply model H_5.

As we noted earlier, model H_5 was equivalent to the system of equations (19)–(22). Thus, if model H_5 is true, then each of these equations will also be true. We can test the hypothesis that equation (19) is true using model H_{20} in table 5 (see, e.g., Bishop 1969; Goodman 1970, 1972*a*). Since $X^2(H_{20}) = 1.45$, with 5 df, we see that this hypothesis is congruent with the data. In a similar way, we can test separately the hypothesis that equation (20) is true (using H_{21}), the hypothesis that equation (21) is true (using H_{22}), and the hypothesis that equation (22) is true (using

[40] From a formula similar to (6), we see that the "expected odds-ratio" between variables A and D is equal to $(\tau^{AD})^4$, when the level of the remaining variables (B, C) is held constant, Since the τ^{AD} parameter is equal to one under model H_5 in table 5, the corresponding expected odds-ratio will also be equal to one. Because of this, the hypothesis $[\bar{A} \otimes \bar{D} | \bar{B}, \bar{C}]$ will be true. For further details, see Goodman 1970, 1972*b*.

[41] The null hypothesis that there is no three-factor interaction in the three-way table $\{ABC\}$ can be tested by the usual likelihood-ratio χ^2 statistic for testing this null hypothesis in a three-way table (see, e.g., Goodman 1970). This likelihood ratio χ^2 can be calculated either from the three-way table $\{ABC\}$, or as the difference $X^2(H_{23}) - X^2(H_{19})$ in table 5. (We can use the difference here because [*a*] H_{19} and H_{23} differ only to the extent that H_{19} includes the $\{ABC\}$ table among the set of fitted marginals and H_{23} does not, *and* [*b*] both H_{19} and H_{23} are "Markov-type" models; see Goodman [1971*b*].) Similarly, the likelihood-ratio χ^2 for testing the null hypothesis that there is no three-factor interaction in the three-way table $\{BCD\}$ can be calculated either from the three-way table $\{BCD\}$ or as the difference $X^2(H_{20}) - X^2(H_{19})$ in table 5. (We can use the difference here because [*a*] H_{19} and H_{20} differ only to the extent that H_{19} includes the $\{BCD\}$ table among the set of fitted marginals and H_{20} does not; *and* [*b*] both H_{19} and H_{20} are Markov-type models.) These likelihood-ratio χ^2s are $X^2 = 0.01$ and $X^2 = 0.71$ for tables $\{ABC\}$ and $\{BCD\}$, respectively. (If the corresponding goodness-of-fit statistic is applied to the three-way tables, we obtain 0.01 and 0.77, respectively.)

H_{23}). From the corresponding χ^2 values in table 5, we see that each of these hypotheses is congruent with the observed data.

When we introduced model H_{14} in section 2.5 (see fig. 3), we noted that it fit the data rather well, but that a statistically significant improvement in the fit could be obtained by the insertion of both the τ^{AB} and τ^{BC} parameters in model H_{14}, thus yielding model H_5. Because of this, we would not expect the various consequences of model H_{14} to be borne out by the data to the same extent as the various consequences of model H_5 considered earlier in this section.

Since model H_{14} implies model H_5, all of the consequences of model H_5 are also consequences of model H_{14}. But there are many additional consequences of model H_{14}. We shall mention only a few of them here.

We shall now use the symbol $[\bar{B}\otimes\bar{C}|\bar{A},\bar{D}]$ to denote the hypothesis that variable B is statistically independent of variable C when the joint level of the remaining variables (A, D) is held constant; and we shall use the symbol $[\bar{A}\otimes\overline{BD}|\bar{C}]$ to denote the hypothesis that variable A is statistically independent of the joint variable (B, D) when the level of variable C is held constant. In other words, $[\bar{B}\otimes\bar{C}|\bar{A},\bar{D}]$ is the hypothesis of *conditional* independence between variables B and C, given the level of (A, D); and $[\bar{A}\otimes\overline{BD}|\bar{C}]$ is the hypothesis of *conditional* independence between variables A and the joint variable (B, D), given the level of C. In a similar way, we can define the hypotheses $[\bar{A}\otimes\bar{B}|\bar{C},\bar{D}]$ and $[\bar{B}\otimes\overline{AC}|\bar{D}]$. The hypotheses $[\bar{B}\otimes\bar{C}|\bar{A},\bar{D}]$, $[\bar{A}\otimes\bar{B}|\bar{C},\bar{D}]$, $[\bar{A}\otimes\overline{BD}|\bar{C}]$, and $[\bar{B}\otimes\overline{AC}|\bar{D}]$ are equivalent to models H_{24}, H_{25}, H_{26}, and H_{27}, respectively (see Goodman 1970). If model H_{14} is true, then each of the four models (H_{24}–H_{27}), and the corresponding four hypotheses, will also be true. We can use each of the four models (H_{24}–H_{27}) to test the corresponding hypotheses. From $X^2(H_{24})$ to $X^2(H_{27})$, we see that none of these hypotheses is rejected by the data.[42]

We shall use the symbol $[\bar{A}\otimes\bar{D}|\bar{C}]$ to denote the hypothesis that variables A and D are conditionally independent, given the level of variable C in the three-way marginal table $\{ACD\}$. In a similar way, we can define the hypotheses $[\bar{B}\otimes\bar{C}|\bar{D}]$, $[\bar{A}\otimes\bar{B}|\bar{D}]$, and $[\bar{A}\otimes\bar{B}|\bar{C}]$. If H_{14} is true, then each of these hypotheses will also be true.[43] The usual methods for testing

[42] The above statement is true when the χ^2 statistic (8) is used, as we do here. However, if the statistic (7) is used instead, then the numerical values obtained would suggest that model H_{25} is rejected at the 5% level, and model H_{27} at the 10% level.

[43] For example, by collapsing variable D in (39), we see that the following kind of model is obtained for the three-way marginal table $\{ABC\}$:

$$F^{ABC}_{ijk} = \eta\tau^{A}_{i}\tau^{B}_{j}\tau^{C}_{k}\tau^{AC}_{ik}\tau^{BC}_{jk},$$

where the value of the parameters τ^{B}_{j} and τ^{C}_{k} may differ from the corresponding values in (39), and where the new parameter τ^{BC}_{jk} is introduced (see Goodman 1972*b*). The model for F^{ABC}_{ijk} given above is equivalent to the hypothesis $[\bar{A}\otimes\bar{B}|\bar{C}]$

the null hypothesis of conditional independence in a three-way table can be applied to each of the above hypotheses (see, e.g., Goodman 1970). When the four hypotheses ($[\bar{A} \otimes \bar{D}|\bar{C}]$, $[\bar{B} \otimes \bar{C}|\bar{D}]$, $[\bar{A} \otimes \bar{B}|\bar{D}]$, and $[\bar{A} \otimes \bar{B}|\bar{C}]$) are tested using the likelihood-ratio χ^2 statistic, we obtain $X^2 = 0.81$, $X^2 = 0.74$, $X^2 = 1.92$, and $X^2 = 7.14$, respectively, each with 2 df.[44] Thus, the first three hypotheses are accepted, and the fourth hypothesis is rejected by the data.

Each of the four hypotheses considered in the preceding paragraph was a hypothesis that pertained to a three-way marginal table formed from the four-way table. They pertained to the three-way tables $\{ACD\}$, $\{BCD\}$, $\{ABD\}$, $\{ABC\}$, respectively. We shall now use the symbol $[(ACD) = 0]$ to denote the hypothesis that there is zero three-factor interaction in the three-way table $\{ACD\}$. In other words, $[(ACD) = 0]$ is the hypothesis that the three-factor interaction is nil in the three-way table $\{ACD\}$.[45] In a similar way, we can define the hypotheses $[(BCD) = 0]$, $[(ABD) = 0]$, $[(ABC) = 0]$. If model H_{14} is true, then each of these hypotheses will also be true.[46] We can test each of these hypotheses using the usual methods for testing the null hypothesis of zero three-factor interaction in a three-way table (see, e.g., Goodman 1970). When the four hypotheses ($[ACD) = 0]$, $[(BCD) = 0]$, $[(ABD) = 0]$, $[(ABC) =$

(see Goodman 1970). Thus, if model (39) (i.e., model H_{14}) is true, then hypothesis $[\bar{A} \otimes \bar{B}|\bar{C}]$ will also be true.

44 Each of these four χ^2 values can be calculated either from the corresponding three-way marginal table, or as a difference between certain χ^2 values in table 5. Corresponding to hypotheses $[\bar{A} \otimes \bar{D}|\bar{C}]$, $[\bar{B} \otimes \bar{C}|\bar{D}]$, $[\bar{A} \otimes \bar{B}|\bar{D}]$, and $[\bar{A} \otimes \bar{B}|\bar{C}]$, we can calculate the following differences:

$$X^2(H_{26}) - X^2(H_{25}), X^2(H_{27}) - X^2(H_{25}),$$

$$X^2(H_{27}) - X^2(H_{24}), X^2(H_{26}) - X^2(H_{19}),$$

respectively. (If the goodness-of-fit statistic is applied to the corresponding three-way tables, we obtain 0.98, 0.76, 2.33, and 8.69, respectively.)

45 The hypothesis $[(ACD) = 0]$ about the three-way table $\{ACD\}$ is different from the hypothesis $[(ACD) = 0|B]$ about the four-way table. The latter hypothesis states that, *given* the level of variable B, the three-factor interaction among variables A, C, D is nil. In other words, it states that the three-factor interaction $\tau^{ACD}{}_{ikl}$, and the four-factor interaction $\tau^{ABCD}{}_{ijkl}$, are not included in the model for the four-way table (see Goodman 1970). From (39) we see that, if model H_{14} is true, then *all* three-factor interactions, $\tau^{ACD}{}_{ikl}$, $\tau^{BCD}{}_{jkl}$, $\tau^{ABD}{}_{ijl}$, $\tau^{ABC}{}_{ijk}$, and the four-factor interaction, $\tau^{ABCD}{}_{ijkl}$, are not included in the model for the four-way table. The hypothesis that *all* three- and four-factor interactions are not included is equivalent to model (4) (i.e., model H_1 of table 5); and from $X^2(H_1)$ we see that this hypothesis fits the data very well.

46 For example, the model obtained for the three-way marginal table $\{ABC\}$ by collapsing variable D in (39) (see n. 43) did not include the three-factor interaction $\tau^{ABC}{}_{ijk}$; and so this model implies that $[(ABC) = 0]$. Thus, if model (39) (i.e., model H_{14}) is true, then the hypothesis $[(ABC) = 0]$ will also be true.

0]) are tested using the likelihood-ratio χ^2 statistics, we obtain $X^2 = 0.04$, $X^2 = 0.71$, $X^2 = 0.61$, $X^2 = 0.01$, respectively, each with 1 df.[47] Thus, each of these hypotheses is congruent with the data.

We have considered above some of the consequences of model H_{14}. This model has other consequences as well; but to save space, we shall not pursue this matter further here.

2.8. A New Model for These Data: A Recursive Model

Model H_5 has the advantage that it fits the data very well, but it has the possible disadvantage that it leads to a negative *B-C* effect, which may at first sight seem peculiar. (The fact that the effect turned out to be negative is actually not so surprising when the meaning of the effect is clearly understood: It is the effect of *B* on *C* [or *C* on *B*] when the joint level of variables (A, D) is held constant.) Model H_{14} has the advantage of parsimony (it includes two less parameters than does model H_5),[48] but it has the possible disadvantage that it leads to path diagrams in which there appears to be time reversal, which may at first seem peculiar. (The apparent time reversal is actually not so surprising when its meaning, as discussed in sec. 2.6, is clearly understood.) We shall now present a new model which has the advantages listed above for both models H_5 and H_{14}, and it does not have the possible disadvantages. In addition, the new model will help to further explain why there was a negative *B-C* effect in model H_5.

The path diagram for the new model is presented in figure 5. From this diagram we can see that the model states that variable *D* is posterior (in a certain sense) to the other three variables,[49] that equation (19) is true,[50] *and* that variable *B* is statistically independent of variable *C*, given the level of variable *A* (i.e., that $[\bar{B} \otimes \bar{C} | \bar{A}]$).[51] We noted in section 2.7 that

[47] With respect to the second and fourth X^2 values, see n. 41. Similarly, the likelihood-ratio χ^2 statistic for testing the null hypothesis $[(ACD) = 0]$ can be calculated either from the three-way table $\{ACD\}$, or as the difference $X^2(H_{24}) - X^2(H'_{21})$, where H'_{21} is a modification of H_{21} in which the τ^{BC} parameter has been deleted. In addition, the likelihood-ratio χ^2 statistic for testing the null hypothesis $[(ABD) = 0]$ can be calculated either from the three-way table $\{ABD\}$, or as the difference $X^2(H_{24}) - X^2(H'_{22})$, where H'_{22} is a modification of H_{22} in which the parameter τ^{BC} has been deleted. (If the goodness-of-fit statistic is applied to the corresponding three-way tables $\{ACD\}$ and $\{ABD\}$, we obtain 0.04 and 0.64, respectively.)

[48] Model H_{14} has 8 df whereas model H_5 has 2 less.

[49] In fig. 5, there are arrows pointing to *D* (from *B* and *C*), and there are no arrows going back from *D* to any of the other variables.

[50] The arrows pointing to *D* from *B* and *C* correspond to $\beta^{B\bar{D}}$ and $\beta^{C\bar{D}}$ in (30), which is equivalent to (19).

[51] In fig. 5, since *D* is posterior to *A*, *B*, and *C* (i.e., there are no arrows pointing to *A*, *B*, or *C* from *D*), the arrows among *A*, *B*, and *C* pertain to the analysis of the

the model described by equation (19) was equivalent to model H_{20} in table 5, and so our new model is a modification of H_{20} which we shall refer to as model H^*_{20}.[52]

The condition that $[\bar{B} \otimes \bar{C} | \bar{A}]$ can be expressed as follows in terms of a formula that somewhat resembles formula (20). Letting F_{ijk} denote the expected frequency in cell (i, j, k) of the marginal table $\{ABC\}$, we first define the "expected odds" $\Omega^{\bar{C}}_{ij.}$ with respect to variable C (at the joint level (i, j) on variable (A, B)) by the following formulae:

$$\Omega^{\bar{C}}_{ij.} = F_{ij1}/F_{ij2} \,. \tag{44}$$

The condition that $[\bar{B} \otimes \bar{C} | \bar{A}]$ is equivalent to the following equation:

$$\Omega^{\bar{C}}_{ij.} = \gamma^{\bar{C}} \gamma^{A\bar{C}}_{i} \,, \tag{45}$$

where the numerical value of the γ's need not be equal to that of the corresponding γ's in (20).[53] In the same way that (20) was rewritten as (31), we can rewrite (45) as[54]

$$\Phi^{\bar{C}}_{ij.} = \beta^{\bar{C}} + \beta^{A\bar{C}}_{i} \,. \tag{46}$$

Figure 5 now can be expressed as follows: It views variable D as posterior to A, B, and C, with equation (19) holding true; and it views variable C as posterior to A and B, with equation (45) holding true. As in figure 2, variables A and B can be viewed here as "exogenous" variables that are not statistically independent (see model [36]), or as "endogenous" variables that affect each other (see model [37]–[38] and nn. 22 and 25).

---

three-way marginal table $\{ABC\}$. With respect to the arrows among A, B, and C, the diagram suggests that $[\bar{B} \otimes \bar{C} | \bar{A}]$ (see Goodman 1972*b* and the discussion below).

52 At first sight, the path diagram in fig. 5 may appear to correspond to model H_{10} in table 5, but it actually does not. Model H^*_{20} is not equivalent to model H_{10}, nor is it equivalent to any other model of the kind listed in table 5. For the analysis of the four-way table, model H^*_{20} is actually a new kind of model which was not included among the 166 different models for the four-way table listed in Goodman (1970, table 4). This new kind of model is discussed further in Goodman 1973*a*.

53 Both fig. 5 and equation (45) view variable C as posterior to A and B. The condition that $[\bar{B} \otimes \bar{C} | \bar{A}]$ is also equivalent to the following equation

$$\Omega^{\bar{B}}_{i \cdot k} = \gamma^{\bar{B}} \gamma^{A\bar{B}}_{i} \,;$$

but this equation differs from fig. 5 in that it views variable B as posterior to A and C.

54 The arrow pointing to C from A corresponds to $\beta^{A\bar{C}}$ in (46), which is equivalent to (45). The $\beta^{A\bar{C}}$ parameter could be estimated from the three-way table $\{ABC\}$ under the hypothesis $[\bar{A} \otimes \bar{B} | \bar{C}]$ using the same general methods used earlier for the four-way table, but the estimate thus obtained is actually equal to the corresponding value obtained from the saturated model for the two-way table $\{AC\}$. This equality is due to the fact that the B-C effect is nil under model (46); and we see from (45) that the $\gamma^{A\bar{C}}$ parameter (and thus the $\beta^{A\bar{C}}$ parameter) is not affected when variable B is collapsed to obtain the two-way table $\{AC\}$. A similar comment applies also to the estimation of the β^{AB} parameter.

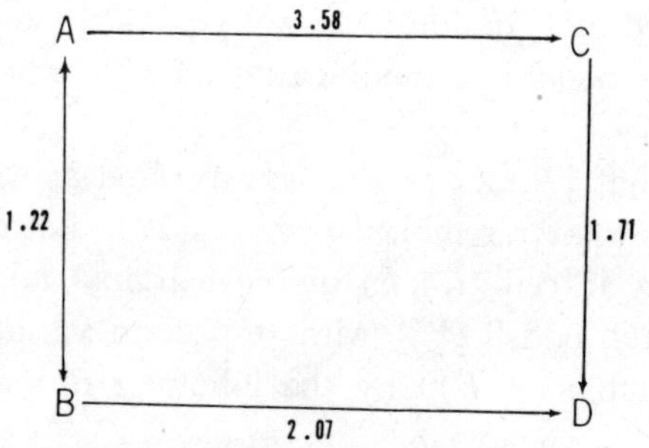

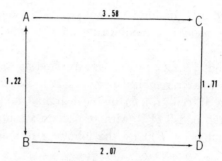

FIG. 5.—Path diagram for the system of eqq. (30) and (46) applied to table 1

With variable C posterior to A and B, and variable D posterior to A, B, and C, we can view figure 5 as a "recursive" model (see also figs. 10 and 13 below). We shall provide a more general discussion of recursive models in section 5.

As in section 2.7, the hypothesis $[\bar{B} \otimes \bar{C} | \bar{A}]$ can be tested by applying the usual methods for testing the null hypothesis of conditional independence in a three-way table. When the hypothesis $[\bar{B} \otimes \bar{C} | \bar{A}]$ is tested in this way using the likelihood-ratio χ^2 statistic, we obtain $X^2 = 0.15$, with 2 df.[55] From table 5 we see that $X^2(\mathrm{H}_{20}) = 1.45$, with 5 df. The hypothesis H^*_{20}, which stated both that $[\bar{B} \otimes \bar{C} | \bar{A}]$ is true and that H_{20} is also true, can be tested by calculating the sum of the above two χ^2 statistics. Since this sum is $0.15 + 1.45 = 1.61$,[56] using the χ^2 distribution with 7 df (i.e., $2 + 5 = 7$ df), we see that model H^*_{20} fits the data very well indeed.

Since H^*_{20} states that $[\bar{B} \otimes \bar{C} | \bar{A}]$ is true and that H_{20} is also true, the expected frequency F^*_{ijkl} in cell (i, j, k, l) of the four-way table $\{ABCD\}$ under H^*_{20} can be written as follows:

$$F^*_{ijkl} = F_{ijk} F_{ijkl} / f_{ijk} , \tag{47}$$

where f_{ijk} is the observed frequency in cell (i, j, k) of the three-way table $\{ABC\}$, where F_{ijk} is the corresponding expected frequency under the hypothesis $[\bar{B} \otimes \bar{C} | \bar{A}]$, and where F_{ijkl} is the expected frequency in cell (i, j, k, l) under H_{20}. (For further details, see Goodman 1973*a*.) Since the three-way table $\{ABC\}$ was included in the set of fitted marginals under H_{20}, we see that

$$f_{ijk} = \sum_{l=1}^{2} F_{ijkl} . \tag{48}$$

The F_{ijkl} can be estimated as we did earlier for each of the hypotheses in

[55] If the corresponding goodness-of-fit statistic is calculated from the three-way table $\{ABC\}$, we obtain 0.16.

[56] See n. 13. The above sum is 1.61, rather than 1.60, when more significant digits are used than are reported here.

table 5; and the F_{ijk} can be estimated by the usual methods for estimating the expected frequency in the three-way table $\{ABC\}$ under $[\bar{B}\otimes\bar{C}|\bar{A}]$ (see, e.g., Goodman 1970). Using (47) to estimate the F^*_{ijkl}, with the F_{ijkl} and F_{ijk} estimated as described above, we obtain table 8. Using (8) to compare the estimated F^*_{ijkl} with the corresponding observed f_{ijkl}, we obtain a value of 1.61 (with 7 df),[57] which agrees with the sum calculated in the preceding paragraph, as it should.

TABLE 8

ESTIMATE OF EXPECTED FREQUENCIES IN FOUR-WAY CROSS-CLASSIFICATION TABLE (TABLE 1) UNDER MODEL H^*_{20} DESCRIBED BY FORMULAE (30) AND (46)

	Second Interview				
	Vote Intention	+	+	−	−
	Candidate Opinion	+	−	+	−
First Interview: Vote Intention	Candidate Opinion				
+	+	128.10	3.72	1.68	1.50
+	−	12.11	22.07	0.01	0.81
−	+	0.73	0.02	12.30	10.95
−	−	0.80	1.45	1.23	68.52

We have described model H^*_{20} as a modification of H_{20}. It can also be described as a modification of model H_5. Model H_5 states that both H_{20} is true and that the three-factor interaction in the three-way table $\{ABC\}$ is nil.[58] Model H^*_{20} states that model H_{20} is true, that the three-factor interaction in table $\{ABC\}$ is nil, and that the two-factor B-C parameter (β^{BC}) in table $\{ABC\}$ is also nil.[59] Thus, model H^*_{20} states both that model H_5 is true and that the two-factor B-C parameter in table $\{ABC\}$ is nil. If model H^*_{20} is true, then model H_5 will also be true; but not vice versa. The various consequences of model H_5 described in section 2.7 are also consequences of model H^*_{20}.

With respect to model H^*_{20}, it is possible to prove that, since the β^{BC} parameter in the marginal table $\{ABC\}$ is nil, the corresponding β^{BC} parameter in the four-way table $\{ABCD\}$ *must* be negative if both β^{BD}

57 If the corresponding goodness-of-fit statistic is calculated, we obtain 1.61.

58 If model H_5 is true, we noted in the preceding section that model H_{20} will be true and that the three-factor interaction in the three-way table $\{ABC\}$ will be nil. Actually, model H_5 is equivalent to the statement that model H_{20} is true *and* that the three-factor interaction in table $\{ABC\}$ is nil.

59 The hypothesis $[\bar{B}\otimes\bar{C}|\bar{A}]$ is equivalent to the statement that there is no three-factor interaction in the three-way table $\{ABC\}$ and that the two-factor β^{BC} parameter in table $\{ABC\}$ is nil. For further details, see n. 65 below.

and β^{CD} have the same sign, and it *must* be positive if β^{BD} and β^{CD} have opposite signs (see Goodman 1972*b*). Thus, the negative *B-C* effect in figure 4 (model H_5) can be viewed as a consequence of the fact that model H^*_{20} (as described in fig. 5) is true.

2.9. The Saturated Model

For each model in table 5, the estimate of the expected frequency F_{ijkl} under the model can be calculated by fitting a given set of marginal tables (see the list of fitted marginals in table 5) using a straightforward iterative computation. In particular, this was true for models H_5 and H_{20}, but it was not true for H^*_{20}. (Recall the additional calculation of [47] under H^*_{20}.) In the present section, we shall consider all of the models in table 5; and we shall ignore H^*_{20}, which is essentially different from those models (see n. 52).

All of the models presented in table 5 are special cases of the "saturated" model. This model states that the expected frequencies F_{ijkl} in the four-way table can be expressed in terms of the τ parameters in the following way:

$$F_{ijkl} = \eta\tau^A{}_i\tau^B{}_j\tau^C{}_k\tau^D{}_l\tau^{AB}{}_{ij}\tau^{AC}{}_{ik}\tau^{AD}{}_{il}\tau^{BC}{}_{jk}\tau^{BD}{}_{jl}$$
$$\tau^{CD}{}_{kl}\tau^{ABC}{}_{ijk}\tau^{ABD}{}_{ijl}\tau^{ACD}{}_{ikl}\tau^{BCD}{}_{jkl}\tau^{ABCD}{}_{ijkl}\,, \tag{49}$$

where

$$\tau^A{}_1 = 1/\tau^A{}_2\,, \qquad \tau^{AB}{}_{11} = \tau^{AB}{}_{22} = 1/\tau^{AB}{}_{12} = 1/\tau^{AB}{}_{21}\,,$$

$$\tau^{ABC}{}_{111} = \tau^{ABC}{}_{221} = \tau^{ABC}{}_{212} = \tau^{ABC}{}_{122}$$
$$= 1/\tau^{ABC}{}_{112} = 1/\tau^{ABC}{}_{121} = 1/\tau^{ABC}{}_{211} = 1/\tau^{ABC}{}_{222}, \text{ etc.} \tag{50a}$$

(Compare [49] with [4], [9], and [39]; and compare [50a] with [5].) The conditions described in (50a) can also be expressed in the following equivalent form:

$$\prod_i \tau^A{}_i = 1, \qquad \prod_i \tau^{AB}{}_{ij} = \prod_j \tau^{AB}{}_{ij} = 1,$$

$$\prod_i \tau^{ABC}{}_{ijk} = \prod_j \tau^{ABC}{}_{ijk} = \prod_k \tau^{ABC}{}_{ijk} = 1,$$

$$\prod_i \tau^{ABCD}{}_{ijkl} = \prod_j \tau^{ABCD}{}_{ijkl} = \prod_k \tau^{ABCD}{}_{ijkl} = \prod_l \tau^{ABCD}{}_{ijkl} = 1. \tag{50b}$$

In the same way that (9) was rewritten as (26), we can also rewrite (49) to express the G_{ijkl} in additive form in terms of the λ parameters. Methods for estimating the τ parameters, and the corresponding λ parameters, were presented in Goodman (1970, 1972*b*), and a formula was also given there for $S_{\hat{\lambda}}$, the estimate of the standard deviation of each of the estimated λ's. In this earlier work, we divided each estimated λ parameter by its estimated standard deviation $S_{\hat{\lambda}}$, and we called the ratio thus obtained the "standardized value" of the corresponding estimated λ. To save space, we shall not go into these matters here, but instead refer the reader to the earlier work.

In the present article, our attention was focused on the γ parameters and the β parameters, which can be calculated from the corresponding τ and λ parameters by formulae (18), (35a), and similar formulae. In view of (35a), the standardized value of the estimated β parameter will be equal to the standardized value of the corresponding estimated λ parameter, which we showed how to calculate in the earlier work.

Table 9 presents the estimated γ and β parameters and the corresponding standardized value of the estimated β parameters. An examination of the relative magnitudes of the standardized values in table 9 will shed further light on some of the results presented in the preceding sections. Note, for example, the relative magnitudes of the standardized values of the estimated two-factor β's in this table. The standardized values can serve as a partial guide in the selection of models that fit the data, but this kind of guidance is not always foolproof (see Goodman 1970, 1971*a*, 1973*b*).

TABLE 9

ESTIMATE OF γ PARAMETERS AND CORRESPONDING β PARAMETERS IN THE SATURATED MODEL (49), FOR FOUR-WAY CROSS-CLASSIFICATION TABLE (TABLE 1)

Variable	γ Parameters	β Parameters	Standardized Value of Estimated β
A	1.30	0.26	0.72
B	1.11	0.11	0.29
C	1.15	0.14	0.37
D	0.90	−0.11	−0.28
AB	1.53	0.43	1.15
AC	12.60	2.53	6.86
AD	1.53	0.43	1.16
BC	0.78	−0.25	−0.68
BD	3.24	1.18	3.18
CD	3.02	1.10	2.99
ABC	0.98	−0.02	−0.06
ABD	1.05	0.05	0.14
ACD	1.02	0.02	0.06
BCD	1.20	0.18	0.49
ABCD	2.13	0.75	2.04

For the data in table 1, the estimated standard deviation of each of the estimated β's was $S_{\hat{\beta}} = 0.369$. This numerical value is the estimated standard deviation of each of the estimated β's in table 9, and it can also be viewed as an estimate of an upper bound for the standard deviation of each of the estimated β's in table 7. Since each of the estimated β's will be approximately normally distributed (when the sample size is large), the above numerical value of $S_{\hat{\beta}}$, taken together with the estimated β's can be used to obtain tests of hypotheses about the β parameters and also confidence intervals for these parameters.[60]

Since the numerical values of the effects presented in figures 1 and 2 (except for the *A-B* effect in fig. 2) were obtained from the estimated β's in table 7, we see that $S_{\hat{\beta}} = 0.369$ also serves as an estimate of an upper bound for the standard deviation of these effects.[61] A similar remark applies also to the effects presented in figures 3 and 4 (except for the *C-D* effect in figure 4),[62] and to the *B-D* and *C-D* effects in figure 5.[63]

Before closing this section, we shall comment briefly upon the relation between the saturated model (49) and the various "unsaturated" models presented in the preceding sections. Models (4), (9), and (39) were obtained from model (49) by specifying that certain τ parameters in (49) would be set equal to one. (For example, with respect to model [4], all three- and four-factor τ's in [49] were set equal to one.) Similarly, each of the models (19)–(22), or the equivalent models (30)–(33), can also be obtained from model (49) by setting certain τ's in (49) equal to one. Table 10 lists the τ's that are set equal to one for each of the models (19)–(22), and also for each of the models (40)–(43).

If both models (19) and (20) are true, then we see from table 10 that the following set of τ parameters in (49) are taken equal to one: $[AD]$, $[ABC]$, $[ABD]$, $[ACD]$, $[BCD]$, $[ABCD]$. Since this set of τ parameters was also taken equal to one in model (9), we see that model (19)–(20) is

60 Since we use here an estimate of an upper bound for the standard deviation of each estimated β, both the tests of hypotheses and the confidence intervals proposed above will be conservative ones. These methods can be used to supplement the procedures used earlier for determining whether a given estimated parameter is statistically significant.

61 With respect to the estimated β^{AB} parameter in fig. 2, we can estimate its standard deviation by applying to the saturated model for the two-way table $\{AB\}$ the same general methods used here with the four-way table (see Goodman 1970). We thus obtain $S_{\hat{\beta}} = 0.150$ for the estimates in table $\{AB\}$.

62 Since each of the estimated effects in figs. 3 and 4 could be obtained from the corresponding two-way marginal table, we can estimate the corresponding standard deviations by applying the methods developed for the analysis of the "saturated model" for each two-way table. For the estimated β^{AC}, β^{BD}, and β^{CD} parameters, we estimate the standard deviations to be 0.362, 0.192, and 0.173, respectively.

63 Since the *A-C* and *B-C* effects in fig. 5 were obtained from an analysis of the three-way table $\{ABC\}$, the corresponding analysis of the "saturated model" for this table provides the following estimate of an upper bound for the standard deviation of these effects: $S_{\hat{\beta}} = 0.363$.

TABLE 10

The τ Parameters in (49) That Are Set Equal to One, When Equations Pertaining to Expected Odds Are Expressed as Unsaturated Models Obtained from (49)

Equation for Expected Odds	Equation Number	Parameters That Are Set at One
$\Omega^{\bar{D}}$	(19)	[*AD*], [*ABD*], [*ACD*], [*BCD*], [*ABCD*]
$\Omega^{\bar{C}}$	(20)	[*ABC*], [*ACD*], [*BCD*], [*ABCD*]
$\Omega^{\bar{B}}$	(21)	[*ABC*], [*ABD*], [*BCD*], [*ABCD*]
$\Omega^{\bar{A}}$	(22)	[*AD*], [*ABC*], [*ABD*], [*ACD*], [*ABCD*]
$\Omega^{\bar{D}}$	(40)	Same as (19)
$\Omega^{\bar{C}}$	(41)	[*BC*], [*ABC*], [*ACD*], [*BCD*], [*ABCD*]
$\Omega^{\bar{B}}$	(42)	[*AB*], [*BC*], [*ABC*], [*ABD*], [*BCD*], [*ABCD*]
$\Omega^{\bar{A}}$	(43)	[*AB*], [*AD*], [*ABC*], [*ABD*], [*ACD*], [*ABCD*]

equivalent to model (9). In addition, from table 10 we find that model (9) will be equivalent to any other pair of equations from (19) to (22) (except for the pair [20]–[21]), to any triplet of equations from (19) to (22), and of course to the four equations (19)–(22). Similarly, from table 10 we find that model (39) will be equivalent to model (42)–(43), to model (40) and (42), to model (41) and (43), to any triplet of equations from (40) to (43), and of course to the four equations (40)–(43).

3. THE ANALYSIS OF TABLE 2

Having already presented a rather detailed analysis of table 1, we shall only outline here the analysis of table 2. Although the same general methods will be used, our conclusions about table 2 will differ from those obtained about table 1.

In our analysis of table 2, we shall use the letters *A, B, C, D* to denote the four variables in the table (membership at the first interview, attitude at the first interview, membership at the second interview, and attitude at the second interview, respectively). Corresponding to table 5 pertaining to table 1, we now present table 11 pertaining to table 2.

As in table 5, the hypothesis described by model (4) is listed as H_1 in table 11. Since $X^2(H_1) = 1.21$, with 5 df, we see that this model fits the data (table 2) very well indeed. However, comparison of $X^2(H_5)$ with $X^2(H_1)$ indicates that the inclusion of the τ^{AD} parameter in model H_1 does not contribute in a statistically significant way; and comparison of $X^2(H_7)$ with $X^2(H_1)$ indicates that the inclusion of the τ^{AB} in model H_1 also does not contribute in a statistically significant way. If both τ^{AD} and τ^{AB} are deleted from model H_1, we then obtain model H_{12}.

Comparison of $X^2(H_{12})$ with $X^2(H_5)$ and $X^2(H_7)$ indicates that the insertion of either τ^{AD} or τ^{AB} in model H_{12} would contribute in a statisti-

TABLE 11

χ^2 VALUES FOR SOME MODELS PERTAINING TO TABLE 2

Model	Fitted Marginals	df	Likelihood-Ratio χ^2	Goodness-of-Fit χ^2
H_1	$\{AB\}, \{AC\}, \{AD\}, \{BC\}, \{BD\}, \{CD\}$	5	1.21	1.21
H_2	$\{AB\}, \{AC\}, \{AD\}, \{BC\}, \{BD\}$	6	15.71	15.76
H_3	$\{AB\}, \{AC\}, \{AD\}, \{BC\}, \{CD\}$	6	262.54	261.28
H_4	$\{AB\}, \{AC\}, \{AD\}, \{BD\}, \{CD\}$	6	16.70	16.90
H_5	$\{AB\}, \{AC\}, \{BC\}, \{BD\}, \{CD\}$	6	4.06	4.10
H_6	$\{AB\}, \{AD\}, \{BC\}, \{BD\}, \{CD\}$	6	989.68	967.49
H_7	$\{AC\}, \{AD\}, \{BC\}, \{BD\}, \{CD\}$	6	3.84	3.82
H_8	$\{AB\}, \{AC\}, \{BC\}, \{BD\}$	7	35.20	34.94
H_9	$\{AB\}, \{AC\}, \{BC\}, \{CD\}$	7	267.48	265.48
H_{10}	$\{AB\}, \{AC\}, \{BD\}, \{CD\}$	7	17.91	18.27
H_{11}	$\{AB\}, \{BC\}, \{BD\}, \{CD\}$	7	1,009.16	982.64
H_{12}	$\{AC\}, \{BC\}, \{BD\}, \{CD\}$	7	8.78	8.82
H_{13}	$\{AC\}, \{AD\}, \{BC\}, \{BD\}$	7	16.82	16.98
H_{14}	$\{AC\}, \{AD\}, \{BC\}, \{CD\}$	7	267.26	265.36
H_{15}	$\{AC\}, \{AD\}, \{BD\}, \{CD\}$	7	36.24	36.21
H_{16}	$\{AD\}, \{BC\}, \{BD\}, \{CD\}$	7	1,009.22	981.09

cally significant way. We therefore focus our attention here on both model H_5 (obtained from H_1 by the deletion of τ^{AD} and from H_{12} by the insertion of τ^{AB}) and model H_7 (obtained from H_1 by the deletion of τ^{AB} and from H_{12} by the insertion of τ^{AD}). Each of the two-factor τ's in model H_5 contributes in a statistically significant way, as can be seen by comparing $X^2(H_5)$ with $X^2(H_8)$ to $X^2(H_{12})$; and each of the two-factor τ's in model H_7 contributes in a statistically significant way as can be seen by comparing $X^2(H_7)$ with $X^2(H_{12})$ to $X^2(H_{16})$. (For further comments on H_{12}, see Goodman 1973*c*.)

Table 12 presents the estimated expected frequencies under model H_5, and figures 6 and 7 present path diagrams corresponding to this model.

TABLE 12

ESTIMATE OF EXPECTED FREQUENCIES IN FOUR-WAY CROSS-CLASSIFICATION TABLE (TABLE 2), UNDER MODEL H_5 (IN TABLE 11) DESCRIBED BY FORMULA (9)

		Second Interview			
	Membership	+	+	−	−
	Attitude	+	−	+	−
First Interview					
Membership	Attitude				
+	+	447.71	151.16	104.60	53.53
+	−	169.36	182.77	54.59	89.29
−	+	192.98	65.15	537.71	275.16
−	−	87.95	94.92	338.10	553.03

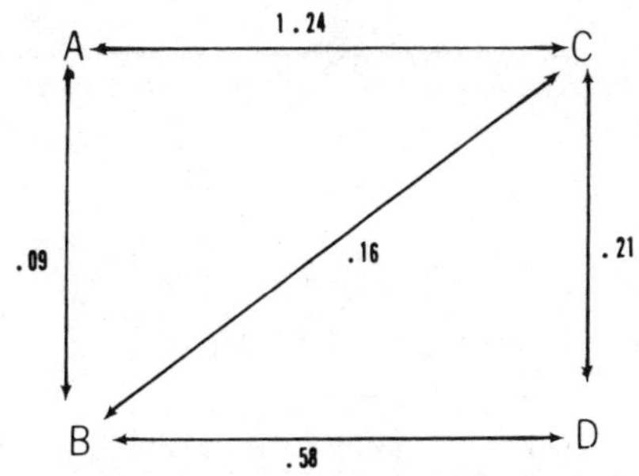

FIG. 6.—Path diagram for the system of eqq. (30)–(33), applied to table 2. See model H_5 in table 11.

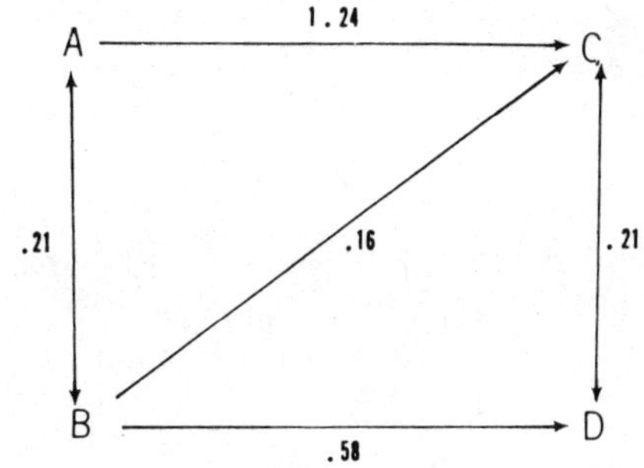

FIG. 7.—Path diagram for the system of eqq. (30)–(31), applied to table 2. See model H_5 in table 11.

Comparison of figures 6 and 7 (for table 2) with the corresponding figures 1 and 2 (for table 1) indicates that the negative *B-C* effect in table 1 is replaced by a positive *B-C* effect in table 2, and that each of the effects in the path diagrams for table 2 is less pronounced than the corresponding effects in the path diagrams for table 1.[64]

Figures 8 and 9 present path diagrams corresponding to model H_7. Comparison of figures 8 and 9 (for table 2) with the corresponding figures 3 and 4 (for table 1) indicates that the *A-D* effect and the *B-C* effect are statistically significant in table 2, although they are relatively small in magnitude there; whereas each of these effects is not statistically significant in table 1.

[64] For a given effect in these path diagrams, we can obtain (conservative) confidence intervals and tests of hypotheses for the difference between the corresponding values for tables 1 and 2 (say, $\beta_1 - \beta_2$), estimating an upper bound for the standard deviation of the difference using the usual formula

$$S_{\hat{\beta}_1 - \hat{\beta}_2} = (S^2_{\hat{\beta}_1} + S^2_{\hat{\beta}_2})^{1/2},$$

with the appropriate numerical values of the $S_{\hat{\beta}}$'s given here. For each effect in these path diagrams, we find the difference to be statistically significant. The technique presented above can also be used when comparing a given set of variables in two different populations, or at two different points in time, or in cases where different sets of variables are comparable.

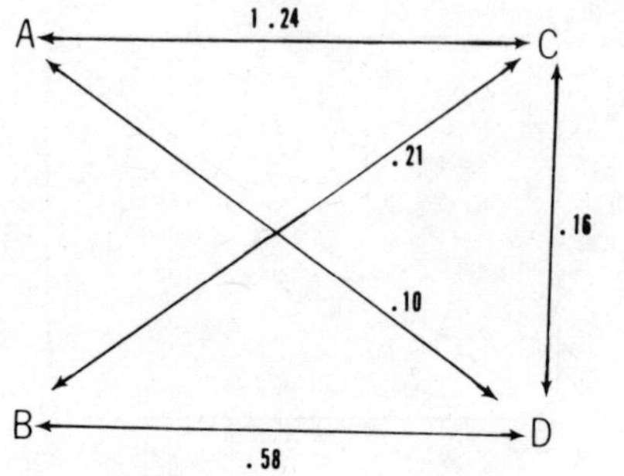

FIG. 8.—Path diagram for the system of four equations corresponding to model H_7 in table 11 applied to table 2.

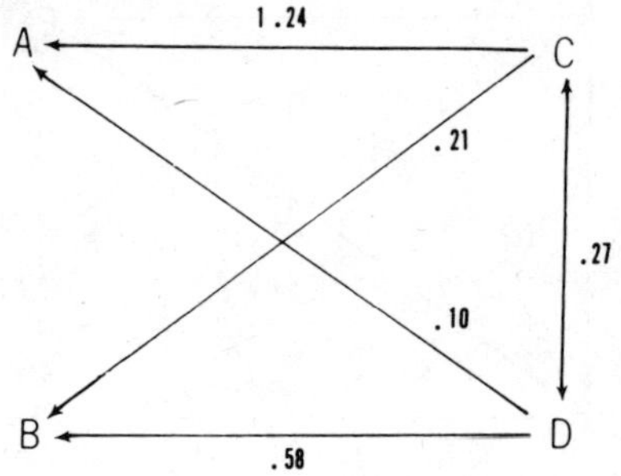

FIG. 9.—Path diagram for a system of two equations corresponding to model H_7 in table 11 applied to table 2.

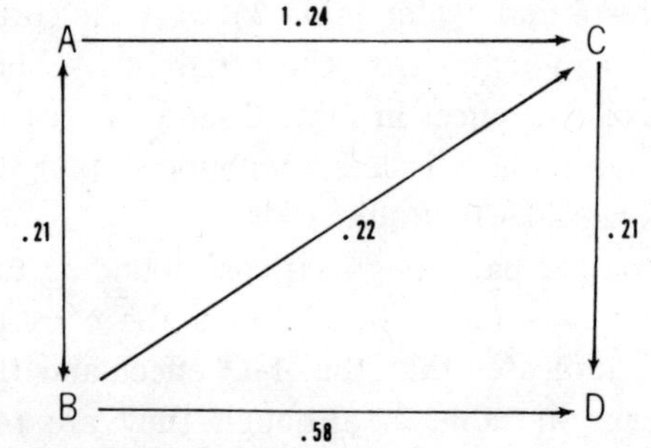

FIG. 10.—Path diagram for the system of eqq. (30) and (51), applied to table 2. See model H_5 in table 11.

Figure 10 presents a path diagram for a model that is somewhat analogous to model H^*_{20} discussed in section 2.8. However, instead of using the equation (46) as in H^*_{20}, we replace it by the following equation:

$$\Phi^{\bar{C}}_{ij\cdot} = \beta^{\bar{C}} + \beta^{A\bar{C}}{}_i + \beta^{B\bar{C}}{}_j\,. \tag{51}$$

Equation (51) states that there is no three-factor interaction in the three-way table $\{ABC\}$.[65] Figure 10 presents the path diagram for the system

[65] Note that equation (46) can be obtained from (51) by setting the β^{BC} parameter equal to zero. Thus, the hypothesis $[\bar{B}\otimes\bar{C}|\bar{A}]$, which was described by (46), states

of equations (30) and (51). This model is actually equivalent to model H_5 in table 11, which fit the data well.

The β parameters in (51) were estimated from the three-way marginal table {*ABC*} using the same kinds of methods as we used earlier in our analysis of the four-way table.[66] Since there are no arrows pointing to *A* or *B* from *C* or *D* in figure 10, the *A-B* effect in this diagram was estimated from the two-way marginal table {*AB*}.

Corresponding to table 9 for the saturated model for table 1, we have table 13 for the saturated model for table 2. An examination of the relative

TABLE 13

ESTIMATE OF γ PARAMETERS AND CORRESPONDING β PARAMETERS IN THE SATURATED MODEL (49), FOR FOUR-WAY CROSS-CLASSIFICATION TABLE (TABLE 2)

Variable	γ Parameters	β Parameters	Standardized Value of Estimated β
A	0.60	−0.51	−11.36
B	1.15	0.14	3.08
C	0.86	−0.15	−3.27
D	1.35	0.30	6.64
AB	1.07	0.07	1.49
AC	3.40	1.22	27.20
AD	1.08	0.07	1.62
BC	1.19	0.17	3.78
BD	1.79	0.58	12.88
CD	1.19	0.17	3.82
ABC	0.99	−0.01	−0.33
ABD	1.04	0.04	0.92
ACD	1.02	0.02	0.37
BCD	0.99	−0.01	−0.28
ABCD	1.01	0.01	0.30

magnitudes of the standardized values in table 13 will shed further light on the results presented earlier in this section. Note, for example, the relative magnitudes of the standardized values of the estimated two-factor β's in this table.

For the data in table 2, the estimated standard deviation of each of the estimated β's was $S_{\hat{\beta}} = 0.045$.[67] This numerical value can be used in the

both that there is no three-factor interaction in the three-way table {*ABC*} and that the β^{BC} parameter is zero.

66 Note that the effects in fig. 10 are equal to the corresponding effects in fig. 7 except for the *B-C* effect. The estimate of the *A-C* effect in (51) will be equal to the corresponding effect in (31), but the estimate of the *B-C* effect in (51) will be larger than the corresponding effect in (31). (For the reason why this is so, see Goodman 1972*b*.)

67 For the saturated models for the three-way table {*ABC*} and the two-way tables {*AB*} and {*CD*}, we obtain the following estimated standard deviations: 0.043, 0.036, and 0.036, respectively. These estimated standard deviations (rather than the

same ways that were described in section 2.9 for the corresponding value for table 1. We note, in passing, that the numerical value of $S_{\hat{\beta}}$ for table 2 is very much smaller than the corresponding value for table 1, as would be expected considering the size of the samples described in the two tables.[68]

4. THE ANALYSIS OF TABLE 3

In our analysis of table 3, we shall now use the letters *P, N, C, S* (rather than *A, B, C, D*) to denote the four variables in the table (proximity, norms, contact, and sentiment, respectively). Corresponding to the earlier tables 5 and 11, we now present table 14.

TABLE 14

χ^2 VALUES FOR SOME MODELS PERTAINING TO TABLE 3

Model	Fitted Marginals	df	Likelihood-Ratio χ^2	Goodness-of-Fit χ^2
H_1	$\{PN\}, \{PC\}, \{PS\}, \{NC\}, \{NS\}, \{CS\}$	5	2.25	2.29
H_2	$\{PN\}, \{PC\}, \{PS\}, \{NC\}, \{NS\}$	6	30.00	30.52
H_3	$\{PN\}, \{PC\}, \{PS\}, \{NC\}, \{CS\}$	6	22.71	22.92
H_4	$\{PN\}, \{PC\}, \{PS\}, \{NS\}, \{CS\}$	6	12.31	12.86
H_5	$\{PN\}, \{PC\}, \{NC\}, \{NS\}, \{CS\}$	6	2.53	2.62
H_6	$\{PN\}, \{PS\}, \{NC\}, \{NS\}, \{CS\}$	6	45.91	46.44
H_7	$\{PC\}, \{PS\}, \{NC\}, \{NS\}, \{CS\}$	6	14.97	15.10
H_8	$\{PN\}, \{PC\}, \{NC\}, \{NS\}$	7	34.31	34.31
H_9	$\{PN\}, \{PC\}, \{NC\}, \{CS\}$	7	24.22	24.21
H_{10}	$\{PN\}, \{PC\}, \{NS\}, \{CS\}$	7	12.37	13.04
H_{11}	$\{PN\}, \{NC\}, \{NS\}, \{CS\}$	7	50.21	49.87
H_{12}	$\{PC\}, \{NC\}, \{NS\}, \{CS\}$	7	16.48	16.46

As in tables 5 and 11, the hypothesis described by model (4) is listed as H_1 in table 14. Since $X^2(H_1) = 2.25$, with 5 df, we see that this model fits the data (table 3) very well indeed. However, comparison of $X^2(H_5)$ with $X^2(H_1)$ indicates that the inclusion of the τ^{PS} parameter in model H_1

one obtained for the four-way table $\{ABCD\}$) can be used when the corresponding estimated β pertains to the three- or two-way tables.

[68] It is possible that some of the effects that we found to be not statistically significant in table 1 might turn out to be statistically significant if a larger sized sample is used. To a rather large extent, the difference between the likelihood-ratio and goodness-of-fit statistics in table 5 was also due to the small frequencies in some of the cells of table 1. (Note that there is much less difference between the corresponding statistics in table 11; see related comments in the second paragraph of n. 31.) The expected frequency in each cell of the four-way table need not be very large in order for the chi-square asymptotic sampling theory to serve as an adequate approximation, but to determine how small they can be in the present context requires further work.

does not contribute in a statistically significant way. We therefore focus our attention on model H_5 (obtained from H_1 by the deletion of τ^{PS}). Each of the two-factor τ's in model H_5 contributes in a statistically significant way as can be seen by comparing $X^2(H_5)$ with $X^2(H_8)$ to $X^2(H_{12})$ in table 14.

Table 15 presents the estimated expected frequencies under model H_5, and figures 11–14 present path diagrams corresponding to this model. Figure 11 corresponds to the system of three equations (30)–(32), and it

TABLE 15

ESTIMATE OF EXPECTED FREQUENCIES IN FOUR-WAY CROSS-CLASSIFICATION TABLE (TABLE 3) UNDER MODEL H_5 (IN TABLE 14) DESCRIBED BY FORMULA (9)

			Sentiment	
Proximity	Norms	Contact	+	−
+	+	+	74.80	34.48
+	+	−	14.75	17.98
+	−	+	32.28	33.45
+	−	−	11.30	30.97
−	+	+	42.94	19.79
−	+	−	28.51	34.76
−	−	+	35.99	37.29
−	−	−	42.44	116.29

represents in more quantitative terms the verbal thesis proposed by Wilner et al. (1955) and Davis (1971). (For further details, see Goodman 1972*b*.) Figure 12 corresponds to the system of two equations (30)–(31), and it represents one of the points of view presented earlier in the present article (see figs. 2 and 7) applied now to table 3. Figure 13 corresponds to the system of equations (30) and (51), and it also represents one of the points of view presented earlier (see fig. 10). Figure 14 presents a modification of the points of view presented in figures 11 and 13, which we shall discuss next.

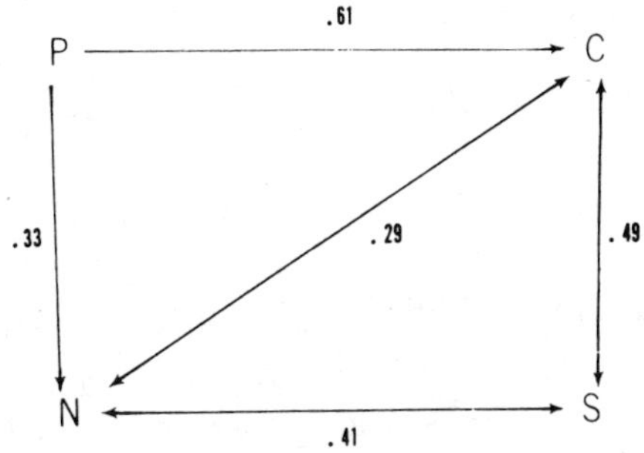

FIG. 11.—Path diagram for the system of eqq. (30)–(32), applied to table 3. See model H_5 in table 14.

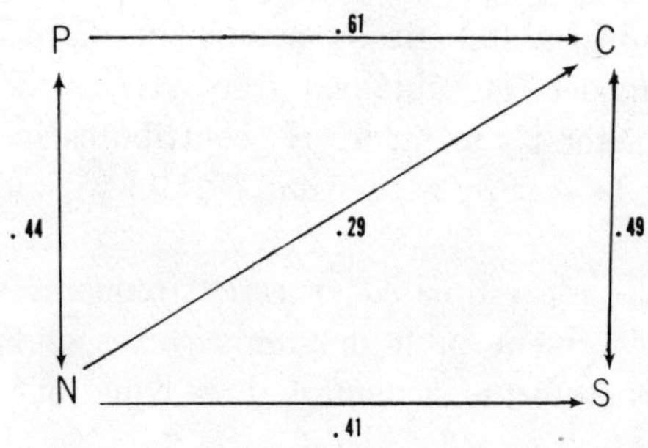

FIG. 12.—Path diagram for the system of eqq. (30)–(31), applied to table 3. See model H_5 in table 14.

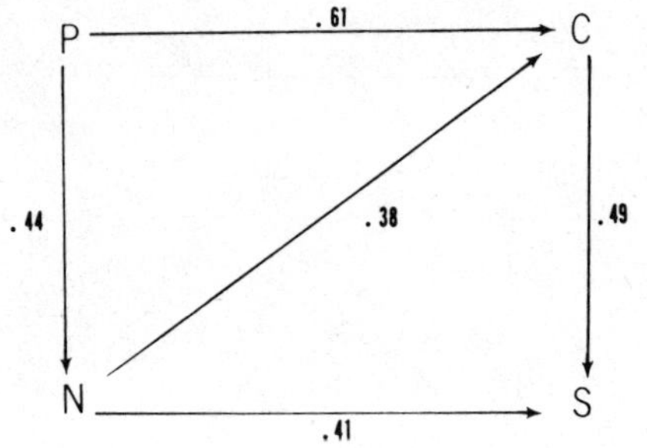

FIG. 13.—Path diagram for the system of eqq. (30) and (51), applied to table 3. See model H_5 in table 14.

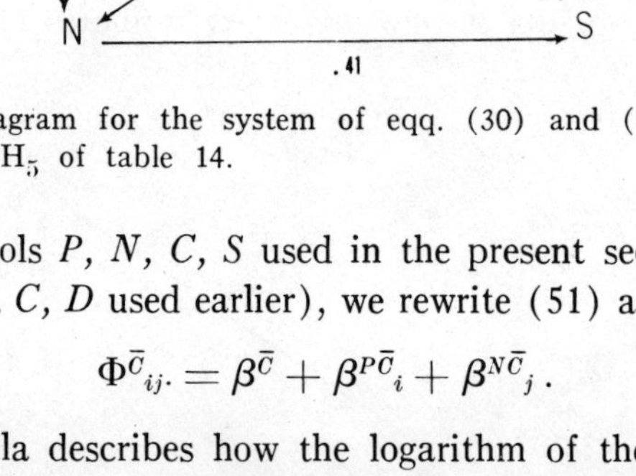

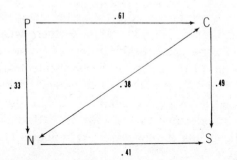

FIG. 14.—Path diagram for the system of eqq. (30) and (52)–(53), applied to table 3. See model H_5 of table 14.

With the symbols P, N, C, S used in the present section (rather than the symbols A, B, C, D used earlier), we rewrite (51) as follows:

$$\Phi^{\bar{C}}_{ij\cdot} = \beta^{\bar{C}} + \beta^{P\bar{C}}_{i} + \beta^{N\bar{C}}_{j}\,. \tag{52}$$

The above formula describes how the logarithm of the "expected odds" with respect to variable C is affected by variables P and N in the three-way table $\{PNC\}$, in the case where there is no three-factor interaction in this three-way table. In this case, we also obtain the following equation which describes how the logarithm of the "expected odds" with respect to variable N is affected by variables P and C in the three-way table $\{PNC\}$:[69]

$$\Phi^{\bar{N}}_{i\cdot k} = \beta^{\bar{N}} + \beta^{P\bar{N}}_{i} + \beta^{C\bar{N}}_{k}\,. \tag{53}$$

Figure 14 presents the path diagram for the system of equations (30) and (52)–(53).[70] This system of equations is actually equivalent to model H_5 of table 14, as was also the case for the systems of equations presented in figures 11–13 (see Goodman 1973*a*).

The β parameters in (52)–(53) were estimated from the three-way marginal table $\{PNC\}$ (under the hypothesis that the three-factor interaction is nil), using the same kinds of methods as we used earlier in our analysis of the four-way table. Note that the numerical values of the effects in figure 14 were the same as the corresponding values in figure 13 except for the β^{PN} effect.[71] The β^{PN} effect in figure 13 was the same as that in figure 12, and the β^{PN} effect in figure 14 was the same as that in figure 11.

As we noted earlier, figure 12 corresponds to the earlier figures 2 and 7, and figure 13 corresponds to the earlier figure 10.[72] In addition, figure 11 corresponds to the earlier figures 1 and 6, except that the arrowheads pointed to A from B and C in figures 1 and 6 do not appear in the corresponding place in figure 11. These arrowheads could be deleted from figures 1 and 6 (or the corresponding ones could be inserted in figure 11) without any changes occurring to the numerical values of the effects in these path diagrams. (Of course the interpretation of the path diagrams will change when arrowheads are deleted or inserted.)

The path diagram in figure 14 does not correspond directly to any of those in the earlier figures. Path diagrams like figure 14 could also be applied to tables 1 and 2 in the same way that we have done here for table 3. To save space, we shall not include this here.

Corresponding to tables 9 and 13 for the saturated model for tables 1 and 2, we have table 16 for the saturated model for table 3. Here too an examination of the relative magnitudes of the standardized value in table

69 The hypothesis that there is no three-factor interaction in the three-way table $\{PNC\}$ is equivalent to (52) or (53) or the system of equations (52)–(53) (see Goodman 1972*b*).

70 The two arrows pointing to S correspond to $\beta^{N\bar{S}}$ and $\beta^{C\bar{S}}$ in (30); the two arrows pointing to C correspond to $\beta^{P\bar{C}}$ and $\beta^{N\bar{C}}$ in (52); the two arrows pointing to N correspond to $\beta^{P\bar{N}}$ and $\beta^{C\bar{N}}$ in (53).

71 Note that the P-N effect in fig. 13 is larger than in fig. 14. For the reason why this is so, see Goodman 1972*b*.

72 The arrowhead pointing to A from B in fig. 10 does not appear in the corresponding place in fig. 13, but this arrowhead could be deleted from fig. 10 (or inserted in fig. 13) without any changes occurring to the numerical values of the effects in these path diagrams.

TABLE 16

ESTIMATE OF γ PARAMETERS AND CORRESPONDING β PARAMETERS IN THE SATURATED MODEL (49), FOR FOUR-WAY CROSS-CLASSIFICATION TABLE (TABLE 3)

Variable	γ Parameters	β Parameters	Standardized Value of Estimated β
P	0.70	−0.36	−3.89
N	0.82	−0.20	−2.18
C	1.24	0.22	2.37
S	0.91	−0.10	−1.06
PN	1.35	0.30	3.28
PC	1.76	0.57	6.14
PS	1.05	0.05	0.58
NC	1.35	0.30	3.23
NS	1.43	0.36	3.88
CS	1.57	0.45	4.89
PNC	1.01	0.01	0.09
PNS	0.97	−0.03	−0.30
PCS	0.94	−0.06	−0.67
NCS	1.10	0.10	1.06
PNCS	1.10	0.09	1.03

16 will shed further light on the results presented earlier in this section. For the data in table 3, the estimated standard deviation of the estimated β's was $S_{\hat{\beta}} = 0.092$.[73]

5. RECURSIVE MODELS AND SOME MORE GENERAL MODELS

In figure 13, we began with a set of two variables (P, N) that were prior to the other variables, and we analyzed the corresponding two-way table. Then we had another variable (C) that was posterior to (P, N) but prior to the other variable (S), and we analyzed the three-way table $\{PNC\}$ considering models for the three-way table in which the marginal $\{PN\}$ is included among the fitted marginals. (For example, under model [52] the marginal $\{PN\}$ is included among the fitted marginals.) Next we had another variable (S) that was posterior to the other three variables, and we analyzed the four-way table $\{PNCS\}$ considering models for the four-way table in which the marginal $\{PNC\}$ is included among the fitted marginals. (For example, under model [30], the marginal $\{PNC\}$ [i.e., $\{ABC\}$] is included among the fitted marginals.)

The above paragraph about figure 13 can be applied directly to figures 5 and 10, with the letters P, N, C, S replaced by A, B, C, D, respectively. The formula used to calculate the expected frequencies F^*_{ijkl} for figure 5

[73] For the saturated models for the three-way table $\{PNC\}$ and the two-way table $\{PN\}$, we obtain the following estimated standard deviations: 0.090, 0.084. For related comments, see n. 67.

(see [47]) can also be expressed in the following more general form which would be applicable also to figures 10 and 13:[74]

$$F^*_{ijkl} = F_{ij}(F_{ijk}/f_{ij})(F_{ijkl}/f_{ijk}), \tag{54}$$

where f_{ij} is the observed frequency in cell (i, j) of the two-way table $\{AB\}$, and where F_{ij} is the corresponding expected frequency in this table under a specified model (e.g., the saturated model as in figs. 5, 10, and 13, or the independence model $[\bar{A}\otimes\bar{B}]$, or any other model listed in Goodman [1970, table 2]); where f_{ijk} is the observed frequency in cell (i, j, k) of the three-way table $\{ABC\}$, and where F_{ijk} is the corresponding expected frequency in this table under a specified model that includes $\{AB\}$ among the fitted marginals (e.g., the model $[\bar{B}\otimes\bar{C} \mid \bar{A}]$ as in fig. 5, or the model $[(ABC) = 0]$ as in figs. 10 and 13, or any other appropriate model listed in Goodman [1970, table 3]); where F_{ijkl} is the expected frequency in cell (i, j, k, l) of the four-way table under a specified model that includes $\{ABC\}$ among the fitted marginals (e.g., model [30] as in figs. 5, 10, and 13 or any other appropriate model listed in Goodman [1970, table 4]). (For related matters, see Goodman 1973*a*.) As in (48), we see that

$$f_{ij} = \sum_{k=1}^{2} F_{ijk}\,, \; f_{ijk} = \sum_{l=1}^{2} F_{ijkl}\,. \tag{55}$$

To calculate the β's in the path diagram, we would express the specified model for table $\{AB\}$ as an equation for $\Phi^{\bar{B}}_{i.}$ or $\Phi^{\bar{A}}_{.j}$ or both (see, e.g., [37]–[38]); the specified model for table $\{ABC\}$ as an equation for $\Phi^{\bar{C}}_{ij.}$ (see, e.g., [46], [51], or [52]); the specified model for table $\{ABCD\}$ as an equation for $\Phi^{\bar{D}}_{ijk.}$ (see, e.g., [30]).[75] Each of the β's in

[74] As we noted earlier, figs. 10 and 13 were actually equivalent to H_5 of tables 11 and 14, respectively; but fig. 5 was not equivalent to any of the models in table 5. We needed (47) or the more general (54) to calculate F^*_{ijkl} for fig. 5, but we actually did not need this kind of formula for figs. 10 and 13 since the F^*_{ijkl} could be calculated under model H_5 using the usual iterative procedure. If an arrow pointing from A to D (or P to S) is interested in figs. 10 or 13, then the model thus obtained would not be equivalent to any of the kinds of models listed in tables 5, 11, and 14 (see Goodman 1973*a*); and the general formula (54) would be needed to calculate the F^*_{ijkl}.

[75] Since the specified models for tables $\{AB\}$, $\{ABC\}$, and $\{ABCD\}$ are of the kind discussed in Goodman (1972*a*), the modified multiple-regression technique presented in that article can be applied to each equation in the recursive system. In this respect, the recursive systems introduced here are analogous to recursive systems for quantitative variables. On the other hand, since the modified regression technique cited above does not need (or make use of) the usual system of "normal equations" in regression analysis, it also does not need (or make use of) the Wright multiplication theorem of path analysis, which is closely related to the usual normal equations. The particular structure of the system of equations in the recursive system for quantitative variables differs somewhat from the structure for the corresponding system of equations con-

the equations for $\Phi^{\bar{B}}_{i\cdot}$ (and/or $\Phi^{\bar{A}}_{\cdot j}$), $\Phi^{\bar{C}}_{ij\cdot}$, and $\Phi^{\bar{D}}_{ijk\cdot}$ would be represented by arrows in the path diagram.[76] The estimated F_{ij}, F_{ijk}, and F_{ijkl} can then be used to estimate the τ parameters in each specified model (see n.11), the corresponding λ parameters (see n.17), and then the β parameters used in the path diagram.

To test whether the overall model described by the path diagram fits the data, we compare the observed f_{ijkl} with the estimated F^*_{ijkl} using the χ^2 statistic (8). This statistic can be partitioned into components that test in turn the specified models for tables $\{AB\}$, $\{ABC\}$, and $\{ABCD\}$, comparing the observed f_{ij}, f_{ijk}, and f_{ijkl} with the corresponding estimated F_{ij}, F_{ijk}, and F_{ijkl}, respectively.

The procedures described above can be extended in order to analyze the following more general situation. Suppose we have a set of $a + b + c$ variables $A_1, A_2, \ldots, A_a, B_1, B_2, \ldots, B_b, C_1, C_2, \ldots, C_c$, where the subset of variables $(A_1, A_2, \ldots, A_a)$ is prior to the other variables, where the subset of variables $(B_1, B_2, \ldots, B_b)$ is posterior to the subset of A's but prior to the subset of C's, and where the subset of variables of $(C_1, C_2, \ldots, C_c)$ is posterior to the A's and B's. In this case, we first consider models for the a-way marginal table $\{A_1A_2 \ldots A_a\}$; we then analyze the $(a + b)$-way marginal table $\{A_1A_2 \ldots A_aB_1B_2 \ldots B_b\}$ considering models for the $(a + b)$-way table in which the a-way marginal table $\{A_1A_2 \ldots A_a\}$ is included among the fitted marginals; and we next analyze the $(a + b + c)$-way table $\{A_1A_2 \ldots A_aB_1B_2 \ldots B_bC_1C_2 \ldots C_c\}$ considering models for the $(a + b + c)$-way table in which the $(a + b)$-way marginal table is included among the fitted marginals. The situation described above can, of course, be generalized further to include a fourth subset of variables $(D_1, D_2, \ldots, D_d)$, a fifth subset of variables $(E_1, E_2, \ldots, E_e)$, etc.

Methods for analyzing a given a-way table (for $a = 2, 3, \ldots$) were presented in, for example, Goodman (1970, 1971*a*), and these articles also presented methods for analyzing a given $(a + b)$-way table (for $a = 1, 2, \ldots$; $b = 1, 2, \ldots$) considering models for the $(a + b)$-way table in which the a-way marginal table is included among the fitted marginals. These methods can be applied directly in the general situation described above. (These methods are not limited to the analysis of dichotomous variables; they can be applied to polytomous variables as well.)

---

sidered in the present article (in the former structure, each "dependent" variable can appear also as an "independent" variable in the equations for the other "dependent" variables that are posterior to it; whereas, in the latter structure, the actual "dependent" variables [i.e., the Φ's] do not appear also as "independent" variables); and because of this the Wright multiplication theorem will not be applicable here, since this theorem can be viewed as a consequence of the former structure.

[76] This sentence does not apply to the "general mean" in these equations (eg., $\beta^{\bar{C}}$ in equation [51]).

The general situation described above includes the path diagrams in figures 1–14. For example, the A, B, C, D variables in these diagrams can be represented by A_1, A_2, A_3, A_4 in figures 1 and 3; by A_1, A_2, B_1, B_2 in figure 2; by B_1, B_2, A_1, A_2, in figure 4; by A_1, A_2, B_1, C_1 in figure 5; etc.[77] For the $A_1, A_2, \ldots, A_a$ in a given path diagram, the specified model for the a-way table can be described by expressing the expected frequencies in this table in terms of the τ parameters (see, e.g., [9], [36], or [39]), or equivalently by a system of a equations expressing the a log-odds (i.e., the Φ's) in terms of the β parameters (see, e.g., [30]–[33], [37]–[38], or [40]–[43]). For the $A_1 A_2, \ldots, A_a, B_1, B_2, \ldots, B_b$ in the given path diagram, the specified model for the $(a+b)$-way table can be described by expressing the expected frequencies in this table in terms of the τ parameters (see, e.g., [9] or [39]), or equivalently by a system of b equations expressing the b log-odds in terms of the β parameters (see, e.g., [30]–[31] or [42]–[43]). Etcetera.

Each of the β's in the a equations for the a-way table, in the b equations for the $(a+b)$-way table, etc., would be represented by arrows in the path diagram.[78] The β parameters can be estimated by the general methods noted above, using the estimated expected frequencies (i.e., the F's) under the specified models for the a-way table, the $(a+b)$-way table, etc. These estimated expected frequencies can also be used to estimate the expected frequencies (i.e., the F^*s) for the overall model described by the path diagram, by applying (54) expressed in somewhat more general form (see Goodman 1973*a*). To test whether the overall model fits the data, we compare the observed frequencies in the full table with the corresponding estimated F^*s, and we can partition the χ^2 statistic (8) thus obtained into components that test in turn the specified models for the a-way table, the $(a+b)$-way table, etc.

In the path diagrams presented here, the models under consideration were such that each β parameter in the corresponding equations had no more than two letters in its superscript (see, e.g., [30]–[33]). However, the general methods proposed are not limited to models for which this is the case. When each of the β parameters has no more than two letters in its superscript, the corresponding path diagram is less complicated than in cases where some superscripts have more than two letters (see, e.g., Goodman 1973*a*). In cases where the path diagrams appear to be too compli-

[77] Single-headed arrows can move from A's to B's, and from A's and B's to C's; but *not* from B's to A's or from C's to A's or B's. Double-headed arrows can appear among the A's (if $a > 1$), among the B's (if $b > 1$), among the C's (if $c > 1$); but *not* between A's and B's, between A's and C's, or between B's and C's. The A's may be viewed as either exogenous or endogenous variables; and the B's and C's are endogenous. When the A's are viewed as exogenous, the straight double-headed arrows appearing among them might be replaced by curved double-headed arrows (see n. 25).

[78] See the exception noted in n. 76.

cated to be readily understood, attention could be focused instead on the corresponding system of equations.

6. COMMENTS ON SOME RELATED WORK

6.1. The Earlier Analyses of Tables 1, 2, and 3

In the earlier analyses of table 1, Lipset et al. (1954) and Lazarsfeld et al. (1968) introduce the concept of a "harmonizing process," which Boudon (1968) interprets as being representable by a path diagram such as figure 15. This figure corresponds to model H_2 in table 5 of the present article,

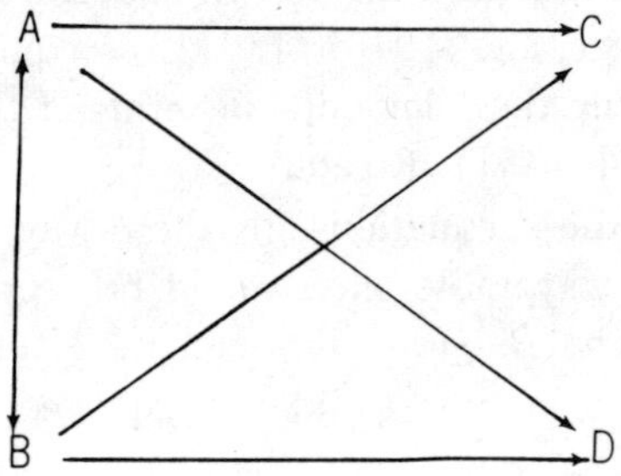

FIG. 15.—Path diagram corresponding to a "harmonizing process"

and since $X^2(H_2) = 16.98$, with 6 df, we see that this model is not congruent with the data. (Recall that $X^2[H_5] = 1.46$, with 6 df, for these data.) Actually, Boudon's quantitative model representing the path diagram is somewhat different from the quantitative model representing this path diagram in the present article, and his method of analysis is also quite different. He does present a table of expected frequencies estimated under his model, and a comparison of that table with our table 6 (or table 8) indicates that his model does not fit the data as well.

If the estimated expected frequencies in the table under a given model are maximum-likelihood estimates, or estimates that are asymptotically equivalent to maximum-likelihood estimates, then these estimates can be compared with the observed frequencies using the χ^2 statistics (7) or (8) If the estimates are not of this kind, then use of the χ^2 statistics can *not* be justified.[79] All of the estimates used herein are maximum-likelihood

[79] If the estimates are not of this kind (i.e., not "asymptotically efficient"), then the distribution of the statistic (7) or (8) will *not* be the corresponding tabulated χ^2 distribution (even when the sample size is large). Thus, under these circumstances, the usual procedure for assessing the statistical significance of an observed numerical value of the statistic (7) or (8), by comparing it with the percentiles of the corresponding tabulated χ^2 distribution, cannot be justified. On the other hand, since use of the maximum-likelihood estimates will minimize the statistic (8) under a given model, the numerical value of this statistic obtained with asymptotically inefficient

estimates (see, e.g., Goodman 1970, 1971*a*), whereas Boudon's methods do not provide maximum-likelihood estimates or asymptotically equivalent ones. Therefore, the use of the χ^2 statistics can be justified with the estimation methods proposed in the present article but not with Boudon's methods.

With respect to the interpretation of table 1, Boudon states that "as regards the *cross effects,* basic attitudes [about party preferences] have an influence on the attitude toward the candidate in the following period (.329), but the effect of the attitude toward the candidate on the party preferences is close to zero (.011)." Somewhat similar kinds of interpretations were presented, for example, by Campbell (1963) and Lazarsfeld et al. (1968). These interpretations did not take explicitly into account the *synchronous* effects (at the time of the second interview) of (*a*) party preferences upon attitude toward the candidate and (*b*) attitude toward the candidate upon party preferences. When these *synchronous* effects are taken into account (see, e.g., our figs. 2 and 5), different interpretations of table 1 are obtained. For example, from figure 5 (where the synchronous effect of party preferences upon attitude toward the candidate is taken into account) we see that, as regards the cross-effects, party preferences do *not* influence attitude toward the candidate in the following period, and attitude toward the candidate does not influence party preferences in the following period.

Contrary to the interpretation presented in some of the earlier literature, we view as spurious the *apparent* effect of party preferences upon attitude toward the candidate in the following period. This *apparent* effect disappears when we take into account, as in our figure 5, the synchronous effect of party preferences in the second period upon attitude toward the candidate. With the observed persistence of party preferences from the first period to the second, followed by the synchronous effect noted above, the *apparent* effect reported in the earlier literature might seem (at first sight) to be genuine even when it actually is spurious.[80]

The consequences that can be deduced from Boudon's path diagram for the "harmonizing process" (see fig. 15 above) are quite different from the

---

estimates will serve as an upper bound for the corresponding numerical value that could be obtained when maximum-likelihood estimates are used. Thus, if we were to assess how well a given model fit the data using the numerical value of the statistic (8) obtained with asymptotically inefficient estimates (with the tabulated χ^2 percentiles serving as a gauge), then if the fit is assessed to be good (i.e., if the numerical value is considered to be small enough), it would have been assessed similarly if the maximum-likelihood estimates had been used instead; *but* if the fit is not assessed to be good (i.e., if the numerical value is not considered to be small enough), we would not know what the assessment would have been if maximum-likelihood estimates had been used instead.

80 In our fig. 5, the *direct* effect of variable A on D is nil (as we have noted), but there is an indirect effect (from A to C to D or from A to B to D).

consequences that can be deduced from, say, our figure 5. For example, according to figure 15, attitude toward the candidate in the second interview would *not* be affected by party preferences in the second period (given the level of variables A and B); whereas, according to figure 5, it would be. Similarly, as we noted earlier, according to figure 15, attitude toward the candidate in the second interview would be affected by party preferences in the first interview (given the level of variables B and C); whereas, according to figure 5, it would *not* be.

To further assess how attitude toward the candidate in the second interview (i.e., variable D) is affected by the other three variables in table 1 (i.e., variables A, B, and C), we return now to model H_{20} in table 5, and also to models H_{28}, H_{29}, and H_{30}. As we noted earlier (see equation [19]), model H_{20} states that the odds pertaining to variable D are affected by variables B and C but not A. Similarly, model H_{28} states that these odds are affected by variables A and B but not C; model H_{29} states that these odds are affected by variables A and C but not B; and model H_{30} states that these odds are affected by variables A, B, and C. Comparison of $X^2(H_{30})$ with $X^2(H_{20})$, $X^2(H_{28})$, and $X^2(H_{29})$ indicates that variable A (party preferences at the first interview) does *not* have a statistically significant effect,[81] but both variables B and C (attitude toward candidate at the first interview and party preferences at the second interview) do have significant effects. Again we note that model H_{20}, which ignored the possible effect of variable A, fit the data very well indeed.

Next we consider the earlier analysis of table 2 by Coleman (1964). His model for this table is quite different from those presented in the present article. As was the case with Boudon, Coleman's methods did not provide maximum-likelihood estimates or estimates that were asymptotically equivalent to maximum-likelihood estimates. Thus, the comment earlier in this section pertaining to the use of the χ^2 statistics (7) and (8) will apply to Coleman's work too (see n. 79).

Coleman did present a table of expected frequencies estimated under his model, and a comparison of that table with the corresponding tables obtained under the models considered in the present article for the analysis of table 2 (models H_5 and H_7 in table 11) indicates that his model did not fit the data as well. Indeed, since his model had 4 df (he set four "transition rates" equal to zero) whereas our models had 6 df, his model

[81] The first sentence of n. 68 should be kept in mind when interpreting this result and the remarks in the second paragraph preceding this one. (Of course, this sentence also applies more generally whenever statistical significance is being assessed.) In addition to the assessment of statistical significance, we can compare the magnitudes of the estimated effects (β^{AD}, β^{BD}, β^{CD}) in the saturated model (see table 9), or in model H^*_{20} (see fig. 5), or in model H_{30}.

was both less parsimonious (in a certain sense) and also less congruent with the data.

With respect to the interpretation of table 2, Coleman states that "the strongest effects relative to the random shocks are the effect of a positive attitude toward the crowd's behavior on locating oneself within the crowd . . . , and the effect of a negative attitude in moving out of the crowd." In contrast to this, we see, for example, from our figure 7, that the *weakest* effect upon membership in the crowd at the second interview is the effect of attitude toward the crowd's behavior at the first interview; the next weakest effect (upon membership at the second interview) is the synchronous effect of attitude toward the crowd's behavior at the second interview; and the strongest effect (upon membership at the second interview) is the effect of membership in the crowd at the first interview. Indeed, in contrast to Coleman's statement quoted at the beginning of this paragraph, we find that the *weakest* effect among the various effects included in our model (fig. 7) is the effect of attitude toward the crowd's behavior upon membership in the crowd in the following period; and the *strongest* effect (among the various effects in our model) is the effect of membership in the crowd upon membership in the following period. (For somewhat different interpretations of table 2, which are also consistent with these data, see our figs. 9 and 10.)

With respect to the analysis of table 3, we have already noted in section 4 that figure 11 represents in more quantitative terms the verbal thesis proposed by Wilner et al. (1955) and Davis (1970). In section 4 we also introduced alternative theses represented by figures 12–14. Further comparisons among these theses can be made by the interested reader.

6.2. The Usual Models of Causal Analysis and Multiple Regression

The usual models of causal analysis for quantitative variables express a set of quantitative variables (say, variables Y_1, Y_2, . . .) as linear functions of the variables that affect them, plus certain disturbance terms. The models assume homogeneity; that is, that the variance of the disturbance term pertaining to variable Y_i is a constant which is independent of the particular values of the variables that affect Y_i (for $i = 1, 2, \ldots$). These models also assume that variable Y_i is not restricted to a specified range (for example, the range between 0 and 1). Both the assumption of homoscedasticity and the assumption that the range is restricted are violated when dealing with dichotomous or polytomous variables. On the other hand, the models and methods proposed in the present article do not require such assumptions. These models and methods were designed specifically to take into account the special properties of dichotomous and polytomous variables.

In the various systems of equations introduced in the present article (e.g., [30]–[33]), the equation for a given log-odds (say, equation [30] for $\Phi^{\bar{D}}{}_{ijk\cdot}$) was expressed in a form that is more reminiscent of the analysis of variance than of multiple-regression analysis. On the other hand, each of these equations can be expressed in an equivalent multiple-regression form, using dummy variables: Let us introduce a dummy variable X_1 that takes on the value 1 when variable A is at level 1, and 0 otherwise; and similarly let us introduce dummy variables X_2 and X_3 corresponding to variables B and C, respectively. Then equation (30) can be rewritten as

$$\Phi^{\bar{D}}{}_{ijk\cdot} = \beta^* + \beta^*{}_1 X_1 + \beta^*{}_2 X_2 + \beta^*{}_3 X_3\,, \tag{56}$$

where

$$\begin{gathered} \beta^* = \beta^{\bar{D}} - \beta^{A\bar{D}} - \beta^{B\bar{D}} - \beta^{C\bar{D}}, \\ \beta^*{}_1 = 2\beta^{A\bar{D}}, \quad \beta^*{}_2 = 2\beta^{B\bar{D}}, \quad \beta^*{}_3 = 2\beta^{C\bar{D}}, \end{gathered} \tag{57}$$

and where $\beta^{A\bar{D}} = 0$.

Formula (56) expresses the $\Phi^{\bar{D}}{}_{ijk\cdot}$ (i.e., the logarithm of the expected odds $\Omega^{\bar{D}}{}_{ijk\cdot}$) as a linear function of the dummy variables. We can also rewrite (56), introducing a "disturbance" term as in regression analysis:[82]

$$\phi^{\bar{D}}{}_{ijk\cdot} = \beta^* + \beta^*{}_1 X_1 + \beta^*{}_2 X_2 + \beta^*{}_3 X_3 + \epsilon, \tag{58}$$

where $\phi^{\bar{D}}{}_{ijk\cdot}$ is the logarithm of the observed odds $\omega^{\bar{D}}{}_{ijk\cdot}$, and where

$$\epsilon = \phi^{\bar{D}}{}_{ijk\cdot} - \Phi^{\bar{D}}{}_{ijk\cdot}\,. \tag{59}$$

Formulae (56) and (58) provide expressions that are equivalent to equation (30) for $\Phi^{\bar{D}}{}_{ijk\cdot}$; and the equations for the other Φ's in this article can be expressed in similar terms.

6.3. Comparison of Multiplicative-Effects Models with Various Kinds of Additive-Effects Models

The equation for a given log-odds (say, equation [30] for $\Phi^{\bar{D}}{}_{ijk\cdot}$) was also equivalent to an equation for the corresponding expected odds (equation [20] for $\Omega^{\bar{D}}{}_{ijk\cdot}$). The former equation was additive in form, whereas

82 In contrast to the usual regression analysis, the variance of the disturbance term ϵ in (58) will not be independent of $\Phi^{\bar{D}}{}_{ijk\cdot}$. This disturbance term is, in part, a consequence of the "chance mechanism" that determines, for each individual, his actual response with respect to variable D. This chance mechanism will have a variance that depends upon the expected value of the individual's response with respect to variable D, given his response on the other variables; but the variance of the corresponding chance mechanism in the usual regression analysis does not depend upon this expected value.

the latter was multiplicative. The odds $\Omega^{\bar{D}}_{ijk\cdot}$ were defined in terms of the expected frequencies (see [14]); but they could also have been defined in terms of the probability $P^{\bar{D}}_{ijk1}$ that an individual will be at level 1 with respect to variable D, given that he was at level (i, j, k) with respect to the joint variable (A, B, C). From (14) we see that

$$\Omega^{\bar{D}}_{ijk\cdot} = P^{\bar{D}}_{ijk1}/(1 - P^{\bar{D}}_{ijk1}); \tag{60}$$

and from (60) we see that

$$P^{\bar{D}}_{ijk1} = \Omega^{\bar{D}}_{ijk\cdot}/(1 + \Omega^{\bar{D}}_{ijk\cdot}). \tag{61}$$

By replacing the $\Omega^{\bar{D}}_{ijk\cdot}$ in (61) by the expression for the $\Omega^{\bar{D}}_{ijk\cdot}$ in (20), we obtain a model for the $P^{\bar{D}}_{ijk1}$ expressed in terms of the γ's in (20). This model for the $P^{\bar{D}}_{ijk1}$ is equivalent to the multiplicative-effects model (20) for the $\Omega^{\bar{D}}_{ijk\cdot}$ and the additive-effects model (30) for the $\Phi^{\bar{D}}_{ijk\cdot}$.

We shall now compare this model with a different model for the $P^{\bar{D}}_{ijk1}$: the model obtained by replacing the $\Phi^{\bar{D}}_{ijk\cdot}$ on the left side of (30) by $P^{\bar{D}}_{ijk1}$. In other words, we shall compare a model for the $P^{\bar{D}}_{ijk1}$ that is equivalent to model (30), which expressed $\Phi^{\bar{D}}_{ijk\cdot}$ in terms of additive effects, with a different model for the $P^{\bar{D}}_{ijk1}$, which expresses $P^{\bar{D}}_{ijk1}$ in terms of additive effects. Since $\Phi^{\bar{D}}_{ijk\cdot}$ and $P^{\bar{D}}_{ijk1}$ are quite different, these two models are different. (As we have noted, the former model is equivalent to the multiplicative-effects model [20] for the expected odds $\Omega^{\bar{D}}_{ijk\cdot}$ and the additive-effects model [30] for the log-odds $\Phi^{\bar{D}}_{ijk\cdot}$, whereas the latter model is an additive-effects model for the corresponding probability $P^{\bar{D}}_{ijk1}$.) On the other hand, from table 17 we find that the relation between the probability $P^{\bar{D}}_{ijk1}$ and the corresponding log-odds $\Phi^{\bar{D}}_{ijk\cdot}$ is approximately linear for values of $P^{\bar{D}}_{ijk1}$ in the range .25–.75, and even in a

TABLE 17

RELATIONSHIP BETWEEN PROBABILITY P AND CORRESPONDING LOG-ODDS Φ, WHERE $\Phi = \text{Log}\,[P/(1 - P)]$

P	Log-Odds Φ	Difference between Successive Log-Odds
.50	0.000	
		0.20
.55	0.201	
		0.20
.60	0.405	
		0.21
.65	0.619	
		0.23
.70	0.847	
		0.25
.75	1.099	
		0.29
.80	1.386	
		0.35
.85	1.735	
		0.46
.90	2.197	
		0.75
.95	2.944	

NOTE.—For values of P that are less than .5, the corresponding Φ value is equal to the negative of the Φ value corresponding to $1 - P$. For example, for $P = .45$, the corresponding Φ is -0.201.

somewhat wider range.[83] Thus, in cases where the $P^{\bar{D}}{}_{ijk1}$ are in this range, if an additive-effects model for the $\Phi^{\bar{D}}{}_{ijk.}$ fits the data then we might reasonably expect that a corresponding additive-effects model for the $P^{\bar{D}}{}_{ijk1}$ would also fit these data approximately. The above remark concerning the relationship between additive-effects models for the $\Phi^{\bar{D}}{}_{ijk.}$ and $P^{\bar{D}}{}_{ijk1}$ will also apply when comparing a system of equations for the Φ's (e.g., [30]–[33]) with the corresponding system for the *P*'s.

REFERENCES

Birch, M. W. 1963. "Maximum Likelihood in Three-Way Contingency Tables." *Journal of the Royal Statistical Society,* ser. B 25:220–33.

Bishop, Y. M. M. 1969. "Full Contingency Tables, Logits, and Split Contingency Tables." *Biometrics* 25:383–400.

Boudon, R. 1968. "A New Look at Correlation Analysis." In *Methodology in Social Research,* edited by H. M. Blalock, Jr., and A. Blalock. New York: McGraw-Hill.

Campbell, D. T. 1963. "From Description to Experimentation: Interpreting Trends as Quasi-Experiments." In *Problems in Measuring Change,* edited by C. W. Harris. Madison: University of Wisconsin Press.

Coleman, J. S. 1964. *Introduction to Mathematical Sociology.* New York: Free Press.

Cox, D. R. 1970. *Analyzing Binary Data.* London: Methuen.

Davis, J. A. 1971. *Elementary Survey Analysis.* Englewood Cliffs, N.J.: Prentice-Hall.

Duncan, O. D. 1966. "Path Analysis: Sociological Examples." *American Journal of Sociology* 72:1–16.

———. 1969. "Some Linear Models for Two-Wave, Two Variable Panel Analysis." *Psychological Bulletin* 72:177–82.

Duncan, O. D., and D. L. Featherman. 1972. "Psychological and Cultural Factors in the Process of Occupational Achievement." In *Structural Equation Models in the Social Sciences: Proceedings of a Conference,* edited by A. S. Goldberger and O. D. Duncan. New York: Seminar, forthcoming.

Goodman, L. A. 1968 "The Analysis of Cross-classified Data: Independence, Quasi-Independence, and Interactions in Contingency Tables with or without Missing Entries." *Journal of the American Statistical Association* 63:1091–1131.

———. 1969. "On Partitioning χ^2 and Detecting Partial Association in Three-Way Contingency Tables." *Journal of the Royal Statistical Society,* ser. B 31:486–98.

———. 1970. "The Multivariate Analysis of Qualitative Data: Interactions among Multiple Classifications." *Journal of the American Statistical Association* 65:225–56.

———. 1971*a*. "The Analysis of Multidimensional Contingency Tables: Stepwise Procedures and Direct Estimation Methods for Building Models for Multiple Classifications." *Technometrics* 13:33–61.

———. 1971*b*. "Partitioning of Chi-Square, Analysis of Marginal Contingency Tables, and Estimation of Expected Frequencies in Multidimensional Contingency Tables." *Journal of the American Statistical Association* 66:339–344.

———. 1972*a*. "A Modified Multiple Regression Approach to the Analysis of Dichotomous Variables." *American Sociological Review* 37:28–46.

———. 1972*b*. "A General Model for the Analysis of Surveys." *American Journal of Sociology* 77:1035–86.

[83] From table 17, we see that the Φ values can be approximated by $\Phi \doteq 4\,(P - \frac{1}{2})$, when *P* is near ½. This approximation will give numerical values that are in error by less than 10% for values of *P* in the range .25–.75; and if the slope of 4 in the above formula is changed to 4.2, the modified approximation will give numerical values that are in error by less than 10% for values of *P* in the range .20–.80 (for some related results, see Cox 1970).

———. 1973*a*. "The Analysis of Multidimensional Contingency Tables When Some Variables Are Posterior to Others: A Modified Path Analysis Approach." *Biometrika* 60 (in press).
———. 1973*b*. "Guided and Unguided Methods for the Analysis of a Set of *T* Multidimensional Contingency Tables." *Journal of the American Statistical Association*, vol. 68 (in press).
———. 1973*c*. "The Analysis of Systems of Qualitative Variables When Some of the Variables Are Unobservable: A Modified Latent Structure Approach," forthcoming.
Ku, H. H., and S. Kullback. 1968. "Interaction in Multidimensional Contingency Tables: An Information Theoretic Approach." *Journal of Research of the National Bureau of Standards* 72B:159–99.
Lazarsfeld, P. F. 1948. "The Use of Panels in Social Research." *Proceedings of the American Philosophical Society* 92:405–10.
———. 1970. "A Memoir in Honor of Professor Wold." In *Scientists at Work*, edited by T. Dalenius, G. Karlsson, S. Malmquist. Uppsala: Almquist & Wiksells.
Lazarsfeld, P. F., B. Berelson, and H. Gaudet. 1968. *The People's Choice*. 3d ed. New York: Columbia University Press.
Lipset, S. M., P. F. Lazarsfeld, A. H. Barton, and J. Linz. 1954. "The Psychology of Voting: An Analysis of Political Behavior." In *Handbook of Social Psychology*, edited by G. Lindzey. Vol. 2. Reading, Mass.: Addison-Wesley.
Wilner, D. M., R. P. Walkley, and S. W. Cook. 1955. *Human Relations in Interracial Housing: A Study of the Contact Hypothesis*. Minneapolis: University of Minnesota Press.
Wright, S. 1954. "The Interpretation of Multivariate Systems." In *Statistics and Mathematics in Biology*, edited by O. Kempthorne, T. A. Bancroft, J. W. Gowen, J. L. Lush. Ames: Iowa State College Press.

Leo A. Goodman

Reprinted from: THE AMERICAN JOURNAL OF SOCIOLOGY, Vol. 78, No. 5, March 1973.

Part Three

Davis on Goodman's Approach

Taking the analogy of the logit model with the usual regression model one step further, Davis describes the hypothesis-testing process for hierarchical log-linear models using the familiar terms *correlations* and *partial correlations.* He suggests some relatively simple methods for structuring and interpreting higher-order interaction terms.

Association or correlation between two variables within the log-linear framework is measured by ratios of odds. A single odds ratio suffices for dichotomies; more than one odds ratio is necessary when either variable is polytomous. To illustrate their use Davis examines the relationship between *(R)* region of origin (North or South), *(J)* belief in employment equality for blacks and whites (yes or no), and *(E)* education level (grade school, high school, or college).

Some readers may prefer to begin the book with this nontechnical exposition before tackling the other articles. The relatively few equations in this section are clearly stated and carefully illustrated with well-chosen examples.

Chapter 7

Hierarchical Models for Significance Tests in Multivariate Contingency Tables: An Exegesis of Goodman's Recent Papers

James A. Davis

I wish to thank Stephen Fienberg, Leo Goodman, Avery Guest, and Howard Schuman for their comments, corrections, and criticisms. I am also most grateful to Edmund D. Meyers, Jr., and the Kiewit Computation Center of Dartmouth College for generous assistance.

In a flurry of recent papers (1970, 1971, 1972a, 1972b, 1972c), Leo Goodman presents and elaborates a system for the analysis of contingency tables (cross-classifications of nominal and ordinal variables) that promises to be extremely useful to sociologists. The papers are however, extremely compressed, heavily salted with the cumbersome notation of contingency analysis, and are not easily accessible to the student and average research worker. The aim of this chapter is to explain the logic and procedures of the system in terms the reader (and writer) may find more comfortable.

The Goodman system consists of two parts: a scheme for making significance tests by means of "hierarchical models" and an extensive discussion of a set of techniques known as "log linear models." The two parts are logically and practically distinct. I shall treat only the former.

Among the important uses of hierarchical models are the following: tests for the significance of partial correlations; tests for interactions (specifications) where the control variable has as many categories as one pleases; tests for higher-order (three or more variable) interactions; succinct statements of what is and what is not going on in a contingency table. Since none of these tools has been easily available to the average sociologist, the Goodman system is well worth learning—especially because it provides considerable insight into the properties of cross-classifications and the logic of significance tests.

My explanation of the Goodman system is nontechnical but it does assume some exposure to contingency table analysis. In particular the reader should be familiar with the following concepts: marginals, expected cell values, statistical independence, the chi square test for differences between expected and observed frequencies, and degrees of freedom (see Davis, 1971, Chap. 3, or any elementary statistics textbook).

BACKGROUND

To understand how to use the Goodman system, the reader should be familiar with three concepts: *odds ratios*, *effects*, and *models*.

Odds Ratios

An odds ratio is the ratio of frequencies for two categories of some variable. The concept, which may be unfamiliar, is extremely useful in defining the models that are the heart of Goodman's system, and it also gives us a common language for talking about a variety of statistical properties, such as skewed distributions, partial associations, specifications, and interactions.

Consider the hypothetical fourfold table shown in Table 1. Looking at the column totals, we observe that 30 cases are Yes on variable B and 70 cases are No. The odds are 30 to 70 that a case in Table 1 is a Yes on variable B and the odds are 70 to 30 that a case in Table 1 is a No on variable B. Taking the ratio $30{:}70 = 0.429$, we can say that the

TABLE 1
Hypothetical Fourfold Table

		Variable *B* Yes	No	Total
	High	15	35	50
Variable *A*	Low	15	35	50
	Total	30	70	100

odds ratio for Yes versus No on variable *B* is 0.429. We can symbolize such calculations (although this is not Goodman's notation) as follows:

$$\begin{pmatrix}\text{Yes}\\ B\\ \text{No}\end{pmatrix} = \text{Odds ratio for } B = 30{:}70 = 0.429$$

$$\begin{pmatrix}\text{High}\\ A\\ \text{Low}\end{pmatrix} = \text{Odds ratio for } A = 50{:}50 = 1.000$$

The category frequency appearing above the letter is divided by the category frequency below the letter. Which category frequency we choose to put above and which we choose to put below is perfectly arbitrary. We can decide to say that

$$\begin{pmatrix}\text{Yes}\\ B\\ \text{No}\end{pmatrix} = 0.429$$

or to say that

$$\begin{pmatrix}\text{No}\\ B\\ \text{Yes}\end{pmatrix} = 2.333$$

but the decision is irrevocable in a particular analysis.

The odds ratio is 1.000 for a situation in which the two categories are equal (such as variable *A* in Table 1) regardless of which ratio we calculate. Its minimum is zero (for example, 0:70 = 0) but it has no defined upper limit. At the extreme, where a frequency is divided by zero (say, 70:0), the odds ratio has no specific numerical value. This property makes it necessary to make certain adjustments in the data when one is calculating Goodman's measures of association, but it has no importance for the significance tests discussed in this chapter.

The concept of odds ratios can be extended to conditional odds ratios and then to relative odds ratios. The odds ratio

$$\begin{pmatrix} \text{Yes} \\ B \\ \text{No} \end{pmatrix}$$

was based on all the cases in Table 1, that is, on the marginals for item B. One can also calculate these ratios for particular categories of another variable; for example, calculating

$$\begin{pmatrix} \text{Yes} \\ B \\ \text{No} \end{pmatrix}$$

when A is high, using the frequencies in the top row of Table 1. We call such results *conditional odds ratios* and use the conventional vertical line to indicate a condition. Thus

$$\left[\begin{pmatrix} \text{Yes} \\ B \\ \text{No} \end{pmatrix} \mid A = \text{high}\right] = 15{:}35 = 0.429$$

in Table 1. Conditional odds ratios lead directly to the notion of independence and association. Many of us would evaluate independence in Table 1 by means of the *cross-product difference* (Davis, 1971, pp. 34–50). Thus $(35 * 15 - 35 * 15) = (525 - 525) = 0$. Since equality of the cross-products is a definition of independence, we would conclude that Table 1 shows independence for A and B. Exactly the same destination can be reached through conditional odds ratios and the notion of relative odds ratios.

TABLE 2
Cell Frequencies in Fourfold Table

		y −	y +
x	+	a	b
x	−	c	d

Table 2 represents a diagram of cell frequencies in a fourfold table. Designating the cell frequencies in a fourfold table as in Table 2, we call $(b/d)/(a/c)$, the ratio of two conditional odds ratios, a *relative* odds ratio:

$$\underset{-}{\overset{+}{X}}\ \underset{-}{\overset{+}{Y}} = \frac{\underset{-}{\overset{+}{X}} \mid Y = +}{\underset{-}{\overset{+}{X}} \mid Y = -} = \frac{b/d}{a/c} \qquad (1)$$

From high school algebra:

$$\frac{b/d}{a/c} = \frac{bc}{ad} = \frac{b/a}{d/c} = \frac{\overset{+}{X}}{\underset{-}{}}\frac{\overset{+}{Y}}{\underset{-}{}} = \frac{\overset{+}{Y}}{\underset{-}{}}\frac{\overset{+}{X}}{\underset{-}{}} \tag{2}$$

Equation (2) tells us two things. First it says that the ratio of the cross-products (not their difference) is identical in value to the ratio of the two conditional odds ratios. In particular, whenever the cross-products are equal in magnitude, their ratio, the relative odds ratio, must equal 1.000. Second it says that one gets the same number whether one takes the ratio of the two X conditionals or the ratio of the two Y conditionals. The relative odds ratio is symmetrical. (But remember, one does not get the same number if one reverses plus and minus for one of the variables. The effect of reversing the signs for one of the items is to change the odds ratio to its reciprocal.)

Thus there are two new definitions of lack of independence in a fourfold table: the conditional odds ratios are not equal; the relative odds ratio does not equal 1.000. The identical concepts apply to larger tables but the number of possibilities increases rapidly. Taking the odds ratio for B in Table 3, we see that there are six ratios we can examine (ignoring which letter is at the top and which is at the bottom):

$$\begin{pmatrix} i \\ B \\ j \end{pmatrix}\begin{pmatrix} i \\ B \\ k \end{pmatrix}\begin{pmatrix} i \\ B \\ l \end{pmatrix}\begin{pmatrix} j \\ B \\ k \end{pmatrix}\begin{pmatrix} j \\ B \\ l \end{pmatrix}\begin{pmatrix} k \\ B \\ l \end{pmatrix}$$

one for each pair of categories. And each has a relative odds ratio for each possible pair of categories on the other variable:

High	High	Medium
A	A	A
Medium	Low	Low

TABLE 3
Hypothetical 3 × 4 Table

		B			
		i	j	k	l
	High				
A	Medium				
	Low				

Since a variable with K categories has $(K)(K-1)/2$ pairs of categories, we can put it more generally. For an $R \times C$ table (where R

is rows and C is columns):

$$\left(\frac{(R)(R-1)}{2}\cdot\frac{(C)(C-1)}{2}\right)=\left(\frac{(R)(C)(R-1)(C-1)}{4}\right) \tag{3}$$

$$= \text{Number of relative odds ratios}$$

Table 3 has $(4)(3)(3)(2)/4 = 18$ different (though not mathematically independent) relative odds ratios. A fourfold table has $(2)(2)(1)(1)/4 = 1$. Equation (3) and its discussion leads to a generalization of the second of our new definitions of lack of independence: an $R \times C$ table lacks independence if any relative odds ratios do not equal 1.000.

This concept can also be extended to tables with more than two variables, which leads us to "relative relative" or higher-order odds ratios. To avoid cumbersome phrases like "super" or "relative relative," we define the order of an odds ratio as the number of variables involved in its calculation. $\underset{-}{\overset{+}{A}}$ is a first-order odds ratio, $\underset{--}{\overset{++}{AB}}$ is a second-order odds ratio, $\underset{---}{\overset{+++}{ABC}}$ is a third-order odds ratio, and so on. The phrase "in its calculation" implies that conditional variables are not counted in assessing order. $\underset{--}{\overset{++}{BC}}|A = +$ is a second-order conditional odds ratio.

Let us consider the third-order odds ratio $\underset{---}{\overset{+++}{ABC}}$ in this hypothetical table (Table 4). We may think of it as calculating the second-order odds ratio (cross-product ratio) for A and B within each category of C and then finding the ratio of these two conditional results. In algebra:

$$\frac{\underset{-}{\overset{+}{A}}\ \underset{-}{\overset{+}{B}}|C=+}{\underset{-}{\overset{+}{A}}\ \underset{-}{\overset{+}{B}}|C=-} = \frac{fg/eh}{bc/ad} = \frac{fgad}{ehbc} = \underset{-}{\overset{+}{A}}\ \underset{-}{\overset{+}{B}}\ \underset{-}{\overset{+}{C}} \tag{4}$$

TABLE 4
Diagram of Third-Order Odds Ratio

$C = -$

		B −	B +
A	+	a	b
	−	c	d

$C = +$

		B −	B +
A	+	e	f
	−	g	h

The third-order and all higher-order results are symmetrical in the sense that one gets the same answer whatever order one picks for the ratios. Thus $\overset{+++}{\underset{---}{BCA}}$ implies that we calculate the $\overset{++}{\underset{--}{BC}}$ odds ratios within levels of A and find their ratio as in Equation (5).

$$\frac{\overset{+}{\underset{-}{B}}\ \overset{+}{\underset{-}{C}} \mid A = +}{\overset{+}{\underset{-}{B}}\ \overset{+}{\underset{-}{C}} \mid A = -} = \frac{fa/be}{ch/dg} = \frac{fadg}{bech} = \frac{fgad}{ehbc} \tag{5}$$

A third-order odds ratio of 1.000 means that all the two-variable associations are identical in magnitude across categories of the third item. (It does not imply that all three—AB, AC, BC—have the same strength or that any of them depart from independence.)

A third-order odds ratio other than 1.000 means that for each of the three associations (AB, AC, BC) the magnitudes in the two conditions differ. It also implies that at least one of the two conditional tables for each association shows lack of independence. If both conditional tables showed independence, the two conditional odds ratios would be 1.000 and so would the third-order coefficient.

When interpreting third-order odds ratios other than 1.000, we usually view the result as a *specification* or an *interaction* (Davis, 1971, pp. 99–103), saying, for example, that the degree of correlation between A and B varies, depending on the category of C. This is a useful way to wrench some sense from the data, but it is important to remember that higher-order effects are interchangeable. When the third-order odds ratio for ABC differs from 1.000, it is equally correct to say that the correlation between A and B varies, depending on the category of C; or that the correlation between A and C varies, depending on the category of B; or that the correlation between B and C varies, depending on the category of A.

A specific example may help clarify the concept of higher-order odds ratios or interactions. Table 5 gives multivariate data on employed persons from the 1960 U.S. Census rounded to the nearest thousand (figures are from Davis, 1971, p. 157). All variables are dichotomies: A is 1959 earnings, dichotomized as $+$ = \$4000 or more, $-$ = less than \$4000; B is occupation, with $+$ = white collar and $-$ = blue collar and farm; C is education, with $+$ = high school graduate and $-$ = less than high school graduate; and D is sex, with $+$ = male and $-$ = female.

Column I lists eight conditional first-order odds ratios for A, the high-income frequencies divided by low-income frequencies within the various combinations of the other three variables.

TABLE 5
Sex, Education, Occupation, and 1959 Earnings

Frequencies					Odds Ratios			
A		*B*	*C*	*D*	I	II	III	IV
Earnings		Occupation	Education	Sex	$\overset{+}{\underset{-}{A}} \mid B,C,D$	$\overset{+\,+}{\underset{-\,-}{AB}} \mid C,D$	$\overset{+\,+\,+}{\underset{-\,-\,-}{ABC}} \mid D$	$\overset{+\,+\,+\,+}{\underset{-\,-\,-\,-}{ABCD}}$
3,451	7,294	+	+	+	2.114	2.321	1.235	1.043
4,188	3,817	−	+	+	0.911			
2,397	1,986	+	−	+	0.829	1.880		
13,303	5,871	−	−	+	0.441			
6,982	1,150	+	+	−	0.165	4.342	1.184	
2,005	76	−	+	−	0.038			
2,538	168	+	−	−	0.066	3.667		
6,063	107	−	−	−	0.018			

Column II lists four conditional second-order odds ratios, $\overset{++}{\underset{--}{AB}}|C,D$. In each the result in Column I for plus on occupation was divided by the result for minus on occupation. All four exceed 1.000. This is equivalent to saying that there is a nonzero partial correlation between occupation and income, controlling for education and sex. If there were no partial association, the second-order odds ratios would all be 1.000.

Column III compares the second-order results for plus on education with those for minus on education. It asks whether the occupation-income relationship is the same within levels of education. Because both third-order odds ratios exceed 1.000, we infer that the association between occupation and income is stronger among the better-educated. There is a specification or interaction.

Column IV is the fourth-order odds ratio. Its value, 1.043, is quite close to 1.000, and without a significance test one would not wish to draw a firm conclusion (more on this later). Assuming for the sake of discussion only that it is greater than 1.000, it could be described as a specification of a specification or as an interaction of an interaction. One English translation would be: "The tendency for high education to accentuate the occupation-income correlation is greater among men."

The concept of higher-order interactions can be extended indefinitely, although plausible substantive interpretations seem to decrease geometrically with the order of the effect. We can summarize our iscussion and illustrate the utility of the odds ratio concept by the following statements:

Odds Ratio Statement	Statistical Property
1. Odds ratio	1. Ratio of two category frequencies for some variable
2. The first-order (nonconditional) odds ratio $\neq$ 1.000	2. The marginal frequencies for the two categories are not identical
3. The second-order odds ratio $\neq$ 1.000	3. The two items are associated (not independent) at the zero-order level
4. The second-order conditional odds ratio $\neq$ 1.000	4. The two items have a nonzero partial association
5. The third-order odds ratio $\neq$ 1.000	5. The three items show an interaction or specification
6. The fourth-order odds ratio $\neq$ 1.000	6. The four items show an interaction (the third-order interaction is specified)

Effects

The second concept in Goodman's system discussed here is *effects*, another term that may be unfamiliar. It is a point of view rather than a statistical property but it helps glue the system together. In a manner of speaking, the analyst working with the Goodman system views the cell frequencies in his cross-tabulation as a dependent variable. His data analysis is an attempt to account for variation in cell frequencies. The broad concept is a familiar one. We are all used to the idea that in multiple regression the analyst attempts to account for variation in scores on a dependent variable through scores on predictor variables. But the analogy is rough and can be misleading. In Goodman analysis we are not accounting for scores on a particular variable but for joint frequencies on a number of variables: our predictors are not scores on a variable but statistical properties involving sets of variables. The specific statistical properties are defined precisely later in this chapter, but for an introduction we can sketch them as follows.

A *single*-variable effect is a difference in cell frequencies that reflects the marginal distributions for one or more items. For example, in a cross-tabulation involving race the cells associated with race = black may be systematically smaller because there are many fewer blacks than whites in the total sample.

A *two*-variable effect is a difference in cell frequencies that reflects an association between a pair of variables. For example, in a cross-tabulation involving occupation and income, cells for the combination of high occupation and high income may be relatively larger because occupation is correlated with income.

A *three*-variable effect is a difference in cell frequencies that reflects an interaction or specification of three variables. For example, in a cross-tabulation involving race, region, and attitudes toward civil rights, cells for the combination of white, Southern, and anti-civil rights may be especially large if it is the case that race differences in attitudes toward civil rights are stronger in the South.

Four-, five-, . . ., N-variable effects can be described in a similar fashion. (The reader has undoubtedly noticed a similarity between the kinds of effects listed and the kinds of odds ratios defined above. They are really the same thing. One uses odds ratios to talk concretely about various kinds of effects. *Effect* is the abstract concept and *odds ratios* are its concrete language.)

Readers familiar with analysis of variance will see an obvious analogy and may find it comfortable to think of the Goodman procedure as akin to carrying out analysis of variance on cell frequencies. There are some potentially confusing semantic problems with this analogy also. In analysis of variance, the term *interaction* applies to effects involving two or more variables. In the statistical tradition of contingency table analysis, two-variable effects are called *associations* or *correlations* and the term *interaction* is reserved for higher-order effects. We shall take the latter position and speak of *single-variable or marginal effects, two-variable or association effects,* and *three or more variable or interaction effects.*

Models

The final concept necessary to understand the Goodman system is a more familiar word, *models.* The central operation in the system is constructing and testing models—sets of semiartificial data created by the analyst and having some statistical property. The analyst creates a series of models, compares them with actual data, and uses the discrepancies or lack of discrepancies to assess the presence of various statistical effects.

The reader is already familiar with the concept of models, but not the word, from standard tests for association. Although we do not usually think of it this way, the standard test for significant association between two variables is an instance of modeling. We create a set of semiartificial data that has some properties identical to the real data (the same marginals and same grand total) and one big difference—in our model the two variables are forced to be statistically independent. We then compare our model with the real data, and if there is a significant discrepancy—if our model does not fit—we say that there is an association between the variables. Goodman's system extends this idea backward and forward: backward to an even more primitive model and forward to models involving many variables and their interactions.

The three background concepts—odds ratios, effects, and models—all fit together. We can summarize this section as follows: in the Goodman system the analyst attempts to account for the cell frequencies in his cross-tabulation by hypothesizing effects, by building models that embody the presence and absence of various effects through setting the odds ratios to appropriate values, and by comparing the model with the actual data.

SPECIFIC MODELS

This section considers the specific models one can build, how such models are constructed and tested, and the inferences that follow from various outcomes. The broader questions of when an analyst should use various models, what the results mean, and how all of this fits into conventional data analysis are considered in the following section. I shall treat the following kinds of models, classified in terms of the effects at issue: no effects, single-variable effects, two-variable effects, three-variable effects, higher-order effects, and mixed models.

Equal-Probability Model: No Effects

Is there a contingency table so devoid of interest, a concatenation of figures so nonproblematical that it could not even arouse the interest of a statistics graduate student seeking a thesis topic? One never knows until he tries, but one candidate is Table 6. The crucial statistical property in Table 6 is not very subtle. Each of the 24 cells in the table has exactly 149 cases. Is the table interesting? If it turned up as the result of a run on actual data, it would be very interesting—one would immediately ask what went wrong with the computer. But as a sort of Platonic "ideal table," it is the essence of nothingness since there is absolutely nothing going on of any statistical interest. Table 6 is an example of the first and most primitive of Goodman's nested set of models, the model assuming equal probability or no effects.

There is more (or perhaps less) to equal-probability tables than meets the eye when we consider their properties in terms of odds ratios and effects. Thus:

1. All first-order odds ratios are 1.000. If we sum the cell frequencies in Table 6 to obtain the item marginals, we find that all the dichotomies have marginal frequencies of 1788 in each category and that the trichotomous variable has a frequency of 1192 in each of its three categories. Equal sizes of all marginal categories is equivalent to saying that all first-order odds ratios are 1.000.
2. All second-order odds ratios are 1.000. A second-order odds ratio is a fourfold table, and we see that any fourfold table we define within Table 6 has cell frequencies of 149, 149, 149, and

TABLE 6
Equal-Probability $2 \times 3 \times 2 \times 2$ Table ($N = 3576$)

Variable A	Something						Something Else					
Variable *B*	This		That		The Other		This		That		The Other	
Variable *C*	Yes	No	Yes	No	Yes	No	Yes	No	Yes	No	Yes	No
Variable *D*												
High	149	149	149	149	149	149	149	149	149	149	149	149
Low	149	149	149	149	149	149	149	149	149	149	149	149

149. The fourfold table shows independence, which is equivalent to saying that the second-order odds ratio is 1.000.

3. All higher-order odds ratios are 1.000. Since all fourfold tables have odds ratios of 1.000, all second-order odds ratios are 1.000:1.000 or 1.000. And, by a similar argument, it follows that odds ratios of any order are all 1.000.
4. Therefore all odds ratios are 1.000 in an equal-probability table.

Since an effect is defined as an odds ratio that departs from 1.000 (to be discussed later), the equal-probability table has no effects. Both intuition and the properties of odds ratios tell us that in an equal-probability table (no-effect table) absolutely nothing is going on.

To construct an equal-probability model one finds the total number of cells in the table (here 24) and then divides N (the total cases) by that number (3576/24 = 149). More abstractly the cell values in an equal-probability model with variables A, B, C, and so on such that A has a total of A categories, B has a total of B categories, and so on, are

$$\text{Cell value} = N/A * B * C \ldots \qquad (6)$$

The values in an equal-probability model tell us nothing until we compare them with actual data. The figures in Table 6 are in fact the equal-probability model for a particular table adapted from M. A. Schwartz (1967, Table 8, p. 133). Schwartz compares a number of surveys from 1942 to 1963 to trace trends in white attitudes toward Negroes. There is no reason why we cannot treat time (year of study) as a variable—provided that each study contains the same questions—and the Goodman system is especially useful for this purpose. Schwartz's variables are as in Table 7. The raw frequencies in her real-data table

TABLE 7
Schwartz's Variables

Variable	Content	Categories	
Time (T)	Year of study	1946	1963
Education (E)	Respondent's educational attainment	Grade school, high school, college	
Region (R)	Respondent's region	South	Other (North)
Jobs (J)	"Do you think Negroes should have as good a chance as white people to get any kind of job or do you think white people should have the first chance at any kind of job?"	Negroes should have as good a chance	Other

(calculated from the percentages and bases she reports) are presented in Table 8. The actual data seem pretty far from the model. Only a few of the cells are anywhere near the predicted value of 149, some are as high as 300 or 400, and some are as small as 50.

The discrepancies between model and data might be accounted for by chance fluctuations—sampling or measurement error. The well-known chi square test enables us to check this possibility. For each cell we first compare the observed (raw data) and expected (model) values (for example, 52 versus 149 in the upper left-hand corner). We next square the discrepancy ($52 - 149 = -97$; $-97^2 = 9409$), divide the squared discrepancy by the expected (model) value ($9409/149 = 63.15$), and then sum the cell chi square values to give a total chi square for the table. In practice the model data should be run out a decimal or two. For example, 3579/24 is really 149.12. I used whole numbers here for simplicity of explanation only.

As is usual in chi square tests, we need to know the degrees of freedom (DF) or number of independent pieces of information. When 23 cell discrepancies have been computed, we can obtain the 24th by subtraction—since the discrepancies in the table must sum to zero. Only one of the chi square values is redundant. Here, and in general, the degree of freedom for the equal-probability model equals

$$(A * B * C * D \ldots) - 1$$

[see Equation (7)]. In our example $24 - 1 = 23$. In a fourfold table we would get $(2 * 2) - 1 = 3$.

$$\text{DF for variables } A, B, C \ldots = (A * B * C \ldots) - 1 \qquad (7)$$

We obtained a chi square value of 2798.49 for the discrepancy between Table 6 (the model) and Table 8 (the data). The standard chi square table (at the back of any statistics book) tells us that for 23 degrees of freedom, a chi square value of 40.3 would be significant at the 0.01 level. This exceeds the alpha level of 0.01 that I shall use throughout this chapter. We conclude that our model does not fit.

What have we learned? It depends on how the test comes out. If the equal-probability model fits (if the chi square is not significant): there is nothing going on in the table. There are no effects to analyze. We cannot reject the hypothesis that all odds ratios are 1.000. If the equal-probability model does not fit (if the chi square is significant): something is going on in the table. The cell frequencies in the data differ from each other more than chance can account for. However, we do not know *what* is going on. It is not the case that all odds ratios are 1.000, but we cannot point to any specific odds ratios that differ from 1.000.

TABLE 8
Cell Frequencies in Schwartz's Data

	1946						1963					
	Grade		High		College		Grade		High		Other	
	South	Other	South	Other	South	Other	South	Other	South	Other	South	Other
As good	52	311	64	439	42	204	71	148	126	410	55	201
Other	206	365	163	374	51	83	44	61	31	56	14	8

Single-Variable Effects: No Associations

The failure of the equal-probability model tells us that there are some effects in the data. Following the principle of parsimony, we next try to fit the data with the simplest possible alternative. It can be described alternatively as:

1. A model in which first-order odds ratios are allowed to depart from 1.000, but all higher-order odds ratios are set to 1.000
2. A model in which cell frequencies are not necessarily identical, but there are no partial associations or interactions
3. A model in which the cell frequencies are exactly proportional to the marginal frequencies for the items
4. A model that allows single-variable effects only
5. A model that tests the hypothesis that all differences in cell frequencies can be accounted for by the single-variable category distributions of the items

The familiar chi square test for significant association in an $R \times C$ table is an example of a marginal-only model. We do not usually dignify such calculations with the awesome term *model*, but testing for a significant association can be viewed as building a model in which cell frequencies are proportional to the marginal frequencies but no higher-order (two-variable, correlational) effects are permitted.

Less familiar is the extension of this idea to any number of variables, as in our example of the Schwartz data. Table 9 gives us her marginals. How many cases should our model have for a particular cell, say, 1946/high school/South/as good, under the hypothesis that cell frequencies are a function of marginal frequencies? Using a not unfamiliar logic, we multiply odds or probabilities like this:

$$\frac{2354}{3579} * \frac{1663}{3579} * \frac{919}{3579} * \frac{2123}{3579} * 3579 = \frac{7{,}637{,}728{,}445{,}974}{164{,}076{,}654{,}996{,}081} * 3579$$

$$= 0.0466 * 3579 = 166.8 = 0.658 * 465 * 0.257 * 0.593 * 3579$$
$$= 0.0467 * 3579 = 167.1$$

Using the actual frequencies we get an expected value of 166.8 for that cell; using proportions we get 167.1; the difference is due to rounding. When working by hand, the proportional approach keeps the number of figures within practical limits, but with computer programs the frequency approach is more accurate.

The same calculations are made for each cell, giving the following general formula (we use here the conventional summation sign Σ to indicate a sum or total, and use the notation $A_iB_jC_k$ to name the cell

TABLE 9
Single-Variable Marginal Frequencies for Data in Table 8

Variable	Categories	Frequency	Proportion
Time (T)	1963	1225	0.342
	1946	2354	0.658
		3579	1.000
Education (E)	Grade	1258	0.351
	High school	1663	0.465
	College	658	0.184
		3579	1.000
Region (R)	South	919	0.257
	Other	2660	0.743
		3579	1.000
Jobs (J)	As good	2123	0.593
	Other	1456	0.407
		3579	1.000

defined by the ith category of variable A, the jth category of variable B, and so on):

$$\Sigma(A_iB_jC_k\cdots) = \left(\frac{\Sigma A_i}{N} * \frac{\Sigma B_j}{N} * \frac{\Sigma C_k}{N} \cdots\right) * N \tag{8}$$

Canceling the Ns we obtain an alternate version that is easier for calculations:

$$\Sigma(A_iB_jC_k\cdots) = \frac{(A_i * B_j * C_k \cdots)}{N^{v-1}} \tag{9}$$

where N means the number of variables.

As before, we compare each cell in the model with its mate in the actual data, square the discrepancy, divide by the expected (model) value, and sum the terms to obtain a total chi square for the model.

How many degrees of freedom do we use? The answer is simple. Starting with $(A * B * C \ldots) - 1$, we subtract for each variable $K - 1$, where K equals the total number of categories in the variable. Because matters become more complex as we move on to more elaborate models, it is worth the time to consider the reasoning behind this magic number. Consider first a single variable A with three categories I, II, and III. If we were to test the equal-probability model in the previous section, we could make our calculations for any two of the three cells in the distribution and then fill in the observed and expected in the last cell by subtraction, since both the observed and expected cell values must sum to N. It makes no difference which cell we treat as the last or redundant cell, but call it III and diagram the situation in Figure 1, where numbers indicate an empirical result and cross-hatching the redundant cell.

Can we apply the single-variable model to our hypothetical A? Of course not. The cell value *is* the category total ($\Sigma A_i = \Sigma A$) and there cannot be discrepancy between observed and predicted. The degrees-of-freedom rule for single-variable effects gives us the same answer. In Figure 1, $(A * B * C \ldots) - 1$ equals $(A) - 1$. We get $3 - 1 = 2$. Since $K = 3$, $K - 1 = 2$. And there is general agreement that $2 - 2 = 0$. When we apply the single-variable model to a lone item, the degrees of freedom are zero. This is an instance of a general property of the Goodman system. To test a model one must have at least one more variable than the level being tested. A single-variable model requires at least two variables in the data. So we add a second variable B with categories i, j, k, and l (see Figure 2). As before, the expected cases must sum to N as well as the observed cases. Consequently there will again be a redundant cell. I have arbitrarily put it down in the lower right-hand corner.

In moving from the equal-probability model to the single-variable model, I have introduced some new restrictions. Not only must both observed and expected cases sum to N but the expecteds in each category also must sum to the actual marginals. In other words: the model always has the same marginals and total as the data, whereas the previous equal-probability model had the same total but different category marginals (except in the special case where it fit). Figure 3 adds these new restrictions.

Let us arbitrarily begin our analysis in the upper left-hand corner and move from left to right across rows. After calculating chi squares for the cells marked 1 and 2, we find that the next result (B_iA_{III}) can be obtained by subtraction since the three cells are required to sum to ΣB_i. I have shaded that cell to indicate that it is redundant. The same thing

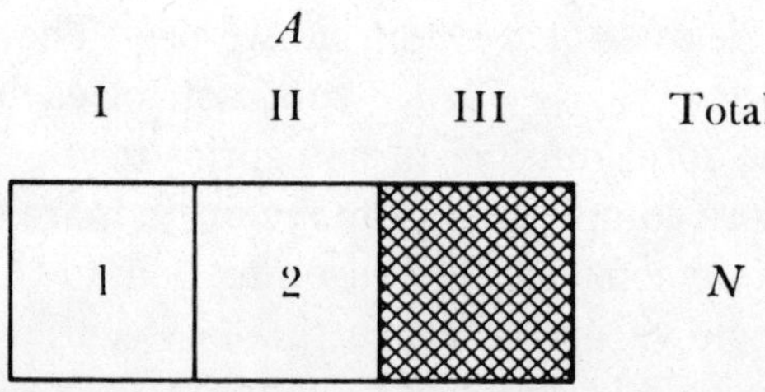

Figure 1. Schematic notion of degrees of freedom in an equal-probability model.

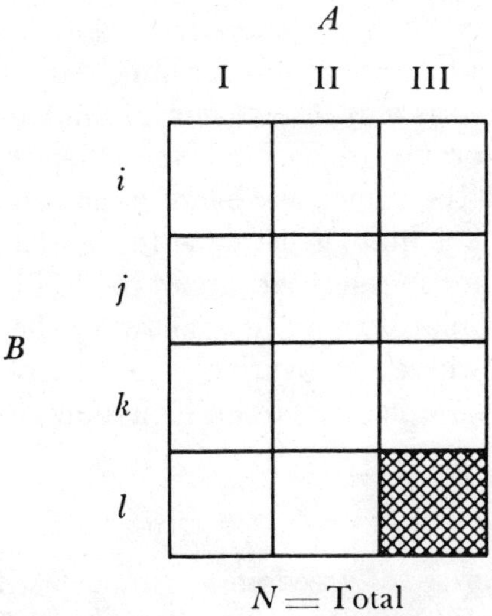

Figure 2. Two-variable model with an arbitrarily chosen redundant cell.

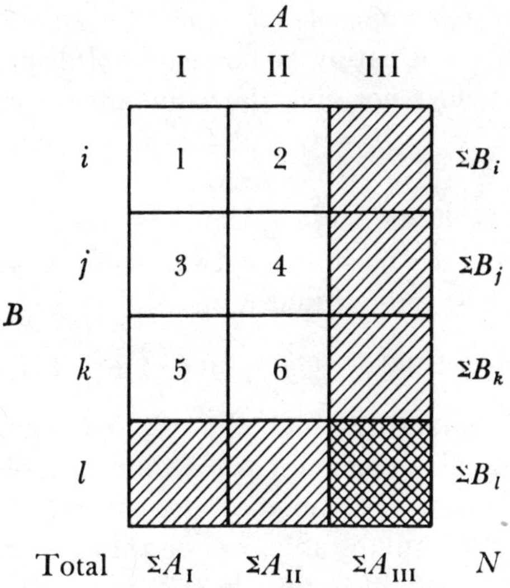

Figure 3. Schematic notion of Degrees of Freedom in a single-variable model.

happens in the second and third rows giving us two more shaded redundant cells in the right-hand column of Figure 3. Moving down to the fourth row we find that the first two cells can also be obtained by subtraction because of the column sum restrictions given by ΣA_{I} and ΣA_{II}.

Finally we reach the lower right-hand corner where we find the redundant cell from the equal-probability model. This cell is triply redundant. We could obtain its results from the row restriction ΣB_l or the column restriction ΣA_{III} or from N. The cell is surely redundant but the single-variable restrictions do not make it more redundant. This is the reason for the minus one in the $K - 1$ formula. Any variable added in a single-variable model adds only $K - 1$ new redundant cells.

In fact the diagrams are illustrative rather than rigorous, since one need not start at the upper left-hand corner and work across the rows. But the principle is quite general, as the reader will see if he tries various schemes for filling in cells that cannot be obtained by subtraction.

Stating the rule more explicitly to allow for the fact that one may wish to create models where some variables are set for equal-probability and others for single-variable, we obtain Equation (10):

$$\begin{aligned} \text{DF} = (A * B * C * D * E * F) - (1) \\ - [(A - 1) + (B - 1) + (C - 1)] \end{aligned} \tag{10}$$

where DF equals degrees of freedom for model with single-variable effects A, B, C and equal-probability predictions D, E, F. For Figure 3 we would get $\text{DF} = (4 * 3) - (1) - (3 + 2) = 6$.

We are all familiar with a different formula for degrees of freedom in a two-variable contingency table $(R - 1)(C - 1)$, and $(4 - 1)(3 - 1)$ does equal 6. But this is a happy coincidence. Multiplication of $K - 1$ beyond two variables does not give the same answer as Equation (10). For example when $A = 3$, $B = 4$, and $C = 5$; $(A - 1)(B - 1)(C - 1)$ equals 24, whereas Equation (10) gives 50.

In the Schwartz data (Table 9) time has two categories, education has three, region has two, and jobs has two. The degrees of freedom for a single-variable model using all four are

$$\text{DF} = (2 * 3 * 2 * 2) - (1) - (1 + 2 + 1 + 1) = 18$$

After constructing a model in which the cell frequencies are given by multiplying the marginal frequencies of the four variables, we compared the model with the data and obtained a total chi square value of 823.53. Looking this up in a chi square table, we see that for 18 degrees of freedom a value of 34.805 is significant at the 0.01 level. The model does not fit.

How shall we interpret the result? If the single-variable model fits (if the chi square is not significant): (1) There are no two-variable or higher-order effects. (2) There may or may not be single-variable effects. If the single-variable model does not fit (if the chi square is significant): (1) There are some two-variable or higher-order effects, but we do not know which ones unless we are analyzing a two-variable table. (2) There may or may not be single-variable effects.

The results of testing a single-variable model are not terribly informative by themselves. Failure to fit, as in the failure of an equal-probability model, may be construed as a hunting license but this time it is a little more specific. The difference between the two failures may be put this way: failure of the equal-probability model indicates that something is going on; failure of the single-variable model indicates that something is going on that cannot be accounted for by the marginals of the variables—therefore it must involve pairs, triplets, or larger combinations of variables.

And, contrary to expectations, testing a single-variable model tells us nothing about the presence or absence of single-variable effects. If the reader refers to the five alternative definitions at the beginning of this section, he will see important qualifiers such as "allowed" in (1), "not necessarily" in (2), and "allows" in (4). The explanation is simple. We build the model from the single-item marginals. If the item has unequal category frequencies (first-order odds ratios differing from 1.000), so will the model. But if the item should have equal category frequencies, equal probability for the item will be built into the model.

The point is a general property of the Goodman system and can be summarized by the following generalization: an X-variable model is one that allows but does not require effects of level X, $X - 1$, $X - 2$, and so on, whereas it forbids effects of level $X + 1$, $X + 2$, and so forth.

Two-Variable Models: No Interactions

If a single-variable model permits single-variable effects but prohibits higher-order differences, the suspicion arises that a two-variable model permits two-variable effects but prohibits all higher-order effects. Correct. A two-variable model can be described alternatively as:

1. A model in which first- and second-order odds ratios are allowed to depart from 1.000 but all higher-order odds ratios are set to 1.000
2. A model in which partial associations between pairs of variables may be present but there are no interactions or higher-level variations in the associations
3. A model that allows only single- and two-variable effects
4. A model that tests the hypothesis that all the differences in cell frequencies can be accounted for by partial correlations constant in magnitude from condition to condition.

How does one construct a two-variable model? The rules are as follows. The data must sum to N. The item marginals must agree with the raw data. The zero-order cross-tabulations for the pairs of variables in question must agree with the raw data. All third-order and higher odds ratios must be 1.000.

It is not obvious how such a model might be formed but there is an algorithm or mathematical strategy for constructing these tables by iteration (successive approximation). The procedure is not very difficult—three or four variables can be done by hand in about 30 minutes once one learns the technique. This task is clearly appropriate for a computer, however, and no one will go broke computing the model, since it has been proved that the iterations must converge at a rapid rate.

The process itself is not illuminating and I have relegated it to the appendix. Instead I wish to give the results of hand calculations for the Schwartz data to show what a three-item, two-variable model looks like. For simplification, in Table 10 I arbitrarily excluded time from the analysis, leaving a three-item, twelve-celled table [(3) * (2) * (2) = 12]. To see whether the modeled data meet the criteria, we first check whether we can reproduce the zero-order correlations for the variables. Taking region and education, for example, we get Table 11.

Although the model and raw data results are very close, if the reader looks back to Table 10 he will see that some cells there differ quite a bit between the raw data and the model. Did we also succeed in meeting the single-variable marginal and *N* restrictions? Of course. If we can reproduce the zero-order cross-tabulation to a desired level of accu-

TABLE 10
Original Data and Two-Variable Model for Schwartz Variables of Region, Education, and Jobs

(a) Raw Data ($N = 3597$)

		Grade School		High School		College	
		South	Other	South	Other	South	Other
Jobs	+	123	459	190	849	97	405
	–	250	426	194	430	65	91

(b) Two-Variable Model

		Grade School		High School		College	
		South	Other	South	Other	South	Other
Jobs	+	122.3	459.8	183.2	856.4	104.5	397.5
	–	250.6	425.4	200.7	422.8	57.5	98.5

TABLE 11
Zero-Order Association between Region and Education in Raw Data and Two-Variable Model

	Grade School		High School		College	
Region	Raw	Model	Raw	Model	Raw	Model
South	373	372.9	384	383.9	162	162.0
Other	885	885.2	1279	1279.2	496	496.0

racy, we must match the marginal and N restrictions as well. Let us see whether the data are free of higher-order interactions. Table 12 gives the figures.

For region and jobs, the top three lines in Table 12, we observe values close to 0.45 for all three education conditions. This is odds ratio language for saying that there is a negative partial between region (South) and jobs, or, in English, Southerners are less likely to endorse equality of employment. If the partials had vanished, the odds ratios would have been 1.000 or close to it. The reader may find the following rule helpful: in a second-order odds ratio, a value less than 1.000 is a negative correlation; a value greater than 1.000 is a positive correlation—under the convention that plus is assigned to the top category of the first item (for example, S for region) and the left category of the second item (for example, plus for jobs). Since all three conditional values are close

TABLE 12

Higher-Order Odds Ratio for Modeled Data in Table 10

Second-Order Odds Ratio	Condition		Third-Order Odds Ratios	
Region (R) and Jobs (J)				
S + R $J \mid E$ O −	Grade school	0.452	1.002 (Grade school vs. High school)	1.002
	High school	0.451	1.002 (High school vs. College)	
	College	0.450		
Education (E) and Jobs (J)				
G + E $J \mid R$ H −	South	0.535	1.002	
	Other	0.534		
G + E $J \mid R$ C −	South	0.268	1.000	
	Other	0.268		
H + E $J \mid R$ C −	South	0.502	1.000	
	Other	0.502		
Region (R) and Education (E)				
S G R $E \mid J$ O H	Jobs −	1.241	0.998	
	Jobs +	1.243		
S H R $E \mid J$ O C	Jobs −	0.813	0.999	
	Jobs +	0.814		
S G R $E \mid J$ O C	Jobs −	4.319	0.998	
	Jobs +	4.359		

to 0.45, the three higher-order odds ratios are all very close to 1.000. The degree of correlation between region and jobs is the same in each level of schooling in this model.

Turning to the second pair of variables, education and jobs, we see that each of the three possible odds ratios is negative and the higher-order odds ratios are all close to 1.000. The reason education has a negative correlation with job attitudes is that we assigned the *lower* level of education to the top of the notation. The matter is quite arbitrary, but it would probably have been better craftsmanship to reverse the categories in these odds ratios.

The last pair, region and education, shows the same pattern but is of some interest in that one of the conditional odds ratios is negative whereas the others are positive. Inspection of the original data shows that when considering only the high school and college groups, Southerners have a slightly greater proportion of persons reporting some college. The point has no particular importance for the Goodman system, but it does illustrate how decomposing a relationship into all possible odds ratios can show us properties we would otherwise miss.

Tables 11 and 12 have now convinced us that our iteration did deliver the goods: a model where all zero-order relationships are identical with those in the raw data, where nonzero partial associations can appear, and where all third-order odds ratios are 1.000.

To test the model we compare the observed and expected values in each cell and calculate the total chi square, just as in the previous examples. For the total four-variable Schwartz data, a computer program generated the complete model and obtained a total chi square value of 31.35.

How many degrees of freedom do we use? The restrictions are given by the cell frequencies in each of the zero-order tables, and we have one redundant calculation for each such cell. Referring to Figure 3, however, we see that not all of these $R \times C$ restrictions are new since the N and the single-variable restrictions are included. Thus each two-variable hypothesis adds only $(R - 1)(C - 1)$ new restrictions, just as each marginal added only $K - 1$ new restrictions when we moved from the equal-probability to the single-variable model.

Goodman provides a handy scheme for talking about his higher-order models and I shall use it to specify the degrees of freedom. He simply assigns a letter to each variable, describes a two-variable hypothesis by a pair of letters in parentheses, describes a single-variable hypothesis by a single letter in parentheses but only if it does not appear in a two-variable hypothesis, and ignores a variable treated only as equal-probability.

Consider, for example, a five-variable model with items A, B, C, D, and E in which A and C are set for two-variable, A and D for two-variable, C and D for two-variable, B for single-variable, and E for equal-probability. In shorthand we can describe the model in this way:

$$(A,C)\ (A,D)\ (C,D)\ (B)$$

Our two-variable model for the Schwartz data can be described like this:

$$(T,E)\ (T,R)\ (T,J)\ (E,R)\ (E,J)\ (R,J)$$

To calculate degrees of freedom, apply the following rules for degrees of freedom in two-variable models:
1. Obtain the total by multiplying $A * B * C * D * E \ldots$
2. Subtract one degree of freedom for N regardless of the number of equiprobable variables
3. Subtract $(K - 1)$ for each different letter that appears anywhere in the list [in our hypothetical example:
 $(A - 1) + (B - 1) + (C - 1) + (D - 1)$]
4. Subtract $(R - 1)(C - 1)$ for each pair of letters within parentheses [in our hypothetical example:
 $(A - 1)(C - 1) + (A - 1)(D - 1) + (C - 1)(D - 1)$]

Remembering that in these data $K = 2$ for all items except education (which has three categories), the degrees of freedom are as follows:
1. $(2 * 3 * 2 * 2) = 24$
2. 1
3. $(1 + 2 + 1 + 1) = 5$ $\Big\}\ 15$
4. $(1 * 2) + (1 * 1) + (1 * 1) + (2 * 1) + (2 * 1) + (1 * 1) = 9$
 DF $= 24 - 15 = 9$

The chi square table tells us that for nine degrees of freedom a total chi square of 21.666 is significant at the 0.01 level. Our chi square value of 31.35 is statistically significant. This third model does not fit either.

The interpretation rule for a two-variable model goes like this. If the two-variable model fits (if the chi square is not significant): (1) There are no significant three-variable or higher-level interactions (no specifications). (We cannot reject the hypothesis that all third-order or higher odds ratios are 1.000.) (2) There may or may not be item skews and nonzero partial associations among the variables. If the two-variable model does not fit (if the chi square is significant): (1) It is not the case that all third-order and higher odds ratios are 1.000, but we do not know whether any particular higher-order effects are significant. (2) There may or may not be item skews and nonzero partial associations among the variables. Special situation: In a three-variable table, a significant chi square for the two-variable model implies that the lone possible higher-order effect, the three-way interaction, is significant.

The significant chi square for our two-variable model means that we must proceed to higher-order models because we cannot account for the cell frequencies by the simple effects so far introduced.

Higher-Order Models

The logic of increasingly complicated models including all prior effects should be clear by now, so it is not necessary to consider higher-order models in detail. Rather I shall review the properties of three-variable models and then sketch the rationale for higher-order models. A three-variable model may be defined alternatively as one that permits skews, partial associations, and three-variable interactions but no higher-order effects; or as one that sets fourth-order and higher odds ratios to 1.000 but allows lower-order odds ratios to depart from 1.000.

Construction of three-variable models is a surprisingly simple extension of the iteration scheme described in the appendix of this chapter. One proceeds in exactly the same fashion as before, except for this data restriction: the model cell frequencies must sum to the cell frequencies of the relevant three-variable cross-tabulations.

Consider variables *A*, *B*, *C*, and *D*. The complete two-variable model required that the model data match the cell frequencies for the zero-order cross-tabulations *AB*, *AC*, *AD*, *BC*, *BD*, and *CD*. For these four variables we can make four three-variable cross-tabulations: *ABC*, *ABD*, *ACD*, and *BCD*. For the data in the Schwartz table we could tabulate region by time by jobs, region by education by jobs, time by jobs by education, and time by region by education. The first would have $2 \times 2 \times 2 = 6$ cells, the second $2 \times 3 \times 2 = 12$ cells, and so on. Following the steps given in the appendix, we proceed to create a set of data such that we can sum the cell frequencies in the model data and obtain the raw data cell values for each of the three-way tables. Such a model has all odds ratios of fourth-order or higher equal to 1.000, just as the two-variable model had all odds ratios of third-order or higher equal to 1.000. As usual the model can be compared with the data, provided of course that there are four or more variables. With three variables, a three-variable model *is* the raw data and no discrepancy can occur.

Degrees of freedom are calculated by a straightforward extension of the rules presented in the preceding section. I shall present them in a general form that should enable the reader to calculate easily the degrees of freedom for models of any order. Given models stated in parentheses, for example (I,J,K,L) (R,S,T) (P,Q):

1. Obtain the total by multiplying all distinct letters; for example, $I * J * K * L * R * S * T * P * Q$.
2. Subtract 1 for the N restriction.
3. Subtract $K - 1$ for each different letter that appears anywhere in the set.
4. Subtract $(R - 1)$ $(C - 1)$ for each pair of letters within parentheses; for example, IJ, IK, IL, JK, JL, KL, RS, RT, ST, PQ. Note: Subtract a particular product only once, no matter how many sets of parentheses include it. Do not subtract a product if the letters are not in the same parentheses (for example, L and R in the sample above).
5. Subtract $(R - 1)$ $(S - 1)$ $(T - 1)$ for each triplet within parentheses, applying the note in rule 4.
6. Subtract $(I - 1)$ $(J - 1)$ $(K - 1)$ $(L - 1)$ for each set of four letters within parentheses, applying the note in rule 4.
7. And so on, for each subset of increasing size up to the total number of variables minus 1.

Degrees of freedom for a complete three-variable model of the Schwartz data can be calculated as follows, using $E = 3$, $R = 2$, $J = 2$, and $T = 2$ for the four variables education, region, jobs, and time. The complete three-variable model is: (E,R,J) (E,R,T) (E,J,T) (R,J,T)—in other words, all possible combinations of three variables. We proceed as follows:

1. Total DF $= 3 * 2 * 2 * 2 = 24$.
2. Subtracting 1 for N gives 23.
3. Since each letter appears, we subtract $(E - 1 = 2) + (R - 1 = 1) + (J - 1 = 1) + (T - 1 = 1) = 5$ for single-variable effect restrictions, giving $23 - 5 = 18$.
4. Since each pair appears at least once (for example, R and T occur in the second and fourth set of parentheses) we further subtract:

$$
\begin{aligned}
(E - 1)\ (R - 1) &= 2 \\
(E - 1)\ (J - 1) &= 2 \\
(E - 1)\ (T - 1) &= 2 \\
(R - 1)\ (J - 1) &= 1 \\
(R - 1)\ (T - 1) &= 1 \\
(J - 1)\ (T - 1) &= 1 \\
\hline
18 - 9 &= 9
\end{aligned}
$$

5. And for the four triplets:

$$(E-1)(R-1)(J-1)=2$$
$$(E-1)(R-1)(T-1)=2$$
$$(E-1)(J-1)(T-1)=2$$
$$(R-1)(J-1)(T-1)=1$$

$$9-7=2$$

We end up with two degrees of freedom for the model.[1]

Our computer program constructed and tested the three-variable model, obtaining a total chi square of 6.91, which is below the 0.01 criterion value of 9.210. Thus the three-variable model fits—the discrepancy between model and actual data is not statistically significant at the 0.01 alpha level we have arbitrarily chosen for this example.

The interpretation is a natural extension of previous argument. Satisfactory fit (as in our example) implies that all higher-order effects (here, the four-variable interaction) are zero; significant discrepancy implies that not all higher-order effects are zero.

The procedure can be applied to models of progressively higher order, provided that the cross-tabulation has enough variables. Remember that to test an X-level hypothesis (that is, to rule out $X+1$ level and higher effects) one must have at least $X+1$ variables in the cross-tabulation.

We shall not continue for two reasons. First the rationale for the scheme should now be clear. Second our example will no longer be useful since we have only four variables. Note further that even if we had more variables, there is no point in continuing with the data in the example. Of mathematical necessity, if a model at level X fits, all higher-order models must also fit. In sum we continue to test increasingly higher-order models until we obtain fit or run out of variables.

Mixed Models and Testing Specific Effects

Assume for purposes of explanation that we have tested progressively complex models and have finally obtained fit with an L-variable (L for level) model. In the Schwartz data, for example, we fitted the three-variable ($L=3$) model after failing with the equal-probability, single-variable, and two-variable hypotheses.

What do we know? We know two things. First we need no effects more complex than level L to describe the cell frequencies. Second we need at least one L-level effect. The first result follows from the success

[1] The remaining degrees of freedom are associated with the lone four-variable relationship $(E-1)(R-1)(J-1)(T-1) = 2 * 1 * 1 * 1 = 2$.

in fit at level L, the second from the previous failure to fit at level $L - 1$. But we may not need all possible L-level effects. In the Schwartz data, for example, our model included four distinct three-variable effects: (E,R,J) (E,R,T) (E,J,T), and (R,J,T). But it could be that only one of them is nonzero (or that two or three or all four of them are nonzero).

Invoking the principle of scientific parsimony, we say that the analyst's aim is to find the least complex set of effects that are necessary and sufficient to account for the data.

Knowing that a model with all possible L-level effects is sufficient, we turn to the question of whether all of them are necessary. This leads us to the construction of mixed models with effects of differing levels. I do not know of any algorithm that provides the best way to find a good mixed model and I do not know whether there is always a unique solution—a single model that meets the criteria—but the following commonsense steps provide one approach.

As a first step, try a series of models with just one L-level effect. In the Schwartz data, for example, we might try a model with only one three-variable effect, for example (education, region, jobs). But taken literally the model (E,R,J) says to allow an interaction effect for the three items but to set the other three relationships (ET), (RT), and (JT) to zero. We have no warrant for such an excision. Rather, when testing to see whether we can fit a model containing only the single interaction (E,R,J) we have to supplement that effect with all possible lower-order effects. Putting it a little more abstractly, here is step 1: begin by testing all possible models with a single L-level effect and all $L - 1$ level effects not subsumed in it. Following the rule we would add two-variable effects for (ET), (RT), and (JT), but we would not build in two-variable effects for (ER), (EJ), and (RJ) since they are subsumed in (E,R,J).

Table 13 gives the results of step 1. The bottom half of Table 13 gives good news. Three of the new models fail and one fits. We can describe the data in our table by the model (ER) (RJ) (RT) (EJT) as well as the model (ERJ) (ERT) (EJT) (RJT). Which is more parsimonious? Goodman does not discuss the matter but the following are suggested as rules: The level of parsimony of a model is the highest-order effect present; among models at the same level L, those with fewer L-level effects are more parsimonious; among models with the same number of L-level effects, the one with fewer $L - 1$ level effects is more parsimonious; and so on. Our new model is more parsimonious from the second of these rules. These results came out neatly but such an event is not inevitable. Thus more steps.

Step 2: if only one model constructed in step 1 fits, go to step 3.

TABLE 13
Some Mixed Models for Schwartz Data

Effects										Chi		
ER	*EJ*	*ET*	*RJ*	*RT*	*JT*	*ERJ*	*ERT*	*EJT*	*RJT*	Square	DF	Criterion[a]
Previous Results												
						( )				2798.49	23	40.289*
(*E*)	(*R*)	(*J*)	(*T*)							823.53	18	34.805*
(*ER*)	(*EJ*)	(*ET*)	(*RJ*)							31.35	9	21.666*
						(*ERJ*)	(*ERT*)	(*EJT*)	(*RJT*)	6.91	2	9.210
		(*ET*)		(*RT*)	(*JT*)	(*ERJ*)				28.52	7	18.475*
	(*EJ*)		(*RJ*)		(*JT*)		(*ERT*)			27.97	7	18.475*
(*ER*)			(*RJ*)	(*RT*)				(*EJT*)		16.4	7	18.475
(*ER*)	(*EJ*)	(*ET*)							(*RJT*)	27.5	8	20.090*

[a] Value of chi square for significance at the 0.01 level for the number of degrees of freedom. Asterisks indicate significant values in data.

If none of the models in step 1 fits, try models with one more L-level effect in each and return to step 1. If two or more fit, go to step 3 and analyze each.

Step 3: next we ask whether the model can be further reduced. Steps 1 and 2 establish the minimum number of L-level effects that must be present in the final model, but we have learned nothing yet about the $L - 1$ or still lower-level effects. It is possible that we could fit an even simpler model that deletes one or more lower-level effects. To test for possible deletions we next build a series of models containing the necessary L-level effects but removing the $L - 1$ effects one at a time. In our Schwartz example this amounts to three models each containing (E,J,T) and two out of the trio (ER) (RJ) (RT).

Table 14 gives the results when we drop particular two-variable effects in our example. None of the three models fits, as the chi squares each exceed the criterion value for the 0.01 level. And that is the end of our hunt for a satisfactory model. We now know that the model (ER) (RJ) (RT) (EJT) fits the data, and that no less complex model does.

TABLE 14
Mixed Models for Schwartz Data Deleting Two-Variable Effects One at a Time

Model				Chi Square	DF	Criterion[a]
	(RJ)	(RT)	(EJT)	25.53	9	21.666*
(ER)	(RJ)		(EJT)	58.50	8	20.090*
(ER)		(RT)	(EJT)	148.07	8	20.090*

[a] Value of chi square for significance at the 0.01 level for the number of degrees of freedom. Asterisks indicate significant values in data.

In other situations, of course, one or more of the new models might fit the data. If so, one should continue the process of simplification through models deleting more than one $L - 1$ level effect or those deleting $L - 2$ and successively lower-level effects.

To summarize: in step 3, test simplified models by deleting $L - 1$ and lower-level effects while maintaining the L-level effects determined in step 2.[2] The process continues until all simpler effects have either been eliminated or shown to be necessary to achieve fit.

At the end of step 3—which might involve a more extensive sequence of model-testing than in our lucky example—one has a model

[2] We do not test models deleting L-level effects since—under the assumption at the beginning of this section—the model denying any L-level effects failed to fit.

(or possibly more than one model) that meets the parsimony criterion. It is the least complex set of effects that are necessary and sufficient to fit the raw data. More effects are not necessary because the model fits; any simpler model with fewer effects is not sufficient since we tried all of them and they failed to fit.

The final step in the analysis is to test the significance of individual effects. The model as a whole is significant in the sense explained above, but it does not follow that each part standing alone is significant. From the logic of the whole enterprise one may expect that almost all the component effects are individually significant, but it is worthwhile to make the tests to nail the point down.

To test the significance of a particular effect, one attempts to fit a model containing "everything else." A significant failure is equivalent to rejecting the null hypothesis that the effect is nil. A model excluding one specific effect may be constructed as follows. First list all $V - 1$ level effects, where V is the number of variables. For example, with four variables list all the three-variable effects. Next delete any that include (or are) the effect in question. For example, if the effect in question is (AC) then the third-order effect (ABC) would be deleted. (This is why the V-level effect is not listed; it includes all other effects.) Then check whether lower-level effects included in the one in question are included in the results of step 2 above. If not, add them.

Let us find a model to test the significance of (ER) in the Schwartz example. Since there are four variables, $V - 1 = 3$ and we first list all the three-variable effects: (EJT), (RJT), (ERT), and (ERJ). Second we delete (ERT) and (ERJ) since they include the pair ER. This leaves (EJT) and (RJT) as our model. Third we see that the lower-level components of ER—E and R—appear in (EJT) and (RJT). Thus we test the model (EJT) (RJT).

Table 15 gives the results for the four effects in our final model for the Schwartz data. All four effects are statistically significant at the 0.01 level. In each case the model that allows all other effects save the one in question has a more than chance discrepancy when compared with the raw data.

Step 4 in the all-purpose plan of analysis is to test the significance of the individual effects in the final model. What if one or more effects are not significant? Such an outcome may seem contradictory since all effects have been shown to be significant in steps 1 or 3. The difference is that the tested models are not exactly the same. Consider the effect (ER). In the top line of Table 14 we concluded that it was significant because the model (RJ) (RT) (EJT) failed; in the top line of Table 15, we concluded that it was significant because the model (EJT) (RJT) failed.

TABLE 15
Test for Specific Effects in the Model (*ER*) (*RJ*) (*RT*) (*EJT*)

Effect Tested	Model				Chi Square	DF	Criterion[a]
(*ER*)			(*EJT*)	(*RJT*)	21.94	8	20.090*
(*RJ*)			(*ERT*)	(*EJT*)	146.25	6	16.812*
(*RT*)			(*ERJ*)	(*EJT*)	54.65	6	16.812*
(*EJT*)	(*ERJ*)	(*ERT*)	(*RJT*)		20.55	4	13.277*

[a] Value of chi square for significance at the 0.01 level for the number of degrees of freedom. Asterisks indicate significant values in data.

Although similar, the two models are not identical. The Table 15 model allows the interaction (*RJT*); in Table 14 it is set to zero (for good reasons, of course). Since the models are not identical, their conclusions do not have to be identical.

Generally the results of the two forms of significance-testing should be close because the logic of the system guarantees that the two models will be quite similar. In the (*ER*) example in the previous paragraph, the only difference between (*RJ*) (*RT*) (*EJT*) and (*RJT*) (*EJT*) is that the latter permits the interaction (*RJT*). We know that (*RJT*) is not a strong effect since the next-to-last line of Table 13 tells us that we can fit the data without it. Thus any play between the two results is because one model allows an effect so small that it is not required to obtain fit.

In sum it is quite unlikely that the two significance tests disagree sharply, but it is quite possible that effects that achieve significance in steps 1 or 3 may turn out to be borderline when specific tests are made.

Substantive Conclusions

At last we are ready to state the substantive conclusions of this investigation. The following propositions are necessary and sufficient to account for the cell frequencies in Table 8. First there is a significant partial association between education and region, controlling for time and jobs, that is essentially similar in magnitude within each control condition. Second there is a statistically significant partial association between region and jobs, controlling for time and education, that is essentially similar in magnitude within each control condition. Third there is a statistically significant partial association between region and time, controlling for education and jobs, that is essentially similar in magnitude within each control condition. And fourth there is a statistically significant interaction involving the three variables education, jobs, and time, such that the degree of association for any two varies with the level of the third. This pattern is essentially similar in each region. The sign and degree of the effects remain to be seen and must be stated before the results can be interpreted (this is discussed in the next section), but these four propositions are an English translation of the conclusions in this part of the analysis.

DISCUSSION

Here I shall discuss the uses and limitations of the system. The uses may be divided into the intellectual and the practical; the limitations into the intrinsic and the extrinsic.

From a purely intellectual point of view, the major utility of the system is that it is a system. Starting with the seemingly trivial concept of odds ratio, it builds a consistent and flexible language for talking about almost any aspect of contingency tables. It is thus possible to see how such concepts as marginal skews, associations, partial associations, interactions, independence, and specification fit together. I believe that anyone who masters the vocabulary and concepts will find a deeper understanding of the logic and properties of contingency tables.

Practically speaking, the system gives us an important general tool and a number of useful specific tests. At the general level, the search for a final model, if successful, gives the analyst a clear and concise statement of what is going on in his contingency tables. With more than two or three variables, it is easy to thrash around endlessly running this-that-and-the-other concatenation of statistics. A final model cuts through all of this and tells us what is there and what is not there. Consider, for example, how a hypothetical final model such as (AB) (CDE) (F) neatly summarizes the dozens of possible relationships in the 64 or more cells of a six-variable table.

Turning to specific practical applications, the system provides at least three significance tests that are not commonly known or used by research workers. First it provides a test for the significance of a partial association, regardless of the number of categories or level of measurement in the associated variables or control items. The question of whether the partial vanishes when such-and-such is controlled is a strategic one in much social research, and the system gives us a significance test for it. Second the system gives us a general test for the significance of interactions and specifications. Goodman's test for a significant difference in Q when the control variable has two levels (Goodman, 1965, p. 291) is well known (Davis, 1971, p. 100), but we can now extend the test to a control variable with more than two categories or to a combination of control variables. Third the fit or failure to fit of the single-variable model provides a simple answer to the question, "Is *anything* going on?" Particularly when working with small samples, an investigator can save time and money by making this test before trying specific hypotheses.

Shifting to the limitations of the system, we shall consider one limitation intrinsic to this particular approach and others that stem from the extrinsic properties of the general problem. The only limitation intrinsic to the system of which I am aware is the possibility of an ambiguous result. I strongly suspect, but cannot prove, that it is possible to fail in the attempt to find a final model. The problem seems to occur in a set of data with a large number of effects of borderline strength. I suspect, for example, that in a set of data with many borderline associations, it could happen that the single-variable model would not fit but that no particular association would be strong enough that its deletion would produce lack of fit for a final model. Putting the same idea another way, the failure of a general model such as the single-variable or the two-variable model means "it is not the case that nothing is going on," which is not quite the same as saying "something definite is going on." In a strict sense the failure is in the empirical data, not in the system; but analysts should be alert to the possibility that there may be no clearcut answer to the problem of finding an unambiguous final model.

Other limitations arise not from the logic of this system itself but from familiar problems in any multivariate significance-testing enterprise. First one should remember that in any analysis all relationships are partials. Any effect has the implicit phrase "all other variables held constant." In practice the investigator—especially if he is working in the tradition of causal models—may wish to control different variables at different stages in his analysis. For example we noted earlier that it seemed awkward to control for jobs when examining the correlation between education and region. This problem is not serious; it is always possible to collapse a multivariate table into a smaller one and carry out the analysis with the collapsed variable(s) ignored. However, it means that the analysis is not self-contained. One needs further assumptions to carry out multivariate analysis. One such set of assumptions is given in Davis (1971, chap. 5 and 6). Goodman does not treat this issue in his early publications, but he does address it in an unpublished paper that has arrived too recently for discussion here (Goodman, 1972c).

Second one should not forget the elementary principle that significance tests are sensitive to sample size as well as to effect magnitudes. When working with extremely large samples, even the most trivial effect can be significant, and the analyst may end up with a model that includes a number of significant higher-order interactions maddening to interpret but trivial in magnitude. One should always remember that unless the data have odds ratios of exactly 1.000, there is always some number by which the frequencies could each be multiplied to obtain significance. Conversely the analyst working with small samples obtains significance only for very large effects.

Third there is the perpetual problem that when generalizing beyond the sample in hand, the chi square test assumes that the cases are a simple random sample (SRS). This is seldom or never the case with social science data. The Schwartz data, for example, come from multistage cluster samples that probably have quota features in the final stage. Since all the tests used in the most sophisticated journals also assume simple random sample, this problem is not unique to the Goodman system.

Fourth, significance tests need to be supplemented by descriptive statistics that convey the magnitude and, when appropriate, the direction of the relationship. There are many possibilities—percentage tables, multivariate *Q*s and gammas (Davis, 1971), the new measures proposed in Goodman's recent papers but not discussed here, or the odds ratios themselves. Since the odds ratios may be unfamiliar, I shall describe the Schwartz data in terms of odds ratios to show how they look.

Our final model is (ER) (RJ) (RT) (EJT), which was translated into English at the end of the previous section. We take each of the four English conclusions in turn.

The first association, between education and region (ER), is interpreted as saying that there is some association between education and region and that it does not vary significantly in magnitude across the control conditions formed by cross-tabulating jobs and time. Since education is a trichotomy, we need to examine (3)(2)(2)(1)/4, or three odds ratios, according to Equation (3). Table 16 gives the results.

The results turn out to be a little complicated—which may be useful for our purposes, as they illustrate a number of issues that arise in interpreting the results in the Goodman system. We begin with the odds ratio

$$\begin{matrix} \text{High} & \text{South} \\ E & R \\ \text{Grade} & \text{Other} \end{matrix}$$

and note that all four conditional figures are less than 1.000 (0.641, 0.767, 0.871, and 0.772) with a weighted average of 0.767. The weighted average of the odds ratios, a useful summary statistic, is obtained by multiplying each conditional odds ratio by its case base (the sum of the four cells involved), summing these products, and dividing the sum by

TABLE 16
Odds Ratios for Education and Region

Control Conditions Time	Jobs	High *E* Grade	South *R* Other	College *E* Grade	South *R* Other	College *E* High	South *R* Other
1963	As good	0.641	(775)[a]	0.570	(475)	0.890	(792)
	Other	0.767	(192)	2.426	(127)	3.161	(109)
1946	As good	0.871	(866)	1.231	(609)	1.412	(749)
	Other	0.772	(1108)	1.089	(705)	1.410	(671)
Weighted average[b]		0.767		1.094		1.315	
Zero order		0.712		0.775		1.410	

[a] Total cases in the fourfold subtable.
[b] See text for description.

the total for the constituent case bases. The purpose of all this is to give those odds ratios based on more cases a greater contribution to the overall average. This says that for the two education levels (high school and grade school) Southerners are less likely than others to be high school graduates. (An odds ratio less than 1.000 may be thought of as a negative association if the categories on top of the letters are assigned to the higher values of ordinal or dichotomous variables.) The value of 0.712 in the bottom line tells us the value in the zero-order table for education and region.

Shifting to the middle column, we see a weighted average value of 1.094 and what appears to be considerable fluctuation from condition to condition. Since the two-variable hypothesis (*ER*) fits the data, however, we do not consider these variations as more than chance. (Note: it is possible that if we had run only the two education levels, college versus grade school, a significant interaction—failure of the two-variable hypothesis—might have occurred. Our significance test applies to the total set of 12 conditional odds ratios in Table 16, not necessarily to a particular subgrouping. Those familiar with analysis of variance will see an analogy between this property and the ANOVA property that significance for a given effect does not mean that particular pairs of means are significantly different.) The zero-order correlation is negative (0.775 is less than 1.000), whereas most of the conditionals are positive with an average of 1.329. This is a familiar property of multivariate analysis and can be interpreted as meaning that when time and jobs are controlled, the association reverses sign. Among persons similar in time and jobs, Southerners are relatively more often college graduates and others are more often grade school graduates.

The pattern for college versus high school is as follows: both the zero-order and conditional associations tend to be positive with a weighted average of 1.315 for the conditionals. Controlling for time and jobs, Southerners are relatively better educated in terms of college versus high school.

As mentioned in the previous section, one would not generally control for an item such as racial attitudes when analyzing the relationship between region and education, and the results here are both complex and somewhat artificial. But for didactic purposes we can summarize the interpretation as follows: when time and attitudes toward jobs are controlled, Southerners tend to be less educated in terms of high school graduation versus grade school only, and better educated in terms of college versus high school or grade school.

The second effect (*RJ*) tells us to look at the association between region and jobs within time and education groups, as in Table 17. This one is a lot simpler. In each of the six conditions the odds ratio is less than 1.000 with a weighted average of 0.379. Within time and education levels, Southerners are less likely to endorse job equality.

The third effect (*RT*), a partial association between region and time, controlling education and jobs, is another finding that would probably not be run in actual research work and I shall not report the figures here.

Finally we turn to the three-variable effect (*EJT*). It says that the three variables have an interaction effect that is about the same in each region. One may interpret it using the formulation that the correlation between two of the items varies according to the level of the third.

TABLE 17
Odds Ratios for Region and Jobs

Time	Education	South *R* Other / As Good *J* Other	Time, Education (*N*)
1946	Grade	0.296	934
	High school	0.334	1040
	College	0.335	380
1963	Grade	0.665	324
	High school	0.555	623
	College	0.156	278
Weighted average		0.379	
Zero order		0.445	

The choice of the specifying variable is not dictated by anything in the Goodman system. One could try: the correlation between education and time varies with category of jobs; the correlation between time and jobs varies with education level; the correlation between education and jobs varies with time. I chose the second version because it is intuitively agreeable and also because a nondichotomous variable is easier to handle as a condition than as one of the correlated items. Table 18 presents the results.

TABLE 18
Association between Time (1946–1963) and Jobs (as good, other) within Education Level and Region

Education Level	South	(*N*)	Other	(*N*)	Weighted Average
College	4.770	(162)	10.222	(496)	8.880
High school	10.352	(384)	6.237	(1279)	7.187
Grade school	6.392	(373)	2.848	(885)	3.899

All the odds ratios are positive. Attitudes toward jobs became more favorable from 1946 to 1963 in all education and region groups. Looking at the weighted averages it seems that change has been greater in the better-educated groups. The odds ratios increase with educational attainment. This is clearly the case among the others, but the college-educated Southerners do not quite fit the pattern. However—and here we see one of the advantages of the Goodman system—if the college-South combination was really out of line, the result would be a four-variable interaction; but this was rejected during significance-testing. A more reasonable interpretation is that this group is small (162 cases in contrast to 496 for college-other) and the discrepancy is explainable by chance fluctuations. Thus our summary: job attitudes became more favorable from 1946 to 1963 in both regional groups, especially among the better-educated.

APPENDIX: ITERATION TECHNIQUE

Three-or-more-variable models permitting effects of level 2 or higher must be constructed by iteration (successive approximations). Here I explain how to carry out the operation, following Goodman (1970, p. 237, Equation 3.8). Readers interested in the history and rationale for the procedure will find a number of useful references in Goodman's article. My approach is purely descriptive and gives no rationale or interpretation. The technique has eight steps:

1. List the effects to be permitted in some arbitrary order, for example *AB*, *CDE*, *FG*.
2. For trial number zero, enter the value 1.00 in each cell of the *ABCDEFG* table.
3. Find the frequencies for the collapsed table involving only the first effect by summing the appropriate frequencies in step 2. (If the first effect is *AB*, we build the zero-order, two-variable table *A* by *B*; if the first effect is *CDE*, we build the three-variable table *C* by *D* by *E*.)
4. Calculate weights for the next trial. For each cell of the sub-table in step 3:

$$\frac{\text{Actual frequency in raw data}}{\text{Frequency in most recent trial}} = \text{Weight}$$

5. Next trial. Multiply each cell frequency in step 2 (or previous trial) by the appropriate weight from step 4. (For example, if *AB* was the first effect, every cell that is $A+\ B-$ in the larger table *ABCDEFG* in the previous trial would be multiplied by the weight calculated for the $A+\ B-$ cell in step 4.)
6. Find the next effect in the list of step 1 (for example, after doing *AB* do *CDE*).

TABLE 19
Raw Data for Iteration of (*JR*) (*JE*) (*RE*)
($N = 3579$)

(a) Jobs by Region by Education

	Grade		High School		College	
Jobs	South	Other	South	Other	South	Other
As good	123	459	190	849	97	405
Other	250	426	194	430	65	91

(b) Jobs by Region

Jobs	South	Other
As good	410	1713
Other	509	947

(c) Jobs by Education

Jobs	Grade	High School	College
As good	582	1039	502
Other	676	624	156

(d) Region by Education

Region	Grade	High School	College
South	373	384	162
Other	885	1279	496

7. Go through steps 3 to 5 substituting the effect in step 6 for first effect.
8. When all effects in the list in step 1 have been used, compare the raw data tables for the effects with those produced by the last iteration. If each cell agrees to within one or two decimals, stop. If not, repeat the cycle starting at step 3 with the first effect in the list.

The technique can be illustrated with three of the variables in our Schwartz data, *J* (jobs), *R* (region), and *E* (education), and the two-variable model (*JR*) (*JE*) (*RE*). Table 19 gives the necessary raw data calculations.

The following figures are numbered according to the steps in the iteration scheme described in Table 19 and are presented with minimal text. By working back and forth from the description to the examples the reader should be able to see how it all works.

Step 1. *JR, JE, RE*

Step 2.

	Grade		High School		College	
	South	Other	South	Other	South	Other
As good	1	1	1	1	1	1
Other	1	1	1	1	1	1

Step 3.

	South	Other
As good	3	3
Other	3	3

Step 4.

	South	Other
As good	410/3 = 136.7	1713/3 = 571.0
Other	509/3 = 169.7	947/3 = 315.7

Step 5.

	Grade		High School		College	
	South	Other	South	Other	South	Other
As good	1×136.7 =136.7	1×571.0 =571.0	1×136.7 =136.7	1×571.0 =571.0	1×136.7 =136.7	1×571.0 =571.0
Other	1×169.7 =169.7	1×315.7 =315.7	1×169.7 =169.7	1×315.7 =315.7	1×169.7 =169.7	1×315.7 =315.7

Step 6. *JE*

Step 3 (second trial).

	Grade	High School	College
As good	707.7	707.7	707.7
Other	485.4	485.4	485.4

Step 4 (second trial).

	Grade	High School	College
As good	582/707.7 = 0.822	1039/707.7 = 1.468	502/707.7 = 0.709
Other	676/485.4 = 1.393	624/485.4 = 1.286	156/485.4 = 0.321

TABLE 20
Iterated and Raw Data Values for Region and Education

Cell		Trial							
Region	Education	0	1	2	3	4	5	6	Raw Data
South	Grade school	2	306.4	348.8	372.9	372.5	371.8	372.9	373
	High school	2	306.4	418.9	384.1	384.2	385.1	383.9	384
	College	2	306.4	151.4	162.0	162.3	162.0	162.0	162
Other	Grade school	2	886.7	909.2	884.6	885.0	886.1	885.2	885
	High school	2	886.7	1244.2	1279.1	1279.0	1278.0	1279.2	1279
	College	2	886.7	506.1	496.0	495.8	496.0	496.0	496

Step 5 (second trial).

	Grade		High School	
	South	Other	South	Other
As good	136.7 × 0.822 = 112.4	571.0 × 0.822 = 469.4	136.7 × 1.468 = 200.7	571.0 × 1.468 = 838.2
Other	169.7 × 1.393 = 236.4	315.7 × 1.393 = 439.8	169.7 × 1.286 = 218.2	315.7 × 1.286 = 406.0

	College	
	South	Other
As good	136.7 × 0.709 = 96.9	571.0 × 0.709 = 404.8
Other	169.7 × 0.321 = 54.5	315.7 × 0.321 = 101.3

Step 6 (second trial). *RE*

Step 3 (third trial).

	Grade	High School	College
South	348.8	418.9	151.4
Other	909.2	1244.2	506.1

And so on. The work continued through two complete cycles, giving six trials. Table 20 shows how the figures converge rapidly on the raw data for one of the three effects, region and education.

REFERENCES

DAVIS, J. A.

1971 *Elementary Survey Analysis.* Englewood Cliffs, N.J.: Prentice-Hall.

GOODMAN, L. A.

1965 "On the multivariate analysis of three dichotomous variables." *American Journal of Sociology* 71:290–301.

1970 "The multivariate analysis of qualitative data: Interactions among multiple classifications." *Journal of the American Statistical Association* 65:226–256.

1971 "The analysis of multidimensional contingency tables: Stepwise procedures and direct estimation methods for building models for multiple classifications." *Technometrics* 13:33–61.

1972a "A modified multiple regression approach to the analysis of dichotomous variables." *American Sociological Review* 37:28–46.

1972b "A general model for the analysis of surveys." *American Journal of Sociology* 77:1035–1086.

1972c "Causal analysis of panel study data and other kinds of survey data." (memorandum)

SCHWARTZ, M. A.

1967 *Trends in White Attitudes Toward Negroes.* University of Chicago, NORC Report No. 119.

James A. Davis

Reprinted from: SOCIOLOGICAL METHODOLOGY, Chapter 8, pp. 189-231, 1974.

Part Four
Latent-Structure and Scaling Models

Social scientists are often interested in phenomena not subject to direct measurement. For example, the precise measurement of the cohesion of a small group, the classification of an authoritarian personality, or the diagnosis of a particular disease require indirect observations on a set of indicators or symptoms of the phenomenon to be measured. Often the indicators are qualitative variables. For example, Stouffer and Toby (1951) used the following four items to classify individuals according to their tendency toward universalistic or particularistic values in situations of role conflict.

A. You are riding in a car driven by a close friend, and he hits a pedestrian. You know he was going at least 35 miles an hour in a 20-mile-an-hour speed zone. There are no other witnesses. His lawyer says that if you testify under oath that the speed was only 20 miles an hour, it may save him from serious consequences. What do you think you'd probably do in view of the obligations of a sworn witness and the obligation to your friend?

Check one:
__________Testify that he was going 20 miles an hour. (–)
__________Not testify that he was going 20 miles an hour. (+)

B. You are a doctor for an insurance company. You examine a close friend who needs more insurance. You find that he is in pretty good shape, but you are doubtful on one or two minor points which are difficult to diagnose. Would you shade the doubts in his favor in view of your obligations to the insurance company and your obligation to your friend?

Check one:
__________Yes (–)
__________No (+)

C. You are a New York drama critic. A close friend of yours has sunk all his savings in a new Broadway play. You really think the play is no good. Would you go easy on his play in your review in view of your obligations to your readers and your obligation to your friend?

Check one:
________ Yes (–)
________ No (+)

D. You have just come from a secret meeting of the board of directors of a company. You have a close friend who will be ruined unless he can get out of the market before the board's decision becomes known. You happen to be having dinner at that friend's home this same evening. Would you tip him off in view of your obligations to the company and your obligation to your friend?

Check one:
________ Yes (–)
________ No (+)

Several methods for scaling such data have been recommended. Likert (1932) proposed the method of summated ratings in which response categories are assigned simple weights (1 or 2 for dichotomous items). A respondent's scale score is then the sum of scores on the separate items. Guttman (1950a) suggested modifying Likert's approach to allow some items (some categories) to have a greater weight in a respondent's scale score. His approach is to select weights for the categories that maximize internal consistency. Both these approaches assume that the items are all measuring the same phenomenon, yet neither provides a statistical test for deciding whether this assumption holds.

Guttman (1944, 1950b) invented scalogram analysis and defined a pure scale. One of the goals of scalogram analysis is to determine whether a set of items does in fact constitute a pure scale. However, Guttman's index of reproducibility is drastically affected by the distributions of the items (Festinger, 1947) and is sometimes large even when items are unrelated.

Lazarsfeld (1950) provided a general mathematical model relating the probability of responding in each category of each item (and the joint probability associated with the response pattern for all items) to an underlying latent variable. This approach, called latent-structure analysis, is very general; it can even isolate response tendencies, such as the tendency for some individuals to agree with all items (acquiescent response set) or to virtually ignore the extreme categories when the number of categories is five or more (extreme response set). See, for example, Lazarsfeld (1954, 1960). Unfortunately, latent-structure analysis has rarely been applied because of the difficulty of developing an efficient estimation algorithm to use with the general latent-structure model.

In a methodological breakthrough Goodman (1974a, 1974b, 1975) solved the estimation problem by showing how latent-structure models can be incorporated into the general framework of the log-linear model. Part IV contains Goodman's approach to latent-structure models. Chapter 8 presents the general approach, followed by specializations to latent-structure measurement models in chapters 9 and 10.

A wide range of causal models is explored in chapter 8. These models permit both unobserved causative factors and imperfectly measured outcomes. They can be used to analyze data or to construct tests and indices for measurement and prediction. To illustrate their wide applicability and flexibility, they are applied to several different sets of data analyzed earlier by others. They result in conclusions that are very different from those presented by the other researchers.

In chapter 9 Goodman reanalyzes the Stouffer-Toby role conflict data and two other data sets using latent-structure models. He shows how Guttman's scaling model can be formulated as a special case of a new latent class model and obtains maximum likelihood estimates and a chi-square goodness-of-fit statistic for Guttman's model, which does not fit the data well. Goodman's new model postulates an additional classification for individuals who are "intrinsically unscalable" with respect to the four Stouffer-Toby items. This new model fits the data exceptionally well, with a large proportion (68%) of the population estimated as intrinsically unscalable with respect to these items. Some other models are also considered.

In chapter 10 Goodman considers a wide range of latent-structure measurement models including identifiable and unidentifiable models. For each model he presents a simple method for determining whether the parameters are identifiable. For parameters that are not identifiable, he describes some restrictions that can be imposed in certain cases (if they are reasonable) to identify the parameters. Again the Stouffer-Toby data are used for illustrative purposes. A simple two-class unrestricted latent-structure model fits the Stouffer-Toby data very well. This model estimates that 28% of the population tends toward universalistic values in situations of role conflict. The class of models considered in chapter 10 is somewhat analogous to the general factor analytic model for quantitative data presented by Jöreskog (1969).

References

Festinger, L. 1947. The treatment of qualitative data by "scale analysis." *Psychological Bulletin* 44:146-61.

Goodman, L.A. 1974a. Exploratory latent structure analysis using both identifiable and unidentifiable models. *Biometrika* 61:215-31.

. 1974b. The analysis of systems of qualitative variables when some of the variables are unobservable. Part I: A modified latent structure approach. *American Journal of Sociology* 79:1179-1259.

. 1975. A new model for scaling response patterns: An application of the quasi-independence concept. *Journal of the American Statistical Association* 70:755-68.

Green, B.F. 1952. Latent structure analysis and its relation to factor analysis. *Journal of the American Statistical Association* 47:71-76.

Guttman, L. 1950a. The principal components of scale analysis. In *Measurement and Prediction,* Edited by S.A. Stouffer, pp. 312-361. John Wiley & Sons, Inc., New York.

. 1950b. The basis for scalogram analysis. In *Measurement and Prediction,* Edited by S.A. Stouffer, pp. 60-90. John Wiley & Sons, Inc., New York.

. 1944. A basis for scaling qualitative data. *American Sociological Review* 9:139-50.

Jöreskog, K.G. 1969. A general approach to confirmatory maximum likelihood analysis. *Psychometrika* 34:183-202.

Lazarsfeld, P.F. 1950. The logical and mathematical foundation of latent structure analysis. In *Measurement and Prediction,* Edited by S.A. Stouffer, pp. 362-412. John Wiley & Sons, Inc., New York.

. 1954. A conceptual introduction to latent structure analysis. In *Mathematical Thinking in the Social Sciences,* Edited by P.F. Lazarsfeld, pp. 349-387. Russell & Russell, New York.

. 1960. Latent structure analysis and test theory. In *Psychological Scaling,* Edited by H. Gulliksen and S. Messick, pp. 83-95. John Wiley & Sons, Inc., New York.

Likert, R. 1932. A technique for the measurement of attitudes. *Arch. Psychol.,* No. 140.

Stouffer, S.A., and Toby, J. 1951. Role conflict and personality. The *American Journal of Sociology* 56:395-406.

Chapter 8

The Analysis of Systems of Qualitative Variables When Some of the Variables Are Unobservable. Part I: A Modified Latent Structure Approach[1]

This article presents methods for analyzing the relationships among a set of qualitative variables when some of these variables are specified manifest (i.e., observed) variables and others are latent (i.e., unobserved or unobservable) variables. We shall show how to estimate the magnitude of the various effects represented in path-diagram models that include both the manifest and latent variables, and also how to test whether this kind of path-diagram model is congruent with the observed data. These methods can be applied in order to analyze data obtained in various kinds of surveys (including panel studies), and also in order to construct tests and indices for purposes of measurement and prediction. To illustrate their wide applicability and flexibility, we shall use these methods to reanalyze several different sets of data which were analyzed earlier by Coleman (1964), Lazarsfeld (1948, 1970), Goodman (1973*a*), and others. Except for some related conclusions in Goodman (1973*a*), the methods introduced herein lead to conclusions that are very different from those presented by the other researchers who had analyzed these data earlier.

1. INTRODUCTION

Let us begin by considering the simple situation where a researcher wishes to obtain a better understanding of the relationship between two observed dichotomous variables (say, variables A and B). In this situation, the researcher will often try to find another dichotomous (or polytomous) variable (say, variable X) that can "explain" this relationship. We say that variable X explains the relationship between variables A and B if this relationship disappears when the level of variable X is "held constant" (i.e., if variables A and B become independent of each other when we examine the conditional relationship between these two variables in the two-way cross-classification table $\{AB\}$ at *each* level of variable X). Similarly, in the analysis of, say, four observed dichotomous variables (A, B, C, D), we say that variable X explains the relationships among the

[1] This research was supported in part by research contract no. NSF GS 31967X from the Division of the Social Sciences of the National Science Foundation. For helpful comments, the author is indebted to O. D. Duncan, R. Fay, S. Haberman, J. Jenkins, A. Madansky, J. Murray, J. Nelson, S. Schooler, A. Stinchcombe, and J. W. Tukey.

four variables if these relationships disappear when the level of variable X is held constant (i.e., if the four variables (A, B, C, D) become mutually independent when we examine the conditional relationships among the four variables in the four-way cross-classification table $\{ABCD\}$ at *each* level of variable X). If these *apparent* relationships disappear, we say they are "spurious." In this situation, the four variables (A, B, C, D) are not *directly* related to each other, but each of the four variables may be affected by variable X; and when the level of variable X is not held constant, then these effects may produce the *apparent* relationships among the four variables. This situation can be described by a simple path diagram (see fig. 1 below).

The above remarks pertain to the situation where there is a dichotomous (or polytomous) variable X that can explain the relationships among the observed variables (A, B, C, D). On the other hand, in some situations where the relationships among the four variables (A, B, C, D) can*not* be explained in this way, it may be possible to find two (or more) dichotomous (or polytomous) variables (say, variables Y and Z) that can explain these relationships when the explanatory variables (Y and Z) are considered jointly. We say that variables Y and Z explain the relationships among the four observed variables (A, B, C, D) if these relationships disappear when the joint level of variables Y and Z is held constant.[2] In this situation, the four variables (A, B, C, D) are not *directly* related to each other, but each of the four variables may be affected by variable Y and/or variable Z and/or an interaction effect between variables Y and Z. This situation also can be described by a path diagram (see, e.g., fig. 2 below, where variables A and C are affected by variable Y [but not by variable Z], and variables B and D are affected by variable Z [but not by variable Y]).

In the preceding two paragraphs, we considered the four variables (A, B, C, and D) to be observed variables, but we did not indicate there whether the possible explanatory variables (X, Y, and Z) were observed or not. If the possible explanatory variables are observed, then elementary methods are available for determining whether or not these variables actually do explain the relationships among the four variables (A, B, C, D) (see, e.g., Goodman 1970). On the other hand, if the possible explanatory variables are not observed (they may be unobservable or latent), then the methods that are needed will not be so elementary. In the present article, we shall be concerned primarily with the latter situation.

The data that we shall analyze here for illustrative purposes will lead us to consider, for example, models in which there is a single explanatory latent

[2] Of course, the joint variable (Y, Z) can be viewed as a single polytomous variable. This will be discussed more fully in sec. 2.1.

variable (see, e.g., fig. 1 below), or two explanatory latent variables (see, e.g., figs. 2 and 8 below), or one explanatory latent variable *and* one explanatory observed variable (see, e.g., figs. 3 and 4 below), or two explanatory observed variables (see, e.g., fig. 6 below). For these particular data, it was usually sufficient to consider models in which there were no more than two explanatory variables, and each of these variables was dichotomous; but the general methods that we shall present here can be applied also to the more general situation where there may be any given number of explanatory variables, and each of these variables may be polytomous (not necessarily dichotomous). (For some examples of this, see the discussion of models H_{11}, H_{12}, H_{14}, H_{15}, H_{16}, H_{22}, H_{23} below, and also additional models in Goodman 1973*d,* 1974.) These methods are applicable in situations where there are multiple indicators of an unobserved variable, multiple indicators of causally related unobserved variables, single or multiple observed causes and multiple indicators of unobserved variables, etc. (see fig. 10 below).

The general methods introduced in the present article can be applied in order to analyze data obtained in various kinds of surveys (including panel studies) and also in order to construct tests and indices for purposes of measurement and prediction. To illustrate the wide applicability and flexibility of these methods, we shall reanalyze three different sets of data, the first of which is table 1.

Table 1 is a two-attribute turnover table (a 16-fold table) presented earlier by Coleman (1964). This table cross-classifies the responses of

TABLE 1

Observed Cross-Classification of 3,398 Schoolboys, in Interviews at Two Successive Points in Time, with Respect to Two Dichotomous Variables*

		Second Interview			
	Membership	+	+	−	−
	Attitude	+	−	+	−
First Interview					
Membership	Attitude				
+	+	458	140	110	49
+	−	171	182	56	87
−	+	184	75	531	281
−	−	85	97	338	554

Source.—Coleman 1964.

Note.—With respect to self-perceived membership in the "leading crowd," being in it is denoted + and being out of it is denoted −; with respect to attitude concerning the "leading crowd," a favorable attitude is denoted + and an unfavorable attitude is denoted −. Membership and attitude will be denoted by the letters *A* and *B*, respectively, in the first interview; and by the letters *C* and *D,* respectively, in the second interview.

* Variables = (1) self-perceived membership in the "leading crowd"; (2) favorableness of attitude concerning the "leading crowd."

3,398 schoolboys, each interviewed at two successive points in time, with respect to questions about their self-perceived membership in the "leading crowd" (in it [+] or out of it [−]) and their attitude concerning it (whether they think that membership in it does not require going against one's principles sometimes [+] or whether they think that it does [−]). This table can be viewed as a four-way cross-classification table in which four dichotomous variables (say, variables A, B, C, and D) are cross-classified. Variables A, B, C, and D can denote the response to the question about (A) self-perceived membership at the first interview, (B) attitude at the first interview, (C) self-perceived membership at the second interview, and (D) attitude at the second interview, respectively. In the present article, we shall show that the observed relationships among the four variables in table 1 can be explained simply by the introduction of two underlying latent variables, say, variables Y and Z (see fig. 2 below), where variable Y pertains to an individual's "latent self-perceived membership in the leading crowd" and variable Z pertains to his "latent attitude toward the leading crowd." We shall show how to estimate the magnitude of the relationship between the latent variables Y and Z, and the magnitude of the effects of the latent variables upon the manifest variables A, B, C, and D; and we shall also provide methods for testing various hypotheses about these effects. In addition, on the basis of a given individual's response to the four questions considered in table 1, we shall show how he can be assigned to one of two classes with respect to latent variable Y, to one of two classes with respect to latent variable Z (in this particular example, both variables Y and Z are dichotomous), and to one of four classes with respect to his joint level on both latent variables Y and Z (see table 10 below).

The explanation introduced in the present article for the relationships observed among the variables in table 1 (see fig. 2 below) is very different from that proposed earlier by Coleman (1964). Our explanation fits the data very well indeed, and it is more congruent with these data than is Coleman's proposed model.

In the present article, we shall actually provide several different models that can serve as explanations of the relationships observed among the variables in table 1 and that are congruent with these data. The particular models that turn out to be appropriate for the Coleman data (table 1) will be different from the particular ones that will be appropriate for the other sets of data which we shall analyze here, but each of these models will serve as an example of the kinds of models that can be obtained with the general methods introduced in the present article (see figs. 1–4, 6, 8, 10, and models H_1–H_{23} listed in tables 5, 15, and 18 below). The three sets of data which we shall analyze will be used to introduce different models and to illustrate different points about the analysis of systems of qualitative

TABLE 2

OBSERVED CROSS-CLASSIFICATION IN THREE DIFFERENT SETS OF DATA

VARIABLE				DATA SET		
A	*B*	*C*	*D*	Coleman Panel (1)	Lazarsfeld Panel (2)	McHugh Test (3)
+	+	+	+	458	129	23
+	+	+	−	140	3	5
+	+	−	+	110	1	5
+	+	−	−	49	2	14
+	−	+	+	171	11	6
+	−	+	−	182	23	3
+	−	−	+	56	0	2
+	−	−	−	87	1	4
−	+	+	+	184	1	8
−	+	+	−	75	0	2
−	+	−	+	531	12	3
−	+	−	−	281	11	8
−	−	+	+	85	1	9
−	−	+	−	97	1	3
−	−	−	+	338	2	8
−	−	−	−	554	68	34

SOURCES.—(1) Coleman 1964; (2) Lazarsfeld 1948, 1970; (3) McHugh 1956.

variables. These sets of data are presented table 2, and they can be described as follows:

1. The Coleman panel data on student attitudes (table 1) viewed now as a four-way cross-classification (col. 1 of table 2). For expository purposes, we shall focus our attention in this article mainly on this one example, but we shall also comment on the other two examples listed below.

2. The Lazarsfeld panel data on voting intention viewed as a four-way cross-classification (see Lazarsfeld 1948, 1970; Lipset et al. 1954; Campbell 1963; Boudon 1968). This table cross-classifies the responses of 266 people, each interviewed at two successive points in time, with respect to their vote intention (Republican [+] versus Democrat [−]) and their opinion of a particular candidate (favorable toward the Republican candidate [+] or unfavorable toward him [−]). We use the letters *A, B, C,* and *D* here to denote vote intention at the first interview, opinion of the particular candidate at the first interview, vote intention at the second interview, opinion of the particular candidate at the second interview, respectively.

3. The McHugh test data on creative ability in machine design (see McHugh 1956; Schumacher, Maxson, and Martinek 1953). This table cross-classifies 137 engineers with respect to their dichotomized scores (above the subtest mean [+] or below [−]) obtained on each of four different subtests that were supposed to measure creative ability in machine

design. We use the letters *A, B, C,* and *D* here to denote the dichotomized scores on the four subtests.

The data sets 1 and 2 in table 2 were also reanalyzed in Goodman (1973*a*), and the present article may be viewed as a sequel to that earlier article.[3] In the earlier article, we referred briefly to explanations of the kind presented herein (cf., e.g., fig. 6 below with fig. 4 in Goodman 1973*a*), but we did not discuss them in any detail (except in some special cases) because the techniques that were necessary for dealing more fully with the analysis of latent variables were beyond the scope of that article.

For ease of exposition, we shall focus our attention in the present article on the case where there are four manifest variables (as in each of the sets of data in table 2), but we shall also take note of the fact that our methods can be applied more generally to the case where there are m manifest variables ($m = 3, 4, \ldots$). We shall be concerned mainly with the case where the manifest variables are dichotomous (as in each of the sets of data in table 2), but some of our results will pertain also to the more general case where the m manifest variables are polytomous (not necessarily dichotomous). The m manifest variables could represent observations made at a single point in time, as in the usual survey or testing situation (e.g., data set 3 in table 2); or they could represent observations made at two or more successive points in time, as in the usual panel study (e.g., data sets 1 and 2 in table 2).

Each set of data in table 2 will be considered in a separate section (secs. 2, 3, and 4 for data sets 1, 2, and 3, respectively), and the more general methods and results will also be presented in a separate section (sec. 5). The reader who is interested primarily in the more general methods and results can turn directly to section 5 below.

For each set of data in table 2, we shall present several different latent structure models that can serve as explanations of the relationships observed in the data and that are congruent with the data. Let us return now for a moment to the explanation introduced earlier in this "Introduction" for the Coleman data (table 1). This explanation stated that (1) the two observed indicators of self-perceived membership in the leading crowd (variables *A* and *C*), at different points in time, are measures of a single latent dichotomous variable (variable *Y*—"latent self-perceived membership"), which remains unchanged over time; (2) the two observed indicators of attitude toward the leading crowd (variables *B* and *D*), at different points in time, are measures of a single latent dichotomous

[3] However, the present article does not presuppose that the reader is familiar with the earlier article.

variable (variable Z—"latent attitude"), which remains unchanged over time; (3) the two latent dichotomous variables (Y and Z) are related to each other (and this relationship also remains unchanged over time); (4) the four observed indicators (A, B, C, D) are mutually independent when the level of the joint latent variable (Y, Z) is taken into account. Thus, the relationships among the four observed indicators in the 16-fold table (table 1) can be accounted for by the two unchanging latent variables. We can replace the 16-fold table by the simpler 2×2 table describing the relationship between the latent variables (see table 6 and the corresponding entries in table 8 below), and by indices that describe the relationship between each observed indicator and the corresponding latent variable (see the corresponding entries in table 8).

The above explanation will be referred to below as model H_2 (fig. 2). In addition to H_2, we shall provide other models that can also serve as explanations of the relationships in table 1 and that are congruent with the data (e.g., H_6 and H_{12} below). The statistics that will be used in the assessment of this congruence will appear in a single table (table 5), and we shall also provide a map (fig. 5) that can guide the reader to see how the various models are related to each other with respect to their implications. This will be done separately for each set of data (see secs. 2, 3, and 4 below).

The methods introduced in the present article can be viewed as modifications and generalizations of the techniques of latent structure analysis developed earlier by Lazarsfeld, Anderson, and others (see, e.g., citations in Lazarsfeld and Henry 1968), and our development of this subject owes much to these earlier important contributions. Our methods are also related to recent work by Haberman (1974*a*).

2. THE ANALYSIS OF THE COLEMAN PANEL DATA

2.1. Some Possible Models

As we have already noted, the letters A, B, C, and D will be used in our analysis of the Coleman data (see table 1 or col. 1 of table 2) to denote the four variables in the table (membership at the first interview, attitudes at the first interview, membership at the second interview, and attitudes at the second interview, respectively). A positive (+) response on a given variable will be denoted by the number 1, and a negative (—) response by the number 2. Let (i, j, k, l) denote the cell in the four-way table corresponding to the case where the respondent is classified as i on variable A ($i = 1$ or 2), j on variable B ($j = 1$ or 2), k on variable C ($k = 1$ or 2), and l on variable D ($l = 1$ or 2). Let f_{ijkl} denote the observed frequency in cell (i, j, k, l). From table 1 or column 1 of table 2, we see, for example, that $f_{1111} = 458$, $f_{1112} = 140$, $f_{1121} = 110$, etc.[4]

Let n denote the total sample size (e.g., $n = 3{,}398$ in table 1). From the definition of the f_{ijkl}, we just see that

$$\sum_{i=1}^{2} \sum_{j=1}^{2} \sum_{k=1}^{2} \sum_{l=1}^{2} f_{ijkl} = n. \tag{1}$$

Let p_{ijkl} denote the observed proportion of individuals in cell (i, j, k, l). Thus,

$$p_{ijkl} = f_{ijkl}/n. \tag{2}$$

From (1) and (2), we see that

$$\sum_{i=1}^{2} \sum_{j=1}^{2} \sum_{k=1}^{2} \sum_{l=1}^{2} p_{ijkl} = 1. \tag{3}$$

For some specified hypothesis about the system of variables in the four-way table, we shall let π_{ijkl} denote the probability that an individual will be in cell (i, j, k, l), and we shall let F_{ijkl} denote the expected frequency in cell (i, j, k, l), calculated under the assumption that the specified hypothesis is true. Thus,

$$F_{ijkl} = n\pi_{ijkl}. \tag{4}$$

Note that π_{ijkl} is the expected proportion of individuals in cell (i, j, k, l) under the assumption that the specified hypothesis is true.

Let us first consider the hypothesis H_1 that there is a latent dichotomous variable X that can explain or "account for" the relationships among the manifest variables (A, B, C, D) in table 1. For the tth latent class of variable X $(t = 1, 2)$, let $\pi^{X}{}_{t}$ denote the probability that an individual will be in class t, and let $\pi^{ABCDX}{}_{ijklt}$ denote the probability that an individual will be at level (i, j, k, l, t) with respect to the joint variable (A, B, C, D, X).[5] Let $\pi^{\bar{A}X}{}_{it}$ denote the *conditional* probability that an individual will be at level i with respect to variable A, given that he is at

[4] Note that a symbol like f_{ijkl} here and in all that follows will denote a set of numbers. Since there are two possible values here for i, j, k, and l, the symbol will denote $2 \times 2 \times 2 \times 2 = 16$ numbers.

[5] Similarly, in the preceding paragraph, we could have used the symbol $\pi^{ABCD}{}_{ijkl}$ instead of π_{ijkl}. However, in order to simplify the notation, we used the latter symbol there. In the present paragraph, we could also have replaced the symbol $\pi^{ABCDX}{}_{ijklt}$ by π_{ijklt}, but we did not do this because we wanted the symbol to remind the reader that it pertains to the four manifest variables *and* the one latent variable—it does not pertain here to five manifest variables.

level t on variable X; and let the *conditional* probabilities $\pi^{\bar{B}X}_{jt}$, $\pi^{\bar{C}X}_{kt}$, $\pi^{\bar{D}X}_{lt}$ be similarly defined.[6] The hypothesis H_1 states that

$$\pi^{ABCDX}_{ijklt} = \pi^{X}_{t}\pi^{\bar{A}X}_{it}\pi^{\bar{B}X}_{jt}\pi^{\bar{C}X}_{kt}\pi^{\bar{D}X}_{lt}, \tag{5}$$

and

$$\pi_{ijkl} = \sum_{t=1}^{2} \pi^{ABCDX}_{ijklt}. \tag{6}$$

Formula (5) states that, given that an individual is in latent class t with respect to variable X, his responses on the manifest variables (A, B, C, D) will be mutually independent. (In other words, when the level t of the latent variable X is held constant, the relationships among the manifest variables disappear; i.e., the latent variable explains the relationships among the manifest variables.) Formula (6) describes the fact that each individual in the population under consideration is in one (and only one) of the two latent classes.

The hypothesis H_1 can also be represented by a simple path diagram (fig. 1).[7] This diagram shows that the manifest variables (A, B, C, D) are affected by the latent variable X, but they do not have any direct effects upon each other.[8]

To test whether hypothesis H_1 is congruent with the data (table 1), we would first need to estimate the parameters π^{X}_{t}, $\pi^{\bar{A}X}_{it}$, $\pi^{\bar{B}X}_{jt}$, $\pi^{\bar{C}X}_{kt}$,

[6] In the present article, we shall use a bar over a letter (say, letter A) in the superscript in order to indicate that the probability pertains to the corresponding variable (i.e., variable A) *conditional* upon the level of the other variable (or variables) that appear in the superscript without a bar. Because we are dealing here with *conditional* probabilities, they will satisfy condition (8) presented below.

[7] In fig. 1, and in the other path diagrams presented below for the analysis of data sets 1 and 2 in table 2, we could have indicated that variables A and B were temporally antecedent to variables C and D, respectively, by moving the latter letters to the right of the former letters in the path diagrams (see, e.g., Goodman 1973*a*, 1973*b*, and fig. 10.D–10.F below), with the corresponding arrows moved accordingly so that all the effects (e.g., in fig. 1, the effects from X to A, B, C, and D) remain unchanged. The reader who wants the path diagrams to remind him of this temporal ordering of the variables can redraw the diagrams as indicated. We have not done so here because we also want the diagrams to remind the reader that all of the methods presented in the present article can be applied both in the case where there is this temporal ordering (or some other temporal ordering) and in the case where there is none.

[8] The numerical entries in fig. 1 measure the magnitude of the "main effect" of the latent variable X upon each of the manifest variables (A, B, C, D). The numerical entries in this path diagram, and in the other path diagrams below, will be explained more fully in sec. 2.2, where we shall introduce indices that measure the effects of the latent variable (or variables) upon the manifest variables. These indices will be particularly appropriate for path analysis in the present context where the latent and manifest variables are dichotomous (or polytomous).

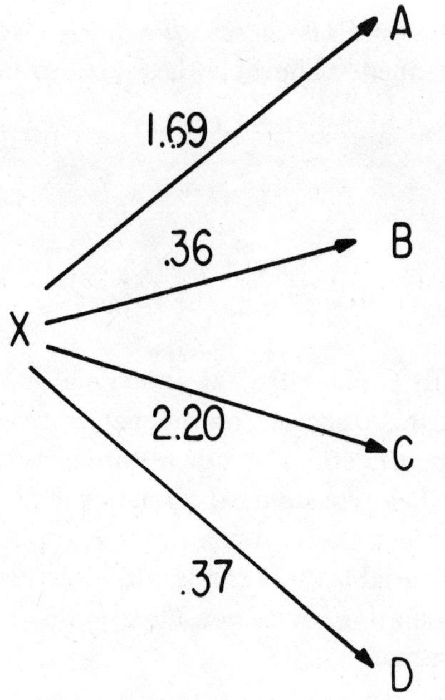

FIG. 1.—Path diagram for the hypothesis H_1 that the latent variable X explains the relationships among the manifest variables (A, B, C, D) in the Coleman panel data (table 1).

$\pi^{\bar{D}X}{}_{lt}$ in formula (5). Recalling the definition of the parameters $\pi^{X}{}_{t}$ (for $t = 1, 2$) presented earlier in this section, we see that

$$\sum_{t=1}^{2} \pi^{X}{}_{t} = 1. \qquad (7)$$

Similarly, we find that

$$\sum_{i=1}^{2} \pi^{\bar{A}X}{}_{it} = 1, \sum_{j=1}^{2} \pi^{\bar{B}X}{}_{jt} = 1, \sum_{k=1}^{2} \pi^{\bar{C}X}{}_{kt} = 1, \sum_{l=1}^{2} \pi^{\bar{D}X}{}_{lt} = 1, \qquad (8)$$

for $t = 1, 2$ (see n. 6). Thus, we need only estimate the following nine parameters, since the other parameters can be expressed in terms of these nine:[9]

$$\pi^{X}{}_{1}, \pi^{\bar{A}X}{}_{1t}, \pi^{\bar{B}X}{}_{1t}, \pi^{\bar{C}X}{}_{1t}, \pi^{\bar{D}X}{}_{1t} \quad (\text{for } t = 1, 2). \qquad (9)$$

The nine parameters could be estimated by the usual method for the latent class model (the Anderson-Lazarsfeld-Dudman method), but there

[9] The nine parameters can also be listed as follows: $\pi^{X}{}_{1}$, $\pi^{\bar{A}X}{}_{11}$, $\pi^{\bar{A}X}{}_{12}$, $\pi^{\bar{B}X}{}_{11}$, $\pi^{\bar{B}X}{}_{12}$, $\pi^{\bar{C}X}{}_{11}$, $\pi^{\bar{C}X}{}_{12}$, $\pi^{\bar{D}X}{}_{11}$, $\pi^{\bar{D}X}{}_{12}$. See table 3 and the last sentence of n. 16 below.

are different ways of carrying out this method,[10] and the numerical values that are actually obtained for the estimates will depend upon the way in which the method is carried out (see, e.g., Anderson 1954). For example, for the data in table 1, depending upon the way in which the method is carried out, we can obtain estimates for π^{X}_{1} that range from 0.39 to 0.52, estimates of $\pi^{\bar{A}X}_{11}$ that range from 0.46 to 1.42, estimates of $\pi^{\bar{A}X}_{12}$ that range from —0.77 to 0.28, etc. (With this method we can obtain estimates that are obviously not permissible; e.g., the estimate of 1.42 for the probability $\pi^{\bar{A}X}_{11}$, or the estimate of —0.77 for the probability $\pi^{\bar{A}X}_{12}$.) As noted in the earlier literature (e.g., Anderson 1959), the estimates obtained by this method are not as efficient as the corresponding estimates obtained by the maximum-likelihood method for model (5)–(6). In other words, the usual method provides estimates that have a larger asymptotic variance in the present context than that obtained with the maximum-likelihood method.

In the present article, we shall present a new method for calculating the maximum-likelihood estimate for the parameters in various kinds of latent structures. This method is simpler and more general than the procedure proposed earlier by McHugh (1956) for the latent class model (see sec. 5 below). The method introduced in the present article can be applied to the latent class model *and* also to a wide variety of other latent structures (see, e.g., figs. 1–4, 6, 8, 10, and models H_1–H_{23} listed in tables 5, 15, and 18 below); whereas McHugh's formulas could be applied only to the (unrestricted) latent class model. Even for the latent class model, the method introduced here is easier to apply than are McHugh's formulas.

For ease of exposition, we shall apply the general method introduced here to each set of data in the particular section devoted to the analysis of that set of data, and then in section 5 we shall describe the method itself in some detail. Applying this method now to the data in table 1, we obtain the maximum-likelihood estimates presented in table 3.

By inserting the estimated parameters obtained from table 3 into formulas (5)–(6), we obtain the maximum-likelihood estimate of π^{ABCDX}_{ijklt} and π_{ijkl} under hypothesis H_1, and we can then estimate the expected frequency F_{ijkl} under this hypothesis by inserting the estimate of π_{ijkl} into formula (4). The estimated F_{ijkl} under hypothesis H_1 are presented in the column so designated in table 4, and they can be compared directly with the corresponding observed frequencies f_{ijkl} in table 1, which

[10] See, e.g., Lazarsfeld and Henry 1968, pp. 71–75. In carrying out this method, any one of the manifest variables could be selected as the "stratifier," and different subsets of the manifest variables could be selected to form the "basic matrix"; but there are no general rules for determining which particular selection is preferable, or how to combine the estimates of a given parameter that are obtained with different selections.

TABLE 3

Maximum-Likelihood Estimate of the Parameters in Model H_1 (Fig. 1), Described by Formulas (5)–(6), for the Four-Way Cross-Classification (Table 1)

Latent Class t	π^{X}_{t}	$\pi^{\bar{A}X}_{1t}$	$\pi^{\bar{B}X}_{1t}$	$\pi^{\bar{C}X}_{1t}$	$\pi^{\bar{D}X}_{1t}$
1	.40	.77	.64	.89	.67
2	.60*	.10	.47	.09	.50

* The estimated π^{X}_{2} can be determined from the estimated π^{X}_{1} using (7). Because of (7), we do not include π^{X}_{2} among the nine parameters listed under (9).

we have listed (to facilitate this comparison) in the initial column of numbers in table 4.[11]

TABLE 4

Estimated Expected Frequency F_{ijkl} under Various Models Applied to the Four-Way Cross-Classification (Table 1)

Variable				Observed	Estimated Expected Frequency			
A	*B*	*C*	*D*	Frequency	Under H_1	Under H_2	Under H_3	Under H_5
+	+	+	+	458	408.29	454.77	438.61	439.62
+	+	+	−	140	199.73	144.23	146.89	145.51
+	+	−	+	110	94.34	109.12	96.50	96.12
+	+	−	−	49	68.49	48.86	49.68	50.02
+	−	+	+	171	227.71	172.30	174.90	173.41
+	−	+	−	182	112.71	179.70	190.61	192.23
+	−	−	+	56	77.92	58.26	59.32	59.69
+	−	−	−	87	63.80	85.76	96.50	96.21
−	+	+	+	184	159.65	188.59	203.39	203.86
−	+	+	−	75	97.12	68.82	68.11	67.48
−	+	−	+	531	403.04	530.52	544.50	542.37
−	+	−	−	281	397.34	283.09	280.32	282.26
−	−	+	+	85	110.58	82.14	81.10	80.41
−	−	+	−	97	76.22	101.45	88.39	89.14
−	−	−	+	338	451.47	337.30	334.68	336.78
−	−	−	−	554	449.60	553.09	544.50	542.88

To test whether hypothesis H_1 is congruent with the data, we compare the observed f_{ijkl} with the corresponding estimated F_{ijkl} under H_1 using either the usual goodness-of-fit χ^2 statistic,

$$\sum_{i=1}^{2}\sum_{j=1}^{2}\sum_{k=1}^{2}\sum_{l=1}^{2}(f_{ijkl}-F_{ijkl})^2/F_{ijkl}, \tag{10}$$

[11] The other columns of table 4 will be discussed later in this section and in sec. 2.4 below.

or the corresponding χ^2 based upon the likelihood-ratio statistic:

$$2\sum_{i=1}^{2}\sum_{j=1}^{2}\sum_{k=1}^{2}\sum_{l=1}^{2} f_{ijkl}\log(f_{ijkl}/F_{ijkl}). \tag{11}$$

The χ^2 value obtained from (10) or (11) can be assessed by comparing its numerical value with the percentiles of the tabulated χ^2 distribution. The degrees of freedom for testing hypothesis H_1 will be $16 - 1 - 9 = 6$ (since [*a*] there are 16 cells in table 1, [*b*] there is the restriction of the form [1] to which the f_{ijkl} are subject, and [*c*] there are nine parameters in [9] that are estimated under the hypothesis). More generally, the degrees of freedom can be calculated by subtracting the number of independent parameters from the number of cells in the table minus one.[12]

Using (10) and (11), we obtain the χ^2 values 251.17 and 249.50, respectively. Since there were 6 df under H_1, we see that this hypothesis is ***not*** congruent with the data.

We shall next consider a number of different hypotheses (models) pertaining to table 1, and shall list in table 5 the χ^2 values obtained in testing

TABLE 5

χ^2 VALUES FOR SOME MODELS PERTAINING TO COLEMAN PANEL DATA (TABLE 1 AND COL. 1 IN TABLE 2)

Model	df	Likelihood-Ratio χ^2	Goodness-of-Fit χ^2
H_1	6	249.50	251.17
H_2	4	1.27	1.28
H_3	6	8.75	8.78
H_4	6	36.11	36.03
H'_3	6	8.75	8.78
H''_3	7	8.78	8.82
H_5	7	8.85	8.89
H_6	6	2.71	2.71
H_7	1	0.03	0.03
H_8	1	0.03	0.03
H_9	2	0.21	0.21
H_{10}	3	1.04	1.06
H_{11}	5	2.29	2.32
H_{12}	6	4.64	4.80

12 This rule for calculating the degrees of freedom can be applied in a straightforward way in the analysis of each of the models considered in this article, except for model H_7 in sec. 2.5. This rule applies whenever the parameters in the model are "identifiable" (see, e.g., Goodman 1974). (In each of the models in this article, except for H_7, the parameters are identifiable.) To analyze latent structures in which the parameters are not identifiable, see Goodman (1974).

these hypotheses. We include both the χ^2 values (10) and (11) in table 5; but in the present context, (11) has some advantages (see, e.g., Goodman 1968, 1970). In the remaining discussions, we shall use only the χ^2 value based upon (11), which we shall denote by the X^2 symbol.[13]

Hypothesis H_1 in table 5 was discussed above. We next consider hypothesis H_2. This hypothesis states that the relationships among the manifest variables (A, B, C, D) in table 1 can be explained by introducing two latent dichotomous variables, say, Y and Z, where Y denotes latent self-perceived membership in the leading crowd and Z denotes latent attitude toward the leading crowd. We now replace figure 1 for H_1 by figure 2 for H_2. Figure 2 shows that variables A and C are affected by

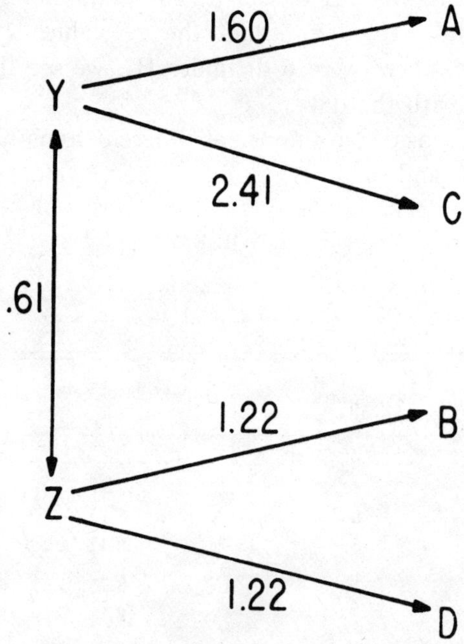

FIG. 2.—Path diagram for the hypothesis H_2 that the latent variables Y and Z explain the relationships among the manifest variables (A, B, C, D) in the Coleman panel data (table 1).

latent variable Y, and that variables B and D are affected by latent variable Z, but variables A, B, C, D do not have any direct effects upon each other.

We included in figure 2 both the direct effects of the latent variables upon the corresponding manifest (indicator) variables *and* the direct effects

[13] The X^2 symbol, which we shall use to denote the χ^2 value based upon (11), should not be confused with the symbol X, which we have used to denote a particular latent variable (as in, e.g., [5]–[6]).

of the latent variables upon each other, but it is important to keep in mind that the effects between variables Y and Z included in the path diagram pertain to the relationship between *latent* variables while the other effects in the diagram pertain to the relationship between each latent variable and its corresponding indicator variables. (With respect to the double-headed arrow between variables Y and Z in fig. 2, see the last two sentences in n. 27 below; with respect to the meaning of the numerical entry associated with this arrow and with the other arrows in fig. 2, see n. 8 above and sec. 2.2 below.)

Let π^{Y}_{r} denote the probability that an individual will be in latent class r on variable Y ($r = 1, 2$), let π^{Z}_{s} denote the probability that an individual will be in latent class s on variable Z ($s = 1, 2$), and let π^{YZ}_{rs} denote the probability that an individual will be at level (r, s) on the joint variable (Y, Z). These probabilities are displayed in the usual 2×2 table format in table 6. (Later in this section, we shall provide estimates of the probabilities displayed in table 6 [see corresponding entries in table 8],

TABLE 6

JOINT DISTRIBUTION OF LATENT VARIABLES Y AND Z

		Variable Z 1	Variable Z 2	
Variable Y	1	π^{YZ}_{11}	π^{YZ}_{12}	$\pi^{Y}_{1} = \pi^{YZ}_{11} + \pi^{YZ}_{12}$
	2	π^{YZ}_{21}	π^{YZ}_{22}	$\pi^{Y}_{2} = \pi^{YZ}_{21} + \pi^{YZ}_{22}$
		$\pi^{Z}_{1} = \pi^{YZ}_{11} + \pi^{YZ}_{21}$	$\pi^{Z}_{2} = \pi^{YZ}_{12} + \pi^{YZ}_{22}$	

and from these estimates we shall be able to calculate the direct effects of the latent variables [Y and Z] upon each other, as indicated by the double-headed arrow in fig. 2.)

Given that an individual is at level (r, s) on the joint variable (Y, Z), let $\pi^{\bar{A}YZ}_{irs}$ denote the *conditional* probability that he will be at level i on variable A, and let the *conditional* probabilities $\pi^{\bar{B}YZ}_{jrs}$, $\pi^{\bar{C}YZ}_{krs}$, $\pi^{\bar{D}YZ}_{lrs}$ be similarly defined. Hypothesis H_2 as described in figure 2 states that the *conditional* probabilities $\pi^{\bar{A}YZ}_{1rs}$ and $\pi^{\bar{C}YZ}_{1rs}$ depend upon the level r on variable Y (but not upon the level s on variable Z) and that the *conditional* probabilities $\pi^{\bar{B}YZ}_{1rs}$ and $^{\bar{D}YZ}_{1rs}$ depend upon the level s on variable Z (but not upon the level r on variable Y).[14] We can express this as follows:

[14] In the above statement, we could have replaced $\pi^{\bar{A}YZ}_{1rs}$, $\pi^{\bar{B}YZ}_{1rs}$, $\pi^{\bar{C}YZ}_{1rs}$, $\pi^{\bar{D}YZ}_{1rs}$ by $\pi^{\bar{A}YZ}_{irs}$, $\pi^{\bar{B}YZ}_{jrs}$, $\pi^{\bar{C}YZ}_{krs}$, $\pi^{\bar{D}YZ}_{lrs}$, respectively. This is a consequence of (8) and the fact that all the parameters in the model can be expressed in terms of the parameters listed in (9).

$$\pi^{\bar{A}YZ}{}_{1r1} = \pi^{\bar{A}YZ}{}_{1r2} = \pi^{\bar{A}Y}{}_{1r};\ \pi^{\bar{C}YZ}{}_{1r1} = \pi^{\bar{C}YZ}{}_{1r2} = \pi^{\bar{C}Y}{}_{1r} \text{ (for } r = 1, 2)$$

and (12)

$$\pi^{\bar{B}YZ}{}_{11s} = \pi^{\bar{B}YZ}{}_{12s} = \pi^{\bar{B}Z}{}_{1s};\ \pi^{\bar{D}YZ}{}_{11s} = \pi^{\bar{D}YZ}{}_{12s} = \pi^{\bar{D}Z}{}_{1s} \text{ (for } s = 1, 2).$$

We shall now express hypothesis H_2 in terms of the usual kind of latent class model. We can view the four cells of the 2×2 table displayed in table 6 as four latent classes of a single latent variable X. In other words, we can use the letter X now to denote the joint variable (Y, Z), and we can refer to the four latent classes as numbered from 1 to 4. The probabilities $\pi^{YZ}{}_{rs}$ in the four cells of the 2×2 table in table 6 can be represented by the symbol $\pi^{X}{}_{t}$ (for $t = 1, 2, 3, 4$), as indicated in table 7, and the

TABLE 7

DISTRIBUTION OF JOINT LATENT VARIABLE (Y, Z) EXPRESSED IN TERMS OF LATENT VARIABLE X

Latent Class t	$\pi^{X}{}_{t}$	Latent Class (r, s)	$\pi^{YZ}{}_{rs}$
1	$\pi^{X}{}_{1}$	(1, 1)	$\pi^{YZ}{}_{11}$
2	$\pi^{X}{}_{2}$	(1, 2)	$\pi^{YZ}{}_{12}$
3	$\pi^{X}{}_{3}$	(2, 1)	$\pi^{YZ}{}_{21}$
4	$\pi^{X}{}_{4}$	(2, 2)	$\pi^{YZ}{}_{22}$

conditional probabilities introduced in the preceding paragraph can be represented by the symbols $\pi^{\bar{A}X}{}_{it}$, $\pi^{\bar{B}X}{}_{jt}$, $\pi^{\bar{C}X}{}_{kt}$, $\pi^{\bar{D}X}{}_{lt}$ (for $t = 1, 2, 3, 4$).

When we consider the latent class model in which variable X has four latent classes, then formulas (5) and (6) still hold true, but t in (5) now applies when $t = 1, 2, 3, 4$ (rather than when $t = 1, 2$), and t in (6) is summed now from 1 to 4 (rather than from 1 to 2). In other words, we now replace (6) by

$$\pi_{ijkl} = \sum_{t=1}^{4} \pi^{ABCDX}{}_{ijklt}. \tag{13}$$

Similarly, formula (7) is replaced now by

$$\sum_{t=1}^{4} \pi^{X}{}_{t} = 1, \tag{14}$$

but formula (8) remains unchanged except that t in (8) now applies when $t = 1, 2, 3, 4$ (rather than $t = 1, 2$). Since the *conditional* probability $\pi^{\bar{A}X}{}_{it}$ represents the *conditional* probability $\pi^{\bar{A}YZ}{}_{irs}$ described earlier (with X representing (Y, Z), and t representing (r, s) as in table 7), and the *conditional* probabilities $\pi^{\bar{B}X}{}_{jt}$, $\pi^{\bar{C}X}{}_{kt}$, $\pi^{\bar{D}X}{}_{lt}$ also represent the corresponding *conditional* probabilities described earlier, we find that condition (12) can now be rewritten in the following equivalent form:[15]

$$\pi^{\bar{A}X}{}_{11} = \pi^{\bar{A}X}{}_{12},\ \pi^{\bar{A}X}{}_{13} = \pi^{\bar{A}X}{}_{14};\ \pi^{\bar{C}X}{}_{11} = \pi^{\bar{C}X}{}_{12},\ \pi^{\bar{C}X}{}_{13} = \pi^{\bar{C}X}{}_{14};$$
$$\pi^{\bar{B}X}{}_{11} = \pi^{\bar{B}X}{}_{13},\ \pi^{\bar{B}X}{}_{12} = \pi^{\bar{B}X}{}_{14};\ \pi^{\bar{D}X}{}_{11} = \pi^{\bar{D}X}{}_{13},\ \pi^{\bar{D}X}{}_{12} = \pi^{\bar{D}X}{}_{14}. \quad (15)$$

Thus, hypothesis H_2 states that the latent variable X has four latent classes ($t = 1, 2, 3, 4$), *and* that the *conditional* probabilities $\pi^{\bar{A}X}{}_{1t}$, $\pi^{\bar{B}X}{}_{1t}$, $\pi^{\bar{C}X}{}_{1t}$, $\pi^{\bar{D}X}{}_{1t}$ are subject to condition (15).

As we noted earlier, the parameters $\pi^{X}{}_{t}$ are subject to (14), and the parameters $\pi^{\bar{A}X}{}_{it}$, $\pi^{\bar{B}X}{}_{jt}$, $\pi^{\bar{C}X}{}_{kt}$, $\pi^{\bar{D}X}{}_{lt}$ are subject to (8) (for $t = 1, 2, 3, 4$) as well as to condition (15). Thus, for purposes of estimation, we need only consider the following 11 parameters, since the other parameters can be expressed in terms of these eleven:[16]

$$\pi^{X}{}_{1},\ \pi^{X}{}_{2},\ \pi^{X}{}_{3}$$
$$\pi^{\bar{A}X}{}_{1t},\ \pi^{\bar{C}X}{}_{1t} \quad (\text{for } t = 1, 3), \quad (16)$$
$$\pi^{\bar{B}X}{}_{1t},\ \pi^{\bar{D}X}{}_{1t} \quad (\text{for } t = 1, 2).$$

The maximum-likelihood estimates of these parameters (under hypothesis H_2) are presented in table 8.[17]

As we did earlier with the estimated parameters in table 3, we can now use the estimated parameters in table 8 to estimate $\pi^{ABCDX}{}_{ijklt}$, π_{ijkl}, and F_{ijkl} under hypothesis H_2 (see [5], [13], and [4]). The estimated F_{ijkl} under H_2 were listed in the column so designated in table 4. To test whether hypothesis H_2 is congruent with the data, we compare the observed f_{ijkl} with the corresponding estimated F_{ijkl} under H_2 (see appropriate columns in table 4). From table 5 we see that, under H_2, the corresponding χ^2 value based upon (11) is 1.27, with 4 df.[18] Thus, we find that hypothesis H_2 fits the data very well indeed.

[15] Recall that condition (12) is a statement of the restrictions imposed by H_2 upon the *conditional* probabilities described earlier.

[16] These 11 parameters pertain to the corresponding 11 entries that are not asterisked in table 8. (The asterisked entries in this table can be expressed in terms of the entries that are not asterisked.) A similar remark applies to the nine entries that are not asterisked in table 3 (see the corresponding nine parameters listed in [9]), and to the entries in the corresponding tables presented below for other models.

[17] Although we are concerned here with a latent class model, we cannot use McHugh's formulas (1956) in the present context since they are not applicable to the case where the parameters are subject to condition (15), or to any other condition of this general kind.

[18] There are 4 df for testing H_2 since $16 - 1 - 11 = 4$. Recall that there were 11 parameters listed in (16).

TABLE 8

MAXIMUM-LIKELIHOOD ESTIMATE OF PARAMETERS IN MODEL H_2 (FIG. 2), DESCRIBED BY FORMULAS (5), (13), AND (15), FOR FOUR-WAY CROSS-CLASSIFICATION (TABLE 1)

LATENT CLASS						
t	(r, s)	$\pi^{X}_{t} = \pi^{YZ}_{rs}$	$\pi^{\bar{A}X}_{1t} = \pi^{\bar{A}YZ}_{1rs}$	$\pi^{\bar{B}X}_{1t} = \pi^{\bar{B}YZ}_{1rs}$	$\pi^{\bar{C}X}_{1t} = \pi^{\bar{C}YZ}_{1rs}$	$\pi^{\bar{D}X}_{1t} = \pi^{\bar{D}YZ}_{1rs}$
1	(1, 1)	.27	.75	.81	.91	.83
2	(1, 2)	.13	.75*	.27	.91*	.30
3	(2, 1)	.23	.11	.81*	.08	.83*
4	(2, 2)	.37*	.11*	.27*	.08*	.30*

NOTE.—The asterisked entries can be determined from the other entries in this table using (14) and (15). Because of (14) and (15), we do not include the π's corresponding to the asterisked entries among the 11 parameters listed under (16).

Under hypothesis H_2, variables A and C could be viewed as fallible indicators for the latent variable Y, and variables B and D could be viewed as fallible indicators for the latent variable Z (see fig. 2). For example, variables A and C indicate the individual's *observed* level on the self-perceived membership variable (at the first and second interviews), whereas variable Y pertains to his *latent* level on this variable. With respect to this variable, if an individual's "true colors" will show up "in the end" (say, at the second interview), then variable C would be an *accurate* indicator for the latent variable Y. In other words, we would find in this case that the *conditional* probability $\pi^{\bar{C}Y}{}_{1r}$ could be expressed as follows:

$$\pi^{\bar{C}Y}{}_{1r} = \begin{cases} 1 \text{ when } r = 1 \\ 0 \text{ when } r = 2. \end{cases} \tag{17}$$

Expressed in terms of the corresponding *conditional* probabilities $\pi^{\bar{C}X}{}_{1t}$, formula (17) can be rewritten in the following equivalent form:

$$\pi^{\bar{C}X}{}_{1t} = \begin{cases} 1 \text{ when } t = 1, 2 \\ 0 \text{ when } t = 3, 4. \end{cases} \tag{18}$$

Consider next the hypothesis H_3 that the relationships among the manifest variables (A, B, C, D) in table 1 can be explained by introducing the two latent dichotomous variables Y and Z (as in hypothesis H_2), *and* that variable C is an *accurate* indicator of the latent variable Y. We now replace figure 2 for H_2 by figure 3 for H_3. Figure 3 shows that variable A is affected by latent variable Y (which can be replaced by the *accurate* indicator C), and that variables B and D are affected by latent variable Z; but variables B, D, and the joint variable (A, C) do not have any direct effects upon each other. Hypothesis H_3 is the same as H_2 except that it introduces the additional condition (18).[19] Thus, H_3 states that the latent class model described by (5) and (13) is true, *and* conditions (15) and (18) are also true.

Note that all but two of the 11 parameters listed in (16) will also be

[19] Under hypothesis H_2, the probabilities $\pi^{\bar{C}X}{}_{1t}$ will satisfy condition (15); and they are, of course, also subject to the same restriction to which all probabilities are subject; namely, that $0 \leqq \pi^{\bar{C}X}{}_{1t} \leqq 1$, for $t = 1, 2, 3, 4$. Under H_3, these probabilities take on extreme values: $\pi^{\bar{C}X}{}_{11} = \pi^{\bar{C}X}{}_{12} = 1$ and $\pi^{\bar{C}X}{}_{13} = \pi^{\bar{C}X}{}_{14} = 0$. Some theoretical results about probabilities and their maximum-likelihood estimates pertain only to the case where the probabilities are *strictly within* the range between zero and one, but not to the case where they take on the extreme values. In the present context, the model obtained when some probabilities take on extreme values (as in H_3) can be analyzed using the same general method (appropriately modified as in sec. 5.2) that we use with the corresponding model obtained when the probabilities are not restricted in that particular way (as in H_2), but the degrees of freedom for testing each model will be different (as we shall see below).

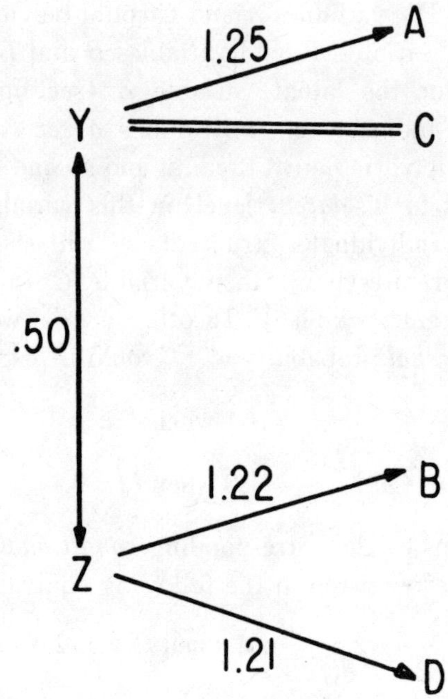

FIG. 3.—Path diagram for the hypothesis H_3 that the latent variables Y and Z explain the relationships among the manifest variables (A, B, C, D) in the Coleman panel data (table 1), *and* that variable C is an accurate indicator for the latent variable Y.

estimated under H_3. The exceptions are the parameters $\pi^{\bar{C}X}{}_{11}$ and $\pi^{\bar{C}X}{}_{13}$, which are not estimated under H_3 since they are completely determined by condition (18). Thus, under H_3 there will be nine (rather than 11) independent parameters to estimate, and the number of degrees of freedom for testing this hypothesis will be $16 - 1 - 9 = 6$. The estimated parameters under H_3 are listed in table 9. The corresponding estimated F_{ijkl} were listed in the column so designated in table 4, and the χ^2 value based on (11) is 8.75, with 6 df (see table 5).

Comparing the χ^2 value obtained for hypothesis H_3 (viz., $X^2(H_3) = 8.75$) with the percentiles of the tabulated χ^2 distribution with 6 df, we see that H_3 fits the data to an acceptable degree. In addition, comparison of this χ^2 value with the corresponding value obtained for hypothesis H_1 (viz., $X^2(H_1) = 249.50$) indicates that H_3 fits the data much better than does H_1.[20] On the other hand, comparison of $X^2(H_3)$ with $X^2(H_2)$ indicates that H_2 fits the data better than does H_3. Indeed, since H_3 states that both hypothesis H_2 is true *and* condition (18) is true, the difference $X^2(H_3) - X^2(H_2)$ can be used to test the hypothesis that condition (18) is true, assuming that H_2 is true. This difference can be assessed by com-

[20] Note that H_1 and H_3 have the same number of degrees of freedom.

TABLE 9

MAXIMUM-LIKELIHOOD ESTIMATE OF PARAMETERS IN MODEL H_3 (FIG. 3), DESCRIBED BY FORMULAS (5), (13), (15), AND (18), FOR FOUR-WAY CROSS-CLASSIFICATION (TABLE 1)

LATENT CLASS						
t	(r, s)	$\pi^{X}{}_{t} = \pi^{YZ}{}_{rs}$	$\pi^{\bar{A}X}{}_{1t} = \pi^{\bar{A}YZ}{}_{1rs}$	$\pi^{\bar{B}X}{}_{1t} = \pi^{\bar{B}YZ}{}_{1rs}$	$\pi^{\bar{C}X}{}_{1t} = \pi^{\bar{C}YZ}{}_{1rs}$	$\pi^{\bar{D}X}{}_{1t} = \pi^{\bar{D}YZ}{}_{1rs}$
1	(1, 1)	.27	.68	.80	1.00**	.82
2	(1, 2)	.14	.68*	.26	1.00**	.29
3	(2, 1)	.25	.15	.80*	.00**	.82*
4	(2, 2)	.34*	.15*	.26*	.00**	.29*

NOTE.—The asterisked entries can be determined from the other entries in this table using (14) and (15). The double asterisked entries are specified by the model using (18).

paring its numerical value with the percentiles of the χ^2 distribution. Since $X^2(H_3)$ had 6 df and $X^2(H_2)$ had 4 df, their difference will have 2 df (see, e.g., Goodman 1968, 1970).[21] Since $X^2(H_3) - X^2(H_2) = 7.48$, with 2 df, we conclude that the extra two parameters in H_2 contribute in a statistically significant way to our explanation of the relationships among the manifest variables in table 1. In other words, viewing variable C as a *fallible* indicator (rather than as an accurate indicator) contributes in a statistically significant way to the analysis of this particular set of data.

For the sake of completeness, we also included in table 5 the χ^2 values for a hypothesis H_4 of the same general form as H_3. Hypothesis H_4 states that both hypothesis H_2 is true *and* variable D is an *accurate* indicator of the latent variable Z. Hypothesis H_4 can be analyzed in the same way as we analyzed H_3. From table 5 we see that this hypothesis is not congruent with the data.

2.2. The Magnitude of the Effects in the Path Diagrams

In the preceding section, we included figures 1, 2, and 3, but we did not explain how the numerical magnitudes in these path diagrams were calculated (see n. 8). We shall now explain this in a way that will also indicate how the magnitudes of the effects are calculated more generally for other models in which some variables are latent and others are manifest.[22] The reader who is not interested in this technical detail can skip this section.

For the sake of simplicity, we first consider hypothesis H_1 as described by figure 1 and formulas (5)–(6). Let $\Omega^{\bar{A}}{}_{.jklt}$ denote the odds that an individual will be at level 1 rather than level 2 on variable A, given that his joint level on the remaining variables (B, C, D, X) is (j, k, l, t). These odds can be defined in terms of the probabilities π^{ABCDX}_{ijklt} which were introduced in the preceding section:

$$\Omega^{\bar{A}}{}_{.jklt} = \pi^{ABCDX}_{1jklt}/\pi^{ABCDX}_{2jklt}. \tag{19}$$

Since $\Omega^{\bar{A}}{}_{.jklt}$ is based upon the probabilities (or "expected proportions") π^{ABCDX}_{ijklt}, we shall refer to $\Omega^{\bar{A}}{}_{.jklt}$ as the "expected odds."[23] From (5) and (19) we see that

[21] Recall that there were two parameters listed in (16) that were estimated under H_2 but *not* under H_3: the parameters $\pi^{\bar{C}X}_{11}$ and $\pi^{\bar{C}X}_{13}$. We are concerned here with determining whether the estimates of these two parameters under H_2 (viz., .91 and .08, respectively, in table 8) can be replaced simply by 1 and 0, respectively.

[22] Both the latent and manifest variables in the path diagrams considered here are dichotomous. For the case where the latent variables are polytomous (not necessarily dichotomous), see Goodman (1973*d*); for the case where the manifest variables are polytomous, see Goodman (1971*a*, 1973*c*).

[23] We use the term "expected odds" usually to distinguish it from the corresponding "observed odds," which are defined in terms of observed proportions. With respect

$$\Omega^{\bar{A}}{}_{\cdot jklt} = \pi^{\bar{A}X}{}_{1t}/\pi^{\bar{A}X}{}_{2t}. \qquad (20)$$

Formula (20) states that the expected odds $\Omega^{\bar{A}}{}_{\cdot jklt}$ pertaining to variable A are affected by the level t of variable X, but *not* by the level (j, k, l) of the other variables (B, C, D).

For the 2×2 table describing the *expected* relationship between variables A and X, let $\Omega^{\bar{A}X}{}_{\cdot t}$ denote the odds that an individual will be at level 1 rather than level 2 on variable A, given that he is at level t on variable X:

$$\Omega^{\bar{A}X}{}_{\cdot t} = \pi^{\bar{A}X}{}_{1t}/\pi^{\bar{A}X}{}_{2t}. \qquad (21)$$

From (20) and (21) we see that

$$\Omega^{\bar{A}}{}_{\cdot jklt} = \Omega^{\bar{A}X}{}_{\cdot t}. \qquad (22)$$

Formula (22) can be rewritten as

$$\Omega^{\bar{A}}{}_{\cdot jklt} = \gamma^{\bar{A}}\gamma^{\bar{A}X}{}_{\cdot t}, \qquad (23)$$

where

$$\gamma^{\bar{A}} = (\Omega^{\bar{A}X}{}_{\cdot 1}\Omega^{\bar{A}X}{}_{\cdot 2})^{1/2}, \qquad (24)$$

$$\gamma^{\bar{A}X} = (\Omega^{\bar{A}X}{}_{\cdot 1}/\Omega^{\bar{A}X}{}_{\cdot 2})^{1/2}, \qquad (25)$$

$$\gamma^{\bar{A}X}{}_{\cdot 1} = \gamma^{\bar{A}X}, \gamma^{\bar{A}X}{}_{\cdot 2} = 1/\gamma^{\bar{A}X}. \qquad (26)$$

From (23) and (26) we see that the effect of variable X upon $\Omega^{\bar{A}}{}_{\cdot jklt}$ is to introduce the multiplicative factor $\gamma^{\bar{A}X}$ when variable X is at level 1, and the multiplicative factor $1/\gamma^{\bar{A}X}$ when variable X is at level 2. The parameter $\gamma^{\bar{A}X}$ is the "main effect" of variable X on the expected odds $\Omega^{\bar{A}}{}_{\cdot jklt}$.

From (25) we see that $\gamma^{\bar{A}X}$ is equal to the square root of the usual "odds-ratio" in the 2×2 table describing the *expected* relationship between variables A and X.

Using (21) we see that (25) can be rewritten as

$$\gamma^{\bar{A}X} = [(\pi^{\bar{A}X}{}_{11}\pi^{\bar{A}X}{}_{22})/(\pi^{\bar{A}X}{}_{21}\pi^{\bar{A}X}{}_{12})]^{1/2}. \qquad (27)$$

Thus, the "expected odds-ratio" in formula (25) is the same as the "expected cross-product ratio" in formula (27).[24] Similarly, we see that (24) can be rewritten as

---

to the expected odds (19), strictly speaking there actually is no corresponding observed quantity since we cannot observe a given individual's latent class t with respect to variable X; and therefore we cannot observe his level (i, j, k, l, t) with respect to the joint variable (A, B, C, D, X). We shall use the term "expected" to refer to the expected value F_{ijkl} or the expected proportion π_{ijkl} (or $\pi^{ABCDX}{}_{ijklt}$) under some specified hypothesis and also to certain quantities (such as the expected odds [19]) that are based upon the F_{ijkl}, π_{ijkl}, or $\pi^{ABCDX}{}_{ijklt}$.

[24] See the end of the last sentence in n. 23 above.

$$\gamma^{\bar{A}} = [(\pi^{\bar{A}X}{}_{11}\pi^{\bar{A}X}{}_{12})/(\pi^{\bar{A}X}{}_{21}\pi^{\bar{A}X}{}_{22})]^{1/2}. \tag{28}$$

To estimate the parameters $\gamma^{\bar{A}}$ and $\gamma^{\bar{A}X}$ in equation (23), we can use formulas (28) and (27), respectively, inserting into these formulae the estimated $\pi^{\bar{A}X}{}_{it}$ obtained from table 3. We shall now use the estimated values of $\gamma^{\bar{A}}$ and $\gamma^{\bar{A}X}$ to obtain further insight into the meaning of equation (23). Inserting the estimated values of $\gamma^{\bar{A}}$ and $\gamma^{\bar{A}X}$ in (23), we find that the estimate of $\Omega^{\bar{A}}{}_{\cdot jklt}$ can be expressed as follows:[25]

$$\Omega^{\bar{A}}{}_{\cdot jkl1} = (0.61)(5.43) = 3.32, \tag{29a}$$

$$\Omega^{\bar{A}}{}_{\cdot jkl2} = (0.61)\left(\frac{1}{5.43}\right) = 0.11. \tag{29b}$$

Comparing (29a) with (29b), we see the effect of $\gamma^{\bar{A}X}$ (i.e., the effect of variable X on the estimate of the expected odds pertaining to variable A).

Equation (23) expresses $\Omega^{\bar{A}}{}_{\cdot jklt}$ as a *product* of certain γ parameters. This formula can also be expressed in additive form via logarithms. Letting $\Phi^{\bar{A}}{}_{\cdot jklt}$ denote the natural logarithm of $\Omega^{\bar{A}}{}_{\cdot jklt}$, we see from (23) that the $\Phi^{\bar{A}}{}_{\cdot jklt}$ can be expressed as

$$\Phi^{\bar{A}}{}_{\cdot jklt} = \beta^{\bar{A}} + \beta^{\bar{A}X}{}_{\cdot t}, \tag{30}$$

where

$$\beta^{\bar{A}} = \log \gamma^{\bar{A}}, \beta^{\bar{A}X} = \log \gamma^{\bar{A}X}. \tag{31}$$

From (26) and (31), we see that

$$\beta^{\bar{A}X}{}_{\cdot 1} = \beta^{\bar{A}X}, \beta^{\bar{A}X}{}_{\cdot 2} = -\beta^{\bar{A}X}. \tag{32}$$

The parameter $\beta^{\bar{A}X}$ is the "main effect" of variable X on the "expected log-odds" $\Phi^{\bar{A}}{}_{\cdot jklt}$; and $\beta^{\bar{A}}$ is the "main effect of the general mean." To estimate the β parameters, we insert the estimated γ's in (31).[26] The estimated $\beta^{\bar{A}X}$ is included in figure 1 as the numerical value corresponding to the arrow from X to A; that is, the effect of variable X on A. The numerical values corresponding to the other arrows in figure 1 were obtained in the same way.

With respect to figure 2, the numerical values corresponding to the arrows from Y to A and C, and from Z to B and D, were obtained in a similar way. For example, the parameter $\gamma^{\bar{A}Y}$ can be written as (27) modified by replacing the letter X there by Y; and this parameter can be

[25] All calculations in this paper were carried out to more significant digits than are reported here.

[26] The β parameters can also be estimated directly, using the estimated $\pi^{\bar{A}X}{}_{it}$, by rewriting (27) and (28) in an equivalent form (via logarithms) expressing the β's in terms of the log $\pi^{\bar{A}X}{}_{it}$.

estimated by inserting in the modified (27) the corresponding estimated $\pi^{\bar{A}Y}{}_{ir}$ obtained from table 8, noting that

$$\pi^{\bar{A}Y}{}_{11} = \pi^{\bar{A}X}{}_{11} = \pi^{\bar{A}X}{}_{12}, \; \pi^{\bar{A}Y}{}_{12} = \pi^{\bar{A}X}{}_{13} = \pi^{\bar{A}X}{}_{14}, \tag{33}$$

as can be seen from (12), (15), and table 8. Having thus estimated $\gamma^{\bar{A}Y}$, the corresponding $\beta^{\bar{A}Y}$ can be estimated using (31) as earlier, but with the letter X replaced there by Y.

The numerical value corresponding to the double-headed arrow between Y and Z in figure 2 can be calculated from the estimates of the probabilities $\pi^{YZ}{}_{rs}$, which were the entries in the 2×2 table presented in table 6. These estimated probabilities are given in the column so designated in table 8. As was the case earlier with formula (27), we can now estimate the parameter γ^{YZ} using the square root of the estimated odds-ratio (i.e., the cross-product ratio) calculated for the 2×2 table presented in table 6; and the parameter β^{YZ} can be estimated using (31) as earlier, but with the letters AX replaced now by YZ.[27]

We have now explained how all the numerical entries in figures 1 and 2 were obtained. The corresponding entries in other path diagrams (e.g., in fig. 3) can be obtained in the same way (but with the estimated parameters in table 8 replaced by the corresponding entries in, e.g., table 9).[28]

2.3. The Assignment of Individuals with Respect to the Latent Variables

For the sake of simplicity, we return again to hypothesis H_1 as described in figure 1. From table 3, we see that the probability $\pi^{X}{}_{t}$ that an individual will be at level t on the latent variable X is estimated to be .40 and .60, for $t = 1$ and 2, respectively. Given that an individual is at level (i, j, k, l) on the joint variable (A, B, C, D), let $\pi^{ABCD\bar{X}}{}_{ijklt}$ denote the *conditional* probability that he will be at level t on variable X. From the definition of the probabilities $\pi^{ABCDX}{}_{ijklt}$ and π_{ijkl} presented in section 2.1 (see [5]–[6]), we find that the *conditional* probability can be expressed as

$$\pi^{ABCD\bar{X}}{}_{ijklt} = \pi^{ABCDX}{}_{ijklt}/\pi_{ijkl}. \tag{34}$$

In section 2.1, we described how the maximum-likelihood estimates of $\pi^{ABCDX}{}_{ijklt}$ and π_{ijkl} could be obtained from the estimated parameters in model H_1, and we can now insert these estimates in (34) to obtain the

27 In (27) variable A was posterior (in a certain sense) to variable X. In the present context, we can view variable Y as posterior to Z, or variable Z as posterior to Y, or simply that these two variables are related to each other without one being posterior to the other. For further details, see the discussion of double-headed arrows in Goodman (1973*a*) and sec. 5 4 below.

28 With respect to fig. 3, the numerical entries in this particular path diagram can also be obtained by an alternative (somewhat simpler) method, which we shall comment upon later in sec. 2.4.

maximum-likelihood estimate of the *conditional* probability $\pi^{ABCD\bar{X}}_{ijklt}$ under this model. On the other hand, since model H_1 was rejected earlier (see table 5), there is not much point in estimating the *conditional* probabilities for a model that is not congruent with the data. We next consider model H_2, which we noted earlier fit the data very well indeed.

For model H_2 as described by figure 2, the probability π^{YZ}_{rs} that an individual will be at level (r, s) on the joint latent variable (Y, Z) was estimated earlier (see the appropriate column in table 8). As in section 2.1, we use the letter X to denote the joint variable (Y, Z), so the estimate of the probability π^{X}_{t} that an individual will be at level t on variable X (for $t = 1, 2, 3, 4$) can be obtained from the same column of table 8. As in the preceding paragraph, we let $\pi^{ABCD\bar{X}}_{ijklt}$ denote the *conditional* probability which we calculate using (34), where the t in (34) now applies when $t = 1, 2, 3, 4$ (rather than when $t = 1, 2$). The maximum-likelihood estimate of the *conditional* probability $\pi^{ABCD\bar{X}}_{ijklt}$ under hypothesis H_2 can be obtained from (34) by using the probabilities π^{ABCDX}_{ijklt} and π_{ijkl} estimated under H_2 as indicated in section 2.1. The estimated *conditional* probabilities $\pi^{ABCD\bar{X}}_{ijklt}$ are presented in table 10. (Note, e.g., that the estimated $\pi^{ABCD\bar{X}}_{1111t}$ are .94, .05, .01, .00 for $t = 1, 2, 3, 4$, respectively.)

TABLE 10

CONDITIONAL PROBABILITY $\pi^{ABCD\bar{X}}_{ijklt}$ THAT AN INDIVIDUAL WILL BE IN LATENT CLASS t WITH RESPECT TO LATENT VARIABLE X (FOR $t = 1, 2, 3, 4$), GIVEN THAT HIS RESPONSE LEVEL IS (i, j, k, l) ON MANIFEST VARIABLES (A, B, C, D)

Variable				Observed Frequency	Estimated Conditional Probability for Latent Class t			
					$t = 1$	$t = 2$	$t = 3$	$t = 4$
A	B	C	D		$(r, s) = (1, 1)$	$(r, s) = (1, 2)$	$(r, s) = (2, 1)$	$(r, s) = (2, 2)$
+	+	+	+	458	.94	.05	.01	.00
+	+	+	−	140	.59	.39	.01	.01
+	+	−	+	110	.39	.02	.50	.09
+	+	−	−	49	.17	.11	.22	.49
+	−	+	+	171	.60	.38	.01	.01
+	−	+	−	182	.11	.85	.00	.03
+	−	−	+	56	.17	.11	.22	.49
+	−	−	−	87	.02	.18	.03	.77
−	+	+	+	184	.73	.04	.19	.04
−	+	+	−	75	.41	.26	.10	.23
−	+	−	+	531	.03	.00	.82	.16
−	+	−	−	281	.01	.01	.31	.68
−	−	+	+	85	.41	.26	.10	.23
−	−	+	−	97	.07	.49	.02	.42
−	−	−	+	338	.01	.01	.31	.67
−	−	−	−	554	.00	.01	.04	.95

NOTE.—The conditional probability is estimated under model H_2 applied to the four-way cross-classification (table 1).

If we know an individual's level (i, j, k, l) on the joint manifest variable (A, B, C, D), then the entries in table 10 can be used to assign him to a latent level with respect to the latent variable X, by selecting the estimated modal latent level t corresponding to his observed level (i, j, k, l). For example, if the individual is at level (1, 1, 1, 1) on the joint variable (A, B, C, D), then he would be assigned to latent level 1 with respect to the latent variable X, since the estimated $\pi^{ABCD\bar{X}}{}_{1111t}$ (for $t = 1, 2, 3, 4$) in table 10 indicate that the corresponding estimated modal level is obtained when $t = 1$.

The letter X denoted the joint variable (Y, Z) in table 10, and so the *conditional* probability $\pi^{ABCD\bar{X}}{}_{ijklt}$ can also be written as $\pi^{ABCD\overline{YZ}}{}_{ijklrs}$. From the entries in table 10, we can also estimate the *conditional* probabilities $\pi^{ABCD\bar{Y}}{}_{ijklr}$ and $\pi^{ABCD\bar{Z}}{}_{ijkls}$. (Thus, for example, the estimated $\pi^{ABCD\bar{Y}}{}_{1111r}$ are .99 [i.e., .94 + .05] and .01 [i.e., .01 + .00] for $r = 1$ and 2, respectively; and the estimated $\pi^{ABCD\bar{Z}}{}_{1111s}$ are .95 [i.e., .94 + .01] and .05 [i.e., .05 + .00] for $s = 1$ and 2, respectively.) The estimated $\pi^{ABCD\overline{YZ}}{}_{ijklrs}$ in table 10 could be used to assign each individual to a latent level with respect to the joint latent variable (Y, Z) as indicated in the preceding paragraph; and, similarly, the estimated $\pi^{ABCD\bar{Y}}{}_{ijklr}$ and $\pi^{ABCD\bar{Z}}{}_{ijkls}$ can be used to assign him to latent levels with respect to the latent variables Y and Z, respectively.

The methods described above, which we applied to hypothesis H_2 (fig. 2), can also be applied directly to hypotheses H_3, H_4, and to other latent structure models.[29] To save space, we shall not include these results here.

2.4. Comments on Some Related Models

This section can be divided into three parts: First, we shall explain how the latent structure H_3 is related to a particular "elementary" model (H'_3), which we shall describe below. Second, we shall modify H'_3 to obtain a more parsimonious model (H''_3). Third, we shall provide a new latent structure (H_5) that is as parsimonious as H''_3. We shall find that the latent structure model H_5 is, in a certain sense, a more fundamental model than H''_3. Since H'_3 and H''_3 are *not* latent structures, the reader who may be concerned about being sidetracked can skip the details in the discussion of these two models and can move on to the last part of the section where the latent structure H_5 is discussed.

As we noted earlier, hypothesis H_3 states that hypothesis H_2 is true *and* that condition (18) is also true. Thus, if H_3 is true, then H_2 will also

[29] With respect to H_3, since the manifest variable C is an accurate indicator in this case for variable Y, the assignment of an individual to a latent class r with respect to variable Y will also be *accurate* under this model. A similar kind of remark applies to model H_4 and to H_{13} below.

be true. In other words, H_3 implies H_2. We noted earlier that H_3 fit the data to an acceptable degree (see table 5). We shall consider next another hypothesis that is implied by H_3. Consideration of this next hypothesis (H'_3) will shed further light on the meaning of H_3.

Consider the hypothesis that variable A is conditionally independent of the joint variable (B, D) when the level of variable C is held constant. This hypothesis was denoted as $[A \otimes BD|C]$ in the notation introduced in Goodman (1970). This hypothesis is an "elementary" hypothesis, and it can be tested using the χ^2 distribution with 6 df (see Goodman 1970). Letting H'_3 denote this hypothesis, the χ^2 values obtained when testing H'_3 were included in table 5.[30] For the data considered here (table 1), we see from table 5 that $X^2(H_3) = X^2(H'_3)$. We shall now explain why this equality was obtained for this particular set of data.[31]

Since variable Y can be replaced by C in the path diagram for hypothesis H_3 (fig. 3), we see from this diagram that H_3 states that $[A \otimes BD|C]$ *and* that variable Z can explain the relationships among variables B, C, and D. The latter hypothesis we denote as $[B \otimes C \otimes D|Z]$; that is, that variables B, C, and D are mutually independent when the level of variable Z is held constant. Since H_3 states that *both* hypotheses H'_3 *and* $[B \otimes C \otimes D|Z]$ are true, we obtain the following relationship between the χ^2 statistics based upon (11):[32]

$$X^2(H_3) \geqq X^2(H'_3). \qquad (35)$$

(See, e.g., the corresponding entries in table 5 and also in table 15 below.)

The expected frequency F_{ijkl} estimated under H_3 and the corresponding quantity F'_{ijkl} estimated under H'_3 will satisfy the following relationship:

$$F_{ijkl} = F_{jkl}F'_{ijkl}/f_{jkl}, \qquad (36)$$

where f_{jkl} is the observed frequency in cell (j, k, l) of the three-way marginal table $\{BCD\}$, and F_{jkl} is the corresponding expected frequency estimated under hypothesis $[B \otimes C \otimes D|Z]$ (see Goodman 1973*b*). For the three-way marginal table $\{BCD\}$ obtained from table 1, hypothesis $[B \otimes C \otimes D|Z]$ fit these data perfectly; the estimated F_{jkl} were equal

30 As we have already noted, models H'_3 and H''_3 in table 5 are *not* latent structure models, although they are closely related to the latent structure H_3. (We shall discuss H''_3 later in this section.) On the other hand, it may also be worth noting that H'_3 is equivalent to a latent structure model (viz., the model in which there is a latent variable Y that can explain the relationships among variables A, C, and the joint variable (B, D), *and* in which variable C is an *accurate* indicator of the latent variable Y, as in [17]); whereas model H''_3 is *not* equivalent to a latent structure model.

31 This equality will *not* be obtained for the data analyzed in sec. 3 (see $X^2(H_3)$ and $X^2(H'_3)$ in table 15 below).

32 The relationship (35) is a particular example of the more general relationship (43) presented below.

to the corresponding f_{jkl}.[33] Thus, for the data in table 1, the estimated F_{ijkl} and F'_{ijkl} were equal (see [36]), and the $X^2(H_3)$ and $X^2(H'_3)$ were also equal.

The latent class model $[B \otimes C \otimes D|Z]$ pertains to *three* manifest variables (B, C, D) and to *two* latent classes (the two classes of variable Z). More generally, suppose there are m manifest variables and T latent classes in the latent class model, where m and T satisfy the following condition:

$$T = 2^m/(m + 1), \text{ for } m = 3, 4, \ldots \qquad (37)$$

(This condition is satisfied, e.g., when $m = 3$ and $T = 2$.) With m manifest (dichotomous) variables, the m-way table will have 2^m cells, and the number of independent parameters that can vary will be $2^m - 1$ in general (i.e., when only the general multinomial model is imposed). Under the latent class model in which there are T latent classes, the number of independent parameters will be $T - 1 + mT = (m + 1)\ T - 1$. When condition (37) is satisfied, the number of independent parameters under the latent class model will be equal to the number in the general multinomial model, since in this case

$$(m + 1)T - 1 = 2^m - 1. \qquad (38)$$

We shall refer to the latent class model in this case as "quasi-tautological" since the model then has as many parameters in it as the general multinomial model. In this case there are, so to speak, zero df for testing the model. If the usual latent class method is used to estimate the parameters in the quasi-tautological case,[34] the fit will be perfect *whenever* the estimates thus obtained are all in the range from zero to one.[35] *If* the estimates thus obtained are all in this range for the quasi-tautological model, then they will be equal to the corresponding estimates obtained by the maximum-likelihood method in this particular case.[36]

33 We shall explain why this was so in the next paragraph.

34 The usual method for estimating the parameters in the latent class model (i.e., the Anderson-Lazarsfeld-Dudman method) can be applied in the quasi-tautological case when $m = 3$ and $T = 2$. We are concerned here with this special case. On the other hand, the usual latent class method for estimating the parameters is limited to situations where $T \leqq (m + 1)/2$ (see Lazarsfeld and Henry 1968); and it therefore cannot be applied in the quasi-tautological case when $m > 3$, since in that case T will be greater than $(m + 1)/2$ (see [37] for $m > 3$). (For a similar reason, the extension of the usual latent class method, which was proposed by Madansky [1960], also cannot be applied in this case. In addition, for other reasons which we shall not go into here, this "extension" cannot actually be used as a method for estimating the parameters in the more general situation where $T > [m + 1]/2$.)

35 The estimates are all in this range for the three-way table $\{BCD\}$ obtained from table 1, but *not* for the corresponding three-way table obtained from the data analyzed in sec. 3.

36 The above sentence pertains to the quasi-tautological case when $m = 3$ and $T = 2$.

We have now explained why model H_3 (which states that $[A \otimes BD|C]$ *and* that $[B \otimes C \otimes D|Z]$) and model H'_3 (which states only that $[A \otimes BD|C]$) produce the same estimated expected frequencies for the data in table 1. In short, the reason is that the latent class model $[B \otimes C \otimes D|Z]$ is quasi-tautological,[37] and it fits these observed data perfectly.

The results presented earlier in this section can also be used to facilitate the calculation of the numerical entries in table 9 and in figure 3 (pertaining to H_3) for table 1. In this calculation, we can take advantage of the fact that $F_{ijkl} = F'_{ijkl}$ for these data, and the maximum-likelihood method of estimation in this particular case can be replaced by the usual method of estimation in the latent class model.[38]

As we have noted, hypothesis H_3 stated that $[A \otimes BD|C]$ *and* that $[B \otimes C \otimes D|Z]$ for the three-way marginal table $\{BCD\}$. Consider next the related hypothesis H''_3 that $[A \otimes BD|C]$ *and* that the three-factor interaction in the three-way marginal table $\{BCD\}$ is zero (see Goodman 1970, 1971*b*). The latter hypothesis we denote as $[(BCD) = 0]$.[39] Hypothesis H''_3 states that *both* hypotheses H'_3 *and* $[(BCD) = 0]$ are true. These hypotheses can be expressed in terms of the marginals that are fitted under them. Since H'_3 is the hypothesis $[A \otimes BD|C]$, the fitted marginals for H'_3 are $\{AC\}$ and $\{BCD\}$; whereas since H''_3 modifies H'_3 by setting the three-factor interaction (BCD) in the three-way table $\{BCD\}$ equal to zero, the fitted marginals for H''_3 are $\{AC\}$, $\{BC\}$, $\{BD\}$, $\{CD\}$ (see Goodman 1971*b*, 1973*b*).[40] With respect to the parameters

For the quasi-tautological case when $m > 3$, see the related comments in n. 34. In this case, the methods presented by Goodman (1974) will be of interest in order to determine whether or not the parameters in the model are "identifiable."

37 With a quasi-tautological model, there is no parsimony, and there are no degrees of freedom available for testing the model. Even when this kind of model fits the data perfectly, the perfect fit cannot serve as *statistical* evidence that the model provides a suitable description of the reality represented by the data, since the description is not parsimonious; nevertheless, there may be *substantive* considerations (and/or additional relevant data) that would lead us to select this model as a suitable description. On the other hand, when this kind of model does not fit the data perfectly, we have *statistical* evidence that the model is not a suitable description.

38 The latent class model referred to here is the model $[B \otimes C \otimes D|Z]$. In view of the comments in nn. 31 and 35, the simplification that is possible in the calculation pertaining to H_3 for these data would lead to incorrect results if applied to the data analyzed in sec. 3.

39 This hypothesis states that the interaction between any pair of variables in the three-way table $\{BCD\}$ does not depend upon the level of the third variable in the table.

40 The results presented in the articles cited above can be used to explain why H''_3 (i.e., $[A \otimes BD|C] \cap [(BCD) = 0]$) is equivalent to the log-linear model in which the fitted marginals are $\{AC\}$, $\{BC\}$, $\{BD\}$, $\{CD\}$. This particular model is listed in Goodman (1973*a*, table 11 [see model H_{12} there]), where various log-linear models are presented for the Coleman panel data.

that need to be estimated under H'_3 and H''_3, there is one less parameter under H''_3 (since the three-factor interaction (BCD) is nil under H''_3 but not under H'_3). There will be one more degree of freedom available for testing H''_3. There were 6 df for testing H'_3, and there will be 7 df for testing H''_3. If H''_3 is true, then H'_3 will also be true.

The χ^2 values for testing H''_3 and H'_3 were included in table 5. By comparing $X^2(H''_3)$ with the percentiles of the χ^2 distribution with 7 df, we see that H''_3 fits the data reasonably well. By comparing $X^2(H''_3)$ with $X^2(H'_3)$, we see in particular that the three-factor interaction (BCD) is not statistically significant.[41]

To recapitulate, we started in this section with the latent structure H_3, and we explained why it gave the same estimated expected frequencies as the "elementary" model H'_3 for the data in table 1. Next we modified H'_3 by setting the three-factor interaction (BCD) in this model equal to zero, thus obtaining model H''_3, a more parsimonious model. Now we shall provide a new latent structure model (H_5) which is somewhat related to H''_3, but which is, in a certain sense, a more fundamental model. We shall introduce the latent structure model H_5 in the paragraph after the next one. A preliminary result will be presented first.

With respect to the three-way table $\{BCD\}$, we have considered in this section two different hypotheses: viz., $[B \otimes C \otimes D|Z]$ and $[(BCD) = 0]$. For the reader who is more familiar with the latent class models, we shall now comment briefly on the meaning of the three-factor interaction (BCD) expressed in terms of the parameters of the latent class model $[B \otimes C \otimes D|Z]$. This three-factor interaction (as defined in, e.g., Goodman 1970) will be positive, negative, or zero according as the following quantity is positive, negative, or zero:

$$(\gamma^{\bar{B}Z} - 1)(\gamma^{\bar{C}Z} - 1)(\gamma^{\bar{D}Z} - 1)(1 - \gamma^{Z}), \tag{39}$$

where $\gamma^{\bar{B}Z}$, $\gamma^{\bar{C}Z}$, and $\gamma^{\bar{D}Z}$ are defined in the same way as we defined $\gamma^{\bar{A}X}$ in (27), and γ^{Z} is defined as follows:

$$\begin{aligned}\gamma^{Z} &= \Big[\prod_{j,k,l} \Omega^{BCD\bar{Z}}{}_{jkl.}\Big]^{1/8} \\ &= \Omega^{\bar{Z}}_{.}\{\psi^{BZ}\psi^{CZ}\psi^{DZ}\},\end{aligned} \tag{40}$$

where

$$\Omega^{BCD\bar{Z}}{}_{jkl.} = \pi^{BCDZ}{}_{jkl1}/\pi^{BCDZ}{}_{jkl2},\ \Omega^{\bar{Z}}_{.} = \pi^{Z}{}_{1}/\pi^{Z}{}_{2},$$

$$\psi^{BZ} = [(\pi^{\bar{B}Z}{}_{11}\pi^{\bar{B}Z}{}_{21})/(\pi^{\bar{B}Z}{}_{12}\pi^{\bar{B}Z}{}_{22})]^{1/2}, \tag{41}$$

[41] The difference $X^2(H''_3) - X^2(H'_3)$ will be equal to the usual χ^2 likelihood-ratio statistic for testing $[(BCD) = 0]$ in the three-way table $\{BCD\}$.

and where ψ^{CZ} and ψ^{DZ} are defined in the same way as we defined ψ^{BZ} above.[42] If the quantity within braces on the second line of (40) is equal to one (or is close to one),[43] then the hypothesis that $\gamma^Z = 1$ corresponds to the hypothesis that $\Omega^{\bar{Z}}_{\cdot} = 1$. From the expression for $\Omega^{\bar{Z}}_{\cdot}$ in (41), we see that the hypothesis that $\Omega^{\bar{Z}}_{\cdot} = 1$ is equivalent to

$$\pi^Z{}_1 = \pi^Z{}_2. \tag{42}$$

Condition (42) states that the classes pertaining to the latent variable Z are equiprobable.

As we noted earlier, hypothesis H''_3 stated that H'_3 is true *and* that the three-factor interaction (BCD) in the three-way table $\{BCD\}$ is zero (i.e., that the quantity defined by [39] is equal to zero). Consider next the related hypothesis H_5 that H_3 is true and that condition (42) is satisfied. With respect to the parameters that need to be estimated under H_5 and H_3, there is one less parameter under H_5 (since the parameter $\pi^Z{}_1$ is a specified numerical constant [viz., 1/2] under H_5 but not under H_3.) In this sense, H_5 is simpler than H_3. There will be one more degree of freedom available for testing H_5. There were 6 df for testing H_3, and there will be 7 df for testing H_5. If H_5 is true, then H_3 will aso be true.

The estimated parameters under H_5 are presented in table 11, and the corresponding path diagram is presented in figure 4.

The χ^2 values for testing H_5 were included in table 5. By comparing $X^2(H_5)$ with $X^2(H_3)$, we see that the data are congruent with the hypothesis that the classes pertaining to the latent variable Z are equiprobable.[44]

As we have noted, H_5 implies H_3. We can denote this as $H_5 \Rightarrow H_3$ (see, e.g., Goodman 1973*c*).[45] Figure 5 describes how the various

[42] The quantities ψ^{BZ}, ψ^{CZ}, and ψ^{DZ} will become more meaningful to the reader after he has read sec. 2.5 where the related "error rates" are discussed (see n. 48).

[43] The quantity ψ^{BZ} will be equal to one if and only if the variable B "error rate" (as defined in sec. 2.5) for individuals who are at level 1 on latent variable Z is equal to the corresponding "error rate" for individuals who are at level 2 on latent variable Z (see n. 48). A similar remark pertains to ψ^{CZ} and ψ^{DZ}. The quantity within braces on the second line of (40) will be equal to one (i.e., $\psi^{BZ} \cdot \psi^{CZ} \cdot \psi^{DZ} = 1$) *if* $\psi^{BZ} = \psi^{CZ} = \psi^{DZ} = 1$; but, of course, the condition that each of the three ψ's is equal to one is not a necessary condition for their product to be equal to one.

[44] The difference $X^2(H_5) - X^2(H_3)$ will be equal to the χ^2 likelihood-ratio statistic for testing condition (42) in the three-way table $\{BCD\}$.

[45] We use the double arrow $\Rightarrow$ in the text to denote the "implication" relationship between two given hypotheses; but for typographical simplicity, the double arrow in the text will be replaced by the corresponding single arrow in fig. 5. The arrows in fig. 5, which tell us that certain *hypotheses imply* certain other *hypotheses,* should not be confused with arrows that appear in path diagrams (e.g., figs. 1–4), which tell us that certain *variables affect* certain other *variables.* The arrows in fig. 5 have a very different meaning from arrows in the path diagrams.

TABLE 11

MAXIMUM-LIKELIHOOD ESTIMATE OF PARAMETERS IN MODEL H_5 (FIG. 4), DESCRIBED BY FORMULAS (5), (13), (15), (18), AND (42) FOR FOUR-WAY CROSS-CLASSIFICATION (TABLE 1)

LATENT CLASS						
t	(r, s)	$\pi^{X}_{t} = \pi^{YZ}_{rs}$	$\pi^{\bar{A}X}_{1t} = \pi^{\bar{A}YZ}_{1rs}$	$\pi^{\bar{B}X}_{1t} = \pi^{\bar{B}YZ}_{1rs}$	$\pi^{\bar{C}X}_{1t} = \pi^{\bar{C}YZ}_{1rs}$	$\pi^{\bar{D}X}_{1t} = \pi^{\bar{D}YZ}_{1rs}$
1	(1, 1)	.26	.68	.81	1.00**	.83
2	(1, 2)	.15	.68*	.27	1.00**	.30
3	(2, 1)	.24*	.15	.81*	.00**	.83*
4	(2, 2)	.35*	.15*	.27*	.00**	.30*

NOTE.—The asterisked entries can be determined from the other entries in this table using (14), (15), and (42). The double asterisked entries are specified by the model using (18).

hypotheses in this section are related to each other with respect to their implications. For any two hypotheses, say, H and H*, we obtain the following general result:

$$\text{If } H^* \Rightarrow H, \text{ then } X^2(H^*) \geqq X^2(H) \quad (43)$$

when the likelihood-ratio χ^2 statistics are used. (Compare the relationships among the hypotheses presented in fig. 5 with the corresponding likelihood-ratio χ^2 values in table 5.)

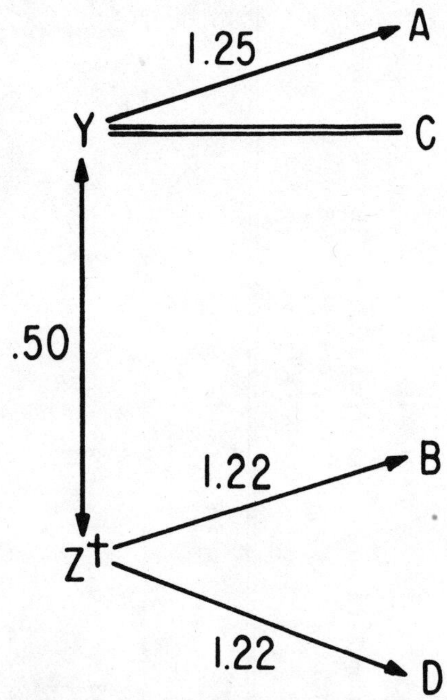

FIG. 4.—Path diagram for the hypothesis H_5 that the latent variables *Y* and *Z* explain the relationships among the manifest variables (*A*, *B*, *C*, *D*) in the Coleman panel data (table 1), *and* that variable *C* is an accurate indicator for the latent variable *Y*, *and* that the latent classes of variable *Z* are equiprobable. The symbol *Z*+ denotes the fact that the two latent classes of variable *Z* are equiprobable under this model.

Hypotheses H_5 and H''_3, each with 7 df, fit the data reasonably well. As we have noted, these two hypotheses are somewhat related to each other (see [39]–[42]). Hypothesis H_5 is, in a certain sense, a more fundamental hypothesis in that it helps to "explain" the various "effects" included in model H''_3 (i.e., the effects of the manifest variables [*A*, *B*, *C*, *D*] upon each other) in terms of the latent variables *Y* and *Z* that are included in model H_5.

2.5. Some Additional Latent Structures

Model H_5, which we introduced in the preceding section, stated that H_3 is true *and* that condition (42) is satisfied. This model can be viewed as a special case of H_3, which can in turn be viewed as a special case of H_2. In fact, all the models in figure 5 that have arrows pointing (either directly or indirectly) to H_2 can be viewed as special cases of H_2.[46] We shall now introduce still another model (H_6) which is also a special case of H_2.

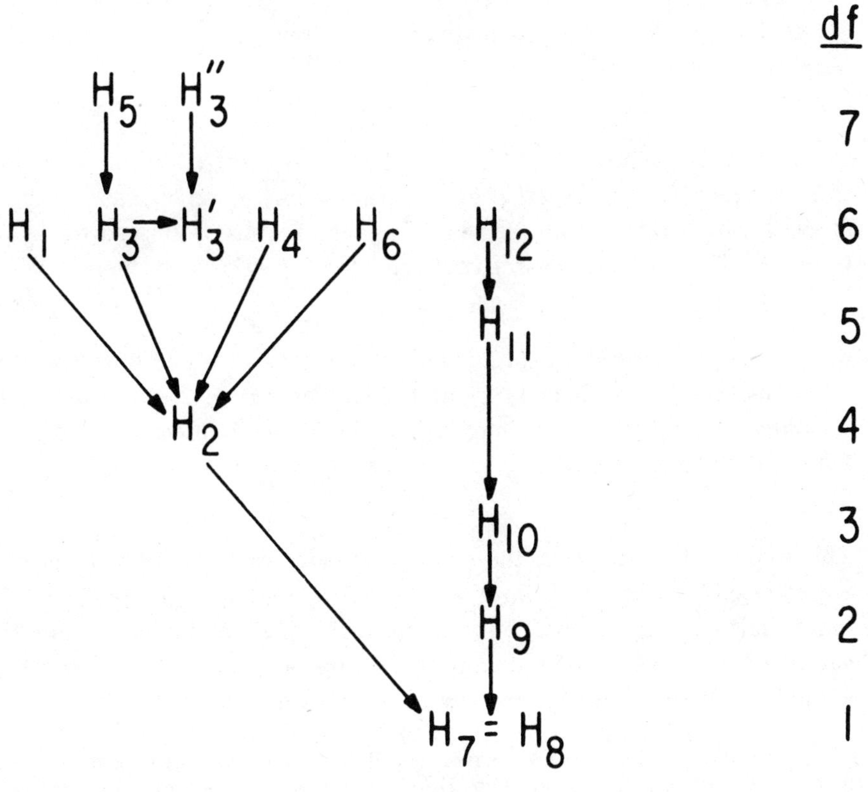

FIG. 5.—The relationship among the models in table 5, with respect to their implications and degrees of freedom. Models H_6–H_{12} will be discussed in sec. 2.5.

Under the parameter values for model H_2 as estimated in table 8, we see that an individual who is at level 1 on the latent variable Z will respond at level 1 on variable B with probability .81, and thus he will respond at level 2 on variable B with probability .19 (note that $.19 = 1 - .81$). For individuals who are at level 1 on the latent variable Z, each response on variable B will be either the "modal response" (level 1) or the

[46] Among the models discussed so far, H'_3 and H''_3 are the only ones in fig. 5 that do not have arrows pointing directly or indirectly to H_2.

"nonmodal response" (level 2).[47] With respect to their responses on variable B, we can view level 1 as the "correct response," and level 2 as the "incorrect response" or "error," for individuals who are at level 1 on latent variable Z. For these individuals, as noted earlier in this paragraph, the variable B "error rate" is estimated to be .19. Similarly, for individuals who are at level 2 on latent variable Z, we see from table 8 that the variable B "error rate" is estimated to be .27, that level 1 is the "incorrect response" ("error") on this variable, and that level 2 is the "correct response."

Letting $\epsilon^{\bar{B}Z}{}_1$ and $\epsilon^{\bar{B}Z}{}_2$ denote the two error rates discussed in the preceding paragraph, we see that they are estimated to be .19 and .27, respectively. Note that

$$\epsilon^{\bar{B}Z}{}_s = \min\,[\pi^{\bar{B}Z}{}_{1s}, \pi^{\bar{B}Z}{}_{2s}], \quad \text{for } s = 1, 2, \tag{44}$$

where $\pi^{\bar{B}Z}{}_{1s}$ is defined by (12), and $\pi^{\bar{B}Z}{}_{2s} = 1 - \pi^{\bar{B}Z}{}_{1s}$. (In other words, $\epsilon^{\bar{B}Z}{}_s$ is equal to the lesser [i.e., the minimum] of $\pi^{\bar{B}Z}{}_{1s}$ and $\pi^{\bar{B}Z}{}_{2s}$.) Similarly, from table 8, we see that $\epsilon^{\bar{C}Y}{}_1$ and $\epsilon^{\bar{C}Y}{}_2$ are estimated to be .09 (i.e., 1 — .91) and .08, respectively; and that

$$\epsilon^{\bar{C}Y}{}_r = \min\,[\pi^{\bar{C}Y}{}_{1r}, \pi^{\bar{C}Y}{}_{2r}], \quad \text{for } r = 1, 2, \tag{45}$$

where $\pi^{\bar{C}Y}{}_{1r}$ is defined by (12), and $\pi^{\bar{C}Y}{}_{2r} = 1 - \pi^{\bar{C}Y}{}_{1r}$. Of course, the estimates of $\epsilon^{\bar{B}Z}{}_s$ (for $s = 1, 2$) and $\epsilon^{\bar{C}Y}{}_r$ (for $r = 1, 2$) are subject to sampling variability, and it might sometimes be of interest to test the following null hypothesis:

$$\epsilon^{\bar{B}Z}{}_1 = \epsilon^{\bar{B}Z}{}_2 \text{ and } \epsilon^{\bar{C}Y}{}_1 = \epsilon^{\bar{C}Y}{}_2. \tag{46}$$

This hypothesis states that the variable B error rate for individuals who are at level 1 on latent variable Z is equal to the corresponding rate for individuals who are at level 2 on latent variable Z; and similarly for the variable C error rates for individuals who are at levels 1 and 2 on latent variable Y. We shall now show how to test this hypothesis.[48]

[47] For the sake of simplicity, we have defined the modal and nonmodal responses in terms of the *estimated* parameters. However, these two kinds of responses could have been defined instead in terms of the actual (unknown) parameter values; in which case, the estimated parameters would have been used to *estimate* which response is modal and which is nonmodal. This estimate of the modal and nonmodal response will be correct with probability approaching one when the sample size is large.

[48] The above discussion of error rates was presented in terms of the latent parameters under model H_2 for the four-way table $\{ABCD\}$, but the basic methods and concepts which we shall present here can also be applied more generally to study the relationship between a given indicator variable and the corresponding latent variable in other latent structure models. For example, with respect to the three-way table $\{BCD\}$ under the latent class model $[B \otimes C \otimes D|Z]$, which was discussed near the end of sec. 2.4, the variable B error rates will be equal to each other (i.e., $\epsilon^{\bar{B}Z}{}_1 = \epsilon^{\bar{B}Z}{}_2$) if and only if the quantity ψ^{BZ} defined in (41) is equal to one. (This is a consequence of [44].) The methods which we shall present here can be applied to the latent class model $[B \otimes C \otimes D|Z]$ and to various other latent structures. (For somewhat different methods, see Murray [1971].)

The remarks in the preceding two paragraphs pertain to the parameter values under model H_2. Let H_6 now denote the hypothesis that model H_2 is true *and* that condition (46) is also true. With respect to the parameters that need to be estimated under H_6 and H_2 there will be two less parameters under H_6, since condition (46) is equivalent to the following condition expressed more directly in terms of these parameters:

$$1 - \pi^{\bar{B}X}{}_{11} = \pi^{\bar{B}X}{}_{12} \text{ and } 1 - \pi^{\bar{C}X}{}_{11} = \pi^{\bar{C}X}{}_{13}, \tag{47}$$

when $\pi^{\bar{B}X}{}_{11}$ and $\pi^{\bar{C}X}{}_{11}$ are greater than .5, and $\pi^{\bar{B}X}{}_{12}$ and $\pi^{\bar{C}X}{}_{13}$ are less than .5 (see table 8). In this sense, model H_6 is simpler than H_2, and there will be two more degrees of freedom for testing H_6. If H_6 is true, then H_2 will also be true (see fig. 5).

The estimated parameters under H_6 are presented in table 12.[49] The corresponding χ^2 values for testing H_6 were included in table 5. By comparing $X^2(H_6)$ with $X^2(H_2)$, we see that the data are congruent with the hypothesis that condition (46) is satisfied.[50] From table 5, we see that H_6 fits the data very well indeed.

Model H_6 stated that H_2 was true *and* that we could ignore the difference between the two estimated error rates for variable *B*, and similarly for variable *C*. On the other hand, the same kind of analysis applied to variables *A* and *D* would have indicated that the difference between the two estimated error rates for variable *A* is statistically significant, and similarly for variable *D*.[51]

The general method applied here to study whether the difference between two estimated error rates is statistically significant can also be applied to study whether the difference between two estimated conditional probabilities (e.g., $\pi^{\bar{A}X}{}_{11} - \pi^{\bar{B}X}{}_{11}$) is statistically significant. We shall illustrate this later in this section.

As we have already noted, the models in figure 5 that have arrows pointing (either directly or indirectly) to H_2 are special cases of H_2: they

49 To save space, we have not included here the corresponding path diagram for model H_6 nor for the other models discussed in this section. The numerical entries in the path diagram for H_6 can be obtained from the estimated parameter values in table 12 in the same way as we had obtained earlier the corresponding numerical entries in fig. 2 from table 8.

50 The difference $X^2(H_6) - X^2(H_2)$ will be equal to the χ^2 likelihood-ratio statistic for testing condition (46), under the assumption that model H_2 is true.

51 To save space, we shall not include these details here. However, it may be of some interest to record the χ^2 values for the model that states that H_2 is true ***and*** that $\epsilon^{\bar{B}Z}{}_1 = \epsilon^{\bar{B}Z}{}_2$, $\epsilon^{\bar{C}Y}{}_1 = \epsilon^{\bar{C}Y}{}_2$, $\epsilon^{\bar{D}Z}{}_1 = \epsilon^{\bar{D}Z}{}_2$. The likelihood-ratio χ^2 is 9.76, and the goodness-of-fit χ^2 is 9.90; each with 7 df. Although these χ^2 values are not statistically significant at the usual levels of significance, comparison of these values with the corresponding χ^2 values for H_6 in table 5 indicates that we would reject the null hypothesis that $\epsilon^{\bar{D}Z}{}_1 = \epsilon^{\bar{D}Z}{}_2$, assuming that model H_6 is true.

TABLE 12

MAXIMUM-LIKELIHOOD ESTIMATE OF PARAMETERS IN MODEL H_6 DESCRIBED BY FORMULAS (5), (13), (15), AND (47) FOR FOUR-WAY CROSS-CLASSIFICATION (TABLE 1)

LATENT CLASS						
t	(r, s)	$\pi^{X}_{t} = \pi^{YZ}_{rs}$	$\pi^{\bar{A}X}_{1t} = \pi^{\bar{A}YZ}_{1rs}$	$\pi^{\bar{B}X}_{1t} = \pi^{\bar{B}YZ}_{1rs}$	$\pi^{\bar{C}X}_{1t} = \pi^{\bar{C}YZ}_{1rs}$	$\pi^{\bar{D}X}_{1t} = \pi^{\bar{D}YZ}_{1rs}$
1	(1, 1)	.29	.76	.77	.92	.80
2	(1, 2)	.10	.76*	.23*	.92*	.27
3	(2, 1)	.28	.12	.77*	.08*	.80*
4	(2, 2)	.33*	.12*	.23*	.08*	.27*

NOTE.—The asterisked entries can be determined from the other entries in this table using (14), (15), and (47).

impose various restrictions (see, e.g., [47]) upon the parameters of H_2, in addition to the restrictions already imposed (see [12] and [15]) by the definition of H_2. Since these models impose additional restrictions upon H_2, the χ^2 values obtained by (11) will be larger for these models,[52] and the number of degrees of freedom for testing the models will also be larger. We shall next introduce a model (H_7) that actually removes some of the restrictions imposed by the definition of H_2 (see fig. 5). This will lead to a reduction in both the χ^2 value and in the degrees of freedom for testing the model (see table 5).[53]

From condition (12) and figure 2, we see that model H_2 stated that variables A and C are directly affected by the latent variable Y (but not Z), and that variables B and D are directly affected by the latent variable Z (but not Y). Recall that Y was interpreted as latent self-perceived membership in the leading crowd, and Z was interpreted as latent attitude toward the leading crowd. Consider next model H_7 in which each of the two manifest variables at the first interview (variables A and B) is directly affected by its corresponding latent variable (A is directly affected by Y [but not Z], and B is directly affected by Z [but not Y]), but the two manifest variables at the second interview (variables C and D) are not restricted in this way. In this model, variable C can be affected by *both* latent variables Y and Z (and by their interaction), and similarly for variable D. In other words, model H_7 leaves unchanged the restrictions in (12) pertaining to variables A and B, but it deletes the restrictions pertaining to variables C and D. Thus, under model H_7, the set of restrictions described by (15) would be modified as follows:

$$\pi^{\bar{A}X}{}_{11} = \pi^{\bar{A}X}{}_{12}, \pi^{\bar{A}X}{}_{13} = \pi^{\bar{A}X}{}_{14},$$
$$\pi^{\bar{B}X}{}_{11} = \pi^{\bar{B}X}{}_{13}, \pi^{\bar{B}X}{}_{12} = \pi^{\bar{B}X}{}_{14}. \qquad (48)$$

The χ^2 values for H_7 were included in table 5. Before commenting upon the number of degrees of freedom for testing H_7, or upon the corresponding estimated parameters under this model, we shall first introduce model H_8, a special case of H_7 in which the following additional restriction is imposed:

$$\epsilon^{\bar{A}Y}{}_1 = \epsilon^{\bar{A}Y}{}_2. \qquad (49)$$

[52] If the additional restrictions are already satisfied by the estimated parameter values, then the χ^2 values will be equal (see [43]).

[53] If model H_2 had not fit the data analyzed here so well, we could have examined the various restrictions imposed by the definition of H_2 (see [15]) in order to determine which of these restrictions is not congruent with the data. There is no real need to do this in the analysis of the present data, since the fit of H_2 is so good. Nevertheless, we shall introduce model H_7, and the subsequent models H_8–H_{12}, because of their substantive interest and also because the reader may wish to consider models of this kind for other sets of data where the fit of H_2 may not be so good.

Condition (49) states that the variable A error rate for those individuals who are at level 1 on latent variable Y is equal to the corresponding error rate for those individuals who are at level 2 on variable Y. As we did earlier when we rewrote (46) as (47), we can now rewrite (49) as

$$1 - \pi^{\bar{A}X}{}_{11} = \pi^{\bar{A}X}{}_{13}. \qquad (50)$$

Model H_8 states both that H_7 is true (i.e., that condition [48] holds true) *and* that (50) is also true. Combining (50) with (48), we see that H_8 states that the following conditions are true:

$$\begin{aligned} \pi^{\bar{A}X}{}_{11} &= \pi^{\bar{A}X}{}_{12} = 1 - \pi^{\bar{A}X}{}_{13} = 1 - \pi^{\bar{A}X}{}_{14}; \\ \pi^{\bar{B}X}{}_{11} &= \pi^{\bar{B}X}{}_{13}, \pi^{\bar{B}X}{}_{12} = \pi^{\bar{B}X}{}_{14}. \end{aligned} \qquad (51)$$

The χ^2 values for H_8 were also included in table 5. From this table, we see that these values are equal to the corresponding values for H_7. The reason for this equality is that the expected frequencies F_{ijkl} estimated under H_7 were equal to the corresponding expected frequencies estimated under H_8. However, the parameters in H_7 actually were not "identifiable," whereas the parameters in H_8 were.[54] The estimated parameters under H_8 are presented in table 13.[55] There are $3 + 1 + 2 + 4 + 4 = 14$ independent parameters estimated under H_8,[56] and so there will be $15 - 14 = 1$ df for testing this model.[57] From table 5, we see that the model fits the data very well indeed.

We shall next introduce models H_9 and H_{10}, which impose additional restrictions upon H_8 (see fig. 5). Model H_9 is a special case of H_8 in which the following additional restriction is imposed:

$$\epsilon^{\bar{B}Z}{}_{1} = \epsilon^{\bar{B}Z}{}_{2}. \qquad (52)$$

Since H_8 was obtained from H_7 by the imposition of (49), we see that H_9

[54] For the analysis of latent structures in which the parameters are not "identifiable," see Goodman (1974). That article also presented methods for determining whether or not the parameters in a given latent structure are identifiable. By applying those methods to H_7 and H_8, we find that the parameters in H_7 are not identifiable; but the imposition of the additional restriction (50) would make the parameters identifiable. The methods in Goodman (1974) should be applied whenever there is any doubt as to whether or not the parameters in the latent structure under investigation are identifiable.

[55] Table 13 also includes the estimated parameters for models H_9–H_{12}. We shall discuss them anon.

[56] From table 13 for model H_8, we see that there are 14 entries that are not asterisked.

[57] Since the imposition of the additional restriction (50) would make the parameters in H_7 identifiable without changing the estimated expected frequencies under the model, the number of degrees of freedom for testing H_7 will be the same as those for testing the model with the additional restriction imposed (model H_8).

TABLE 13

MAXIMUM-LIKELIHOOD ESTIMATE OF PARAMETERS IN FIVE RELATED MODELS FOR FOUR-WAY CLASSIFICATION (TABLE 1)

Model	Latent Class t	Latent Class (r, s)	$\pi^{X}_{t} = \pi^{YZ}_{rs}$	$\pi^{\bar{A}X}_{1t} = \pi^{\bar{A}YZ}_{1rs}$	$\pi^{\bar{B}X}_{1t} = \pi^{\bar{B}YZ}_{1rs}$	$\pi^{\bar{C}X}_{1t} = \pi^{\bar{C}YZ}_{1rs}$	$\pi^{\bar{D}X}_{1t} = \pi^{\bar{D}YZ}_{1rs}$
H_8:	1	(1, 1)	.27	.90	.73	.89	.81
	2	(1, 2)	.06	.90*	.17	.86	.02
	3	(2, 1)	.38	.10*	.73*	.21	.73
	4	(2, 2)	.29*	.10*	.17*	.11	.24
H_9:	1	(1, 1)	.25	.94	.78	.85	.82
	2	(1, 2)	.10	.94*	.22*	.75	.21
	3	(2, 1)	.32	.06*	.78*	.24	.78
	4	(2, 2)	.33*	.06*	.22*	.13	.28
H_{10}:	1	(1, 1)	.22	.87	.87*	.94	.80
	2	(1, 2)	.10	.87*	.13*	1.00	.39
	3	(2, 1)	.33	.13*	.87*	.19	.71
	4	(2, 2)	.35*	.13*	.13*	.12	.34
H_{11}:	1	(1, 1)	.22	.90	.90*	.90*	.79
	2	(1, 2)	.11	.90*	.10*	.90*	.42
	3	(2, 1)	.33	.10*	.90*	.20	.70
	4	(2, 2)	.34*	.10*	.10*	.13	.35
H_{12}:	1	(1, 1)	.22	.89	.89*	.89*	.79
	2	(1, 2)	.11	.89*	.11*	.89*	.42
	3	(2, 1)	.33	.11*	.89*	.21	.70
	4	(2, 2)	.34*	.11*	.11*	.11*	.35

NOTE.—Model H_8 described by formulas (5), (13), and (51); model H_9 described by formulas (5), (13), and (53); model H_{10} described by formulas (5), (13), and (55); model H_{11} described by formulas (5), (13), (55), and (60); model H_{12} described by formulas (5), (13), (55), (60), and (61). The asterisked entries can be determined from the other entries in the table for a given model using (14) and the corresponding restrictions pertaining to the model.

is obtained from H_7 by the imposition of both (49) and (52). Thus, H_9 states that the following conditions are true:

$$\pi^{\bar{A}X}_{11} = \pi^{\bar{A}X}_{12} = 1 - \pi^{\bar{A}X}_{13} = 1 - \pi^{\bar{A}X}_{14};$$
$$\pi^{\bar{B}X}_{11} = \pi^{\bar{B}X}_{13} = 1 - \pi^{\bar{B}X}_{12} = 1 - \pi^{\bar{B}X}_{14} \qquad (53)$$

(cf. [53] with [51]). Model H_{10} is a special case of H_9 in which the following additional restriction is imposed:

$$\epsilon^{\bar{A}Y}_{1} = \epsilon^{\bar{B}Z}_{1}. \qquad (54)$$

Model H_{10} states that the variable A error rates and the variable B error rates are all equal to each other. Thus, model H_{10} states that the following condition is true:

$$\pi^{\bar{A}X}_{11} = \pi^{\bar{A}X}_{12} = 1 - \pi^{\bar{A}X}_{13} = 1 - \pi^{\bar{A}X}_{14}$$
$$= \pi^{\bar{B}X}_{11} = \pi^{\bar{B}X}_{13} = 1 - \pi^{\bar{B}X}_{12} = 1 - \pi^{\bar{B}X}_{14} \qquad (55)$$

(cf. [55] with [53]).

The χ^2 values for testing both H_9 and H_{10} were included in table 5, and we see from these values that these models also fit the data very well indeed. The estimated parameters under these models were included in table 13.[58]

We shall next introduce a latent structure (H'_{10}) that may appear at first sight to be quite different from H_{10}, but we shall see shortly that it is closely related to it. In model H_{10}, each of the manifest variables at the first interview (variables A and B) could be viewed as a "direct" indicator of the corresponding latent variable (variables Y and Z, respectively); but the manifest variables at the second interview (variables C and D) could not be viewed in this simple way since each of these manifest variables was affected by ***both*** latent variables and by their interaction effect. Consider now a latent structure H'_{10} in which, in addition to the two latent dichotomous variables Y and Z (which pertain to the corresponding manifest variables at the first interview), there are two other latent dichotomous variables, say, Y' and Z', which pertain to the corresponding manifest

[58] The same general methods used in sec. 2.2 to obtain the numerical entries in the path diagrams (e.g., fig. 2) can now be applied also to the models in table 13. In addition to the various effects in fig. 2, variable C will now also be directly affected by latent variable Z and by the interaction effect between Y and Z; and variable D will now also be directly affected by latent variable Y and by the interaction effect between Y and Z (see fig. 10.*F.1* below). For various ways to include the interaction effects in path diagrams, see comments in, e.g., sec. 6 of Goodman (1973*b*). The method used to describe the interaction effects in fig. 10.*F.1* (and in some of the other path diagrams in fig. 10) is similar to one of the methods introduced for this purpose in Goodman (1973*b*). It is also somewhat similar to a method used in Duncan (1973).

variables at the second interview. In model H'_{10}, variables A and B are "direct" indicators for the latent variables Y and Z, respectively; and variables C and D are "direct" indicators for the latent variables Y' and Z', respectively.

Let $\pi^{\bar{C}Y'}_{ku}$ denote the *conditional* probability that an individual will be at level k on variable C (for $k = 1, 2$), given that he was at level u on latent variable Y' (for $u = 1, 2$); and let $\pi^{\bar{D}Z'}_{lv}$ be defined in a similar way for variable D (with $l = 1, 2$) and variable Z' (with $v = 1, 2$). (As was the case for $\pi^{\bar{C}Y}_{kr}$ and $\pi^{\bar{D}Z}_{ls}$ in [12], the $\pi^{\bar{C}Y'}_{ku}$ do not depend upon the level v of variable Z'; and the $\pi^{\bar{D}Z'}_{lv}$ do not depend upon the level u of variable Y'.) Similarly, we let $\pi^{\bar{Y}'YZ}_{urs}$ denote the *conditional* probability that an individual will be at level u on the latent variable Y' (for $u = 1, 2$), given that he was at level (r, s) on the joint latent variables (Y, Z); and let $\pi^{\bar{Z}'YZ}_{vrs}$ be defined in a similar way for the latent variable Z' (with $v = 1, 2$). Then the conditional probabilities $\pi^{\bar{C}YZ}_{1rs}$ and $\pi^{\bar{D}YZ}_{1rs}$ (which were estimated for, say, model H_{10} in table 13), can be expressed as

$$\pi^{\bar{C}YZ}_{1rs} = \pi^{\bar{C}Y'}_{11}\pi^{\bar{Y}'YZ}_{1rs} + \pi^{\bar{C}Y'}_{12}\pi^{\bar{Y}'YZ}_{2rs},$$

and (56)

$$\pi^{\bar{D}YZ}_{1rs} = \pi^{\bar{D}Z'}_{11}\pi^{\bar{Z}'YZ}_{1rs} + \pi^{\bar{D}Z'}_{12}\pi^{\bar{Z}'YZ}_{2rs},$$

for $r = 1, 2$; $s = 1, 2$. The first line of (56) states that the latent variables Y and Z (which actually pertain to the first interview) have an *apparent* effect upon variable C that is actually due to the direct effect of these latent variables upon the latent variable Y' (which pertains to the second interview), and subsequent direct effect of the latent variable Y' upon variable C. The second line of (56) can be interpreted in a similar way for variable D and the latent variable Z'.

In the preceding two paragraphs we introduced a model (H'_{10}) which expressed the four parameters $\pi^{\bar{C}YZ}_{1rs}$ (in, say, model H_{10}) in terms of a set of six new parameters ($\pi^{\bar{C}Y'}_{11}$, $\pi^{\bar{C}Y'}_{12}$, and $\pi^{\bar{Y}'YZ}_{1rs}$, for $r = 1, 2$; $s = 1, 2$);[59] and the four parameters $\pi^{\bar{D}YZ}_{1rs}$ were expressed in similar terms. Consider now a model in which the following restriction is imposed upon the $\pi^{\bar{C}Y'}_{1u}$ and $\pi^{\bar{D}Z'}_{1v}$:

$$\pi^{\bar{C}Y'}_{1u} = \pi^{\bar{A}Y}_{1u}, \pi^{\bar{D}Z'}_{1v} = \pi^{\bar{B}Z}_{1v}, \tag{57}$$

for $u = 1, 2$; $v = 1, 2$. Restriction (57) states that the corresponding conditional probabilities for variables A and C are equal, and the corresponding conditional probabilities for variables B and D are equal.[60]

[59] Note that $\pi^{\bar{Y}'YZ}_{2rs}$ can be determined from $\pi^{\bar{Y}'YZ}_{1rs}$, since these conditional probabilities will satisfy the following equation when Y' is a dichotomous variable: $\pi^{\bar{Y}'YZ}_{1rs} + \pi^{\bar{Y}'YZ}_{2rs} = 1$.

[60] Recall that the manifest variables A and C report on the same issue at the first and second interview, and their conditional probabilities pertain to the latent variables

When (57) is imposed, then the $\pi^{\bar{C}Y'}{}_{1u}$ can be determined by the $\pi^{\bar{A}Y}{}_{1r}$, and thus there will be four new parameters ($\pi^{\bar{Y}'YZ}{}_{1rs}$, for $r = 1,2; s = 1, 2$) in the set of parameters that is used to express the $\pi^{\bar{C}YZ}{}_{1rs}$; and a similar remark applies to the set of parameters that is used to express the $\pi^{\bar{D}YZ}{}_{1rs}$. Formula (56) expressed the $\pi^{\bar{C}YZ}{}_{1rs}$ and $\pi^{\bar{D}YZ}{}_{1rs}$ in terms of the $\pi^{\bar{Y}'YZ}{}_{1rs}$ and $\pi^{\bar{Z}'YZ}{}_{1rs}$; and from (56) we obtain the following expressions for the $\pi^{\bar{Y}'YZ}{}_{1rs}$ and $\pi^{\bar{Z}'YZ}{}_{1rs}$ in terms of the $\pi^{\bar{C}YZ}{}_{1rs}$ and $\pi^{\bar{D}YZ}{}_{1rs}$:

$$\pi^{\bar{Y}'YZ}{}_{1rs} = [\pi^{\bar{C}YZ}{}_{1rs} - \pi^{\bar{C}Y'}{}_{12}]/[\pi^{\bar{C}Y'}{}_{11} - \pi^{\bar{C}Y'}{}_{12}],$$

and (58)

$$\pi^{\bar{Z}'YZ}{}_{1rs} = [\pi^{\bar{D}YZ}{}_{1rs} - \pi^{\bar{D}Z'}{}_{12}]/[\pi^{\bar{D}Z'}{}_{11} - \pi^{\bar{D}Z'}{}_{12}],$$

for $r = 1, 2$; $s = 1, 2$.[61] The conditional probabilities $\pi^{\bar{Y}'YZ}{}_{1rs}$ and $\pi^{\bar{Z}'YZ}{}_{1rs}$ cannot lie outside the interval from zero to one, and thus the $\pi^{\bar{C}YZ}{}_{1rs}$ and $\pi^{\bar{D}YZ}{}_{1rs}$ need to satisfy the following inequalities:

$$\pi^{\bar{C}Y'}{}_{11} \geqq \pi^{\bar{C}YZ}{}_{1rs} \geqq \pi^{\bar{C}Y'}{}_{12}, \pi^{\bar{D}Z'}{}_{11} \geqq \pi^{\bar{D}YZ}{}_{1rs} \geqq \pi^{\bar{D}Z'}{}_{12}, \qquad (59)$$

when

$$\pi^{\bar{C}Y'}{}_{11} > \pi^{\bar{C}Y'}{}_{12} \text{ and } \pi^{\bar{D}Z'}{}_{11} > \pi^{\bar{D}Z'}{}_{12} \text{ (see table 13)}.$$

Each of the models H_8–H_{10} in table 13 can be reinterpreted by the introduction of the conditional probabilities $\pi^{\bar{Y}'YZ}{}_{1rs}$ and $\pi^{\bar{Z}'YZ}{}_{1rs}$, and also the conditional probabilities $\pi^{\bar{C}Y'}{}_{1u}$ and $\pi^{\bar{D}Z'}{}_{1v}$ subject to condition (57). In accordance with (57), the estimated parameters $\pi^{\bar{A}Y}{}_{1r}$ and $\pi^{\bar{B}Z}{}_{1s}$ in table 13 can be used to estimate $\pi^{\bar{C}Y'}{}_{1u}$ and $\pi^{\bar{D}Z'}{}_{1v}$, respectively. These estimates, and the estimates of $\pi^{\bar{C}YZ}{}_{1rs}$ and $\pi^{\bar{D}YZ}{}_{1rs}$ in table 13, can now be inserted in (58) in order to obtain estimates of $\pi^{\bar{Y}'YZ}{}_{1rs}$ and $\pi^{\bar{Z}'YZ}{}_{1rs}$. Note that the estimates obtained from table 13 need to satisfy the inequalities (59).

Unfortunately, corresponding to each of the models H_8–H_{10} in table 13, there are some estimates that do not satisfy (59). For example, corresponding to H_{10}, the estimated $\pi^{\bar{C}YZ}{}_{111}$, $\pi^{\bar{C}YZ}{}_{112}$, and $\pi^{\bar{C}YZ}{}_{122}$ do not satisfy (59).[62] To remedy this, we shall now introduce model H_{11}, which is a special case of H_{10} in which the following additional conditions are satisfied:[63]

Y and *Y'*, respectively. A similar remark also applies to the manifest variables *B* and *D*.

[61] Formula (58) can be obtained from (56) using the relationship given at the end of n. 59. We assume here that $\pi^{\bar{C}Y'}{}_{11} \neq \pi^{\bar{C}Y'}{}_{12}$ and $\pi^{\bar{D}Z'}{}_{11} \neq \pi^{\bar{D}Z'}{}_{12}$.

[62] From the estimated parameters under H_{10} in table 13, we see that $\pi^{\bar{C}YZ}{}_{111} > \pi^{\bar{A}YZ}{}_{111}$, $\pi^{\bar{C}YZ}{}_{112} > \pi^{\bar{A}YZ}{}_{112}$, and $\pi^{\bar{C}YZ}{}_{122} < \pi^{\bar{A}YZ}{}_{122}$, for the corresponding estimates. Since $\pi^{\bar{A}YZ}{}_{111} = \pi^{\bar{A}YZ}{}_{112} = \pi^{\bar{A}Y}{}_{11} = \pi^{\bar{C}Y'}{}_{11}$ and $\pi^{\bar{A}YZ}{}_{122} = \pi^{\bar{A}YZ}{}_{121} = \pi^{\bar{A}Y}{}_{12} = \pi^{\bar{C}Y'}{}_{12}$ (see [12] and [57]), we find that (59) is violated due to the three inequalities listed in the preceding sentence.

[63] This special case of H_{10} is somewhat different from the special case of H_{10} in

$$\pi^{\bar{C}YZ}{}_{111} = \pi^{\bar{A}YZ}{}_{111}, \pi^{\bar{C}YZ}{}_{112} = \pi^{\bar{A}YZ}{}_{112}. \tag{60}$$

The estimated parameters under H_{11} were included in table 13. The corresponding χ^2 values for testing H_{11} were presented earlier in table 5, and they indicate that the model fits the data very well indeed. The estimated parameters under H_{11} in table 13 can now be used, as indicated in the preceding paragraph, to estimate the conditional probabilities $\pi^{\bar{Y}'YZ}{}_{1rs}$ and $\pi^{\bar{Z}'YZ}{}_{1rs}$, which we give in table 14.[64] The estimates in table 14 can

TABLE 14

Maximum-Likelihood Estimate of Conditional Probabilities $\pi^{\bar{Y}'YZ}{}_{1rs}$ and $\pi^{\bar{Z}'YZ}{}_{1rs}$ Obtained from Formulas (58) and (57), Applied to Models H_{11} and H_{12} for Four-Way Cross-Classification (Table 1)

Model	Latent Class t	(r, s)	$\pi^{\bar{Y}'YZ}{}_{1rs}$	$\pi^{\bar{Z}'YZ}{}_{1rs}$
H_{11}:	1	(1, 1)	1.00	.87
	2	(1, 2)	1.00	.40
	3	(2, 1)	.12	.75
	4	(2, 2)	.03	.31
H_{12}:	1	(1, 1)	1.00	.87
	2	(1, 2)	1.00	.40
	3	(2, 1)	.13	.75
	4	(2, 2)	.00	.30

which the parameters are limited only to the extent that they are required to satisfy the inequalities (59). However, for the data considered here, the maximum-likelihood estimate of the parameters under the latter special case will be equal to the corresponding estimate under the former special case. As noted in n. 62, the estimated parameters under H_{10} violated (59) due to the three *inequalities* listed in that footnote. Thus, the maximum-likelihood estimate of the parameters under the latter special case of H_{10} (i.e., when [59] must be satisfied) will be such that one or more of the following three *equalities* will be satisfied for the corresponding estimates: $\pi^{\bar{C}YZ}{}_{111} = \pi^{\bar{A}YZ}{}_{111}$, $\pi^{\bar{C}YZ}{}_{112} = \pi^{\bar{A}YZ}{}_{112}$, $\pi^{\bar{C}YZ}{}_{122} = \pi^{\bar{A}YZ}{}_{122}$. (This holds true as long as the likelihood-function under H_{10} does not have a "local maximum" that satisfies [59].) To obtain the maximum-likelihood estimate of the parameters under the latter special case of H_{10}, we first considered each of the three special cases of H_{10} in which the parameters were subject to only one of the three *equalities* listed above, and then we considered each of the special cases in which the parameters were subject to two of the three *equalities*.

[64] In addition to the estimated conditional probabilities under H_{11}, table 14 gives the corresponding estimated values under H_{12}, a model which we shall introduce in the next paragraph. With respect to the particular model H_{11}, estimates corresponding to those presented in table 14 were obtained by different methods in Murray, Wiley, and Wolfe (1972). These authors report a χ^2 value of 2.28 with 3 df for their model. Since their estimated model can be shown to be equivalent to our estimated H_{11}, their reported χ^2 value and number of degrees of freedom should have agreed with the corresponding quantities for H_{11} in table 5. The discrepancy in the number

help us to assess how stable (over time) is an individual's level with respect to the latent variables Y and Z, and how large are the various effects of the latent variables Y and Z upon their subsequent levels.[65]

We shall now comment briefly upon one more model (H_{12}). This model is a special case of H_{11} in which the following additional condition is satisfied:

$$\pi^{\bar{C}YZ}_{122} = \pi^{\bar{A}YZ}_{122}. \tag{61}$$

The χ^2 values for testing H_{12} were presented earlier in table 5, and they indicate that the model fits the data rather well. The estimated parameters under H_{12} were included in table 13, and the corresponding conditional probabilities $\pi^{\bar{Y}'YZ}_{1rs}$ and $\pi^{\bar{Z}'YZ}_{1rs}$ were included in table 14. The estimated conditional probabilities under H_{12} can be used in the same way as indicated earlier for the entries under H_{11} in table 14.

Models H_{11} and H_{12}, when interpreted in terms of the latent variables Y, Z, Y', and Z', state that there are two evolving latent variables (Y and Z evolving into Y' and Z', respectively),[66] and that the latent parameters satisfy certain additional conditions (see [55], [56], [57], [60], [61]); whereas model H_2 (and the related models H_3–H_6) stated that there are two unchanging latent variables. A model in which the latent variables are unchanging (e.g., H_2) is more parsimonious than the corresponding model in which they are evolving. We can make the latter kind of model

of degrees of freedom is due to the fact that these authors should have included an additional 2 df for testing their model, since their estimated model is at a "terminal maximum" where the following two conditions are satisfied: $\pi^{\bar{Y}'YZ}_{111} = 1$, $\pi^{\bar{Y}'YZ}_{112} = 1$ (see Goodman 1974). These two conditions correspond directly to the two conditions presented as (60). Recall that the imposition of the two conditions (60) changed H_{10} to H_{11}, and it also changed the corresponding number of degrees of freedom from 3 to 5. (For a related comment, see last sentence of n. 19.)

[65] Since Y and Y' pertain to the same latent variable at the first and second interview, respectively, we can refer to the level of Y' as the subsequent level of Y. With respect to variable Y', the assessment of the magnitude of the various effects upon this variable can be made by analyzing the three-way table $\{YZY'\}$ (see Goodman 1972*a*, 1973*a*), which can be formed using the estimates for π^{YZ}_{rs} and $\pi^{\bar{Y}'YZ}_{1rs}$ from tables 13 and 14, respectively. Similar comments apply with respect to variable Z'. In the path diagram for model H_{11} expressed in terms of the four latent variables Y, Z, Y', and Z', the variable Y' is directly affected by Y, Z, and by the interaction effect between Y and Z; and the variable Z' is subject to similar kinds of effects (see fig. 10.*E.1* below). In this path diagram, each of the manifest varilables is directly affected by only one latent variable (A by Y, B by Z, C by Y', D by Z').

[66] From the entries in table 14, we see that variable Y is evolving much less than variable Z. If variable Y were unchanging at both latent levels, then $\pi^{\bar{Y}'YZ}_{1rs}$ would be equal to 1 and 0 for $r = 1$ and 2, respectively (for both $s = 1$ and 2). Variable Y is unchanging at latent level 1 but not at latent level 2, under H_{11}. Comparison of $X^2(H_{12})$ with $X^2(H_{11})$ in table 5 shows that the estimated $\pi^{\bar{Y}'YZ}_{122}$ under H_{11} in table 14 does not differ from zero in a statistically significant way; but a similar procedure can be applied to show that $\pi^{\bar{Y}'YZ}_{121}$ under H_{12} (or under H_{11}) does differ from zero in a statistically significant way.

more parsimonious either by imposing restrictions that replace evolving latent variables by unchanging ones (as in H_2–H_6) or by imposing other kinds of restrictions on the latent parameters (as in H_{11} and H_{12}). For the data analyzed here, each of these routes led to models that were congruent with the data.

2.6. Some Further Comments on the Assignment of Individuals with Respect to the Latent Variables

On the basis of an individual's observed joint level (i, j, k, l) on the manifest variables (A, B, C, D), we can assign him to a latent class with respect to the latent variables in a given model, using the general method presented in section 2.3. This method can be applied to any of the latent structures considered in this article. Since the particular models considered in section 2.3 were expressed in terms of one or two latent variables, whereas models H_{11} and H_{12} in the preceding section were expressed in terms of four latent variables, it may be helpful to include here the following addendum to section 2.3. The reader who is not particularly interested in this topic can skip the details here.

For the sake of simplicity let us first consider, say, model H_{12} expressed in terms of the *two* latent variables Y and Z. In this case, we can directly apply (34) with $\pi^{ABCDX}{}_{ijklt}$ and π_{ijkl} estimated by inserting in (5) and (13) the estimated parameters under H_{12} in table 13. The latent variable X in (34) can denote the joint latent variable (Y, Z), as it does in this particular case and in the application to H_2 in section 2.3 (see tables 7 and 13).

For the sake of comparison with the remarks in the preceding paragraph, let us now consider model H_{12} expressed in terms of the *four* latent variables Y, Z, Y', and Z'. In this case, we can also apply (34) with the latent variable X in (34) now denoting the joint latent variable (Y, Z, Y', Z'). This joint latent variable has 16 latent classes, since each of the four individual latent variables has two latent classes, and $2^4 = 16$. The tth latent class $(t = 1, 2, \ldots, 16)$ with respect to latent variable X denotes the latent class (r, s, u, v) in which variables Y, Z, Y', and Z' are at level r, s, u, and v, respectively (for $r = 1, 2$; $s = 1, 2$; $u = 1, 2$; $v = 1, 2$). The π_{ijkl} in (34) can now be estimated using (13) modified so that t is summed from 1 to 16, and the $\pi^{ABCDX}{}_{ijklt}$ in (34) can be estimated using (5) with

$$\begin{gathered}\pi^{X}{}_{t} = \pi^{YZY'Z'}{}_{rsuv} = \pi^{YZ}{}_{rs}\pi^{\bar{Y}'YZ}{}_{urs}\pi^{\bar{Z}'YZ}{}_{vrs},\\ \pi^{\bar{A}X}{}_{it} = \pi^{\bar{A}Y}{}_{ir}, \pi^{\bar{B}X}{}_{jt} = \pi^{\bar{B}Z}{}_{js}, \pi^{\bar{C}X}{}_{kt} = \pi^{\bar{C}Y'}{}_{lu}, \pi^{\bar{D}X}{}_{lt} = \pi^{\bar{D}Z'}{}_{lv},\end{gathered} \tag{62}$$

inserting in (5), (13), and (62) the estimated parameters under H_{12} in table 13 and the estimates for H_{12} in table 14.[67]

Given that an individual is at level (i, j, k, l) on the joint manifest variable (A, B, C, D), let $\pi^{ABCD\overline{YZ}}_{ijklrs}$ denote the *conditional* probability that he will be at level (r, s) on the joint latent variable (Y, Z), and let $\pi^{ABCD\overline{YZY'Z'}}_{ijklrsuv}$ denote the *conditional* probability that he will be at level (r, s, u, v) on the joint latent variable (Y, Z, Y', Z'). In the preceding two paragraphs, we described how these two conditional probabilities should be estimated under model H_{12} expressed in terms of the joint latent variables (Y, Z) and (Y, Z, Y', Z'), respectively. In addition, it may be worthwhile to note that these two conditional probabilities are related in the following way:

$$\pi^{ABCD\overline{YZY'Z'}}_{ijklrsuv} = \pi^{ABCD\overline{YZ}}_{ijklrs}\pi^{\bar{C}Y'}_{ku}\pi^{\bar{D}Z'}_{lv}\pi^{\bar{Y}'YZ}_{urs}\pi^{\bar{Z}'YZ}_{vrs}/[\pi^{\bar{C}Y}_{kr}\pi^{\bar{D}Z}_{ls}]. \quad (63)$$

Thus, from the estimated $\pi^{ABCD\overline{YZ}}_{ijklrs}$ for, say, model H_{12} (corresponding to the entries in table 10), we can also estimate $\pi^{ABCD\overline{YZY'Z'}}_{ijklrsuv}$ by inserting in (63) the estimated parameters under H_{12} in table 13 and the estimates for H_{12} in table 14.

3. THE ANALYSIS OF THE LAZARSFELD PANEL DATA

3.1. Some Possible Models

For the Lazarsfeld panel data in table 2, we present in table 15 the χ^2 values obtained in testing various hypotheses about these data. Table 15 includes some of the hypotheses tested in section 2 (see table 5) and also some new hypotheses. We shall now describe one of the new hypotheses (H_{13}).

Hypothesis H_{13} states that the relationships among the manifest variables (A, B, C, D) can be explained by introducing two latent dichotomous variables Y and Z (as in hypothesis H_2), *and* that variables C and D are *accurate* indicators of latent variables Y and Z, respectively. We now introduce figure 6 for H_{13}.[68] This figure shows that variable A is affected by latent variable Y (which can be replaced by the *accurate* indicator C), and that variable B is affected by latent variable Z (which can be replaced by the accurate indicator D).

[67] With respect to the last two equalities on the second line of (62), note that (57) is used in order to estimate $\pi^{\bar{C}Y'}_{1u}$ and $\pi^{\bar{D}Z'}_{1v}$ from the estimated parameters in table 13. The estimated π_{ijkl} in this paragraph will be equal to the corresponding quantity in the preceding paragraph.

[68] Since H_{13} is a modification of H_2, we can view fig. 6 for H_{13} as a modification of fig. 2 for H_2. Similarly, we can view fig. 6 for H_{13} as a modification of fig. 3 for H_3.

TABLE 15

χ^2 VALUES FOR SOME MODELS PERTAINING TO LAZARSFELD PANEL DATA (COL. 2 IN TABLE 2)

Model	df	Likelihood-Ratio χ^2	Goodness-of-Fit χ^2
H_1	6	105.72	123.86
H_2	4	7.32	11.53
H_3	6	8.27	11.18
H_4	6	7.81	11.83
H'_3	6	7.88	10.48
H_{13}	8	8.62	11.45
H_{14}	4	3.14	2.25
H_{15}	8	5.44	4.93
H_{16}	10	8.38	7.40

Hypothesis H_{13} states that *both* hypotheses H_3 and H_4 are true. Under H_3, condition (18) was imposed; and under H_4, the corresponding condition is imposed:

$$\pi^{\bar{D}X}{}_{1t} = \begin{cases} 1 \text{ when } t = 1, 3 \\ 0 \text{ when } t = 2, 4. \end{cases} \tag{64}$$

As we noted earlier, when H_3 is true, two of the parameters in (16) are

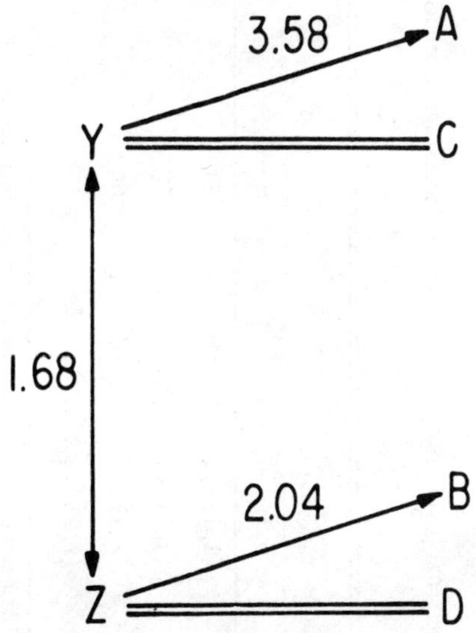

FIG. 6.—Path diagram for the hypothesis H_{13} that the latent variables Y and Z explain the relationships among the manifest variables (A, B, C, D) in the Lazarsfeld panel data, *and* that variables C and D are accurate indicators for the latent variables.

TABLE 16

MAXIMUM-LIKELIHOOD ESTIMATE OF PARAMETERS IN MODEL H_{13} (FIG. 6), DESCRIBED BY FORMULAS (5), (13), (15), (18), AND (64), FOR LAZARSFELD PANEL DATA

LATENT CLASS						
t	(r, s)	$\pi^{X}_{t} = \pi^{YZ}_{rs}$	$\pi^{\bar{A}X}_{1t} = \pi^{\bar{A}YZ}_{1rs}$	$\pi^{\bar{B}X}_{1t} = \pi^{\bar{B}YZ}_{1rs}$	$\pi^{\bar{C}X}_{1t} = \pi^{\bar{C}YZ}_{1rs}$	$\pi^{\bar{D}X}_{1t} = \pi^{\bar{D}YZ}_{1rs}$
1	(1, 1)	.53	.98	.91	1.00**	1.00**
2	(1, 2)	.10	.98*	.15	1.00**	.00**
3	(2, 1)	.06	.04	.91*	.00**	1.00**
4	(2, 2)	.31*	.04*	.15*	.00**	.00**

* The asterisked entries can be determined from the other entries in the table using (14) and (15). The double asterisked entries are specified by the model using (18) and (64).

completely determined ($\pi^{\bar{C}X}{}_{11}$ and $\pi^{\bar{C}X}{}_{13}$); and when H_4 is true, two other parameters in (16) are completely determined ($\pi^{\bar{D}X}{}_{11}$ and $\pi^{\bar{D}X}{}_{12}$). When both H_3 and H_4 are true (i.e., when H_{13} is true), the four parameters listed above are completely determined. Because four parameters in (16) are completely determined under H_{13} but not under H_2, there will be four more degrees of freedom available for testing H_{13} than for testing H_2. There were 4 df for testing H_2, and there will be 8 df for testing H_{13}. By comparing $X^2(H_{13})$ in table 15 with the percentiles of the χ^2 distribution with 8 df, we see that H_{13} fits the data to an acceptable degree. The estimated parameters under H_{13} are presented in table 16.

Since variables *Y* and *Z* can be replaced by *C* and *D*, respectively, in the path diagram for model H_{13} (fig. 6), we see from this diagram that H_{13} states that variable *A* is affected directly by the level of variable *C* (not *B* or *D*), and variable *B* is affected directly by the level of variable *D* (not *A* or *C*).[69] This particular model was introduced in Goodman (1973*a*) for these data. The present development of this model (as a modification of H_2, H_3, and H_4) sheds further light on it. In figure 7 we

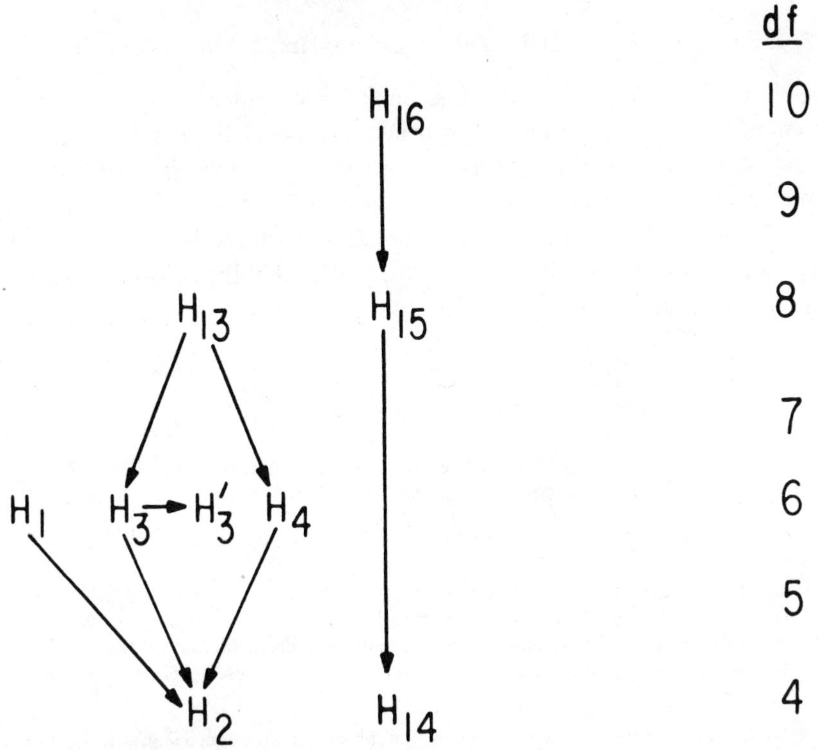

FIG. 7.—The relationship among the models in table 15, with respect to their implications and degrees of freedom. Models H_{14}–H_{16} will be discussed in sec. 3.2.

69 This interpretation of H_{13} can be used to gain further insight into the meaning of

describe how this model (i.e., H_{13}), and the other models considered in this section and section 3.2, are related to each other with respect to their implications and degrees of freedom.

In section 2.5, we showed how H_2 could be modified to obtain H_6, a special case of H_2 in which certain error rates were assumed to be equal (see [46]). Similarly for the data analyzed in the present section, we could consider various special cases of H_2 and H_{13}. Some of these special cases fit the data to an acceptable degree when the χ^2 statistic (11) is used, but the fit was found to be less acceptable when the statistic (10) is used.[70] To save space, we shall not discuss these special cases here.

In section 2.5, we also presented model H_7 and various special cases of it (H_8–H_{12}). Similarly, for the data analyzed in the present section, we could consider these kinds of models. When this was done, we found that the remarks in the preceding paragraph concerning the fit obtained with the statistics (10) and (11) also held true for some special cases of H_9 and H_{10}. To save space, we shall forego a discussion of these special cases too.

3.2. Some New Kinds of Latent Structures and "Quasi-latent-Structures"

We have already discussed some models in which certain conditional probabilities are equal to one and others are equal to zero (see, e.g., [18] and [64]). We next consider a different kind of latent structure in which some of the conditional probabilities are restricted in this way.

Let us consider the latent class model H_{14} in which the latent variable X has four latent classes, and the conditional probabilities pertaining to the first and last latent class are restricted as follows:

$$\pi^{\bar{A}X}{}_{11} = \pi^{\bar{B}X}{}_{11} = \pi^{\bar{C}X}{}_{11} = \pi^{\bar{D}X}{}_{11} = 1, \tag{65a}$$

$$\pi^{\bar{A}X}{}_{14} = \pi^{\bar{B}X}{}_{14} = \pi^{\bar{C}X}{}_{14} = \pi^{\bar{D}X}{}_{14} = 0. \tag{65b}$$

Condition (65a) states that an individual in latent class 1 with respect to latent variable X will be (with probability equal to one) at level (1, 1, 1, 1) with respect to the joint manifest variable (A, B, C, D); and,

the numerical entries in table 16 and in fig. 6. For this particular model, the general method used in the present article to calculate these entries (see sec. 5.2) is particularly easy to apply. This ease of calculation is a consequence of the above interpretation of H_{13}, which yields explicit formulas for these entries (see Goodman 1970, 1973*b*).

[70] As noted in Goodman (1973*a*), when the observed frequencies in the cross-classification are large enough (see, e.g., table 1), the conclusions obtained using statistic (11) will usually agree with those obtained using (10); but when some of the observed frequencies are very small (as in the table analyzed in the present section), this need not be the case. When different conclusions are obtained, it would be prudent to suspend judgment until additional data can be obtained.

similarly, from (65b) we see that an individual in latent class 4 with respect to latent variable X will be (with probability equal to one) at level $(2, 2, 2, 2)$ with respect to the joint manifest variable. The χ^2 values for testing H_{14} were presented in table 16, and they indicate that H_{14} fits the data very well.

The estimated parameters under H_{14} are presented in table 17.[71] As noted above, the individuals who are in latent classes 1 and 4 on latent variable X will be at level (1, 1, 1, 1) and (2, 2, 2, 2), respectively, with respect to the joint manifest variables (A, B, C, D).[72] If the two cells (1, 1, 1, 1) and (2, 2, 2, 2) are deleted from the four-way table $\{ABCD\}$, model H_{14} can be expressed as a "quasi-latent-structure" for the remaining 14 cells in the table.[73] In this quasi-latent-structure, latent classes 1 and 4 are deleted, and the latent variable X is replaced by the dichotomous latent variable X'.[74] When H_{14} is expressed in this way, the model states that a *dichotomous* latent variable X' can explain the relationships among the manifest variables, when cells (1, 1, 1, 1) and (2, 2, 2, 2) are deleted.[75]

Since H_{14} can be expressed in terms of the dichotomous latent variable

71 Table 17 also includes the estimated parameters for models H_{15} and H_{16}, which we shall discuss below.

72 Latent classes 1 and 4 can be viewed in deterministic terms as the two "extreme scale types."

73 The concept of a "quasi-latent-structure" (a modification of the usual concept of a latent structure) is introduced here in order to analyze cross-classification tables in which certain cells have been deleted. In a similar way, Goodman (1968) introduced the concept of "quasi-independence" (a modification of the usual concept of "independence") in order to analyze cross-classification tables in which certain cells have been deleted. The usual model of independence in a cross-classification table is a special case of the latent class model in which there is only one latent class; and similarly the model of quasi-independence is a special case of the quasi-latent-structure model.

74 The remaining two latent classes (viz., latent classes 2 and 3) in table 17 for H_{14} now form the dichotomous variable X'. The *conditional* probabilities associated with these two latent classes are estimated by the corresponding entries in table 17, and the probabilities $\pi^{X'}_t$ for these two classes can be estimated by dividing each of the corresponding entries in table 17 by the sum of the two entries.

75 To estimate the parameters under this quasi-latent-structure model, it is not actually necessary to delete these two cells. The observed frequencies in these two cells will not affect the estimated parameters under the model *if* the procedure in n. 74 is followed, as long as each of the two observed frequencies is not smaller than the corresponding estimated expected frequencies under H_{14}. (In the situation where this condition is not satisfied for the observed frequency in, say, cell (1, 1, 1, 1) then π^X_1 would be zero in table 17 for H_{14}, and cell (1, 1, 1, 1) would *not* be deleted under H_{14}. On the other hand, to analyze the quasi-latent-structure with *both* cells deleted in this situation, we can replace the observed frequency in cell (1, 1, 1, 1) by a sufficiently large number [leaving unchanged the observed frequencies in the remaining 15 cells], then estimate the parameters under H_{14} for the table thus modified, and then apply the procedure in n. 74 to these estimated parameters.)

TABLE 17

MAXIMUM-LIKELIHOOD ESTIMATE OF PARAMETERS IN THREE RELATED MODELS APPLIED TO LAZARSFELD PANEL DATA

Model	Latent Class t	Latent Class (r, s)	$\pi^{X}_{t} = \pi^{YZ}_{rs}$	$\pi^{\bar{A}X}_{1t} = \pi^{\bar{A}YZ}_{1rs}$	$\pi^{\bar{B}X}_{1t} = \pi^{\bar{B}YZ}_{1rs}$	$\pi^{\bar{C}X}_{1t} = \pi^{\bar{C}YZ}_{1rs}$	$\pi^{\bar{D}X}_{1t} = \pi^{\bar{D}YZ}_{1rs}$
H_{14}:	1	(1, 1)	.48	1.00**	1.00**	1.00**	1.00**
	2	(1, 2)	.15	.95	.11	.98	.33
	3	(2, 1)	.12	.12	.86	.04	.51
	4	(2, 2)	.25*	.00**	.00**	.00**	.00**
H_{15}:	1	(1, 1)	.48	1.00**	1.00**	1.00**	1.00**
	2	(1, 2)	.16	.93	.12	.97	.39
	3	(2, 1)	.11	.07*	.88*	.03*	.61*
	4	(2, 2)	.25*	.00**	.00**	.00**	.00**
H_{16}:	1	(1, 1)	.48	1.00**	1.00**	1.00**	1.00**
	2	(1, 2)	.16	.93	.07*	.93*	.38
	3	(2, 1)	.11	.07*	.93*	.07*	.62*
	4	(2, 2)	.25*	.00**	.00**	.00**	.00**

NOTE.—Model H_{14} described by formulas (5), (13), and (65); model H_{15} described by formulas (5), (13), and (66); model H_{16} described by formulas (5), (13), (65), (66), and (67).

* The asterisked entries can be determined from the other entries in the table for a given model using (14) and the corresponding restrictions pertaining to the model. The double asterisked entries are specified by the model using (65).

X', a path diagram of the same form as figure 1 could be used here, with X in figure 1 replaced by X'. The numerical entries in this path diagram would be calculated in the same way as indicated earlier for figure 1.[76]

We next consider a special case of H_{14} (viz., model H_{15}). In section 2.5, we showed how H_2 could be modified to obtain H_6 (see [46]); and similarly we shall now show how H_{14} can be modified to obtain H_{15}.

As in our earlier discussion of error rates in section 2.5, we can now let $\epsilon^{\bar{A}X}{}_t$ denote the variable A error rates for individuals who are at level t (for $t = 1, 2, 3, 4$) with respect to latent variable X in model H_{14}. Under the parameter values for model H_{14} as estimated in table 17, we see that $\epsilon^{\bar{A}X}{}_1 = \epsilon^{\bar{A}X}{}_4 = 0$; and $\epsilon^{\bar{A}X}{}_2$ and $\epsilon^{\bar{A}X}{}_3$ are estimated as .05 (i.e., $1 - .95$) and .12, respectively. In a similar way, we can define and estimate the error rates $\epsilon^{\bar{B}X}{}_t$, $\epsilon^{\bar{C}X}{}_t$, and $\epsilon^{\bar{D}X}{}_t$ (for $t = 1, 2, 3, 4$). The estimates of these error rates are subject to sampling variability (for $t = 2, 3$), and it might sometimes be of interest to test the following null hypothesis:[77]

$$\epsilon^{\bar{A}X}{}_2 = \epsilon^{\bar{A}X}{}_3,\ \epsilon^{\bar{B}X}{}_2 = \epsilon^{\bar{B}X}{}_3,\ \epsilon^{\bar{C}X}{}_2 = \epsilon^{\bar{C}X}{}_3,\ \epsilon^{\bar{D}X}{}_2 = \epsilon^{\bar{D}X}{}_3 \qquad (66)$$

(cf. [66] with [46]). This hypothesis states that each error rate for individuals who are at level 2 on latent variable X is equal to the corresponding rate for individuals who are at level 3 on this latent variable. To test this hypothesis, we can use the same kinds of methods applied earlier in section 2.5 to test hypothesis (46). In the present context, we apply these methods by first introducing model H_{15}, which states that model H_{14} is true *and* that condition (66) is also true, and then by comparing the χ^2 values obtained for testing models H_{15} and H_{14}.

The χ^2 values for testing these models were included in table 15, and we see from these values that the data are congruent with the hypothesis that condition (66) is satisfied. We also see that model H_{15} fits the data very well indeed. The estimated parameter values under H_{15} were included in table 17.

We next consider a special case of H_{15} (viz., model H_{16}). In section 2.5, we showed how H_9 could be modified to obtain H_{10} (see [54]); and similarly we shall now show how H_{15} can be modified to obtain H_{16}.

[76] When model H_{14} is expressed in terms of the dichotomous latent variable X', it pertains to the four-way table $\{ABCD\}$, with the two cells (1, 1, 1, 1) and (2, 2, 2, 2) deleted. On the other hand, when model H_{14} is expressed in terms of the latent variable X, which has four latent classes (see table 17), it pertains to the entire four-way table $\{ABCD\}$; and it will provide a perfect fit for the observed frequencies in the two cells (1, 1, 1, 1) and (2, 2, 2, 2). (This will be the case as long as each of the two observed frequencies is not smaller than the corresponding estimated expected frequencies under the quasi-latent-structure model; see related comments in n. 75.)

[77] For the first and fourth latent class, the "estimates" of the error rates are not subject to sampling variability under model H_{14}, since these "error rates" are set equal to zero under the model.

As in our earlier discussion of H_{10} in section 2.5, we can impose restrictions that state that, say, the variable *A* error rates and the variable *B* error rates are equal to each other (see [54]). Similarly, we can now introduce model H_{16}, a special case of H_{15} in which the following additional restrictions are imposed:

$$\epsilon^{\bar{A}X}{}_2 = \epsilon^{\bar{B}X}{}_2 = \epsilon^{\bar{C}X}{}_2. \tag{67}$$

Model H_{16} states that H_{15} is true *and* that the error rates pertaining to variables *A*, *B*, and *C* are all equal to each other.[78]

The χ^2 values for testing H_{16} were also included in table 15. By comparing the χ^2 values for H_{16} and H_{15}, we see that the data are congruent with the hypothesis that (67) is satisfied. We also see that model H_{16} fits the data well.

4. THE ANALYSIS OF THE McHUGH TEST DATA

4.1. Some Possible Models

The various hypotheses considered in the preceding sections were applied there to panel data. They can also be applied in the usual survey or testing situation. To illustrate this, we shall consider the McHugh test data in table 2. Here variables *A*, *B*, *C*, *D* refer to the dichotomized scores on four subtests that were supposed to measure creative ability in machine design.

From table 18 we see that the simple latent class model H_1 (in which

TABLE 18

χ^2 VALUES FOR SOME MODELS PERTAINING TO MCHUGH TEST DATA
(COL. 3 IN TABLE 2)

Model	df	Likelihood-Ratio χ^2	Goodness-of-Fit χ^2
H_1	6	25.20	27.33
H_2	4	25.20	27.33
$\tilde{H}_2$	4	1.09	1.17
H_{17}	8	6.42	6.72
H_{18}	11	7.19	7.92
H_{19}	8	4.57	4.54
H_{20}	10	5.83	5.71
H_{21}	11	7.14	6.62
H_{22}	5	5.54	5.93
H_{23}	10	8.59	8.42

[78] We could also have considered the hypothesis that the error rates pertaining to the four manifest variables (*A*, *B*, *C*, *D*) are all equal to each other. This hypothesis was

the relationships among the four manifest variables (A, B, C, D) are explained by the introduction of one latent dichotomous variable X) does not fit the observed data.[79]

From table 18 we also see that model H_2 (in which there are two latent dichotomous variables Y and Z, as earlier in fig. 2) does not fit the McHugh data. In fact, comparison of $X^2(H_2)$ with $X^2(H_1)$ in table 18 indicates that replacement of model H_1 by the more general model H_2 does not result in a better fit for these particular data. Model H_1 states both that H_2 is true *and* that the following condition is satisfied by the parameters $\pi^{YZ}{}_{12}$ and $\pi^{YZ}{}_{21}$ in table 6:

$$\pi^{YZ}{}_{12} = \pi^{YZ}{}_{21} = 0. \tag{68}$$

Thus, we see that H_1 is a special case of H_2, that H_1 imples H_2, and that $X^2(H_1) \geqq X^2(H_2)$ (see [43] and fig. 5). The fact that $X^2(H_1)$ and $X^2(H_2)$ were equal for these particular data (table 18) was due to the fact that the maximum-likelihood estimates of the two parameters $\pi^{YZ}{}_{12}$ and $\pi^{YZ}{}_{21}$ under model H_2 turned out to be zero for these data.[80] Neither H_1 nor H_2 was congruent with these data.

Consider next hypothesis $\widetilde{H}_2$ in table 18. This hypothesis fits the data very well indeed. This hypothesis states that the relationships among the manifest variables (A, B, C, D) can be explained by the introduction of two latent dichotomous variables, say, V and W, where the level of variable

---

not congruent with these particular data; and to save space we shall not include it here. This hypothesis could be tested using the same kinds of methods applied here to test model H_{16}.

[79] The χ^2 goodness-of-fit value for H_1 presented in table 18 should have agreed with the corresponding value presented earlier by McHugh (1956), but it does not. This is due to an error in his formulas for calculating maximum-likelihood estimates. The error in the formulas was corrected in his 1958 note, but the error in the numerical results was not.

[80] In accordance with the rule referred to in n. 12 or in the last sentence of n. 19, there are 6 df associated with $X^2(H_1)$, and under usual circumstances there would be 4 df associated with $X^2(H_2)$. However, since certain *estimated* parameters took on extreme values for these particular data, the circumstances are not usual. The usual asymptotic χ^2 distribution theory under H_2 applies with 4 df only when the *parameters* in the model do *not* take on these extreme values. When the *parameters* in the model do *not* take on extreme values, then the *estimated* parameters (with probability approaching one) would also *not* have taken on extreme values. When certain *estimated* parameters actually do take on extreme values, in the case where the corresponding *parameters* in the model were not limited to the extreme values, the correct asymptotic distribution theory is complicated. To simplify matters for the user, we would apply the same procedure here that is appropriate in the case where the corresponding parameters take on the extreme values; viz., adjust the degrees of freedom accordingly. Thus, for these particular data, we would assess congruence using the degrees of freedom associated with H_1. To avoid confusing the reader who may not be concerned with this unusual circumstance, this adjustment was not made in table 18—we make it here instead.

V affects variables A and B and the level of variable W affects variables C and D (see fig. 8). The estimated parameters under $\tilde{H}_2$ are presented in table 19.[81]

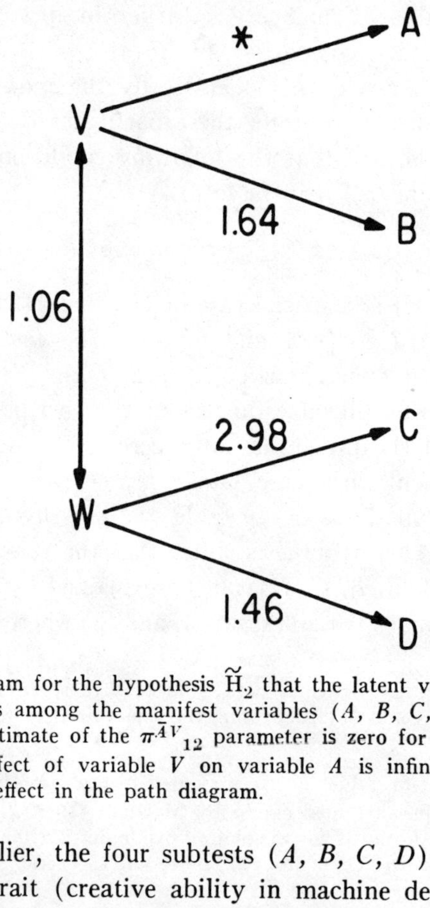

FIG. 8.—Path diagram for the hypothesis $\tilde{H}_2$ that the latent variables V and W explain the relationships among the manifest variables (A, B, C, D) in the McHugh test data. Since the estimate of the $\pi^{\bar{A}V}_{12}$ parameter is zero for these data (see table 19), the estimated effect of variable V on variable A is infinite; and the asterisk indicates this infinite effect in the path diagram.

As we noted earlier, the four subtests (A, B, C, D) were supposed to measure a single trait (creative ability in machine design). The results presented above indicate that there actually are two (correlated) latent traits, and that subtests A and B are directly affected by one of the traits, and subtests C and D are directly affected by the other one.

4.2. Some Additional Latent Structures

We next consider various special cases of model $\tilde{H}_2$ (viz., models H_{17}–H_{23}). Their χ^2 values were included in table 18, and the estimated parameters under these models are presented in table 20. We shall describe in figure 9 how these models, and the other models considered in section 4.1, are

[81] Note that $\tilde{H}_2$ is the same *kind* of hypothesis as H_2, with variables B and C interchanged (cf. fig. 8 with fig. 2, and table 19 with table 8).

TABLE 19

MAXIMUM-LIKELIHOOD ESTIMATE OF PARAMETERS IN MODEL $\tilde{H}_2$ (FIG. 8) APPLIED TO MCHUGH TEST DATA

LATENT CLASS						
t	(r, s)	$\pi^{X}{}_{t} = \pi^{VW}{}_{rs}$	$\pi^{\bar{A}X}{}_{1t} = \pi^{\bar{A}VW}{}_{1rs}$	$\pi^{\bar{B}X}{}_{1t} = \pi^{\bar{B}VW}{}_{1rs}$	$\pi^{\bar{C}X}{}_{1t} = \pi^{\bar{C}VW}{}_{1rs}$	$\pi^{\bar{D}X}{}_{1t} = \pi^{\bar{D}VW}{}_{1rs}$
1	(1, 1)	.40	.75	.76	.89	.79
2	(1, 2)	.20	.75*	.76*	.02	.17
3	(2, 1)	.08	.00	.10	.89*	.79*
4	(2, 2)	.32*	.00*	.10*	.02*	.17*

NOTE.—The asterisked entries can be determined from the other entries in this table using (14) and a formula similar to (15) but with the letters *B* and *C* in (15) interchanged.

TABLE 20

MAXIMUM-LIKELIHOOD ESTIMATE OF PARAMETERS IN SEVEN RELATED MODELS APPLIED TO MCHUGH TEST DATA

Model	Latent Class t	(r, s)	$\pi^{X}_{t} = \pi^{VW}_{rs}$	$\pi^{\bar{A}X}_{1t} = \pi^{\bar{A}VW}_{1rs}$	$\pi^{\bar{B}X}_{1t} = \pi^{\bar{B}VW}_{1rs}$	$\pi^{\bar{C}X}_{1t} = \pi^{\bar{C}VW}_{1rs}$	$\pi^{\bar{D}X}_{1t} = \pi^{\bar{D}VW}_{1rs}$
H_{17}:	1	(1, 1)	.27	1.00**	.76	1.00**	.78
	2	(1, 2)	.18	1.00**	.76*	.00**	.23
	3	(2, 1)	.16	.00**	.28	1.00**	.78*
	4	(2, 2)	.39*	.00**	.28*	.00**	.23*
H_{18}:	1	(1, 1)	.27	1.00**	.76	1.00**	.76*
	2	(1, 2)	.18	1.00**	.76*	.00**	.24*
	3	(2, 1)	.16	.00**	.24*	1.00**	.76*
	4	(2, 2)	.39*	.00**	.24*	.00**	.24*
H_{19}:	1	(1, 1)	.31	.88	.81	.95	.80
	2	(1, 2)	.16	.88*	.81*	.05*	.20*
	3	(2, 1)	.11	.12*	.19*	.95*	.80*
	4	(2, 2)	.41*	.12*	.19*	.05*	.20*
H_{20}:	1	(1, 1)	.40	.80	.80*	.80*	.80*
	2	(1, 2)	.16	.80*	.80*	.08*	.08*
	3	(2, 1)	.11	.08	.08*	.80*	.80*
	4	(2, 2)	.34*	.08*	.08*	.08*	.08*
H_{21}:	1	(1, 1)	.32	.86	.86*	.86*	.86*
	2	(1, 2)	.15	.86*	.86*	.14*	.14*
	3	(2, 1)	.11	.14*	.14*	.86*	.86*
	4	(2, 2)	.42*	.14*	.14*	.14*	.14*
H_{22}:	1	(1, 1)	.42	.66	.71	.91	.84
	2	(1, 2)	.26	.66*	.71*	.08	.20
	3	(2, 2)	.32*	.00	.04	.08*	.20*
H_{23}:	1	(1, 1)	.43	.68	.68*	.86	.86*
	2	(1, 2)	.27	.68*	.68*	.14*	.14*
	3	(2, 2)	.30	.00	.00*	.14*	.14*

NOTE.—Each model imposes additional restrictions upon model $\tilde{H}_2$. For models H_{17}–H_{23} the corresponding restrictions are (70)–(76), respectively. The asterisked entries can be determined from the other entries in the table for a given model using the same formulas as used in table 19 and also the corresponding restrictions pertaining to the model. The double asterisked entries are specified for models H_{17} and H_{18} using (69).

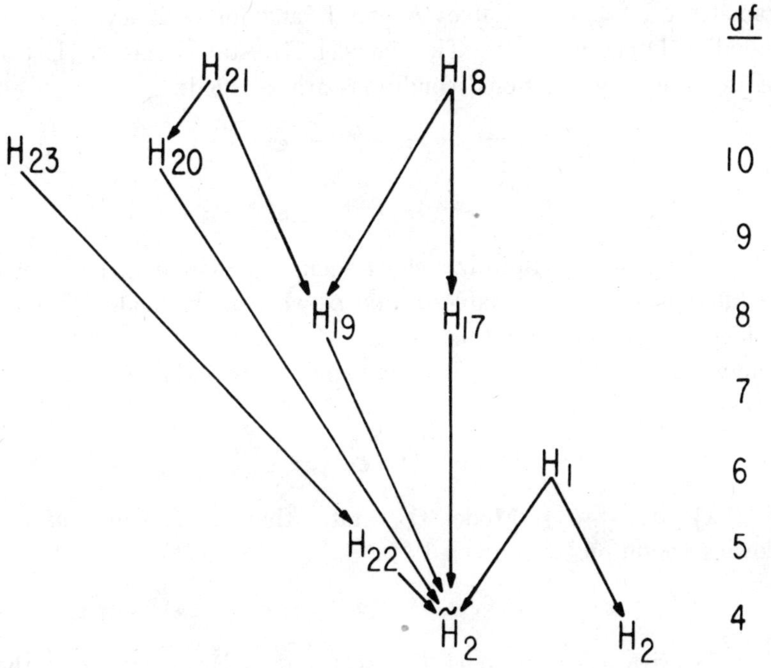

FIG. 9.—The relationship among the models in table 18 with respect to their implications and degrees of freedom.

related to each other with respect to their implications and degrees of freedom.

Model H_{17} states that $\tilde{H}_2$ is true *and* that variables A and C are accurate indicators for the latent variables V and W, respectively.[82] Thus, under H_{17} the following additional conditions are satisfied:

$$\pi^{\bar{A}V}_{1r} = \begin{cases} 1 \text{ when } r = 1 \\ 0 \text{ when } r = 2 \end{cases} \qquad \pi^{\bar{C}W}_{1s} = \begin{cases} 1 \text{ when } s = 1 \\ 0 \text{ when } s = 2 \end{cases} \tag{69}$$

(cf. [69] with [17]). Expressed in terms of error rates, condition (69) states that the error rate is zero for both variables A and C and for both levels of the corresponding latent variables:

$$\epsilon^{\bar{A}V}_1 = \epsilon^{\bar{A}V}_2 = \epsilon^{\bar{C}W}_1 = \epsilon^{\bar{C}W}_2 = 0. \tag{70}$$

Thus, model H_{17} states that $\tilde{H}_2$ is true *and* that condition (70) is satisfied.

Model H_{18} states that H_{17} is true *and* that there is a single error rate

[82] Model H_{17} is of the same form as H_{13}, with variables C, D, A, B in H_{13} replaced by A, C, B, D, respectively, in H_{17} (cf. table 16 with the corresponding entries for H_{17} in table 20). Thus, the comments about H_{13} in sec. 3.1 can be translated into corresponding comments about H_{17}. Some similar comments also apply to H_{18}, which we shall introduce in the next paragraph.

parameter ϵ for both variables B and D and for both levels of the corresponding latent variables. Thus, model H_{18} states that $\tilde{H}_2$ is true *and* that the following additional conditions are satisfied:

$$\epsilon^{\bar{A}V}{}_1 = \epsilon^{\bar{A}V}{}_2 = \epsilon^{\bar{C}W}{}_1 = \epsilon^{\bar{C}W}{}_2 = 0,$$
$$\epsilon^{\bar{B}V}{}_1 = \epsilon^{\bar{B}V}{}_2 = \epsilon^{\bar{D}W}{}_1 = \epsilon^{\bar{D}W}{}_2 = \epsilon \qquad (71)$$

(cf. [71] with [70]). Both models H_{17} and H_{18} viewed variables A and C in a different way from variables B and D, whereas the models that we shall consider next do not have this view.

Model H_{19} states that $\tilde{H}_2$ is true *and* that the following conditions are satisfied:[83]

$$\epsilon^{\bar{A}V}{}_1 = \epsilon^{\bar{A}V}{}_2, \epsilon^{\bar{B}V}{}_1 = \epsilon^{\bar{B}V}{}_2, \epsilon^{\bar{C}W}{}_1 = \epsilon^{\bar{C}W}{}_2, \epsilon^{\bar{D}W}{}_1 = \epsilon^{\bar{D}W}{}_2 \qquad (72)$$

(cf. [72] with [46]). Model H_{20} states that $\tilde{H}_2$ is true *and* that the following conditions are satisfied:[84]

$$\epsilon^{\bar{A}V}{}_1 = \epsilon^{\bar{B}V}{}_1 = \epsilon^{\bar{C}W}{}_1 = \epsilon^{\bar{D}W}{}_1, \epsilon^{\bar{A}V}{}_2 = \epsilon^{\bar{B}V}{}_2 = \epsilon^{\bar{C}W}{}_2 = \epsilon^{\bar{D}W}{}_2 \qquad (73)$$

(cf. [73] with [54]). Model H_{21} states that $\tilde{H}_2$ is true *and* that the following conditions are satisfied:[85]

$$\epsilon^{\bar{A}V}{}_1 = \epsilon^{\bar{B}V}{}_1 = \epsilon^{\bar{C}W}{}_1 = \epsilon^{\bar{D}W}{}_1 = \epsilon^{\bar{A}V}{}_2 = \epsilon^{\bar{B}V}{}_2 = \epsilon^{\bar{C}W}{}_2 = \epsilon^{\bar{D}W}{}_2. \qquad (74)$$

From table 18 we see that these models fit the data very well indeed.

Now let us consider models H_{22} and H_{23}. Model H_{22} states that $\tilde{H}_2$ is true *and* that the following condition is satisfied:[86]

$$\pi^{VW}{}_{21} = 0. \qquad (75)$$

[83] Condition (72) states that, corresponding to each manifest item (viz., variables A, B, C, D), there is an error-rate parameter (viz., $\epsilon^{\bar{A}}$, $\epsilon^{\bar{B}}$, $\epsilon^{\bar{C}}$, $\epsilon^{\bar{D}}$) that applies to both levels of the corresponding latent variable. In other words, the error rates are "item-specific" here, but they are not "latent-level-specific."

[84] Condition (73) states that, corresponding to levels 1 and 2 of the latent variables, there are two error-rate parameters (ϵ_1 and ϵ_2) that apply to all of the manifest items. In other words, the error rates are latent-level-specific here, but they are not item-specific.

[85] Condition (74) states that there is a single error-rate parameter ϵ that applies to all manifest items and to both levels of the corresponding latent variables. In other words, the error rate here is not item-specific nor latent-level-specific. Note that H_{21} can also be expressed as special cases of H_{19} and H_{20} (see fig. 9).

[86] From condition (75), we see that (with probability equal to one) there will be no individuals at the joint level (2, 1) with respect to the joint latent variable (V, W). Thus, the row in table 19 (for $\tilde{H}_2$) pertaining to level (2, 1) with respect to variable (V, W) is deleted in the corresponding table 20 (for H_{22}). For the sake of simplicity, the three latent classes in table 20 (for H_{22}) are renumbered from 1 to 3.

Model H_{23} states that $\widetilde{H}_2$ is true *and* that the following conditions are satisfied:[87]

$$\pi^{VW}{}_{21} = 0, \epsilon^{\bar{A}V}{}_1 = \epsilon^{\bar{B}V}{}_1, \epsilon^{\bar{A}V}{}_2 = \epsilon^{\bar{B}V}{}_2,$$

$$\epsilon^{\bar{C}W}{}_1 = \epsilon^{\bar{D}W}{}_1 = \epsilon^{\bar{C}W}{}_2 = \epsilon^{\bar{D}W}{}_2. \quad (76)$$

From table 18 we see that models H_{22} and H_{23} also fit the data rather well.[88] On the other hand, by comparing $X^2(H_{22})$ with $X^2(\widetilde{H}_2)$, we see that a statistically significant contribution would be made by the insertion of a fourth level (viz., level (2, 1)) into the three-level joint latent variable (V, W) for model H_{22}, thus obtaining the four levels of the joint latent variable (V, W) for model $\widetilde{H}_2$. A somewhat similar result is obtained with respect to the statistical significance of the insertion of a fourth level into the three-level joint latent variable (V, W) for model H_{23}.[89]

In the two-way table describing the relationship between the latent variables V and W,[90] condition (75) states that one of the four cells (viz., cell (2, 1)) will be empty. If condition (75) were replaced by the following more symmetric condition,

$$\pi^{VW}{}_{12} = \pi^{VW}{}_{21} = 0, \quad (77)$$

then two of the four cells (viz., cells (1, 2) and (2, 1)) will be empty.[91] Condition (77) states that variables V and W are equivalent, while condition (75) describes an extreme situation where variables V and W are strongly related to each other but they need not be equivalent.

Models H_{22} and H_{23} can be expressed either in terms of the dichotomous latent variables V and W (see table 20 and the comments in the preceding paragraph) *or* in terms of the trichotomous latent variable X. With the latter expression for these models, the latent classes can be ordered from

[87] Note that H_{23} is a special case of H_{22} (see fig. 9). We could also have considered the special case of H_{22} in which the following conditions are imposed: $\epsilon^{\bar{A}V}{}_1 = \epsilon^{\bar{B}V}{}_1 = \epsilon^{\bar{A}V}{}_2 = \epsilon^{\bar{B}V}{}_2$, $\epsilon^{\bar{C}W}{}_1 = \epsilon^{\bar{D}W}{}_1 = \epsilon^{\bar{C}W}{}_2 = \epsilon^{\bar{D}W}{}_2$. This particular special case was not congruent with the data analyzed in this section.

[88] For these particular data, the maximum-likelihood estimate of the parameter $\pi^{\bar{A}X}{}_3$ was zero under models $\widetilde{H}_2$, H_{22}, and H_{23}. Thus, the corresponding degrees of freedom in table 18 should be adjusted as suggested in n. 80.

[89] With respect to the four-level model corresponding to H_{22}, the comparison of $X^2(H_{22})$ with $X^2(\widetilde{H}_2)$ in table 18 indicates that the insertion of the joint level (2, 1) is statistically significant at the 5% level; whereas, with respect to the four-level model corresponding to H_{23}, a similar comparison would indicate that the insertion is statistically significant at the 10% level.

[90] This two-way table is the same as table 6 with Y and Z replaced there by V and W, respectively.

[91] Condition (77) is the same as condition (68) with Y and Z replaced there by V and W, respectively.

latent class 1 to 2 to 3, and the corresponding modal responses are (1, 1, 1, 1), (1, 1, 2, 2), and (2, 2, 2, 2), respectively (see table 20). These models are related to, but different from, the usual latent distance model. For comments on the relationship between these models and the usual latent distance model, see Goodman (1973*d*).

5. A GENERAL LATENT STRUCTURE MODEL AND RESTRICTED LATENT STRUCTURE MODELS FOR POLYTOMOUS VARIABLES

5.1. A General Latent Structure Model

In this section, we shall introduce a method for obtaining the maximum-likelihood estimates of the parameters in a general latent structure model for polytomous variables. These estimates can then be used to obtain the maximum-likelihood estimates of the expected frequencies under the model, which can in turn be used to calculate the χ^2 statistics that we utilize in testing whether the model is congruent with the observed data.

Let us now consider the case where variables A, B, C, and D are polytomous, and where these variables consist of I, J, K, and L classes, respectively. (Thus, the numerical values for I, J, K, and L can be 2 or more.) Let π_{ijkl} denote the probability that an individual will be at level (i, j, k, l) with respect to the joint variable (A, B, C, D) (for $i = 1, 2, \ldots, I$; $j = 1, 2, \ldots, J$; $k = 1, 2, \ldots, K$; $l = 1, 2, \ldots, L$). Let us suppose for a moment that there is a latent polytomous variable X, consisting of T latent classes that can "explain" the relationships among the manifest variables (A, B, C, D). For the tth latent class $(t = 1, 2, \ldots, T)$, let $\pi^{X}{}_{t}$ denote the probability that an individual will be in class t, and let $\pi^{ABCDX}{}_{ijklt}$ denote the probability that an individual will be at level (i, j, k, l, t) with respect to the joint variable (A, B, C, D, X). Let $\pi^{\bar{A}X}{}_{it}$, $\pi^{\bar{B}X}{}_{jt}$, $^{\bar{C}X}{}_{kt}$, $\pi^{\bar{D}X}{}_{lt}$ denote the *conditional* probabilities as defined in section 2.1. Formula (5) holds true for polytomous variables (as well as for dichotomous variables), and we now replace (6) by

$$\pi_{ijkl} = \sum_{t=1}^{T} \pi^{ABCDX}{}_{ijklt}. \tag{78}$$

We can also replace formulas (7) and (8) by

$$\sum_{t=1}^{T} \pi^{X}{}_{t} = 1, \tag{79}$$

and

$$\sum_{i=1}^{I} \pi^{\bar{A}X}{}_{it} = 1, \sum_{j=1}^{J} \pi^{\bar{B}X}{}_{jt} = 1, \sum_{k=1}^{K} \pi^{\bar{C}X}{}_{kt} = 1, \sum_{l=1}^{L} \pi^{\bar{D}X}{}_{lt} = 1, \tag{80}$$

for $t = 1, 2, \ldots, T$.

Given that an individual is at level (i, j, k, l) with respect to the joint variable (A, B, C, D), let $\pi^{ABCD\bar{X}}{}_{ijklt}$ denote the *conditional* probability that he will be at level t with respect to variable X. The *conditional* probability $\pi^{ABCD\bar{X}}{}_{ijklt}$ can be expressed in terms of the probabilities π_{ijkl} and $\pi^{ABCDX}{}_{ijklt}$ as we did earlier (see [34] and also table 10):

$$\pi^{ABCD\bar{X}}{}_{ijklt} = \pi^{ABCDX}{}_{ijklt}/\pi_{ijkl}. \tag{81}$$

From the definitions of the probabilities $\pi^{X}{}_{t}$ and $\pi^{ABCDX}{}_{ijklt}$, we see that

$$\pi^{X}{}_{t} = \sum_{i=1}^{I} \sum_{j=1}^{J} \sum_{k=1}^{K} \sum_{l=1}^{L} \pi^{ABCDX}{}_{ijklt}, \tag{82}$$

and from the definition of the *conditional* probability $\pi^{\bar{A}X}{}_{it}$, we see that

$$\pi^{X}{}_{t}\pi^{\bar{A}X}{}_{it} = \sum_{j=1}^{J} \sum_{k=1}^{K} \sum_{l=1}^{L} \pi^{ABCDX}{}_{ijklt}. \tag{83}$$

Formulas similar in form to (83) can be obtained for $\pi^{X}{}_{t}$ multiplied by $\pi^{\bar{B}X}{}_{jt}$, $\pi^{\bar{C}X}{}_{kt}$, and $\pi^{\bar{D}X}{}_{lt}$. From (81) we see that (82) can be rewritten as follows:

$$\pi^{X}{}_{t} = \sum_{i=1}^{I} \sum_{j=1}^{J} \sum_{k=1}^{K} \sum_{l=1}^{L} \pi_{ijkl}\pi^{ABCD\bar{X}}{}_{ijklt}. \tag{84}$$

Similarly, formula (83) can be rewritten as

$$\pi^{\bar{A}X}{}_{it} = \Big(\sum_{j=1}^{J} \sum_{k=1}^{K} \sum_{l=1}^{L} \pi_{ijkl}\pi^{ABCD\bar{X}}{}_{ijklt}\Big)/\pi^{X}{}_{t}, \tag{85a}$$

and we can also obtain the following additional formulas:

$$\pi^{\bar{B}X}{}_{jt} = \Big(\sum_{i=1}^{I} \sum_{k=1}^{K} \sum_{l=1}^{L} \pi_{ijkl}\pi^{ABCD\bar{X}}{}_{ijklt}\Big)/\pi^{X}{}_{t}, \tag{85b}$$

$$\pi^{\bar{C}X}{}_{kt} = \Big(\sum_{i=1}^{I} \sum_{j=1}^{J} \sum_{l=1}^{L} \pi_{ijkl}\pi^{ABCD\bar{X}}{}_{ijklt}\Big)/\pi^{X}{}_{t}, \tag{85c}$$

$$\pi^{\bar{D}X}{}_{lt} = \Big(\sum_{i=1}^{I} \sum_{j=1}^{J} \sum_{k=1}^{K} \pi_{ijkl}\pi^{ABCD\bar{X}}{}_{ijklt}\Big)/\pi^{X}{}_{t}\,. \tag{85d}$$

Suppose for a moment that the numerical values of the *conditional* probabilities $\pi^{ABCD\bar{X}}_{ijklt}$ were available.[92] In this case, we could estimate the parameters in the model (viz., π^{X}_{t}, $\pi^{\bar{A}X}_{it}$ $\pi^{\bar{B}X}_{jt}$, $\pi^{\bar{C}X}_{kt}$, $\pi^{\bar{D}X}_{lt}$) in the following way, using the *observed* proportion p_{ijkl} of individuals at level (i, j, k, l) with respect to the manifest variables (A, B, C, D).[93] We would replace (84) and (85a)–(85d) by

$$\hat{\pi}^{X}_{t} = \sum_{i=1}^{I} \sum_{j=1}^{J} \sum_{k=1}^{K} \sum_{l=1}^{L} p_{ijkl}\pi^{ABCD\bar{X}}_{ijklt}, \tag{86}$$

$$\hat{\pi}^{\bar{A}X}_{it} = \Big(\sum_{j=1}^{J} \sum_{k=1}^{K} \sum_{l=1}^{L} p_{ijkl}\pi^{ABCD\bar{X}}_{ijklt}\Big)/\hat{\pi}^{X}_{t}, \tag{87a}$$

$$\hat{\pi}^{\bar{B}X}_{jt} = \Big(\sum_{i=1}^{I} \sum_{k=1}^{K} \sum_{l=1}^{L} p_{ijkl}\pi^{ABCD\bar{X}}_{ijklt}\Big)/\hat{\pi}^{X}_{t}, \tag{87b}$$

$$\hat{\pi}^{\bar{C}X}_{kt} = \Big(\sum_{i=1}^{I} \sum_{j=1}^{J} \sum_{l=1}^{L} p_{ijkl}\pi^{ABCD\bar{X}}_{ijklt}\Big)/\hat{\pi}^{X}_{t}, \tag{87c}$$

$$\hat{\pi}^{\bar{D}X}_{lt} = \Big(\sum_{i=1}^{I} \sum_{j=1}^{J} \sum_{k=1}^{K} p_{ijkl}\pi^{ABCD\bar{X}}_{ijklt}\Big)/\hat{\pi}^{X}_{t}. \tag{87d}$$

For a given set of numerical values for the $\pi^{ABCD\bar{X}}_{ijklt}$, we can use (86) and (87a)–(87d) to obtain tentative estimates of the parameters, and these tentative estimates can be inserted in (5), (78), and (81) to obtain in turn the tentative estimate $\hat{\pi}^{ABCDX}_{ijklt}$, $\hat{\pi}_{ijkl}$, and $\hat{\pi}^{ABCD\bar{X}}_{ijklt}$. We can now replace the initial numerical values of the $\pi^{ABCD\bar{X}}_{ijklt}$ by the tentative $\hat{\pi}^{ABCD\bar{X}}_{ijklt}$ thus obtained, and again use (86) and (87a)–(87d) to estimate the parameters in the model. These new estimates can be inserted in (5), (78), and (81); and the new estimate of $\pi^{ABCD\bar{X}}_{ijklt}$ thus obtained can again be inserted in (86) and (87a)–(87d) to obtain new estimates of the

[92] To apply the iterative procedure which we shall describe below, we only need to have *initial trial* values for the numerical values of the $\pi^{ABCD\bar{X}}_{ijklt}$ (see n. 94 below). Estimates of the $\pi^{ABCD\bar{X}}_{ijklt}$ were presented in table 10. (In actual practice, the true numerical values of the $\pi^{ABCD\bar{X}}_{ijklt}$ would not be known, and the estimates of these quantities [in, e.g., table 10] would not be available until after the iterative procedure described below were applied.) *Initial trial* values of the $\pi^{ABCD\bar{X}}_{ijklt}$ will be discussed more fully in sec. 5.3 below.

[93] The observed proportion p_{ijkl} is the maximum-likelihood estimate of the corresponding probability π_{ijkl} in the situation where the latent structure model may or may not be true. This estimate will differ from the corresponding maximum-likelihood estimate under the latent structure model (except in the special case where the model fits the observed data perfectly). The latter estimate ($\hat{\pi}_{ijkl}$) will be obtained at the end of the iterative procedure described next, using the parameter estimates calculated by the procedure.

parameters in the model. This procedure is repeated until the estimates of the parameters in the model remain unchanged.[94]

The estimates obtained by the procedure described above will provide a solution to the system of equations (86) and (87a)–(87d), where $\pi^{ABCD\bar{X}}_{ijklt}$ in these equations is replaced by $\hat{\pi}^{ABCDX}_{ijklt}/\hat{\pi}_{ijkl}$ (see [81]), and where $\hat{\pi}^{ABCDX}_{ijklt}$ and $\hat{\pi}_{ijkl}$ are defined by (5) and (78), replacing the π's in these formulas by corresponding $\hat{\pi}$'s. If the parameters in this latent structure model have maximum-likelihood estimates, it is possible to prove that the maximum-likelihood estimates will satisfy this same system of equations. In this case, the iterative procedure described above can be used to obtain the maximum-likelihood estimates.[95] (For some related but different results, see Haberman [1973, 1974*a*].)

The procedure described above can be used when the manifest variables (A, B, C, D) are polytomous as well as in the special case where they are dichotomous. The earlier literature on maximum-likelihood estimation in the latent class model (see, e.g., McHugh 1956, 1958) dealt only with the

[94] For expository purposes, the iterative procedure was described above as starting with initial trial values for the numerical values of the $\pi^{ABCD\bar{X}}_{ijklt}$. However, in actual practice, we would usually start the iterative procedure using initial trial values for the parameters (viz., π^{X}_{t}, $\pi^{\bar{A}X}_{it}$, $\pi^{\bar{B}X}_{jt}$, $\pi^{\bar{C}X}_{kt}$, $\pi^{\bar{D}X}_{lt}$), and these trial values would then be inserted in (5), (78), and (81) to obtain the corresponding initial trial values for the $\pi^{ABCD\bar{X}}_{ijklt}$ (see sec. 5.3 below).

[95] The iterative procedure will converge to a solution of the system of equations under quite general conditions. (In the special case where the numerical values obtained for $\hat{\pi}^{X}_{t}$ by [86] approach zero for a given value of t [say, $t = t_0$], we modify the iterative procedure by deleting the latent class t_0 from the latent structure under consideration.) If there is more than one solution (e.g., if there is a "terminal maximum" as well as a "global maximum"), by trying different initial trial values to start the iterative procedure we can determine the various solutions; and then we can compare the corresponding X^2 values to determine which solution provides the maximum-likelihood estimates (see Goodman 1974). Even in situations where the solution to the system of equations includes minima (and/or saddle-points) as well as maxima, the iterative procedure will successfully avoid the former points and it will yield the maxima (except when the initial trial values are exactly at one of the former points). In our experience with the iterative procedure, it has yielded the "global maximum" and "terminal maxima" but no "local maxima"; and when initial trial values are appropriately obtained (see sec. 5.3 below), there was no need to use more than one set of initial trial values. (The "global maximum" gives the value of the parameter vector $(\pi^{X}_{t}, \pi^{\bar{A}X}_{it}, \pi^{\bar{B}X}_{jt}, \pi^{\bar{C}X}_{kt}, \pi^{\bar{D}X}_{lt})$ that maximizes the likelihood over the entire range of possible parameter values where each parameter is *strictly* within the permissible range [e.g., *strictly* within the interval from zero to one]; a "local maximum" gives the value of a parameter vector that maximizes the likelihood within a neighborhood of that parameter vector [but not necessarily over the entire range of possible parameter values] where each parameter is *strictly* within the permissible range; and a "terminal maximum" gives the value of a parameter vector that maximizes the likelihood [either globally or locally] but where one or more of the parameters is on the boundary of the permissible range [e.g., where some parameter values are either zero or one].)

special dichotomous case. In this case, our procedure is simpler to apply than are the formulas presented by McHugh.[96]

From (86) and (87a)–(87d), we find that the estimates ($\hat{\pi}^X{}_t$, $\hat{\pi}^{\bar{A}X}{}_{it}$, $\hat{\pi}^{\bar{B}X}{}_{jt}$, $\hat{\pi}^{\bar{C}X}{}_{kt}$, $\hat{\pi}^{\bar{D}X}{}_{lt}$) will satisfy the same conditions as the corresponding parameters, namely, (79) and (80). Because of this, we can simplify the iterative procedure described above by estimating only $T - 1$ of the $\pi^X{}_t$ (say, $\pi^X{}_t$ for $t = 1, 2, \ldots, T - 1$) using (86); estimating only $I - 1$ of the $\pi^{\bar{A}X}{}_{it}$ (say, $\pi^{\bar{A}X}{}_{it}$ for $i = 1, 2, \ldots, I - 1$) using (87a); etc.

From (87a) we find that, under the latent structure model considered here, the estimated expected proportion of individuals at level i on variable A

$$\left(\text{i.e.,} \sum_{t=1}^{T} \hat{\pi}^{\bar{A}X}{}_{it}\hat{\pi}^X{}_t\right)$$

will be equal to the corresponding observed proportion

$$\left(\text{i.e.,} \sum_{j=1}^{J}\sum_{k=1}^{K}\sum_{l=1}^{L} p_{ijkl}\right).$$

Similarly, from (87b) to (87d) we find that analogous results hold true with respect to levels j, k, and l on variables B, C, and D, respectively.[97] Thus, the method of estimation introduced here for the latent structure model can be viewed as an extension of the usual method of estimating the parameters in the "elementary" model that states that the manifest variables (A, B, C, D) are mutually independent.[98]

In the latent structure model considered here, the number T of latent classes is assumed to be known. When this is not the case, we can proceed

[96] The system of equations (86) and (87a)–(87d) can be used to provide a straightforward interpretation for the system of equations presented by McHugh.

[97] These results can be used to prove that, e.g., under model H_2 (see fig. 2), the estimated expected proportion of individuals at level (i, k) on the joint variable (A, C) will be equal to the corresponding observed proportion, in the case where the three-way table $\{ACZ\}$ can be fit perfectly by the quasi-tautological model in which the latent dichotomous variable Y serves to explain the relationships in the three-way table. Similarly, under H_2, the estimated expected proportion at level (j, l) on the joint variable (B, D) will be equal to the corresponding observed proportion, in the case where the three-way table $\{BDY\}$ can be fit perfectly by the quasi-tautological model in which the latent dichotomous variable Z serves to explain the relationships in this three-way table.

[98] The latter model can be denoted as $[A \otimes B \otimes C \otimes D]$, using the notation introduced in Goodman (1970). Under this model, the set of fitted marginals is $\{A\}$, $\{B\}$, $\{C\}$, $\{D\}$ (see, e.g., Goodman 1970, 1972*b*). Thus, under this model, the estimated expected proportion of individuals at level i on variable A will be equal to the corresponding observed proportion; and analogous results hold true with respect to levels j, k, and l on variables B, C, and D, respectively. The latent structure model can be denoted as $[A \otimes B \otimes C \otimes D|X]$; and in the special case where the latent variable X has only one latent class (i.e., $T = 1$), we obtain the "elementary" model $[A \otimes B \otimes C \otimes D]$.

by considering various possible numerical values for T. When $T = 1$, we have the "elementary" model described in the preceding paragraph (see also n. 98). When $T = 2$, we have model H_1 descibed in section 2.1 (see fig. 1). We could also consider here $T = 3, 4, \ldots$, both in the case where the parameters in the models are subject to certain restrictions (as in H_{22} where $T = 3$, and in H_2 where $T = 4$) *and* also in the case where the parameters are not restricted in these ways.[99] For a given value of T (e.g., for $T = 4$), we could compare the results obtained for the general latent structure model with the corresponding results obtained when the parameters are restricted in various ways.[100] We could also compare latent structure models that have different values of T (e.g., our earlier comparison of H_1 with H_2, or $\breve{H}_2$ with H_{22}) and also models that have the same value of T (e.g., our earlier comparison of H_2 with H_6).

5.2. Restricted Latent Structure Models

The iterative procedure presented in section 5.1 was suitable for the general latent structure model, where the parameters in the model are unrestricted, except for the usual restrictions that pertain to probabilities and conditional probabilities.[101] In the present section, we shall show how to modify this iterative procedure in order to make it suitable also for models in which the parameters are subject to various kinds of restrictions.

Consider first the model in which the latent classes are equiprobable. (This will be the case, for example, in model H_5; see [42].)[102] Under this

[99] When there are four manifest dichotomous variables, problems will arise pertaining to the identifiability of the parameters when $T \geqq 3$ in the general latent structure model and also in some restricted models. More generally, when there are m manifest polytomous variables, problems pertaining to the identifiability of the parameters will arise when the latent structure model does not incorporate enough restrictions and/or when T is too large in relative terms. For the analysis of latent structure models in which the parameters may or may not be identifiable, see Goodman (1974).

[100] When the parameters in the general latent structure model are identifiable, in the case where this model fits the observed data reasonably well, then the difference between the χ^2 value obtained in testing this model and the corresponding χ^2 value obtained in testing the restricted latent structure model could be used to test whether the particular restrictions introduced in the latter model are congruent with the data. When the parameters in the general latent structure model are not identifiable (as, e.g., when $T = 4$ with four manifest dichotomous variables), in the case where this model provides a perfect fit to the data (see Goodman 1974), then the χ^2 value obtained in testing the restricted latent structure model could serve by itself as a test of the restrictions.

[101] Of course, the probabilities π^X_t are subject to the usual restriction (79), the *conditional* probabilities are subject to the corresponding restriction (80), and these parameters cannot be outside the range from zero to one.

[102] Under model H_5, there were really only two latent classes, viz., the classes pertaining to latent variable Z. Variable Y in this model was not really latent, since the

model, with X denoting the latent variable, the $\pi^{X}{}_{t}$ are completely specified, and, in particular,

$$\pi^{X}{}_{t} = 1/T. \tag{88}$$

When the $\pi^{X}{}_{t}$ are completely specified, then equation (86) for $\hat{\pi}^{X}{}_{t}$ would be deleted from our iterative procedure, and the specified numerical values of the $\pi^{X}{}_{t}$ would be used instead.[103]

The remarks in the preceding paragraph can be extended to the case where some of the *conditional* probabilities are completely specified. For example, under our earlier model H_3, the *conditional* probabilities $\pi^{\bar{C}X}{}_{kt}$ were specified by (18). When the $\pi^{\bar{C}X}{}_{kt}$ are completely specified, then equation (87c) for $\hat{\pi}^{\bar{C}X}{}_{kt}$ would be deleted from our iterative procedure, and the specified numerical values of the $\pi^{\bar{C}X}{}_{kt}$ would be used instead.[104] A similar modification of our iterative procedure would be made when some of the other *conditional* probabilities are completely specified (as in models H_4, and H_{13}–H_{18}).[105]

We also considered models in which the *conditional* probabilities are not *completely* specified but are subject to specified restrictions (e.g., condition [15] for H_2). To simplify exposition, consider next the case where the following simple condition is imposed:

$$\pi^{\bar{A}X}{}_{i1} = \pi^{\bar{A}X}{}_{i2}, \quad \text{for } i = 1, 2, \ldots, I. \tag{89}$$

In view of this restriction, the estimates $\hat{\pi}^{\bar{A}X}{}_{i1}$ and $\hat{\pi}^{\bar{A}X}{}_{i2}$ should be equal to each other, and they can be calculated by replacing the formulas for these quantities (i.e., [87a] for $t = 1, 2$) by the following formula at each repetition of the iterative procedure:

$$\hat{\pi}^{\bar{A}X}{}_{i1} = \hat{\pi}^{\bar{A}X}{}_{i2} = \Big(\sum_{t=1}^{2}\sum_{j=1}^{J}\sum_{k=1}^{K}\sum_{l=1}^{L} p_{ijkl}\hat{\pi}^{ABCD\bar{X}}{}_{ijklt}\Big)\Big/\Big(\sum_{t=1}^{2}\hat{\pi}^{X}{}_{t}\Big). \tag{90}$$

---

manifest variable C was an *accurate* indicator for it and thus could serve as its equivalent (see fig. 4).

[103] The specified numerical values of the parameter $\pi^{X}{}_{t}$ would be used in the calculation of the tentative estimate $\hat{\pi}^{ABCDX}{}_{ijklt}$, $\hat{\pi}_{ijkl}$, and $\hat{\pi}^{ABCD\bar{X}}{}_{ijklt}$ at each repetition in the iterative procedure. However, to estimate the other parameters using (87a)–(87d), the denominator on the right side of the equality sign in (87a)–(87d) is obtained at each repetition in the iterative procedure in the same way as earlier in this section, using (86) for $\hat{\pi}^{X}{}_{t}$. (Thus, condition [80] will also hold true as earlier.)

[104] In the special case when the specified numerical values are ones and/or zeros, the iterative procedure described earlier in this section actually need not be changed at all if the specified numerical values are used as the initial trial values of the corresponding parameters.

[105] In models H_{14}–H_{16} (see table 17), the conditional probabilities $\pi^{\bar{A}X}{}_{it}$ are completely specified for $t = 1$ and 4, but not for $t = 2$ and 3; and similarly for $\pi^{\bar{B}X}{}_{jt}$, $\pi^{\bar{C}X}{}_{kt}$, $\pi^{\bar{D}X}{}_{lt}$. The comments in the above paragraph and in n. 104 apply to models of this general kind as well.

Note that the estimates obtained by (90) can also be obtained by taking a weighted average of the estimates $\hat{\pi}^{\bar{A}X}_{i1}$ and $\hat{\pi}^{\bar{A}X}_{i2}$ obtained from (87a) unmodified, using weights proportional to $\hat{\pi}^{X}_{1}$ and $\hat{\pi}^{X}_{2}$, respectively. In other words, the quantity (90) is equal to

$$(\hat{\pi}^{X}_{1}\hat{\pi}^{\bar{A}X}_{i1} + \hat{\pi}^{X}_{2}\hat{\pi}^{\bar{A}X}_{i2})/(\hat{\pi}^{X}_{1} + \hat{\pi}^{X}_{2}), \tag{91}$$

where the $\hat{\pi}^{\bar{A}X}_{it}$ in (91) are obtained from (87a) unmodified. Thus, when condition (89) is imposed, we modify our iterative procedure by deleting equations (87a) for $t = 1, 2$ (corresponding to $\hat{\pi}^{\bar{A}X}_{i1}$ and $\hat{\pi}^{\bar{A}X}_{i2}$), and replacing them (at each repetition in the iterative procedure) by equation (90) or by the equivalent quantity (91).

Condition (89) is similar in kind to the other conditions imposed in models H_2, $\tilde{H}_2$, and other related models (e.g., condition [15]). Furthermore, although condition (89) set $\pi^{\bar{A}X}_{i1}$ equal to $\pi^{\bar{A}X}_{i2}$, this condition is similar in kind to the more general condition in which $\pi^{\bar{A}X}_{i1}$ is set equal to, say, $\pi^{\bar{B}X}_{j2}$ (rather than $\pi^{\bar{A}X}_{i2}$), where the levels i and j may or may not be equal.[106] This more general kind of condition was imposed in models H_{10}–H_{12}, H_{16}, H_{18}, H_{20}, H_{21}, H_{23} (see tables 13, 17, and 20). The modification of our iterative procedure, which we presented in the preceding paragraph for the case where condition (89) is imposed, can be extended in a straightforward way to the cases where these other conditions are imposed instead.

Condition (46), and the other related conditions which we introduced in sections 2.5, 3.2, and 4.2, are also similar to the kinds of conditions discussed in the preceding two paragraphs. Recall, for example, that condition (46) was rewritten as condition (47), and the latter condition sets $\pi^{\bar{B}X}_{11}$ equal to $\pi^{\bar{B}X}_{22}$, and $\pi^{\bar{C}X}_{11}$ equal to $\pi^{\bar{C}X}_{22}$. To further clarify this point, consider next the case where the following simple condition is imposed:

$$\pi^{\bar{A}X}_{11} = 1 - \pi^{\bar{A}X}_{12} = \pi^{\bar{A}X}_{22}, \tag{92}$$

in the special case where variable A is dichotomous. In view of this restriction, the estimates $\hat{\pi}^{\bar{A}X}_{11}$ and $\hat{\pi}^{\bar{A}X}_{22}$ should be equal to each other, and they can be calculated by taking a weighted average of the estimates $\hat{\pi}^{\bar{A}X}_{11}$ and $\hat{\pi}^{\bar{A}X}_{22}$ obtained from (87a) unmodified, at each repetition in the iterative procedure, as in the paragraph before the preceding one. In other words, the estimates $\hat{\pi}^{\bar{A}X}_{11}$ and $\hat{\pi}^{\bar{A}X}_{22}$ are set equal to

[106] Condition (89) was applied at each level i (for $i = 1, 2, \ldots, I$), and the same holds true for the more general condition considered here when the levels i and j are equal. When the levels i and j are not equal, the more general condition will apply at each level i and at the corresponding level j (for $j = 1, 2, \ldots, J$), where there is a one-to-one correspondence between the levels i and j. The more general kind of condition described above could also include the condition in which $\pi^{\bar{A}X}_{i1}$ is set equal to $\pi^{\bar{A}X}_{i'2}$ (rather than $\pi^{\bar{A}X}_{i2}$) where the levels i and i' may or may not be equal.

$$(\hat{\pi}^{X}{}_{1}\hat{\pi}^{\bar{A}X}{}_{11} + \hat{\pi}^{X}{}_{2}\hat{\pi}^{\bar{A}X}{}_{22})/(\hat{\pi}^{X}{}_{1} + \hat{\pi}^{X}{}_{2}), \tag{93}$$

where the $\hat{\pi}^{\bar{A}X}{}_{11}$ and $\hat{\pi}^{\bar{A}X}{}_{22}$ in (93) are obtained from (87a) unmodified. Note that (93) can also be rewritten as

$$[\hat{\pi}^{X}{}_{1}\hat{\pi}^{\bar{A}X}{}_{11} + \hat{\pi}^{X}{}_{2}(1 - \hat{\pi}^{\bar{A}X}{}_{12})]/(\hat{\pi}^{X}{}_{1} + \hat{\pi}^{X}{}_{2}). \tag{94}$$

Thus, when condition (92) is imposed, we modify our iterative procedure by deleting equation (87a) for $\hat{\pi}^{\bar{A}X}{}_{i1}$ and $\hat{\pi}^{\bar{A}X}{}_{i2}$ (for $i = 1, 2$), and replacing them (at each repetition in the iterative procedure) by the quantity (93) or the equivalent quantity (94), for $\hat{\pi}^{\bar{A}X}{}_{11} = \hat{\pi}^{\bar{A}X}{}_{22} = 1 - \hat{\pi}^{\bar{A}X}{}_{12}$.

Condition (92) is similar in kind to condition (47) in H_6 (see table 12), and to related conditions in models H_8–H_{12}, H_{15}, H_{16}, H_{18}, H_{19}, H_{21}, H_{23} (see tables 13, 17, and 20). The modification of our iterative procedure, which we presented in the preceding paragraph for the case where condition (92) is imposed, can be extended in a straightforward way to the cases where these other conditions are imposed instead.

The results noted in section 5.1 pertaining to the maximum-likelihood estimates of the parameters in the general latent structure model can be extended to the models considered in the present section. If the parameters in the *restricted* latent structure have maximum-likelihood estimates, it is possible to prove (using Lagrange multipliers to incorporate the restrictions) that the maximum-likelihood estimates will satisfy the system of equations (86) and (87a)–(87d) modified in the ways described above. In this case, the modified iterative procedures described above can be used to obtain the maximum-likelihood estimates (see n. 95).

Before closing this section, it may be worthwhile to express some of the results presented above in a somewhat different form. In order to do this, let us return for a moment to condition (89). This condition states that the $\pi^{\bar{A}X}{}_{i1}$ and the corresponding $\pi^{\bar{A}X}{}_{i2}$ are equal (for $i = 1, 2, \ldots, I$). More generally, suppose that the T latent classes can be partitioned into α mutually exclusive and exhaustive subsets $\mathcal{T}^{A}{}_{1}, \mathcal{T}^{A}{}_{2}, \ldots, \mathcal{T}^{A}{}_{\alpha}$ (where $\alpha < T$), such that the following condition is satisfied:

$$\pi^{\bar{A}X}{}_{it} = \pi^{\bar{A}X}{}_{it'}, \quad \text{for } i = 1, 2, \ldots, I, \tag{95}$$

whenever t and t' are in subset $\mathcal{T}^{A}{}_{a}$, for $a = 1, 2, \ldots, \alpha$. (Compare [95] with [89].) In this case, we modify our iterative procedure by replacing the T equations (87a) (for $t = 1, 2, \ldots, T$) by a corresponding set of α equations, where the ath equation ($a = 1, 2, \ldots, \alpha$) is

$$\hat{\pi}^{\bar{A}X}{}_{it} = \Big(\sum_{t'}\sum_{j,k,l} p_{ijkl}\hat{\pi}^{ABCD\bar{X}}{}_{ijklt'}\Big)\Big/\Big(\sum_{t'}\hat{\pi}^{X}{}_{t'}\Big), \tag{96}$$

whenever t is in subset $\mathcal{T}^A{}_a$, and where

$$\sum_{t'}$$

denotes summation over each latent class t' in $\mathcal{T}^A{}_a$ (cf. [96] with [90]).

Condition (89) and the generalized form (95) were expressed in terms of the *conditional* probabilities pertaining to variable A. We can also introduce similar kinds of conditions expressed in terms of the *conditional* probabilities pertaining to any of the m manifest variables. (We have been concerned here with the case when $m = 4$.) With respect to variable A, the T latent classes were partitioned into α subsets such that (95) was satisfied; and, similarly, the T latent classes could be partitioned into $\beta, \gamma, \delta, \ldots$ subsets with respect to variables $B, C, D, \ldots$ such that conditions corresponding to (95) are satisfied for each of these variables. Formulas corresponding to (96) can be directly applied in this case.

The above results can be generalized further in the following way. Suppose that the α subsets $\mathcal{T}^A{}_1, \mathcal{T}^A{}_2, \ldots, \mathcal{T}^A{}_\alpha$ and the β subsets $\mathcal{T}^B{}_1, \mathcal{T}^B{}_2, \ldots, \mathcal{T}^B{}_\beta$ are such that for a given pair of subscripts a and b the following condition is satisfied:

$$\pi^{\bar{A}X}{}_{it} = \pi^{\bar{B}X}{}_{it'}, \quad \text{for } i = 1, 2, \ldots, I, \tag{97}$$

whenever t is in $\mathcal{T}^A{}_a$ and t′ is in $\mathcal{T}^B{}_b$. (We are concerned here with the case where the number I of levels of variable A is equal to the number J of levels of variable B.) Even more generally, we can replace (97) by the following condition:

$$\pi^{\bar{A}X}{}_{it} = \pi^{\bar{B}X}{}_{jt'}, \quad \text{for } i = 1, 2, \ldots, I, \tag{98}$$

where there is a one-to-one correspondence between the levels i and j (see n. 106). Still more generally, the above conditions, which were expressed in terms pertaining to variables A and B, can be directly extended so that they are applicable to a given subset of the m manifest variables. Formula (96) can be directly generalized so that it is applicable in the situations described above.

5.3. Initial Trial Values to Start the Iterative Procedure

In order to use the iterative procedure described in sections 5.1 and 5.2, we needed to begin with initial trial values for the $\pi^{ABCD\bar{X}}{}_{ijklt}$ or for the parameters in the model (viz., $\pi^X{}_t$, $\pi^{\bar{A}X}{}_{it}$, $\pi^{\bar{B}X}{}_{jt}$, $\pi^{\bar{C}X}{}_{kt}$, $\pi^{\bar{D}X}{}_{lt}$); see footnotes 92 and 94. We shall now consider this matter in more detail.

Consider first the usual latent class model H_1 in which variables A, B, C, D, and X are dichotomous. As the initial trial values for the parameters in this model, we can take the quantities obtained by the estimation method usually used for the latent class model (the "Anderson-Lazarsfeld-Dudman method").[107] With these initial trial values for the parameters, we can then use the method in footnote 94 to obtain initial values for $\pi^{ABCD\bar{X}}_{ijklt}$, at the start of the iterative procedure introduced in section 5.1.

Next we shall consider model H_2 (see fig. 2). If H_2 is true, then the latent variable Y can be used to explain the relationships among variables A, C, and B; and it can also be used to explain the relationships among variables A, C, and D. Similarly, the latent variable Z can be used to explain the relationships among variables B, D, and A; and it can also be used to explain the relationships among variables B, D, and C. Applying the usual latent class methods to the three-way tables $\{ACB\}$ and $\{ACD\}$, we can obtain initial values for the quantities π^{Y}_{r}, $\pi^{\bar{A}Y}_{ir}$, $\pi^{\bar{C}Y}_{kr}$, $\pi^{\bar{B}Y}_{jr}$, and $\pi^{\bar{D}Y}_{lr}$. Applying these methods to the three-way tables $\{BDA\}$ and $\{BDC\}$, we can obtain initial values for the quantities π^{Z}_{s}, $\pi^{\bar{B}Z}_{js}$, $\pi^{\bar{D}Z}_{ls}$, $\pi^{\bar{A}Z}_{is}$, and $\pi^{\bar{C}Z}_{ks}$. The parameters in model H_2 are $\pi^{\bar{A}Y}_{ir}$, $\pi^{\bar{C}Y}_{kr}$, $\pi^{\bar{B}Z}_{js}$, $\pi^{\bar{D}Z}_{ls}$ (see [12]), and also the probabilities π^{YZ}_{rs} (see table 6). The quantities π^{Y}_{r}, π^{Z}_{s}, $\pi^{\bar{B}Y}_{jr}$, $\pi^{\bar{D}Y}_{lr}$, $\pi^{\bar{A}Z}_{is}$, and $\pi^{\bar{C}Z}_{ks}$ can be expressed in terms of these parameters. For example, from table 6 we see that

$$\pi^{Y}_{r} = \pi^{YZ}_{r1} + \pi^{YZ}_{r2}, \qquad (99)$$

and we also find that

$$\pi^{Y}_{r}\pi^{\bar{B}Y}_{jr} = \pi^{YZ}_{r1}\pi^{\bar{B}Z}_{j1} + \pi^{YZ}_{r2}\pi^{\bar{B}Z}_{j2}. \qquad (100)$$

From (99) and (100), we find that the π^{YZ}_{rs} can be expressed as follows:[108]

$$\pi^{YZ}_{r1} = [(\pi^{\bar{B}Y}_{jr} - \pi^{\bar{B}Z}_{j2})\pi^{Y}_{r}]/(\pi^{\bar{B}Z}_{j1} - \pi^{\bar{B}Z}_{j2})$$

and

$$\pi^{YZ}_{r2} = [(\pi^{\bar{B}Z}_{j1} - \pi^{\bar{B}Y}_{jr})\pi^{Y}_{r}]/(\pi^{\bar{B}Z}_{j1} - \pi^{\bar{B}Z}_{j2}). \qquad (101)$$

By applying the usual latent class methods to the three-way tables $\{ACB\}$ and $\{BDA\}$, we obtained initial values for the quantities on the right side

[107] See n. 10 above. The usual method is based upon an earlier procedure first suggested by Lazarsfeld and Dudman (1951) and independently by Koopmans (1951), which was developed further by Anderson (1954). Extensions of this method were introduced by Gibson (1955, 1962) and Madansky (1960). (With respect to the latter extension, see the last sentence in n. 34.) As noted earlier, the usual method (and its extensions) can yield numerical values that are obviously not permissible (see sec. 2.1). Among the various numerical values that can be obtained by applying the usual method in different ways (see n. 10), we take as initial values here only quantities that are permissible.

[108] An expression similar to (99) for π^{Y}_{r} can be obtained for π^{Z}_{s}, and expressions similar to (100) for π^{Y}_{r} $\pi^{\bar{B}Y}_{jr}$ can be obtained for π^{Y}_{r} $\pi^{\bar{D}Y}_{lr}$, and for π^{Z}_{s} $\pi^{\bar{A}Z}_{is}$ and π^{Z}_{s} $\pi^{\bar{C}Z}_{ks}$, and these expressions can be used (as we used [99]–[100]) to obtain additional expressions for the π^{YZ}_{rs} similar to (101).

of (101), and these initial values can now be inserted in (101) to obtain initial values for the π^{YZ}_{rs} as well.[109] Thus, we obtain initial values for all the parameters in model H_2.

Most of the other models introduced in this article were obtained by imposing certain restrictions on the parameters of H_2 (or the corresponding $\tilde{H}_2$), and initial values for the parameters in these models can be obtained by imposing the corresponding restrictions on the initial values obtained for H_2 (or $\tilde{H}_2$).[110]

In some cases, the restrictions imposed upon the parameters can lead to somewhat simpler ways to obtain initial values for the parameters. For example, when the restriction that $\pi^{YZ}_{21} = 0$ is imposed (corresponding to restriction (75) for model H_{22}), then initial values for the parameters π^{YZ}_{11}, π^{YZ}_{12}, and π^{YZ}_{22} can be obtained directly from the initial values for the marginals π^{Y}_{r} and π^{Z}_{s} (see, e.g., [99]), so that formula (101) is not needed in this case.

In addition to the particular models presented in this article, the general methods introduced in sections 5.1 and 5.2 can be applied to other models as well, and initial values for the latter models can be obtained in ways that are similar to those used here. For example, in some situations where there is a single evolving unobservable variable and a single indicator for it at each of several points in time, the general methods introduced herein can be applied to some of the models considered by Wiggins (1973), and his estimates can be used as initial trial values in our iterative procedure. (The estimates presented by Wiggins are not efficient, and their direct use in the calculation of supposed χ^2 statistics can not be justified [see n.79 in Goodman 1973*a*]; whereas the estimates obtained with our iterative procedure are efficient, and their use in the calculation of the χ^2 statistics [10] or [11] is justified.)

The method described earlier in the present section (see [99]–[101]) for obtaining initial values for the parameters in model H_2 (fig. 2) when there are four manifest variables can be directly extended to the case where there are m manifest variables ($m \geqq 4$), where latent variable Y can affect m' manifest variables ($m' \geqq 2$) and latent variable Z can affect the remaining m'' manifest variables ($m'' \geqq 2$). Instead of using the three-way marginal tables (e.g., $\{ACD\}$ and $\{BDA\}$) as earlier in this section, we would now use an $(m' + 1)$-way marginal table that includes the m'

[109] The initial values obtained from the other three-way tables considered above can be inserted in the additional expressions similar to (101) for the π^{YZ}_{rs} (see n. 108) in order to obtain additional initial values for the π^{YZ}_{rs}.

[110] Except for models H_7–H_{12} and H_{14}–$\tilde{H}_{16}$, all the other latent structures in figs. 5, 7, and 9 are special cases of either H_2 or $\tilde{H}_2$. With respect to the models listed above as exceptions, to obtain initial values for the parameters in these kinds of models, see Goodman (1974).

variables that can be affected by latent variable Y, and an $(m'' + 1)$-way marginal table that includes the m'' variables that can be affected by latent variable Z. (We assume here and earlier that the two-class latent class model applied to the $(m' + 1)$-way and $(m'' + 1)$-way tables has parameters that are identifiable.)

The methods presented above can also be extended to the case where there are three (or more) latent variables, say (Y, Z, U), where variable Y can affect m' manifest variables $(m' \geqq 2)$, variable Z can affect some other m'' manifest variables $(m'' \geqq 2)$, and variable U can affect the remaining m''' manifest variables $(m''' \geqq 2)$. We shall not pursue this further here.

A word of caution is needed when discussing initial values. In the analysis of a given model, say, model H, if we had used (by mistake) initial values that are more appropriate for a different model (say, model H*, where H* implies H), then the iterative procedure for the analysis of model H can lead sometimes to the maximum-likelihood estimates (and corresponding χ^2 value) for H* (rather than H).[111] If there is some doubt as to whether a given set of initial values is appropriate for model H, the researcher can insert other sets of initial values as well in the iterative procedure introduced here for model H, and the χ^2 values (11) that are obtained with the iterative procedure can be used to determine which estimates are appropriate for model H. The maximum-likelihood estimates under model H will minimize the χ^2 statistic (11) under H. (For related comments, see n. 95.)

Before closing this section, a cautionary remark about sample size is also appropriate. When estimating parameters in the latent class model using the usual Anderson-Lazarsfeld-Dudman method, the asymptotic variance of the estimates can be very large unless the sample size is also large (see, e.g., Madansky 1968); and the asymptotic normal distribution

[111] On the other hand, the iterative procedure introduced in the present article for the analysis of model H will lead to the *correct* maximum-likelihood estimates for H under quite general conditions. The *correct* maximum-likelihood estimates for H will be obtained when initial trial values are appropriately chosen (see related comment in n. 95), *and* also in many situations where initial values that are more appropriate for H* (rather than H) are used by mistake. For example, in the analysis of H_2 in sec. 2.1, the following initial values for $\pi^X{}_t$ $(t = 1, 2, 3, 4)$ would have been more appropriate for H_1 than for H_2 (cf. tables 3 and 8): $\pi^X{}_1 = .40 - \epsilon$, $\pi^X{}_2 = \epsilon$, $\pi^X{}_3 = \epsilon$, $\pi^X{}_4 = .60 - \epsilon$. where ϵ is a small number (say, .00001). (Similarly, we can provide initial values for the other parameters in H_2 that would have been more appropriate for H_1 than for H_2.) Nevertheless, if these initial values are used (by mistake) in the iterative program introduced here for the analysis of H_2, the *correct* maximum-likelihood estimates for H_2 will be obtained. (Of course, the initial values used in the iterative program for H_2 need to satisfy conditions [14] and [15].) The above result applies even when ϵ is as small as, say, .00001; but the iterative procedure for H_2 will not work when ϵ is actually zero, since the denominator in (87a)–(87d) will be zero (for $t = 2, 3$) in that case.

theory for the estimates will also require a large-size sample in order for it to be adequate (see Anderson and Carleton 1957). On the other hand, with the use of the maximum-likelihood procedure presented in the present article, the corresponding asymptotic variance will be reduced.[112] With respect to the maximum-likelihood estimates, the adequacy of the asymptotic normal distribution theory requires further investigation. It may also be the case that a smaller sample size is needed for the adequacy of the asymptotic χ^2 distribution theory for the statistics (10) and (11) than for the adequacy of the asymptotic normal distribution theory for estimates of the parameters in the latent structure.[113]

5.4. Some Further Generalizations

For ease of exposition, we viewed the "explanatory" latent variables here as "antecedent" in some sense (see, e.g., the first column of fig. 10), but the same methods can be applied also when these variables are "intervening" (see, e.g., the third column of fig. 10), or when they are both "antecedent" and "intervening" (see, e.g., the second column of fig. 10). Indeed, the statistical analysis of a given cross-classification table (e.g., $\{ABC\}$ or $\{ABCD\}$) cannot distinguish by itself among the corresponding three causal systems in a given row of figure 10. Each row of that figure describes three different causal systems that would be analyzed by the same latent structure model. Note that H_1-type models would be used in cases where there are multiple indicators for a given unobserved variable, where there is an observed cause and multiple indicators for the unobserved variable, and where there are multiple observed causes and multiple indicators for the unobserved variable (see fig. 10. *A.1–A.3; B.1–B.3; G.1–G.3*).[114] Note also that H_2-type models would be used in cases where there

112 This reduction will take place in all situations in which the usual latent class method for estimating the parameters is applicable (see n. 34), except for the quasi-tautological case when $m = 3$ and $T = 2$. Even in this case, although the *asymptotic* variance of the maximum-likelihood estimate will be the same as that of the estimate obtained by the usual method, the *exact* variance will be smaller, since the usual method yields estimates that are not permissible (i.e., that are strictly outside the range from zero to one) in many cases where the maximum-likelihood procedure yields estimates that are permissible.

113 This is the case in the simpler situation where asymptotic distribution theory is applied to a sample from a binomial population (see, e.g., Cochran 1963, table 3.3; Haberman 1974*b*).

114 The pair of letters *AB* in fig. 10.*G.1* and 10*G.2* denote the polytomous joint variable *AB*. The latent structure model $[AB \otimes C \otimes D|X]$ corresponding to fig. 10.*G.1–G.3* can be analyzed by the methods in sec. 5.1. (This model is also related to the H_2-type model, where there are two latent variables *X* and *Y*, where *X* affects *C* and *D* directly [but not A and B], and where *Y* affects *A* and *B* directly [but not *C* and *D*].) For some comments on the interaction effects in fig. 10.*G.3* (and in some of the other path diagrams in fig. 10), see the last two sentences of n. 58.

are multiple indicators for causally related unobserved variables, where there is an observed cause and multiple indicators for causally related unobserved variables, and where there are multiple observed causes and multiple indicators for causally related unobserved variables (see fig. 10. *C.1–C.3*).[115] The corresponding $\widetilde{H}_2$-type model would be used when there

Fig. 10.A.1

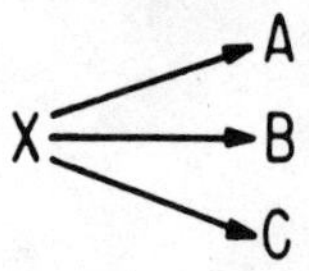

Fig. 10.A.2

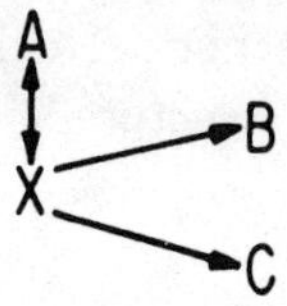

Fig. 10.A.3.

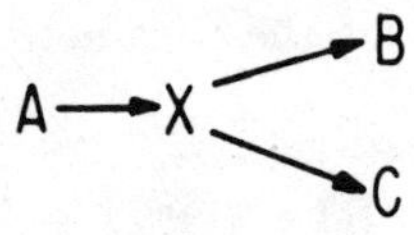

Fig. 10.B.1

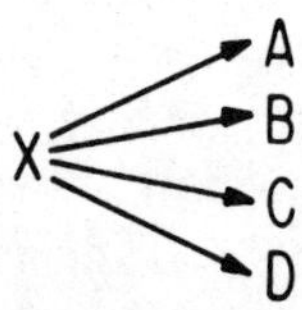

Fig. 10.B.2

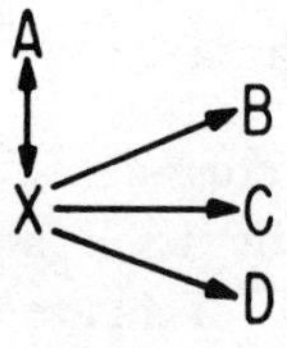

Fig. 10.B.3

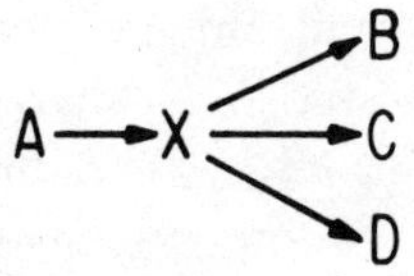

Fig. 10.C.1

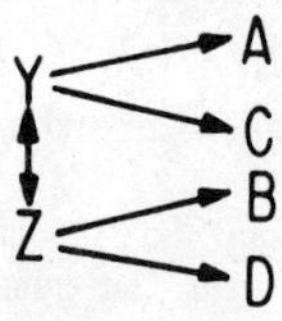

Fig. 10.C.2

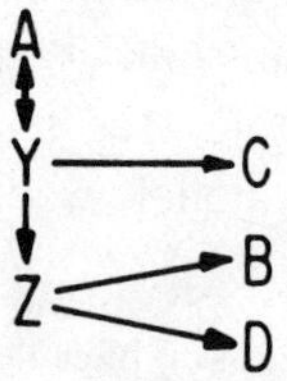

Fig. 10.C.3

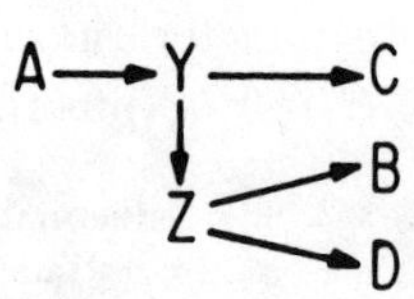

Fig. 10.D.1

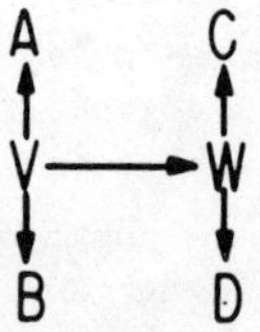

Fig. 10.D.2

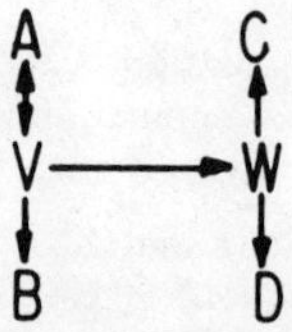

Fig. 10.D.3

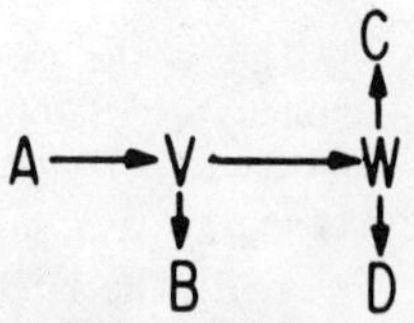

[115] In fig. 10.*C.1* and *C.2*, if variable *A* were replaced by a joint variable (as in fig. 10.*G.1* and *G.2*), then the single observed causal variable *A* in the corresponding fig. 10.*C.3* would be replaced by multiple observed causal variables (as in fig. 10.*G.3*).

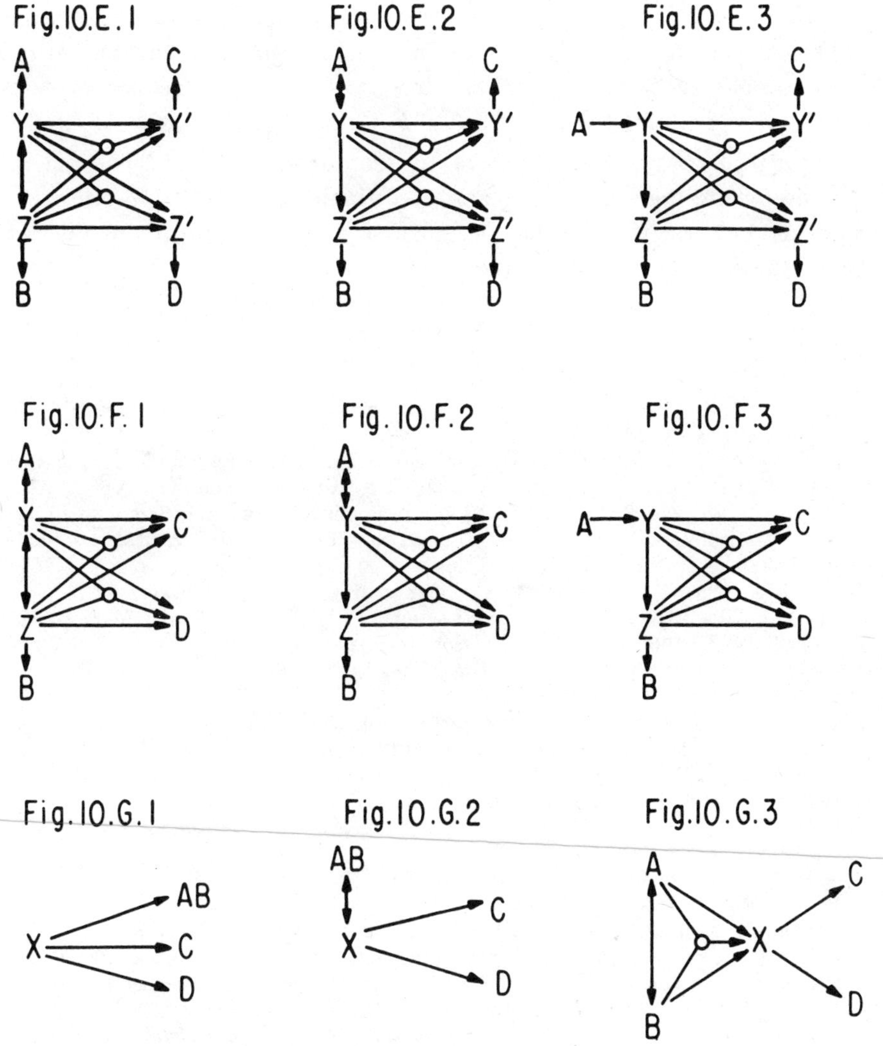

FIG. 10.—Diagrams of some additional causal systems described by a given latent structure model.

are multiple indicators for an evolving unobserved variable, where there is an observed cause and multiple indicators for the evolving unobserved variable, and where there are multiple observed causes and multiple indicators for the evolving unobserved variable (see fig. 10. *D.1–D.3*).[116]

This comment about the three path diagrams (fig. 10.*C.1–C.3*) can also be applied to the other corresponding sets of path diagrams (fig. 10.*A.1–A.3* to fig. 10.*F.1–F.3*).

[116] Model $\tilde{H}_2$, as described in sec. 4, can be applied directly to the analysis of panel data of the kind considered in secs. 2 and 3 (cf. fig. 8 with fig. 10.*D.1*). We could have introduced this model in sec. 2 or 3 (rather than sec. 4), but we did not do so since this model did not fit the particular data sets 1 and 2 in table 2 (but it did fit data set 3).

Models of the H_{11} and H_{12} type would be used in cases where there are indicators for causally related evolving unobserved variables, and where there are observed causes and indicators for the causally related evolving unobserved variables (see fig. 10. *E.1–E.3*).

For expository purposes, we have focused our attention in this article on the case where there are four manifest variables (A, B, C, D). *All* of the methods and results presented here can be extended directly to the case where there are m manifest variables ($m = 3, 4, \ldots$).

REFERENCES

Anderson, T. W. 1954. "On Estimation of Parameters in Latent Structure Analysis." *Psychometrika* 19 (March): 1–10.

———. 1959. "Some Scaling Models and Estimation Procedures in the Latent Class Model." In *Probability and Statistics,* edited by U. Grenander. New York: Wiley.

Anderson, T. W., and R. O. Carleton. 1957. "Sampling Theory and Sampling Experience in Latent Structure Analysis." *Journal of the American Statistical Association* 52 (September): 363.

Boudon, R. 1968. "A New Look at Correlation Analysis." In *Methodology in Social Research,* edited by H. M. Blalock, Jr., and A. Blalock. New York: McGraw-Hill.

Campbell, D. T. 1963. "From Description to Experimentation: Interpreting Trends as Quasi-Experiments." In *Problems in Measuring Change,* edited by C. W. Harris. Madison: University of Wisconsin Press.

Cochran, W. G. 1963. *Sampling Techniques.* 2d ed. New York: Wiley.

Coleman, J. S. 1964. *Introduction to Mathematical Sociology.* New York: Free Press.

Duncan, O. D. 1973. "Preliminary Analysis of Data on Sex-Role Socialization." Unpublished report to the Russell Sage Foundation.

Gibson, W. A. 1955. "An Extension of Anderson's Solution for the Latent Structure Equations." *Psychometrika* 20 (March): 69–73.

———. 1962. "Extending Latent Class Solutions to Other Variables." *Psychometrika* 27 (March): 73–81.

Goodman, L. A. 1968. "The Analysis of Cross-classified Data: Independence, Quasi-Independence, and Interactions in Contingency Tables with or without Missing Entries." *Journal of the American Statistical Association* 63 (December) 1091–1131.

———. 1970. "The Multivariate Analysis of Qualitative Data: Interactions among Multiple Classifications." *Journal of the American Statistical Association* 65 (March): 225–56.

———. 1971*a*. "The Analysis of Multidimensional Contingency Tables: Stepwise Procedures and Direct Estimation Methods for Building Models for Multiple Classifications." *Technometrics* 13 (February): 33–61.

———. 1971*b*. "Partitioning of Chi-Square, Analysis of Marginal Contingency Tables, and Estimation of Expected Frequencies in Multidimensional Contingency Tables." *Journal of the American Statistical Association* 66 (June): 339–44.

———. 1972*a*. "A Modified Multiple Regression Approach to the Analysis of Dichotomous Variables." *American Sociological Review* 37 (February): 28–46.

———. 1972*b*. "A General Model for the Analysis of Surveys." *American Journal of Sociology* 77 (May): 1035–86.

———. 1973*a*. "Causal Analysis of Data from Panel Studies and Other Kinds of Surveys." *American Journal of Sociology* 78 (March): 1135–91.

———. 1973*b*. "The Analysis of Multidimensional Contingency Tables When Some Variables Are Posterior to Others: A Modified Path Analysis Approach:" *Biometrika* 59 (1): 579–96.

———. 1973*c*. "Guided and Unguided Methods for the Selection of Models for a

Set of *T* Multidimensional Contingency Tables." *Journal of the American Statistical Association* 68 (March): 165–75.

———. 1973*d*. "The Analysis of Systems of Qualitative Variables When Some of the Variables Are Unobservable. Part 2—the Use of Modified Latent Distance Models."

———. 1974. "Exploratory Latent Structure Analysis Using Both Identifiable and Unidentifiable Models." *Biometrika,* vol. 61, in press.

Haberman, S. H. 1973. "Non-linear Programming and Frequency Tables Involving Indirect Observation." Unpublished manuscript.

———. 1974*a*. "Log-linear Models for Frequency Tables Derived by Indirect Observation: Maximum Likelihood Equations." *Annals of Statistics,* in press.

———. 1974*b*. *The Analysis of Frequency Data*. Chicago: University of Chicago Press.

Koopmans, T. C. 1951. "Identification Problems in Latent Structure Analysis." Cowles Commission Discussion Paper: Statistics, no. 360.

Lazarsfeld, P. F. 1948. "The Use of Panels in Social Research." *Proceedings of the American Philosophical Society* 92 (5): 405–10. Reprinted in P. F. Lazarsfeld, A. K. Pasanella, M. Rosenberg. 1972. *Continuities in the Language of Social Research*. New York: Free Press.

———. 1970. "A Memoir in Honor of Professor Wold." In *Scientists at Work,* edited by T. Dalenius, G. Karlsson, and S. Malmquist. Uppsala: Almquist & Wiksells.

Lazarsfeld, P. F., and J. Dudman. 1951. "The General Solution of the Latent Class Case." In *The Use of Mathematical Models in the Measurement of Attitudes,* edited by P. F. Lazarsfeld. RAND Research Memorandum no. 455.

Lazarsfeld, P. F., and N. W. Henry. 1968. *Latent Structure Analysis*. Boston: Houghton Mifflin.

Lipset, S. M., P. F. Lazarsfeld, A. H. Barton, and J. Linz. 1954. "The Psychology of Voting: An Analysis of Political Behavior." In *Handbook of Social Psychology,* edited by G. Lindzey. Vol. 2. Reading, Mass.: Addison-Wesley.

McHugh, R. B. 1956. "Efficient Estimation and Local Identification in Latent Class Analysis." *Psychometrika* 21 (December): 331–47.

———. 1958. "Note on 'Efficient Estimation and Local Identification in Latent Class Analysis.'" *Psychometrika* 23 (September): 273–74.

Madansky, A. 1960. "Determinantal Methods in Latent Class Analysis." *Psychometrika* 25 (June) 183–98.

———. 1968. "Latent Structure." In *International Encyclopedia of the Social Sciences,* edited by D. L. Sills. Vol. 9. New York: Macmillan and Fress Press.

Murray, J. R. 1971. "Statistical Models for Qualitative Data with Classification Errors." Ph.D. thesis, University of Chicago.

Murray, J. R., D. E. Wiley, and R. G. Wolfe. 1972. "Models for Lazarsfeld's 16-Fold Table Problem Which Include Errors of Classification." Paper presented at meeting of the Psychometric Society, Princeton, N.J.

Schumacher, C. F., G. R. Maxson, and H. Martinek. 1953. "Tests for Creative Ability in Machine Design." Project ONR 458. Armed Services Technical Information Agency 21 284.

Wiggins, L. M. 1973. *Panel Analysis*. Amsterdam: Elsevier.

Leo A. Goodman

Reprinted from: THE AMERICAN JOURNAL OF SOCIOLOGY,
Vol. 79, No. 5, March 1974.

Chapter 9

A New Model for Scaling Response Patterns: An Application of the Quasi-Independence Concept[1]

To analyze the "scalability" of the observed response patterns for a set of m dichotomous items, we introduce a model in which a given individual in the population is either "intrinsically scalable" or "intrinsically unscalable" (with respect to the m items), and there are d different types of "intrinsically scalable" individuals. With this model, using methods already developed for the study of quasi-independence in contingency tables, we show how to (a) test whether the model fits the observed data, (b) estimate the proportion of intrinsically scalable (and unscalable) individuals, and (c) estimate the distribution of the intrinsically scalable individuals among the d different scale types.

1. INTRODUCTION

Consider first the situation where there are four dichotomous items, say, A, B, C and D, and let (i, j, k, l) denote the response pattern in which the responses on items A, B, C and D are at levels i, j, k and l, respectively (for $i = 1, 2$; $j = 1, 2$; $k = 1, 2$; $l = 1, 2$). In this situation, there are 2^4 possible response patterns; but suppose we consider for the moment the special case in which only five patterns are actually observable,

$$(1, 1, 1, 1),\ (1, 1, 1, 2),\ (1, 1, 2, 2),\ (1, 2, 2, 2),\ (2, 2, 2, 2). \quad (1.1)$$

In this special case, we have a Guttman scale, all respondents are scalable, and there are five different scale types (see, e.g., [8]).

In actual practice, response patterns other than those listed in (1.1) will also occur. To analyze the "scalability" of the observed patterns (when the five patterns in (1.1) correspond to the scale types), we introduce the following model.

> We consider a population in which individuals are classified into one of *six* mutually exclusive and exhaustive categories, viz., a single category for the "intrinsically unscalable" individuals and five other categories for the "intrinsically scalable" individuals in the five scale types corresponding to the response patterns in (1.1). For the "intrinsically unscalable" individuals, the re-

sponses on items A, B, C and D are mutually independent; and the response patterns for the "intrinsically scalable" individuals correspond to their scale type.[1] For the sake of simplicity, the category for the intrinsically unscalable will be labeled category zero, and the other five categories will be numbered in order from one to five.

Let π_t denote the probability that an individual will be in the tth category (for $t = 0, 1, 2, 3, 4, 5$); let $\pi_{it}^{\bar{A}}$ denote the *conditional* probability that his response will be at level i on variable A (for $i = 1, 2$), given that he was in the tth category; and let the conditional probabilities $\pi_{jt}^{\bar{B}}$, $\pi_{kt}^{\bar{C}}$, $\pi_{lt}^{\bar{D}}$ be defined similarly (for $j = 1, 2$; $k = 1, 2$; $l = 1, 2$). Let π_{ijkl}^{ABCD} denote the probability of obtaining response pattern (i, j, k, l); and let $\pi_{ijklt}^{\bar{A}\bar{B}\bar{C}\bar{D}}$ denote the *conditional* probability of obtaining this pattern, given that the individual was in the tth category. Our model states that

$$\pi_{ijkl}^{ABCD} = \sum_{t=0}^{5} \pi_t \pi_{ijklt}^{\bar{A}\bar{B}\bar{C}\bar{D}}, \tag{1.2}$$

where

$$\pi_{ijkl0}^{\bar{A}\bar{B}\bar{C}\bar{D}} = \pi_{i0}^{\bar{A}}\pi_{j0}^{\bar{B}}\pi_{k0}^{\bar{C}}\pi_{l0}^{\bar{D}}, \tag{1.3}$$

$$\pi_{11111}^{\bar{A}\bar{B}\bar{C}\bar{D}} = \pi_{11122}^{\bar{A}\bar{B}\bar{C}\bar{D}} = \pi_{11223}^{\bar{A}\bar{B}\bar{C}\bar{D}} = \pi_{12224}^{\bar{A}\bar{B}\bar{C}\bar{D}} = \pi_{22225}^{\bar{A}\bar{B}\bar{C}\bar{D}} = 1, \tag{1.4}$$

with $\pi_{ijklt}^{\bar{A}\bar{B}\bar{C}\bar{D}} = 0$ otherwise. Formula (1.2) states that the individuals in the population can be classified into six mutually exclusive and exhaustive categories ($t = 0, 1, 2, \cdots, 5$); (1.3) states that the responses on items A, B, C and D are mutually independent for the individuals in the 0th category; and (1.4) states that the response pattern for an individual in the tth category (for $t = 1, 2, 3, 4, 5$) corresponds to his scale type (with probability one).

From (1.2)–(1.4), we see that

$$\begin{aligned}
\pi_{1111}^{ABCD} &= \pi_1 + \pi_0\pi_{10}^{\bar{A}}\pi_{10}^{\bar{B}}\pi_{10}^{\bar{C}}\pi_{10}^{\bar{D}}, \\
\pi_{1112}^{ABCD} &= \pi_2 + \pi_0\pi_{10}^{\bar{A}}\pi_{10}^{\bar{B}}\pi_{10}^{\bar{C}}\pi_{20}^{\bar{D}}, \\
\pi_{1122}^{ABCD} &= \pi_3 + \pi_0\pi_{10}^{\bar{A}}\pi_{10}^{\bar{B}}\pi_{20}^{\bar{C}}\pi_{20}^{\bar{D}}, \\
\pi_{1222}^{ABCD} &= \pi_4 + \pi_0\pi_{10}^{\bar{A}}\pi_{20}^{\bar{B}}\pi_{20}^{\bar{C}}\pi_{20}^{\bar{D}}, \\
\pi_{2222}^{ABCD} &= \pi_5 + \pi_0\pi_{20}^{\bar{A}}\pi_{20}^{\bar{B}}\pi_{20}^{\bar{C}}\pi_{20}^{\bar{D}};
\end{aligned} \tag{1.5}$$

and for any response pattern (i, j, k, l) that does *not* correspond to one of the specified scale types, we obtain

$$\pi_{ijkl}^{ABCD} = \pi_0 \pi_{i0}^{\bar{A}} \pi_{j0}^{\bar{B}} \pi_{k0}^{\bar{C}} \pi_{l0}^{\bar{D}}. \tag{1.6}$$

Formula (1.6) states that a response pattern (i, j, k, l) that does not correspond to one of the specified scale types can be obtained from an individual in the 0th category (with conditional probability $\pi_{i0}^{\bar{A}} \pi_{j0}^{\bar{B}} \pi_{k0}^{\bar{C}} \pi_{l0}^{\bar{D}}$, given the individual in the 0th category); and (1.5) states that a pattern that corresponds to one of the specified scale types can be obtained *either* from an individual in the corresponding scale-type category (with conditional probability one, given the individual in the scale-type category) or from an individual in the 0th category (with conditional probability $\pi_{i0}^{\bar{A}} \pi_{j0}^{\bar{B}} \pi_{k0}^{\bar{C}} \pi_{l0}^{\bar{D}}$, given the individual in the 0th category).

Let S denote the set of possible response patterns that correspond to scale types, and let U denote the set of possible patterns that do not correspond to scale types. (In the particular situation just considered, there are five patterns in S, and there $2^4 - 5 = 11$ patterns in U.) Let us now delete (i.e., blank out) the patterns in S. Considering only the remaining patterns (i.e., those in U), (1.6) states that the responses on items A, B, C and D are quasi-independent (see, e.g., [2]). This formula can be rewritten as

$$\pi_{ijkl}^{ABCD} = \alpha_i \beta_j \gamma_k \delta_l, \quad \text{for} \quad (i, j, k, l) \in U, \tag{1.7}$$

with

$$\begin{aligned} \pi_{i0}^{\bar{A}} &= \alpha_i/(\alpha_1 + \alpha_2), \quad & \pi_{j0}^{\bar{B}} &= \beta_j/(\beta_1 + \beta_2), \\ \pi_{k0}^{\bar{C}} &= \gamma_k/(\gamma_1 + \gamma_2), \quad & \pi_{l0}^{\bar{D}} &= \delta_l/(\delta_1 + \delta_2), \\ \pi_0 &= (\alpha_1 + \alpha_2)(\beta_1 + \beta_2)(\gamma_1 + \gamma_2)(\delta_1 + \delta_2). \end{aligned} \tag{1.8}$$

To calculate the maximum-likelihood estimates, $\hat{\pi}_{i0}^{\bar{A}}$, $\hat{\pi}_{j0}^{\bar{B}}$, $\hat{\pi}_{k0}^{\bar{C}}$, $\hat{\pi}_{l0}^{\bar{D}}$ and $\hat{\pi}_0$, for the corresponding parameters in (1.6), we first calculate $\hat{\alpha}_i$, $\hat{\beta}_j$, $\hat{\gamma}_k$, $\hat{\delta}_l$ corresponding to the parameters in the quasi-independence model (1.7), using the kind of iterative procedure presented for this purpose in [2, p. 1118–19].[2] The maximum-likelihood estimates, $\hat{\pi}_{i0}^{\bar{A}}$, $\hat{\pi}_{j0}^{\bar{B}}$, $\hat{\pi}_{k0}^{\bar{C}}$, $\hat{\pi}_{l0}^{\bar{D}}$ and $\hat{\pi}_0$, can then be obtained from (1.8) by replacing each term on the right side of these formulas (the $\alpha_i, \beta_j, \gamma_k, \delta_l$) by the corresponding quantity just calculated (the $\hat{\alpha}_i, \hat{\beta}_j, \hat{\gamma}_k, \hat{\delta}_l$). To calculate the maximum-likelihood estimate $\hat{\pi}_t$ (for $t = 1, 2, 3, 4, 5$), we use (1.5) to obtain the formulas

$$\begin{aligned} \hat{\pi}_1 &= p_{1111}^{ABCD} - \hat{\pi}_0\hat{\pi}_{10}^{\bar{A}}\hat{\pi}_{10}^{\bar{B}}\hat{\pi}_{10}^{\bar{C}}\hat{\pi}_{10}^{\bar{D}}, \\ \hat{\pi}_2 &= p_{1112}^{ABCD} - \hat{\pi}_0\hat{\pi}_{10}^{\bar{A}}\hat{\pi}_{10}^{\bar{B}}\hat{\pi}_{10}^{\bar{C}}\hat{\pi}_{20}^{\bar{D}}, \\ \hat{\pi}_3 &= p_{1122}^{ABCD} - \hat{\pi}_0\hat{\pi}_{10}^{\bar{A}}\hat{\pi}_{10}^{\bar{B}}\hat{\pi}_{20}^{\bar{C}}\hat{\pi}_{20}^{\bar{D}}, \end{aligned} \tag{1.9}$$

etc., where p_{ijkl}^{ABCD} is the observed proportion of response patterns that are (i, j, k, l). The maximum-likelihood estimates $\hat{\pi}_{ijkl}^{ABCD}$ of the corresponding π_{ijkl}^{ABCD} can then be obtained from (1.5)–(1.6) by replacing each term on the right side of these formulas by the corresponding maximum-likelihood estimate.

We assume here that the probability π_t is greater than zero. In this case, the corresponding maximum-likelihood estimate $\hat{\pi}_t$ will also be greater than zero, with probability approaching one, as the sample size n approaches infinity. On the other hand, if any of the $\hat{\pi}_t$ (for $t = 1, 2, 3, 4, 5$) calculated by (1.9) turn out to be negative (which would happen with probability approaching zero), the method just described would need to be modified slightly. We comment later on this situation (see Section 4), but for the time being we will assume that it does not happen.

For the response patterns that are in set S, we see from (1.5) and (1.9) that

$$\hat{\pi}_{ijkl}^{ABCD} = p_{ijkl}^{ABCD}, \quad \text{for} \quad (i, j, k, l) \in S. \tag{1.10}$$

In other words, for (i, j, k, l) in the set S, the observed proportion p_{ijkl}^{ABCD} is fitted perfectly (except for round-off

[2] Further comments on this iterative procedure, and alternative procedures, will be included in Sections 7 and 8 (see, e.g., (8.5)–(8.7)). To apply the methods proposed here, we assume that the observed frequency f_{ijkl} of response pattern (i, j, k, l) is greater than zero for each pattern (i, j, k, l) in set U; or, more generally, that it is greater than zero for enough of the patterns in U to estimate the parameters in the quasi-independence model. (Some cases where problems arise in the estimation of the parameters will be considered later; see the discussion of Model H_6 in Section 4 and the first model presented in the appendix.)

error) by the corresponding $\hat{\pi}_{ijkl}^{ABCD}$ under the model considered here.

Let f_{ijkl} and F_{ijkl} denote the observed and expected frequency of response pattern (i, j, k, l) in a sample of n individuals, and let $\hat{F}_{ijkl}$ denote the maximum-likelihood estimate of F_{ijkl} under the model. Note that

$$f_{ijkl} = np_{ijkl}^{ABCD}, \quad F_{ijkl} = n\pi_{ijkl}^{ABCD}, \quad \hat{F}_{ijkl} = n\hat{\pi}_{ijkl}^{ABCD}. \tag{1.11}$$

From (1.10) we see that

$$\hat{F}_{ijkl} = f_{ijkl}, \quad \text{for} \quad (i, j, k, l) \in S. \tag{1.12}$$

To test whether the model is congruent with the observed data, we compare the observed f_{ijkl} with the corresponding $\hat{F}_{ijkl}$ for (i, j, k, l) in set U, using either the usual goodness-of-fit chi-square statistic

$$\sum_{U} (f_{ijkl} - \hat{F}_{ijkl})^2/\hat{F}_{ijkl}, \tag{1.13}$$

or the corresponding chi-square based on the likelihood-ratio statistic

$$2 \sum_{U} f_{ijkl} \log (f_{ijkl}/\hat{F}_{ijkl}), \tag{1.14}$$

where $\sum_U$ denotes summation over all (i, j, k, l) in set U.[3] The chi-square value obtained from (1.13) or (1.14) can be assessed by comparing its numerical value with the tabulated percentiles of the chi-square distribution. The number of degrees of freedom for testing the model will be

$$2^4 - 5 - 5 = 6; \tag{1.15}$$

i.e., the number of possible response patterns $(2^4 - 5)$ in set U minus the number of independent parameters estimated under the quasi-independence model for U (i.e., $\pi_{10}^{\bar{A}}, \pi_{10}^{\bar{B}}, \pi_{10}^{\bar{C}}, \pi_{10}^{\bar{D}}$ and π_0).

In this calculation of the degrees of freedom for testing the model, the number of patterns in set S was taken to be five, the full set of scale types as described earlier. More generally, in the situation where there are m dichotomous items, there will be 2^m possible response

[3] From (1.12) we see that the chi-square values (1.13) and (1.14) would remain unchanged if the summation was over *all* (i, j, k, l) in the 2^4 table. On the other hand, in view of (1.12), there is no need to sum over the (i, j, k, l) in set S.

1. Observed Cross-Classification in Three Different Sets of Data

Response pattern				Data set		
Item				Stouffer-Toby questionnaire (1)	McHugh test (2)	Lazarsfeld-Stouffer questionnaire (3)
A	B	C	D			
1	1	1	1	42	23	75
1	1	1	2	23	5	69
1	1	2	1	6	5	55
1	1	2	2	25	14	96
1	2	1	1	6	8	42
1	2	1	2	24	2	60
1	2	2	1	7	3	45
1	2	2	2	38	8	199
2	1	1	1	1	6	3
2	1	1	2	4	3	16
2	1	2	1	1	2	8
2	1	2	2	6	4	52
2	2	1	1	2	9	10
2	2	1	2	9	3	25
2	2	2	1	2	8	16
2	2	2	2	20	34	229

patterns, there will be $m + 1$ possible patterns corresponding to the full set of scale types, there will be $m + 1$ independent parameters estimated under the quasi-independence model for U, and thus the number of degrees of freedom for testing the model will be

$$2^m - 2(m + 1). \tag{1.16}$$

Formula (1.16) applies to the situation where the full set of $m + 1$ scale types can occur. In situations where this is *not* the case, a slight modification of (1.16) will be required (see (4.11)).

To illustrate the general approach proposed, we apply it to the following three sets of data presented in Table 1:

1. The Stouffer-Toby response patterns for respondents to questionnaire items on role conflict (see [5, 11, 17]). This table cross-classifies 216 respondents with respect to whether they tend toward *universalistic* values (+) or *particularistic* values (−) when confronted by each of four different situations of role conflict.[4]

[4] For these data, the letters A, B, C, D in Table 1 denote the dichotomous responses when confronted by the four different role conflicts. In Table 1, level 1 on a given dichotomous response denotes a positive (+) response, and level 2 denotes a negative (−) response.

2. The McHugh test data on creative ability in machine design (see [6, 12, 15]). This table cross-classifies 137 engineers with respect to their dichotomized scores (above the subtest mean (+) or below (−)) obtained on each of four different subtests that were supposed to measure creative ability in machine design.[5]
3. The Lazarsfeld-Stouffer response patterns for noncommissioned officers responding to items on attitude toward the Army (see [10, 16]). This table cross-classifies a sample of 1,000 noncommissioned officers with respect to their dichotomized responses (favorable (+) or unfavorable (−)) obtained on each of four different items on general attitudes toward the Army.[6]

For the sake of simplicity, we focus our attention in this article on the case where there are four dichotomous items (as in each set of data in Table 1), but we also note that our methods can be applied more generally to the case where there are m dichotomous items ($m = 3, 4, \cdots$).[7] The three different sets of data in Table 1 provide an opportunity to illustrate different kinds of considerations that arise in the application of the general approach proposed here. Each set of data will be considered separately (see Sections 2, 3 and 4 for data sets 1, 2 and 3, respectively), as will some additional points that apply more generally (see Sections 5 to 8). We begin with the analysis of the first set of data.

2. THE ANALYSIS OF THE STOUFFER-TOBY DATA

From the Stouffer-Toby data (Column 1 of Table 1), we see that the observed frequencies f_{ijkl}, for the response patterns included in set S described by (1.1), are

$$f_{1111} = 42, \quad f_{1112} = 23, \quad f_{1122} = 25, \quad f_{1222} = 38, \quad f_{2222} = 20. \tag{2.1}$$

[5] For these data, the letters A, B, C, D in Table 1 denote the dichotomized scores on the four subtests. For ease of later exposition, we have interchanged in Table 1 the order of the subtests which were considered first and second in the earlier literature.

[6] For these data, the letters A, B, C, D in Table 1 denote the dichotomized responses on the four items. For ease of later exposition, these items appear in Table 1 in the same order as in [10], which is the opposite order from that used in [16].

[7] The case where $m = 3$ and $m = 2$ will be discussed briefly in the appendix. Further insight into the meaning of the model introduced in this article can be obtained from the results presented in the appendix and from some of the other results presented in the following sections.

Set S consists of five patterns; and the remaining 11 patterns in the 2^4 table form set U. We delete the five patterns in S from the 2^4 table, and consider only the 11 patterns in U. We next consider the quasi-independence model H_1, which states that the responses on items A, B, C and D are quasi-independent with respect to set U; and we apply to set U the method described in Section 1 to obtain the maximum-likelihood estimates of the parameters $\pi_{10}{}^{\bar{A}}$, $\pi_{10}{}^{\bar{B}}$, $\pi_{10}{}^{\bar{C}}$, $\pi_{10}{}^{\bar{D}}$, and π_0 (see comments following (1.8)).

For the Stouffer-Toby data under H_1, we obtain estimates of the parameters,

$$\hat{\pi}_{10}{}^{\bar{A}} = .77, \quad \hat{\pi}_{10}{}^{\bar{B}} = .38, \quad \hat{\pi}_{10}{}^{\bar{C}} = .44, \quad \hat{\pi}_{10}{}^{\bar{D}} = .19, \quad \hat{\pi}_0 = .68, \tag{2.2}$$

which can then be used to calculate the corresponding $\hat{\pi}_{ijkl}^{ABCD}$ for the patterns (i, j, k, l) in set U (by replacing the π's on the right side of (1.6) by the corresponding $\hat{\pi}$'s); and we then obtain the corresponding $\hat{F}_{ijkl}$ under H_1 (using the last formula in (1.11)), which we have presented in Table 2. (All calculations in this table and throughout were carried to more significant digits than reported here.) Next we use (1.9) to obtain

$$\hat{\pi}_1 = .18, \; \hat{\pi}_2 = .03, \; \hat{\pi}_3 = .03, \; \hat{\pi}_4 = .03, \; \hat{\pi}_5 = .05; \tag{2.3}$$

and we use (1.13) and (1.14) to obtain a goodness-of-fit chi-square of 1.01 and a likelihood-ratio chi-square of .99, each with six degrees of freedom (see (1.15)). Thus, the model under consideration here fits the data very well indeed.

For purposes of comparison, we consider next the usual model (H_0) in which responses on items A, B, C and D are mutually independent for the individuals in the population. This model can be viewed as stating that all the individuals are intrinsically unscalable and that there are no scale types; i.e., that $\pi_0 = 1$ and $\pi_t = 0$, for $t = 1, 2, 3, 4, 5$ (see (1.2)–(1.3) and (1.5)–(1.6)). To test Model H_0, the usual chi-square methods of testing for mutual independence can be applied, with $2^4 - 5 = 11$

2. Observed and Estimated Expected Frequencies for the Stouffer-Toby Data Under Two Different Quasi-Independence Models, H_1 and H_2[a]

Response pattern: Item A	B	C	D	Observed frequency	Expected frequency estimated under model H_1	Expected frequency estimated under model H_2
1	1	1	1	42	—	—
1	1	1	2	23	—	20.19
1	1	2	1	6	4.72	5.50
1	1	2	2	25	—	25.96
1	2	1	1	6	5.99	5.92
1	2	1	2	24	24.74	27.93
1	2	2	1	7	7.56	7.61
1	2	2	2	38	—	35.91
2	1	1	1	1	1.14	1.10
2	1	1	2	4	4.73	5.18
2	1	2	1	1	1.44	1.41
2	1	2	2	6	5.97	6.66
2	2	1	1	2	1.83	1.52
2	2	1	2	9	7.57	7.17
2	2	2	1	2	2.31	1.95
2	2	2	2	20	—	—

[a] In H_1, the full set of scale types is deleted. In H_2, only the two extreme scale types are deleted.

degrees of freedom (see, e.g., [4]).[8] The chi-square values obtained thereby, for the Stouffer-Toby data, are included in Table 3a.[9] Comparing the chi-square values for H_0 with those for H_1, we see the dramatic improvement obtained when H_0 is replaced by H_1![10]

[8] Note that there are five more degrees of freedom for testing H_0 than for testing H_1 (see (1.15)); this corresponds to the fact that π_t is set at zero (for $t = 1, 2, 3, 4, 5$) under H_0 but not under H_1. More generally, in the situation where there are m dichotomous items, the number of degrees of freedom for testing H_0 will be $[2^m - (m + 1)]$ (see, e.g., [4]); there are $m + 1$ more degrees of freedom for testing H_0 than for testing H_1 (see (1.16)); and this corresponds to the fact that π_t is set at zero (for $t = 1, 2, \cdots, m + 1$) under H_0 but not under H_1.

[9] Table 3a includes the results for H_0, some of the results for H_1 presented earlier in this section, and some results for Model H_2 which we shall consider next.

[10] This comparison of the chi-square values can be used to test the null hypothesis that $\pi t = 0$ (for $t = 1, 2, 3, 4, 5$), with five degrees of freedom (see Footnote 8). The difference between the chi-squares obtained with statistic (1.14) gives the likelihood-ratio chi-square for testing this null hypothesis, assuming that Model H_1 is true (see, e.g., [2]). Later we show how to test the null hypothesis that $\pi_t = 0$ for a given subset of the five π_t ($t = 1, 2, 3, 4, 5$).

3. Some Estimated Parameters and Chi-Square Values for Various Examples of the New Scaling Model

Model	Number of scale types	Estimated proportion $\hat{\pi}_0$ of intrinsically unscalable individuals	Degrees of freedom	Goodness-of-fit chi-square	Likelihood-ratio chi-square
		a. Stouffer-Toby Data			
H_0	0	1.00	11	104.11	81.08
H_1	5	.68	6	1.01	.99
H_2	2	.78	9	2.28	2.28
		b. McHugh Data			
H_1	5	.49	6	5.63	5.90
H_2	2	.67	9	23.08	22.59
H_3	4	.52	7	7.11	7.42
		c. Lazarsfeld-Stouffer Data			
H_4	4	.67	7	26.09	26.50
H_5	8[a]	.47	3[a]	5.47	5.86
H_6	12	.04[b]	1	.42	.44
H_7	4	.67	7	25.05	25.01

[a] For comments on the results actually obtained for the Lazarsfeld-Stouffer data, see discussion in the text.
[b] Estimate obtained under condition (4.9).

The parameters π_1 and π_5 in Model H_1 pertained to the scale types corresponding to the extreme response patterns (1, 1, 1, 1) and (2, 2, 2, 2), respectively; whereas π_2, π_3 and π_4 pertain to the scale types corresponding to the less extreme response patterns (1, 1, 1, 2), (1, 1, 2, 2) and (1, 2, 2, 2), respectively (see (1.4)–(1.5)). Instead of five different scale types, we next consider the situation where there are only two different scale types; i.e., the types corresponding to the two extreme response patterns,

$$(1, 1, 1, 1) \quad \text{and} \quad (2, 2, 2, 2). \tag{2.4}$$

Consider now the situation where set S consists of the two response patterns in (2.4); and the remaining 14 patterns in the 2^4 table form the set U. We delete the two patterns in S from the 2^4 table, and consider only the 14 patterns in U. The corresponding quasi-independence model H_2 states that responses on items A, B, C and D are quasi-independent with respect to set U; and we now apply to set U the general approach described in Section 1.

For the Stouffer-Toby data under H_2, we obtain estimates of the parameters,

$$\hat{\pi}_{10}{}^{\bar{A}} = .80, \quad \hat{\pi}_{10}{}^{\bar{B}} = .42, \quad \hat{\pi}_{10}{}^{\bar{C}} = .44, \quad \hat{\pi}_{10}{}^{\bar{D}} = .17, \quad \hat{\pi}_0 = .78; \tag{2.5}$$

and the estimates corresponding to (2.3) are now .18 and .05 under H_2, for the two scale types corresponding to the response patterns in (2.4).[11] The $\hat{F}_{ijkl}$ under H_2 are included in Table 2, and the chi-square values for testing H_2 are included in Table 3a.[12] From Table 3a we see that H_2 also fits the data very well.

From the results for H_1 and H_2 in Table 3a, we see that the estimated proportion $\hat{\pi}_0$ of "intrinsically unscalable" individuals is somewhat smaller under H_1 than under H_2. On the other hand, because H_1 has three more scale types than H_2, there are three more parameters to be estimated under H_1 than under H_2; and thus, there are three less degrees of freedom for testing H_1 than for testing H_2. By comparing the corresponding likelihood-ratio chi-squares, we obtain a difference of 1.29 (i.e., $2.28 - .99 = 1.29$), which can be assessed in terms of the tabulated percentiles of the chi-square distribution with three degrees of freedom (i.e., $9 - 6 = 3$). The difference in the chi-squares (1.29) can be used to test the null hypothesis that the three additional scale types under H_1 do not contribute in a statistically significant way to the fit of

[11] The $\hat{\pi}_1$ and $\hat{\pi}_5$ under H_1 in (2.3) agreed to the order of accuracy reported here with the corresponding estimates just presented under H_2, but this need not be the case in general.

[12] For testing H_2, the number of degrees of freedom will be $2^4 - 2 - 5 = 9$ (as indicated in Table 3a); i.e., the number of possible response patterns ($2^4 - 2$) in set U (obtained with H_2) minus the number of independent parameters estimated under the quasi-independence model for U. (See related comment following (1.15), and the more general formulas (4.10)–(4.11) later.)

the model (i.e., that the estimated values of π_2, π_3 and π_4 in (2.3) do not differ from zero in a statistically significant way). This null hypothesis is supported by the data.

The method just described (see also Footnote 10) for comparing two different likelihood-ratio chi-squares can be applied more generally to examine any given subset of the $\hat{\pi}$'s in (2.3) to determine whether they differ from zero in a statistically significant way. To save space, we shall not give the numerical details here, but summarize briefly: Assuming H_1 to be true, we can show that $\hat{\pi}_2$, $\hat{\pi}_3$ and $\hat{\pi}_4$ in (2.3) are not significantly different from zero, but $\hat{\pi}_1$ and $\hat{\pi}_5$ are.[13] Similarly, assuming H_2 to be true, we can also show that the latter two $\hat{\pi}$'s continue to be significantly different from zero.[14]

We note that $\hat{\pi}_0$ under Model H_1 will be larger than the *observed* proportion of individuals whose response patterns do not correspond to the scale types specified by the model (i.e., 68/216 = .31); and, similarly, $\hat{\pi}_0$ under H_2 will be larger than the corresponding *observed* proportion (i.e., 154/216 = .71). This is because the individuals whose observed response patterns correspond to scale types will include both the individuals who are in scale-type categories *and* some individuals who are not.[15] Thus, the proportion of individuals whose observed response

[13] The difference in the chi-squares just presented tested the null hypothesis that $\pi_2 = \pi_3 = \pi_4 = 0$ (with three degrees of freedom), and the difference in the chi-squares considered in Footnote 10 tested the null hypothesis that $\pi_1 = \pi_2 = \pi_3 = \pi_4 = \pi_5 = 0$ (with five degrees of freedom), assuming H_1 to be true. The general method can also be applied to test separately each null hypothesis stating that $\pi_t = 0$ for a specified value of t (with one degree of freedom for each test), or each null hypothesis stating that $\pi_t = 0$ for two specified values of t (with two degrees of freedom for each test), etc.

[14] The difference in the chi-squares for H_0 and H_2 (see Table 3a) can be used to test the null hypothesis that $\pi_1 = \pi_5 = 0$ (corresponding to the two $\hat{\pi}$'s that are relevant here), with two degrees of freedom. We can also test separately the null hypothesis that $\pi_1 = 0$ and the null hypothesis that $\pi_5 = 0$ (with one degree of freedom for each test), using the general method mentioned at the end of Footnote 13.

[15] See (1.15) and its interpretation (included in the paragraph after (1.6)). In the model introduced here, the "intrinsically scalable" individual's observed response pattern is in set S (in accordance with the usual "perfect scale" concept) and it cannot be in set U; whereas the "intrinsically unscalable" individual's responses on items A, B, C and D are mutually independent, and so his observed pattern can be in set S or in set U. For comments on some models in which the "intrinsically scalable" individual's observed pattern can be in either set S or U, see Section 6.

patterns correspond to scale types will tend to overestimate the proportion of individuals who are "intrinsically scalable."[16] The latter proportion is estimated here by $1 - \hat{\pi}_0$.

3. THE ANALYSIS OF THE McHUGH DATA

We consider next the data in Column 2 of Table 1. For the response patterns in set S described in (1.1), we apply the same methods used in Section 2 to study Model H_1, and we obtain estimates of the parameter for the McHugh data,

$$\hat{\pi}_{10}{}^{\bar{A}} = .43, \quad \hat{\pi}_{10}{}^{\bar{B}} = .43, \quad \hat{\pi}_{10}{}^{\bar{C}} = .57, \quad \hat{\pi}_{10}{}^{\bar{D}} = .68, \quad \hat{\pi}_0 = .49. \tag{3.1}$$

The estimates corresponding to (2.3) are now

$$\hat{\pi}_1 = .13, \quad \hat{\pi}_2 = .02, \quad \hat{\pi}_3 = .09, \quad \hat{\pi}_4 = .04, \quad \hat{\pi}_5 = .23; \tag{3.2}$$

and the chi-square values for testing H_1 are included in Table 3b for these data. Here too the model under consideration fits the data well.

In Section 2 on the analysis of the Stouffer-Toby data, we considered both H_1 corresponding to the five scale types and H_2 corresponding to the two extreme scale types. We have also applied these two models to the McHugh data, and the results thus obtained are summarized in Table 3b. From the chi-square values for H_2 in Table 3b, we see that this particular model does not fit the McHugh data.

Instead of having five different scale types as in H_1, or two different scale types as in H_2, consider the situation where there are four different scale types; viz., the types corresponding to the response patterns

$$(1, 1, 1, 1), \quad (1, 1, 2, 2), \quad (1, 2, 2, 2), \quad \text{and} \quad (2, 2, 2, 2). \tag{3.3}$$

In this situation, set S consists of these four patterns, and set U consists of the remaining 12 patterns in the

[16] The magnitude of this overestimation is equal to $\hat{\pi}_0\hat{\pi}_{i0}{}^{\bar{A}}\hat{\pi}_{j0}{}^{\bar{B}}\hat{\pi}_{k0}{}^{\bar{C}}\hat{\pi}_{l0}{}^{\bar{D}}$ summed over all patterns (i, j, k, l) in set S (see (1.9)). This sum will depend on the proportion of individuals whose observed response patterns are in set U and the observed one-way marginal distributions (with respect to items A, B, C and D) obtained for those individuals with patterns in U. (For some related results, see (8.5)–(8.7) and the appendix.)

2^4 table. The corresponding quasi-independence model H_3 states that responses on items A, B, C and D are quasi-independent with respect to set U. Applying the general approach described earlier, we now obtain estimates for the parameters under H_3,

$$\hat{\pi}_{10}{}^{\bar{A}} = .47, \quad \hat{\pi}_{10}{}^{\bar{B}} = .46, \quad \hat{\pi}_{10}{}^{\bar{C}} = .59, \quad \hat{\pi}_{10}{}^{\bar{D}} = .66, \quad \hat{\pi}_0 = .52; \tag{3.4}$$

and the estimates corresponding to (3.2) are .12, .09, .04, and .23 under H_3, for the scale types corresponding to the four patterns listed in (3.3). From the chi-square values for H_3 in Table 3b, we see that this model fits the data well.

From the results for H_1 and H_3 in Table 3b, we see that the estimated proportion $\hat{\pi}_0$ of "intrinsically unscaled" individuals is somewhat smaller under H_1 than under H_3. On the other hand, because H_1 has one more scale type than H_3, there is one more parameter to be estimated under H_1 than under H_3; and thus, there is one less degree of freedom for testing H_1 than for testing H_3. By comparing the corresponding likelihood-ratio chi-squares, we obtain a difference of 1.52 (i.e., $7.42 - 5.90 = 1.52$), which can be assessed in terms of the tabulated percentiles of the chi-square distribution with one degree of freedom (i.e., $7 - 6 = 1$). Here we find that the additional scale type under H_1 does not contribute in a statistically significant way to the fit of the model (i.e., that the estimated value of π_2 in (3.2) does not differ from zero in a statisically significant way).

By applying the same general method used in the preceding paragraph and also near the end of Section 2, we find that, assuming H_1 to be true, the estimate $\hat{\pi}_2$ in (3.2) is not significantly different from zero, but $\hat{\pi}_1$, $\hat{\pi}_3$, $\hat{\pi}_4$ and $\hat{\pi}_5$ are significantly different from zero. Similarly, assuming H_3 to be true, we can also show that the latter $\hat{\pi}$'s continue to be significantly different from zero.

4. THE ANALYSIS OF THE LAZARSFELD-STOUFFER DATA

In Section 3, analyzing the McHugh data, we started with a model (H_1) having five scale types and ended with a model (H_3) having four scale types. We now analyze the Lazarsfeld-Stouffer data, and for reasons which will become clear later, we start with a model that has four scale types; viz., the types corresponding to the scale patterns

$$(1, 1, 1, 1), \quad (1, 1, 1, 2), \quad (1, 2, 2, 2), \quad (2, 2, 2, 2). \tag{4.1}$$

Note that (4.1) deletes from (1.1) the middle pattern (1, 1, 2, 2), but it includes the two most extreme patterns (as in (2.4) for H_2) and also the two next most extreme patterns. Set S now consists of the four patterns in (4.1), and set U consists of the remaining 12 patterns in the 2^4 table. The corresponding quasi-independence model H_4 states that responses on items A, B, C and D are quasi-independent with respect to set U. Applying the general approach described earlier, we obtain estimates for the parameters under H_4,

$$\begin{gathered}\hat{\pi}_{10}{}^{\bar{A}} = .75, \quad \hat{\pi}_{10}{}^{\bar{B}} = .47, \quad \hat{\pi}_{10}{}^{\bar{C}} = .36, \\ \hat{\pi}_{10}{}^{\bar{D}} = .30, \quad \hat{\pi}_0 = .67;\end{gathered} \tag{4.2}$$

and the estimates corresponding to (3.2) are .05, .01, .08, .19 under H_4, for the scale types corresponding to the four patterns listed in (4.1).

Since the four patterns in (4.1) are the first, second, fourth and fifth patterns listed in (1.1) (with the third in (1.1) deleted), the preceding corresponding four estimated parameters can be expressed as

$$\hat{\pi}_1 = .05, \quad \hat{\pi}_2 = .01, \quad \pi_3 = 0, \quad \hat{\pi}_4 = .08, \quad \hat{\pi}_5 = .19. \tag{4.3}$$

For Model H_4, the parameter π_3 in (4.3), which pertains to the scale type corresponding to pattern (1, 1, 2, 2), is assumed to be equal to zero, since this pattern was not included in (4.1); whereas for H_1, all five π's (including π_3) are to be estimated from the data, and all of them are assumed to be positive, or at least nonnegative (see, e.g., (2.3) and (3.2)).

When Model H_1 is applied to the Lazarsfeld-Stouffer data using the general approach described earlier, we find that the estimate of π_3 obtained from (1.8) turns out negative, which indicates that the maximum-likelihood estimate of π_3 must be zero (since $\pi_3 \geqq 0$). In this case, to obtain the maximum-likelihood estimates of the parameters under H_1, we will first need to make π_3 equal to zero. Because of this, the maximum-likelihood estimates of the parameters under H_1 will be equal to the corresponding quantities obtained for H_4.

These comments can be applied more generally whenever the estimate obtained from (1.9) is negative for a subset of the π's. In this case, to obtain the maximum-likelihood estimates of the parameters under the model, we will first need to make one or more of π's in that particular subset equal to zero.

Let us return now to Model H_4. From the chi-square values for this model presented in Table 3c, we see that this particular model does not fit the Lazarsfeld-Stouffer data. From previous remarks, we also see that the same chi-square values would be obtained for H_1 as for H_4, when these values are based on the maximum-likelihood estimates. Since neither model H_4 (with its four scale-type categories) nor H_1 (with its five scale-type categories) fits these data, we consider next a model in which there are still more categories corresponding both to "scale types" and "demi-scale types."

If we consider the set of five patterns listed in (1.1) as corresponding to the "scale types," then three additional patterns could be viewed as corresponding to "demi-scale types,"

$$(1, 1, 2, 1),\ (1, 2, 1, 2),\ (2, 1, 2, 2). \tag{4.4}$$

Pattern 1 in (4.4) would be a scale type if items C and D are interchanged; pattern 2 would be a scale type if items B and C are interchanged; and pattern 3 would be a scale type if items A and B are interchanged. The scale-type categories corresponding to the five patterns in (1.1) were numbered earlier in order from one to five; and now we continue this by using 6, 7 and 8 to denote the demi-scale-type categories corresponding to the three patterns in (4.4).

Now set S consists of the eight patterns in (1.1) and (4.4), and set U consists of the remaining eight patterns in the 2^4 table. The corresponding quasi-independence model H_5 states that responses on items A, B, C, D are quasi-independent with respect to set U. Applying the general approach described here, we obtain estimates of the parameters under H_5,

$$\hat{\pi}_{10}{}^{\bar{A}} = .74, \quad \hat{\pi}_{10}{}^{\bar{B}} = .35, \quad \hat{\pi}_{10}{}^{\bar{C}} = .44, \quad \hat{\pi}_{10}{}^{\bar{D}} = .35, \quad \hat{\pi}_0 = .47; \tag{4.5}$$

and the estimates corresponding to (3.2) are

$$\hat{\pi}_1 = .06, \quad \hat{\pi}_2 = .03, \quad \hat{\pi}_3 = .05, \quad \hat{\pi}_4 = .12, \quad \hat{\pi}_5 = .20, \quad \hat{\pi}_6 = .03, \quad \pi_7 = .00, \quad \hat{\pi}_8 = .04. \tag{4.6}$$

To test H_5 with its eight categories for scale and demi-scale types, there will be three degrees of freedom (i.e., $2^4 - 8 - 5 = 3$);[17] and from the chi-square values for H_5 in Table 3c, we find that this model fits the data.

Let us return now for a moment to Models H_1 and H_4. We noted earlier that to test H_1 with its five scale types there will usually be six degrees of freedom (see, e.g., H_1 in Tables 3a and 3b); but in the special case considered earlier, where we needed to make the parameter π_3 equal to zero (see (4.3)), we actually replaced H_1 by H_4, and so there were seven degrees of freedom (rather than six) for testing the model (see H_4 in Table 3c). Similarly, to test H_5 with its eight categories for scale and demi-scale types, there will usually be three degrees of freedom, as noted previously. But when this model was actually applied to the Lazarsfeld-Stouffer data, we found that the parameter π_7 needed to be equated to zero (see (4.6)), and so there were four degrees of freedom (rather than three) for testing the model.[18]

By applying the same general methods used at the end of Section 3, we can show that, assuming H_5 to be true, the $\hat{\pi}$'s in (4.6) are significantly different from zero (except for π_7, which was set at zero).

Since H_5 fits the data, there is no need to consider models in which there are still more categories. However,

[17] See related comments following (1.15), and the more general formulas (4.10)–(4.11) later.

[18] In order not to confuse the reader who might want to apply H_5 with its eight categories for scale and demi-scale types, Table 3c gives the number of scale (and demi-scale) types, and the number of degrees of freedom, usually appropriate for this model; but in the actual application to the Lazarsfeld-Stouffer data, there really were seven scale (and demi-scale) types (rather than eight), and four degrees of freedom (rather than three).

in situations where H_5 does not fit, we may wish to consider models that include both the scale and demi-scale types listed in (1.1) and (4.4), and also the demi-demi-scale types

$$(1, 2, 1, 1), \quad (2, 1, 1, 2), \quad (1, 2, 2, 1), \quad (2, 2, 1, 2). \quad (4.7)$$

Pattern 1 in (4.7) would be a demi-scale type if items B and C are interchanged, and it would be a scale type if items B and D are interchanged; pattern 2 would be a demi-scale type if items A and B are interchanged, and it would be a scale type if items A and C are interchanged; pattern 3 would be a demi-scale type if items C and D are interchanged, and it would be a scale type if items B and D are interchanged; pattern 4 would be a demi-scale type if items B and C are interchanged, and it would be a scale type if items A and C are interchanged.

There are 12 patterns in (1.1), (4.4) and (4.7). Consider now the case where set S consists of these 12 patterns, and set U consists of the remaining four patterns in the 2^4 table,

$$(2, 1, 1, 1), \quad (2, 2, 1, 1), \quad (2, 1, 2, 1), \quad (2, 2, 2, 1). \quad (4.8)$$

The corresponding quasi-independence model H_6 with respect to this particular set U is actually equivalent to the model that states that responses on items B and C are conditionally independent of each other, given that the responses on items A and D are at levels 2 and 1, respectively.[19] Thus, to test H_6, the usual chi-square test of independence in the 2×2 contingency table (formed by the four response patterns in (4.8)) can be applied. From the chi-square values for H_6 in Table 3c, we see that this model fits the data very well indeed.

As we noted earlier, for the quasi-independence models considered here we first estimated the parameters $\pi_{10}^{\bar{A}}$, $\pi_{10}^{\bar{B}}$, $\pi_{10}^{\bar{C}}$, $\pi_{10}^{\bar{D}}$, π_0 (see, e.g., (4.5)) using the data in set U. However, for the particular set U in (4.8), the parameters $\pi_{10}^{\bar{A}}$, $\pi_{10}^{\bar{D}}$ and π_0 are not identifiable (unless some restrictions are imposed on these parameters), since set U includes only response patterns in which the responses to

[19] In the notation used in [4], this model states that $[B \otimes C | A = 2, D = 1]$.

items A and D are at levels 2 and 1, respectively.[20] In view of the preceding restrictions on the responses to items A and D in set U, we might accordingly introduce restrictions on $\pi_{10}{}^{\bar{A}}$ and $\pi_{10}{}^{\bar{D}}$,

$$\pi_{10}{}^{\bar{A}} = 0, \quad \pi_{10}{}^{\bar{D}} = 1; \tag{4.9}$$

and in that particular case, the parameter π_0 can be estimated by the observed proportion p_U of individuals in the 2^4 table whose response patterns are in set U.[21]

We now use the symbol $\#S$ to denote the number of different response patterns in set S. For the 2^4 table, the corresponding number of patterns in set U will be $2^4 - \#S$. From the data in set U, we need to estimate the five parameters $\pi_{10}{}^{\bar{A}}$, $\pi_{10}{}^{\bar{B}}$, $\pi_{10}{}^{\bar{C}}$, $\pi_{10}{}^{\bar{D}}$ and π_0; and when these are identifiable (which is the case for all the models considered so far, except for H_6), the number of degrees of freedom for testing the corresponding quasi-independence model will be

$$2^4 - \#S - 5. \tag{4.10}$$

(Compare (4.10) with (1.15).) More generally, in the situation where there are m dichotomous items, the corresponding number of degrees of freedom will be

$$2^m - \#S - (m + 1). \tag{4.11}$$

(Compare (4.11) with (1.16).) Formulas (4.10) and (4.11) can be applied to all the models considered so far, except for H_6 (see Table 3). Even in the case of H_6, these formulas can be applied if a simple adjustment is made to take account of the fact that *two* additional restrictions

[20] Another example similar to H_6 in this respect will be included as the first model presented in the appendix. For H_6 and for this model in the appendix, the corresponding set U does not provide sufficient data for the estimation of some of the relevant parameters; there are too few patterns in U in both cases. In the rest of this article, this problem does not arise.

[21] The probability that an "intrinsically unscalable" individual's response pattern will be in set U defined by (4.8) is $\pi_{20}{}^{\bar{A}}\pi_{10}{}^{\bar{D}}$, which will be equal to one under condition (4.9). In this case, the observed proportion p_U is the maximum-likelihood estimate of π_0. On the other hand, if the particular restriction (4.9) is not imposed, but instead the parameters $\pi_{10}{}^{\bar{A}}$ and $\pi_{10}{}^{\bar{D}}$ are equated to two specified numerical values which are such that the product $\pi_{20}{}^{\bar{A}}\pi_{10}{}^{\bar{D}}$ is not equal to one (nor equal to zero), then the corresponding maximum-likelihood estimate of π_0 would be greater than p_U. (In passing, we note that $E\{p_U\} = \pi_0\pi_{20}{}^{\bar{A}}\pi_{10}{}^{\bar{D}}$). For related comments, see the appendix.

have been imposed on the parameters (see (4.9)). In this case, the number of degrees of freedom for testing the model will be $2^4 - 12 - 5 + 2 = 1$, which agrees with the result obtained from our discussion of Model H_6 earlier in this section (see Table 3c).[22]

We now comment briefly on Model H_7 in Table 3c. As noted earlier, if Model H_1 (with its five scale types) is applied to the Lazarsfeld-Stouffer data, we obtain H_4 (with its four scale types). Similarly, if H_1 (with its five scale types) is applied to the Lazarsfeld-Stouffer data with the order of items C and D interchanged in the data, we obtain another model having four scale types which we shall call H_7.[23] Comparing the chi-squares for H_7 and H_4 in Table 3c, we see that H_7 fits the data slightly better than does H_4. Comparing the estimated proportions $\hat{\pi}_0$ in Table 3c, we see that the same numerical value (to the order of accuracy reported here) is obtained for H_7 and H_4.[24]

Since H_7 fits the data slightly better than does H_4, and the estimated proportion $\hat{\pi}_0$ is not greater for H_7 than for H_4, we pursue further the analysis of the Lazarsfeld-Stouffer data with the order of items C and D interchanged (as was done in H_7). Since H_7 does not really fit the data (see Table 3c), we consider next Model H_5 (with its eight categories for scale and demi-scale types) and H_6 (with its 12 categories for scale, demi-scale, and demi-demi-scale types) applied to the data with the order of the items interchanged (as in H_7). The results obtained in these applications did not actually improve the original results presented in Table 3c for Models H_5 and H_6 (when the order of the items was not interchanged), and to save space we do not go into these details here. If the results obtained in these applications would have improved the original results (e.g., if H_5

[22] For further comments on identifiability in models of this general kind, see [5].

[23] The fact that a four scale-type model was obtained when H_1 was applied to a given set of data does not necessarily mean that a four scale-type model would also be obtained when H_1 is applied to the same set of data with the order of two of the items interchanged; but it just happened to turn out that way for the Lazarsfeld-Stouffer data.

[24] The numerical value of $\hat{\pi}_0$ obtained for H_7 does not necessarily have to be the same as that obtained for H_4, but it just happened to turn out that way for these data.

would have fit the data better when the order of the items was interchanged, and if the estimated proportion $\hat{\pi}_0$ were not greater in that case), then we would have suggested that the order of the items should be interchanged when considering scaling models for this particular set of data.[25]

5. UNIFORM, BIFORM AND MULTIFORM SCALES

The scaling model in which set S consists of the five response patterns in (1.1) describes the case (H_1) where the four items (A, B, C, D) are ordered $ABCD$; and the ordering $ABDC$ is obtained when pattern (1, 1, 2, 1) replaces pattern 2 in (1.1). In the former case, all "intrinsically scalable" individuals in the population conform to the ordering $ABCD$; and in the latter case, they conform to the ordering $ABDC$. Consider now the case where some of the "intrinsically scalable" individuals in the population conform to the ordering $ABCD$ and the other "intrinsically scalable" individuals conform to the ordering $ABDC$. In this case, the "scale-type categories" include both the scale types under the ordering $ABCD$ *and* the scale types under the ordering $ABDC$. Set S now consists of six patterns: (1, 1, 1, 1), (1, 1, 1, 2), (1, 1, 2, 2), (1, 2, 2, 2), (2, 2, 2, 2), (1, 1, 2, 1); viz., the five patterns in (1.1) *and* also pattern (1, 1, 2, 1). The general methods applied in Section 4 (see, e.g., discussion of H_5) can be applied directly to the case considered here. (To save space, we shall not go into these details.)

The model obtained when set S consists of the six patterns just listed is an example of a "biform" scale, in which the two forms correspond to the $ABCD$ and $ABDC$ orderings; whereas the usual scaling Model (H_1) is an example of a "uniform" scale corresponding to the single ordering $ABCD$. For any two given orderings of the items, the corresponding "biform" scale can be analyzed using the general methods described here. For the "biform" scales, set S will consist of no fewer than six

[25] More generally, for any given scaling model of the kind introduced in this article, the two criteria just noted (i.e., goodness-of-fit and magnitude of $\hat{\pi}_0$) can be used in attempting to determine the appropriate order of the items, when this order is not clearly established otherwise.

patterns (as in the case just considered) and no more than eight patterns (as would be the case if the two given orderings were, say, *ABCD* and *DABC*). More generally, in the situation where there are m dichotomous items, set S would consist of $m + 1$ response patterns in the "uni-from" scales, and between $m + 2$ and $2m$ response patterns in the "biform" scales.

These comments about "biform" scales can be directly extended to "multiform" scales. These kinds of models will be of substantive interest in various applied contexts, and in addition they will provide us with another way to view the demi-scale and demi-demi-scale models introduced in Section 4: For the demi-scale model H_5, with its eight categories for scale and demi-scale types (see (1.1) and (4.4)), using comments of the kind immediately following (4.4), we find that set S consists of the scale types under the ordering[26]

$$ABCD,\ ABDC,\ ACBD,\ BACD,\ BADC.$$

Thus, the demi-scale model H_5 is an example of a "quinqueform" scale, with the five forms corresponding to the orderings just listed.

For the demi-demi-scale model H_6, with its 12 categories for scale, demi-scale, and demi-demi-scale types (see (1.1), (4.4) and (4.7)), using comments of the kind immediately following (4.7), we find that set S consists of the scale types under the following orderings in addition to the orderings just listed[27]

$$ADCB,\ CBAD,\ ACDB,\ BCAD,\ ADBC,\ CABD,\ CADB.$$

[26] For each of the orderings listed here, the corresponding five scale-type response patterns are all included within the eight patterns in (1.1) and (4.4). Letting d_i denote the difference between the letter in the ith position ($i = 1, 2, 3, 4$) in a given ordering and the corresponding letter in the ordering *ABCD*, we see that $|d_i| \leqq 1$ (for $i = 1, 2, 3, 4$) for each ordering listed here. (In calculating the difference between letters, we replace the letters *A*, *B*, *C*, *D* by the numbers 1, 2, 3, 4, respectively.) Each ordering listed here (except for *ABCD*) is obtained by interchanging the items within a pair of adjacent items, for one or more such disjoint pairs, in the ordering *ABCD*.

[27] For each of the orderings listed here, the corresponding five scale-type response patterns are all included within the twelve patterns in (1.1), (4.4) and (4.7). For d_i defined as in Footnote 26, we see that $|d_i| \leqq 2$ (for $i = 1, 2, 3, 4$), *and* $|\Sigma_{i=1}^{j} d_i| \leqq 2$ (for $j = 1, 2, 3, 4$), for each ordering listed here.

Thus, the demi-demi-scale model H_6 is an example of a "duodecaform" scale, with the 12 forms corresponding to the orderings listed in total in this paragraph.

To illustrate the application of the multiform scales, we return for a moment to the Stouffer-Toby data. We shall now let $\mathfrak{F}_1$ denote the uniform scale model (i.e., the scale model H_1 introduced in Section 2) using the ordering *ABCD*. Let $\mathfrak{F}_2$ denote the biform scale model using the orderings *ABCD* and *ACBD*; let $\mathfrak{F}_5$ denote the quinqueform scale model discussed earlier in this section (i.e., the demi-scale model H_5 introduced in Section 4); let $\mathfrak{F}_8$ denote the octaform scale model using the orderings *ACDB*, *CADB*, *CABD* and the five orderings in $\mathfrak{F}_5$; and let $\mathfrak{F}_{12}$ denote the duodecaform scale model discussed earlier in this section (i.e., the demi-demi-scale model H_6 introduced in Section 4). The scale-type response patterns for $\mathfrak{F}_1$, $\mathfrak{F}_5$, and $\mathfrak{F}_{12}$ were listed earlier, and we display in the figure the corresponding patterns for the other two models just described ($\mathfrak{F}_2$ and $\mathfrak{F}_8$).[28] Table 4 summarizes some of the results obtained with these models.[29] Since Model $\mathfrak{F}_1$ fits the data well (see also H_1 in Table 3a), there was no need to introduce additional scale-type patterns into $\mathfrak{F}_1$ (to obtain $\mathfrak{F}_2$, $\mathfrak{F}_5$, $\cdots$); but we did so here for illustrative purposes, and

[28] For purposes of comparison, Models $\mathfrak{F}_1$ and $\mathfrak{F}_5$ are also included in the figure; the corresponding display of $\mathfrak{F}_{12}$ is somewhat more complicated than those included there. Each of these models could be displayed in various ways; e.g., the patterns in $\mathfrak{F}_5$ could also be displayed as in $\mathfrak{F}_2$ with patterns (1, 1, 2, 1) and (2, 1, 2, 2) inserted to the left of (1, 1, 1, 2) and (1, 2, 2, 2), respectively; or, equivalently, as in $\mathfrak{F}_8$ with patterns (1, 2, 1, 1) and (2, 2, 1, 2) deleted. In the displays for both $\mathfrak{F}_2$ and $\mathfrak{F}_8$, the response patterns that are at levels (1, 2) and (2, 1) on the joint variable (*B*, *C*) are on the left and right sides, respectively, of the display; and the patterns that are at levels (1, 1) and (2, 2) on variable (*B*, *C*) are on a vertical center line. Each of the models considered in this section can be described in terms of its scale-type patterns (as in Section 4), or in terms of the orderings that yield those patterns (as in this section), or in terms of various properties of the corresponding displays.

[29] With respect to Models $\mathfrak{F}_2$, $\mathfrak{F}_5$ and $\mathfrak{F}_8$ in Table 4, a comment of the same kind as in Footnote 18 should be included here. In the actual application to the Stouffer-Toby data, for Model $\mathfrak{F}_2$, there really were five scale types (rather than six), and six degrees of freedom (rather than five); for $\mathfrak{F}_5$, there really were seven scale types (rather than eight), and four degrees of freedom (rather than three); for $\mathfrak{F}_8$, there really were nine scale types (rather than ten), and two degrees of freedom (rather than one). For $\mathfrak{F}_2$ and $\mathfrak{F}_5$, the scale-type category corresponding to response pattern (1, 2, 1, 2) needed to be void; and for $\mathfrak{F}_8$, the scale-type category corresponding to (1, 2, 1, 1) needed to be void.

4. Some Estimated Parameters and Chi-Square Values for the Uniform Scale Model and Some Multiform Scale Models Applied to the Stouffer-Toby Data

Model	Number of scale types	Estimated proportion $\hat{\pi}_0$ of intrinsically unscalable individuals	Degrees of freedom	Goodness-of-fit chi-square	Likelihood-ratio chi-square
$\mathcal{F}_1(=H_1)$	5	.68	6	1.01	.99
$\mathcal{F}_2$	6[a]	.68	5[a]	1.01	.99
$\mathcal{F}_5(=H_5)$	8[a]	.59	3[a]	.11	.11
$\mathcal{F}_8$	10[a]	.59	1[a]	.03	.03
$\mathcal{F}_{12}(=H_6)$	12	.03[b]	1	.00	.00

[a] For comments on the results actually obtained for the Stouffer-Toby data, see Footnote 29.

[b] Estimate obtained under condition (4.9).

to show how $\hat{\pi}_0$ changes with the increase in scale-type patterns.[30]

6. THE NEW SCALING MODEL, THE USUAL SCALING MODEL AND THE USUAL LATENT CLASS MODELS

The scaling model was described at the beginning of this article by formulas (1.5)–(1.6) for the case (H_1), where the response patterns in (1.1) formed set S; and we then saw how to extend this model to the case where a proper subset of the patterns in (1.1) formed set S and to the case where set S consisted of the patterns in (1.1) and some other specified patterns as well. We shall now comment on the relationship between these models, the more usual Guttman scaling model, and the more usual latent class models. For the sake of simplicity, we shall focus our attention on the relationship between H_1 and the other more usual approaches; but these comments can be extended to cover the other scaling models that can be obtained with the general approach introduced here.

[30] The estimate $\hat{\pi}_0$ will depend upon the proportion of individuals whose observed response patterns are in set U *and* the observed one-way marginal distributions (with respect to items A, B, C and D) obtained for those individuals with patterns in U. For some further insight into $\hat{\pi}_0$, see related comments in the appendix.

The Scale-Type Response Patterns in the Uniform Scale and in Some Multiform Scales

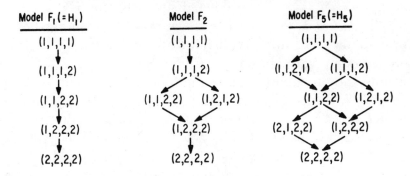

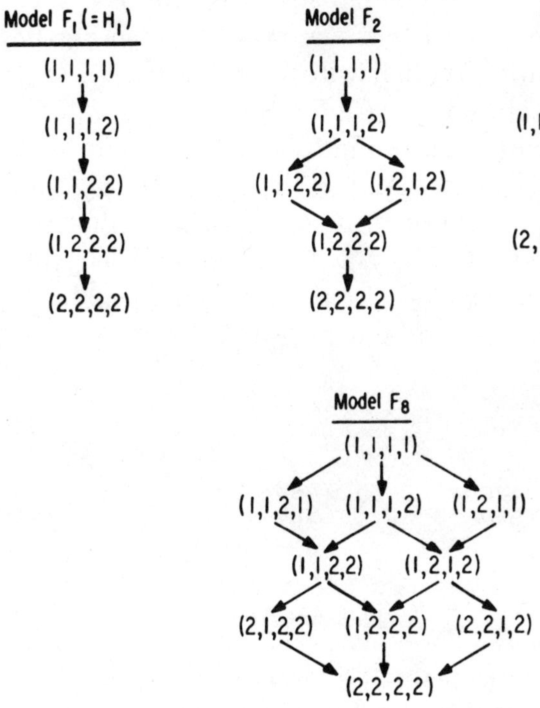

We noted earlier that the usual Guttman scaling model is a special case of Model H_1 in which all individuals in the population under consideration are assumed to be "intrinsically scalable." In terms of the parameters used in (1.5)–(1.6) to describe H_1, the usual Guttman scaling model assumes that $\pi_0 = 0$. This assumption is not supported by the data considered here, and the estimates of π_0 in Tables 3 and 4 shed some light on the magnitude of the discrepancy between the actual value of π_0 and the assumed value of zero.

Consider next the usual latent class model in which there are six latent classes, which we shall number 0, 1, 2, $\cdots$, 5. We shall use the letter X to denote the corresponding latent variable, with its six latent classes (or levels). We let π_t^{X} denote the probability that an individual in the population will be at level t on variable X (for $t = 0, 1, 2, \cdots, 5$); we also let $\pi_{it}^{\bar{A}X}$ denote the *conditional* probability that the individual will be at level i on variable A (for $i = 1, 2$), given that he was at level t on variable X; and we let the *conditional* probabilities $\pi_{jt}^{\bar{B}X}$, $\pi_{kt}^{\bar{C}X}$, $\pi_{lt}^{\bar{D}X}$ be defined similarly (for $j = 1, 2$; $k = 1, 2$; $l = 1, 2$). As earlier, we let π_{ijkl}^{ABCD} denote the probability that an individual will be at level (i, j, k, l) on the joint variable (A, B, C, D); we also let π_{ijklt}^{ABCDX} denote the probability that an individual will be at level (i, j, k, l, t) on the joint variable (A, B, C, D, X); and we let $\pi_{ijklt}^{\bar{A}\bar{B}\bar{C}\bar{D}X}$ denote the *conditional* probability that the individual will be at level (i, j, k, l) on the joint variable (A, B, C, D), given that he was at level t on variable X. The latent class model can be expressed by

$$\pi_{ijkl}^{ABCD} = \sum_{t=0}^{5} \pi_{ijklt}^{ABCDX}, \tag{6.1}$$

$$\pi_{ijklt}^{ABCDX} = \pi_t^X \pi_{ijklt}^{\bar{A}\bar{B}\bar{C}\bar{D}X}, \tag{6.2}$$

and

$$\pi_{ijklt}^{\bar{A}\bar{B}\bar{C}\bar{D}X} = \pi_{it}^{\bar{A}X} \pi_{jt}^{\bar{B}X} \pi_{kt}^{\bar{C}X} \pi_{lt}^{\bar{D}X}. \tag{6.3}$$

Formula (6.1) states that the individuals can be classified into six mutually exclusive and exhaustive latent classes $(t = 0, 1, 2, \cdots, 5)$ with respect to the latent variable X; (6.2) follows directly from the definition of the π's appearing in (6.2); and (6.3) states that within the tth latent class the levels on variables A, B, C and D are mutually independent (for $t = 0, 1, 2, \cdots, 5$).

Since this latent class model has four manifest variables (A, B, C, D) and six latent classes, the parameters in the model will not be identifiable unless some restrictions are imposed (see, e.g., [5]). Various kinds of restrictions could be imposed. Consider now the following set of restrictions imposed on the parameters

$$\begin{aligned}
\pi_{11}^{\bar{A}X} &= \pi_{11}^{\bar{B}X} = \pi_{11}^{\bar{C}X} = \pi_{11}^{\bar{D}X} = 1,\\
\pi_{12}^{\bar{A}X} &= \pi_{12}^{\bar{B}X} = \pi_{12}^{\bar{C}X} = \pi_{22}^{\bar{D}X} = 1,\\
\pi_{13}^{\bar{A}X} &= \pi_{13}^{\bar{B}X} = \pi_{23}^{\bar{C}X} = \pi_{23}^{\bar{D}X} = 1,\\
\pi_{14}^{\bar{A}X} &= \pi_{24}^{\bar{B}X} = \pi_{24}^{\bar{C}X} = \pi_{24}^{\bar{D}X} = 1,\\
\pi_{25}^{\bar{A}X} &= \pi_{25}^{\bar{B}X} = \pi_{25}^{\bar{C}X} = \pi_{25}^{\bar{D}X} = 1.
\end{aligned} \tag{6.4}$$

From (6.1)–(6.3) and (6.4), we see that

$$\begin{aligned}
\pi_{1111}^{ABCD} &= \pi_0{}^X \pi_{10}^{\bar{A}X} \pi_{10}^{\bar{B}X} \pi_{10}^{\bar{C}X} \pi_{10}^{\bar{D}X} + \pi_1{}^X, \\
\pi_{1112}^{ABCD} &= \pi_0{}^X \pi_{10}^{\bar{A}X} \pi_{10}^{\bar{B}X} \pi_{10}^{\bar{C}X} \pi_{20}^{\bar{D}X} + \pi_2{}^X, \\
\pi_{1122}^{ABCD} &= \pi_0{}^X \pi_{10}^{\bar{A}X} \pi_{10}^{\bar{B}X} \pi_{20}^{\bar{C}X} \pi_{20}^{\bar{D}X} + \pi_3{}^X, \\
\pi_{1222}^{ABCD} &= \pi_0{}^X \pi_{10}^{\bar{A}X} \pi_{20}^{\bar{B}X} \pi_{20}^{\bar{C}X} \pi_{20}^{\bar{D}X} + \pi_4{}^X, \\
\pi_{2222}^{ABCD} &= \pi_0{}^X \pi_{20}^{\bar{A}X} \pi_{20}^{\bar{B}X} \pi_{20}^{\bar{C}X} \pi_{20}^{\bar{D}X} + \pi_5{}^X;
\end{aligned} \tag{6.5}$$

and for any other level (i, j, k, l) on the joint variable (A, B, C, D), we obtain

$$\pi_{ijkl}^{ABCD} = \pi_0{}^X \pi_{i0}^{\bar{A}X} \pi_{j0}^{\bar{B}X} \pi_{k0}^{\bar{C}X} \pi_{l0}^{\bar{D}X}. \tag{6.6}$$

Comparing (6.5)–(6.6) with (1.5)–(1.6), we see that the scaling model H_1 is a special case of a latent class model; i.e., it is the latent class model (6.1)–(6.3) with the additional restrictions in (6.4) imposed on the parameters.

For purposes of comparison, we consider next a latent class model that closely resembles the usual Guttman scaling model; viz., Proctor's model [14].[31] It differs from the model introduced here in that it has five latent classes[32] rather than six (i.e., it has no "intrinsically unscalable" class); and under this model an "intrinsically scalable" individual's response on a given item will correspond to his scale type with probability π (rather than one). Thus, the conditional probabilities in (6.4) are equated to π here (rather than to one), and the "response error rate" is $1 - \pi$. This model is a special case of the *restricted* latent structures considered in [5, 6], and the statistical methods developed in those articles can be applied to this special case to obtain the results reported in Table 5.[33] For the three sets of data con-

[31] This model was also discussed briefly in [11]. It is a special case of a latent distance model discussed in [9, 10] and of the more general latent distance model discussed in [11]. For this special case, Proctor [14] provided maximum-likelihood estimates of the parameters in the model, and he showed how the special case could be used to analyze various sets of data.

[32] More generally, when there are m dichotomous items, there will be $m + 1$ latent classes in Proctor's model.

[33] With respect to the results for the McHugh data in Table 5, a comment of the same kind as in Footnote 18 should be included here. In the actual application to the McHugh data, there really were three scale types (rather than five), and twelve degrees of freedom (rather than ten); the scale-type categories corresponding to the response patterns (1, 1, 1, 2) and (1, 2, 2, 2) needed to be void.

5. Some Estimated Parameters and Chi-Square Values for Proctor's Model of Guttman Scaling, Applied to the Data in Table 1

Data	Number of scale types	Response error rate	Degrees of freedom	Goodness-of-fit chi-square	Likelihood-ratio chi-square
Stouffer-Toby	5	15	10	28.03	27.16
McHugh	5[a]	.18	10[a]	25.20	20.31
Lazarsfeld-Stouffer	5	.16	10	71.54	75.60

[a] For comments on the results actually obtained for the McHugh data, see Footnote 33.

sidered, by comparing Table 5 with Tables 3 and 4, we find that the general approach introduced earlier led to models that fit the data better than Proctor's model.[34]

In addition to Proctor's model, other latent class and latent distance models could be applied to these data; but to save space we shall not go into this here. The interested reader can compare the results presented here using the new scaling models with results obtained using various latent class and latent distance models.[35]

7. SOME FURTHER COMMENTS ON ESTIMATION METHODS

We began this article by showing how the methods which had been developed earlier for the analysis of the quasi-independence model could be applied to calculate the maximum-likelihood estimates of the parameters in

[34] The appropriate models in Tables 3 and 4 fit the data better than Proctor's model, but there were fewer degrees of freedom associated with the former models. The scaling models introduced earlier can be modified, by methods described later in Section 8, to obtain new scaling models that have more degrees of freedom associated with them.

[35] See, e.g., [5, 6, 7, 10, 11, 13]. The methods for analyzing the new scaling models, presented earlier, are somewhat easier to apply than the corresponding methods for the latent class and latent distance models. Compare, e.g., the analysis of the Stouffer-Toby data in [5], or the McHugh data in [6], with the analysis of these data presented here. Considering the various models that may fit a given set of data, the choice among them would depend, in part, on which models are more meaningful in the particular substantive context.

the new scaling models introduced here. On the other hand, in view of the equivalence noted in Section 6, between these new scaling models and the corresponding latent class model described there (see (6.1)–(6.4)), an alternative method for estimating these parameters is available; viz., the general procedure presented in [5, 6] for calculating maximum-likelihood estimates of the parameters in a wide variety of latent class models.[36] However, the method described in Section 1 is somewhat easier to apply.

In calculating the maximum-likelihood estimates using the method described in Section 1 (applying the quasi-independence model to set U), we *first* calculated the estimates $\hat{\pi}_{10}{}^{\bar{A}}$, $\hat{\pi}_{10}{}^{\bar{B}}$, $\hat{\pi}_{10}{}^{\bar{C}}$, $\hat{\pi}_{10}{}^{\bar{D}}$, $\hat{\pi}_0$ using the iterative procedure mentioned earlier, and *then* calculated $\hat{\pi}_t$ (for $t = 1, 2, \cdots, 5$) and $\hat{\pi}_{ijkl}$ (see (1.9) and comments following it). Another alternative to this method would be to *first* calculate the estimates $\hat{\pi}_{ijkl}$, under the quasi-independence model applied to set U, using an appropriate kind of iterative procedure (see , e.g., [1; 2, p. 1119–20]);[37] and *then* calculate the estimates $\hat{\pi}_{10}{}^{\bar{A}}$, $\hat{\pi}_{10}{}^{\bar{B}}$, $\hat{\pi}_{10}{}^{\bar{C}}$, $\hat{\pi}_{10}{}^{\bar{D}}$, and $\hat{\pi}_t$ (for $t = 0, 1, 2, \cdots, 5$).

Let $\Omega_{.0}{}^{\bar{A}}$ denote the odds in favor of a level-1 (rather than a level-2) response on item A for the individuals in the population who are in the 0th category, i.e.,

$$\Omega_{.0}{}^{\bar{A}} = \pi_{10}{}^{\bar{A}}/\pi_{20}{}^{\bar{A}}; \tag{7.1}$$

and let the odds $\Omega_{.0}{}^{\bar{B}}$, $\Omega_{.0}{}^{\bar{C}}$, $\Omega_{.0}{}^{\bar{D}}$ be defined similarly. From (7.1) and (1.6), we see that

$$\Omega_{.0}{}^{\bar{A}} = \pi_{1jkl}^{ABCD}/\pi_{2jkl}^{ABCD}, \tag{7.2}$$

for any pair of response patterns $(1, j, k, l)$ and $(2, j, k, l)$ that are included in set U; and similar formulas can be

[36] This general procedure can be applied to latent class models in which restrictions of the kind described by (6.4) are imposed, or in which various other kinds of restrictions are imposed. As noted in Section 6, we used this general procedure to obtain the results reported in Table 5 for Proctor's model, and it can also be used to obtain the results reported in Tables 2, 3 and 4. Other latent class and latent distance models can also be analyzed using this general procedure.

[37] The $\hat{\pi}_{ijkl}$ can be calculated using, e.g., the computer program ECTA (Everyman's Contingency Table Analyzer), prepared by R. Fay and the author, and available from the author.

obtained for the $\Omega_{.0}^{\bar{B}}$, $\Omega_{.0}^{\bar{C}}$, $\Omega_{.0}^{\bar{D}}$. From (7.1) we see that

$$\pi_{10}^{\bar{A}} = \Omega_{.0}^{\bar{A}}/(1 + \Omega_{.0}^{\bar{A}}). \tag{7.3}$$

To obtain the maximum-likelihood estimate for $\pi_{10}^{\bar{A}}$, we insert the corresponding estimates $\hat{\pi}_{1jkl}^{ABCD}$ and $\hat{\pi}_{2jkl}^{ABCD}$ on the right side of (7.2) to obtain $\hat{\Omega}_{.0}^{\bar{A}}$, and we then insert $\hat{\Omega}_{.0}^{\bar{A}}$ on the right side of (7.3) to obtain $\hat{\pi}_{10}^{\bar{A}}$. A similar kind of calculation will also yield $\hat{\pi}_{10}^{\bar{B}}$, $\hat{\pi}_{10}^{\bar{C}}$, $\hat{\pi}_{10}^{\bar{D}}$; and from (1.6) we see that $\hat{\pi}_0$ can be obtained as follows:

$$\hat{\pi}_0 = \hat{\pi}_{ijkl}^{ABCD}/(\hat{\pi}_{i0}^{\bar{A}}\hat{\pi}_{j0}^{\bar{B}}\hat{\pi}_{k0}^{\bar{C}}\hat{\pi}_{l0}^{\bar{D}}), \tag{7.4}$$

for any pattern (i, j, k, l) that is included in set U. Having thus obtained $\hat{\pi}_{10}^{\bar{A}}$, $\hat{\pi}_{10}^{\bar{B}}$, $\hat{\pi}_{10}^{\bar{C}}$, $\hat{\pi}_{10}^{\bar{D}}$ and $\hat{\pi}_0$, we can then use (1.9) to obtain estimates for the remaining parameters π_t, for $t = 1, 2, \cdots, 5$.[38]

We have just noted two alternatives to the estimation method presented in Section 1: first, a method based on the more general procedure appropriate for latent class models; and second, a method based on the quasi-independence model (as in Section 1), but differing in detail from the procedure suggested earlier. Each of these methods yields the same results. The researcher can choose whichever technique he finds most convenient.

8. SOME NEW SCALING MODELS THAT ARE MORE PARSIMONIOUS

In our discussion of the quasi-independence model for set U, we noted in Section 1 that there were five independent parameters estimated under the model, $\pi_{10}^{\bar{A}}$, $\pi_{10}^{\bar{B}}$, $\pi_{10}^{\bar{C}}$, $\pi_{10}^{\bar{D}}$ and π_0. The first four parameters are the *conditional* probabilities of being at level 1 on variables A, B, C and D, respectively, for the individuals in the 0th category (viz., the "intrinsically unscalable" category). In some contexts, there may be good reason to

[38] As noted in Section 4, whenever the estimate of the π_t obtained from (1.9) turns out to be negative for a subset of the π_t's (for $t = 1, 2, 3, 4, 5$), to obtain the maximum-likelihood estimates of the parameters under the model, we first need to make one or more of the π_t's in that particular subset equal to zero, thus deleting the corresponding response pattern from set S.

consider models in which it is assumed. e.g., that the following condition holds true:

$$\pi_{10}^{\bar{A}} = \pi_{10}^{\bar{B}} = \pi_{10}^{\bar{C}} = \pi_{10}^{\bar{D}}; \tag{8.1}$$

or, more generally, that the parameters in some specified subset of the four parameters in (8.1) are equal to each other. With these kinds of restrictions imposed on the parameters in the model, the number of independent parameters estimated under the model is reduced. (For example, under (8.1), there will be two rather than five independent parameters estimated under the model.) With models of this kind, we can assess whether a particular model is congruent with the observed data, and whether the particular restrictions (e.g., (8.1)) are supported by the data.

In Section 7 we noted that the general estimation procedure presented in [5, 6] can be applied to a wide variety of latent class models including the scaling models considered earlier, where restrictions of the kind described in (6.4) are imposed. In addition, this general estimation procedure can also be applied to scaling models of the kind just described where restrictions of the kind described in (8.1) are imposed (see, e.g., [6]).

The kind of restriction described in (8.1) pertained to the same level (level 1) on each of the variables A, B, C and D. On the other hand, in some contexts, we might want to consider the more general kind of restriction

$$\pi_{i0}^{\bar{A}} = \pi_{j0}^{\bar{B}} = \pi_{k0}^{\bar{C}} = \pi_{l0}^{\bar{D}}, \tag{8.2}$$

for a specified level i, j, k, l on variables A, B, C, D, respectively (e.g., level 1, 2, 1, 2, rather than level 1, 1, 1, 1 as in (8.1)). Even more generally, we might want to consider the restriction that the parameters in some specified subset of the four parameters in (8.2) are equal to each other. To analyze scaling models in which these more general kinds of restrictions are imposed, we can also apply the general estimation procedure just mentioned. The results in Table 6 were obtained in this way, for various *restricted* scaling models applied to the data in Table 1.

Model H_2' in Table 6 states that H_2 is true *and* that the parameters in the model satisfy the condition

$$\pi_{10}{}^{\bar{B}} = \pi_{10}{}^{\bar{C}}. \tag{8.3}$$

Because of the restriction imposed in (8.3), model H_2' has one less parameter to estimate than H_2, and there is one more degree of freedom for testing H_2' than for testing H_2. Model H_2'' in Table 6 states that H_2 is true *and* that the parameters in the model satisfy the two conditions

$$\pi_{10}{}^{\bar{B}} = \pi_{10}{}^{\bar{C}} \quad \text{and} \quad \pi_{10}{}^{\bar{A}} = \pi_{20}{}^{\bar{D}}. \tag{8.4}$$

Because of the two restrictions imposed in (8.4), model H_2'' has two less parameters to estimate than H_2, and there are two more degrees of freedom for testing H_2'' than for testing H_2. By comparing the chi-square values for H_2 and H_2', and then the chi-square values for H_2' and H_2'', we see that the restrictions imposed in (8.3) and in (8.4) are supported by the data to which those particular models were applied (i.e., the Stouffer-Toby data). These three models (i.e., H_2, H_2', H_2'') fit the data very well indeed, but H_2' is slightly more parsimonious than H_2, and H_2'' is slightly more parsimonious than H_2'.

The restricted scaling models in Table 6 for the McHugh data and the Lazarsfeld-Stouffer data can be analyzed by the same methods just used for the analysis of the Stouffer-Toby data. We leave these details for the interested reader.

In Section 1 we presented a method for calculating the maximum-likelihood estimates of the parameters in the scaling model introduced there (i.e., the unrestricted scaling model), which was based on the procedure used for the quasi-independence model; and in Section 7 we noted that an alternative method could be based on the more general procedure appropriate for latent class models. In this section we used the general procedure appropriate for latent class models to calculate the maximum-likelihood estimates of the parameters in the *restricted* scaling models introduced here. For those who wish to analyze *restricted* scaling models but prefer to use procedures similar to those used for the quasi-inde-

6. Some Estimated Parameters and Chi-Square Values for Various Restricted Scaling Models, Applied to the Data in Table 1

Model	Number of scale types	Restrictions imposed upon the parameters	$\hat{\pi}_0$	Degrees of freedom	Goodness-of-fit chi-square	Likelihood-ratio chi-square
		a. Stouffer-Toby Data				
H_2'	2	$\pi_{10}^{\bar{B}} = \pi_{10}^{\bar{B}}$	.78	10	2.42	2.39
H_2''	2	$\pi_{10}^{\bar{B}} = \pi_{10}^{\bar{C}}, \quad \pi_{10}^{\bar{A}} = \pi_{20}^{\bar{D}}$	.77	11	2.85	2.72
		b. McHugh Data				
H_3'	4	$\pi_{10}^{\bar{A}} = \pi_{10}^{\bar{B}}$	.52	8	7.11	7.43
H_3''	4	$\pi_{10}^{\bar{A}} = \pi_{10}^{\bar{B}}, \quad \pi_{10}^{\bar{C}} = \pi_{10}^{\bar{D}}$	.52	9	8.00	8.17
H_3'''	4	$\pi_{10}^{\bar{A}} = \pi_{10}^{\bar{B}} = \pi_{10}^{\bar{C}} = \pi_{10}^{\bar{D}}$	.56	10	13.45	12.91
		c. Lazarsfeld-Stouffer Data				
H_5'	7	$\pi_{10}^{\bar{B}} = \pi_{10}^{\bar{D}}$	.48	5	5.46	5.87
H_5''	7	$\pi_{10}^{\bar{B}} = \pi_{10}^{\bar{D}} = \pi_{20}^{\bar{A}}$	.49	6	10.06	9.65

pendence model, we now show how this can be done.

For response pattern (i, j, k, l), let $\delta_{ijkl} = 1$ if this pattern is in set U, and let $\delta_{ijkl} = 0$ otherwise. Under the usual (i.e., unrestricted) quasi-independence model for set U (see (1.7)), the maximum-likelihood estimate $\hat{\pi}_{ijkl}^{ABCD}$ of the corresponding probability π_{ijkl}^{ABCD} (i.e., the probability of obtaining pattern (i, j, k, l)) can be calculated from

$$\hat{\pi}_{ijkl}^{ABCD} = \hat{\alpha}_i\hat{\beta}_j\hat{\gamma}_k\hat{\delta}_l, \quad \text{for} \quad (i, j, k, l) \in U, \tag{8.5}$$

where the $\hat{\alpha}_i$, $\hat{\beta}_j$, $\hat{\gamma}_k$, $\hat{\delta}_l$ satisfy

$$\begin{aligned} p_{i\ldots}^{A} &= \hat{\alpha}_i \sum_{j,k,l} \delta_{ijkl}\hat{\beta}_j\hat{\gamma}_k\hat{\delta}_l, & p_{.j..}^{B} &= \hat{\beta}_j \sum_{i,k,l} \delta_{ijkl}\hat{\alpha}_i\hat{\gamma}_k\hat{\delta}_l, \\ p_{..k.}^{C} &= \hat{\gamma}_k \sum_{i,j,l} \delta_{ijkl}\hat{\alpha}_i\hat{\beta}_j\hat{\delta}_l, & p_{\ldots l}^{D} &= \hat{\delta}_l \sum_{i,j,k} \delta_{ijkl}\hat{\alpha}_i\hat{\beta}_j\hat{\gamma}_k, \end{aligned} \tag{8.6}$$

where $p_{i\ldots}^{A}$ is calculated from the observed proportion p_{ijkl}^{ABCD},

$$p_{i\ldots}^{A} = \sum_{j,k,l} \delta_{ijkl}p_{ijkl}^{ABCD}, \tag{8.7}$$

and where $p_{.j..}^{B}$, $p_{..k.}^{C}$ and $p_{\ldots l}^{D}$ are calculated similarly. On the other hand, under the *restricted* quasi-independence model for set U, where, e.g., the restriction in (8.8) is imposed on the parameters,

$$\pi_{10}^{\bar{A}} = \pi_{10}^{\bar{B}}, \tag{8.8}$$

(1.7) would be replaced by

$$\pi_{ijkl}^{ABCD} = \alpha_i\alpha_j\gamma_k\delta_l, \quad \text{for} \quad (i, j, k, l) \in U, \tag{8.9}$$

since $\beta_j = \alpha_j$ (for $j = 1, 2$) under the restriction in (8.8), (8.5) would be replaced by

$$\hat{\pi}_{ijkl}^{ABCD} = \hat{\alpha}_i\hat{\alpha}_j\hat{\gamma}_k\hat{\delta}_l, \tag{8.10}$$

and (8.6) would be replaced by[39]

[39] Note that the formula on the first line in (8.11) is obtained from the two formulas on the first line in (8.6), by setting $\hat{\beta}_j = \hat{\alpha}_j$ in the two formulas in (8.6), then adding the corresponding terms on the left side of the two formulas, and equating the sum thus obtained on the left with the corresponding sum obtained on the right side of the two formulas.

[40] Letting p_U denote the observed proportion of individuals whose response patterns are in set U, we see that, in this particular case, this proportion is equal to the observed proportion p_{21} of individuals whose response pattern is at level (2, 1) on the joint variable (A, B).

$$\overset{A}{p}_{i\ldots} + \overset{B}{p}_{.i..} = \hat{a}_i\Big[\sum_{j,k,l} \delta_{ijkl}\hat{a}_j\hat{\gamma}_k\hat{\delta}_l + \sum_{t,k,l} \delta_{tikl}\hat{a}_t\hat{\gamma}_k\hat{\delta}_l\Big], \qquad (8.11)$$

$$\overset{C}{p}_{..k.} = \hat{\gamma}_k \sum_{i,j,l} \delta_{ijkl}\hat{a}_i\hat{a}_j\hat{\delta}_l, \quad \overset{D}{p}_{...l} = \hat{\delta}_l \sum_{i,j,k} \delta_{ijkl}\hat{a}_i\hat{a}_j\hat{\gamma}_k.$$

The kind of iterative procedure presented in [2, p. 1118–9] for calculating the $\hat{\alpha}_i$, $\hat{\beta}_j$, $\hat{\gamma}_k$, $\hat{\delta}_l$ in (8.6) can be modified in a straightforward way in order to calculate the corresponding quantities in (8.11). Similar modifications can be made for the other kinds of *restricted* quasi-independence models considered earlier in this section (see, e.g., (8.1)–(8.4)).

APPENDIX

We noted in Section 1 that the model and methods introduced can be applied directly to the case where there are m dichotomous variables, for $m = 3, 4, \cdots$. Further insight into this approach can be obtained by considering in more detail the special case where $m = 2$ and $m = 3$. When $m = 2$, a problem of identifiability arises, and the discussion of this problem here will shed further light on our analysis of H_6 in Section 4 and on related matters. When $m = 3$, we can present explicit formulas for the estimated parameters, and these formulas will shed further light on the meaning of $\hat{\pi}_0$ and related quantities.

Consider first the situation where there are two dichotomous items, say, A and B, and let (i, j) denote the response pattern in which responses on items A and B are at levels i and j, respectively (for $i = 1, 2; j = 1, 2$). Corresponding to the five scale-type patterns in (1), we now have three such patterns:

$$(1, 1),\ (1, 2),\ (2, 2). \qquad (A.1)$$

Here set S consists of the three patterns in (A.1), and set U consists of the remaining pattern (2, 1) in the 2×2 table.[40]

Corresponding to the six categories in the model (H_1) introduced earlier, we now have a total of four categories: the "intrinsically unscalable" category and the three "intrinsically scalable" categories pertaining to the patterns in (A.1). The parameters in the scaling model are now π_t (for $t = 0, 1, 2, 3$) and the conditional probabilities $\pi_{i0}^{\bar{A}}$ and $\pi_{j0}^{\bar{B}}$ pertaining to the 0th category (i.e., the intrinsically unscalable category).

Since set U consists only of one response pattern, we face an identifiability problem here similar to the one faced with H_6 in Section 4. To deal with this problem, we need to impose some restrictions on the relevant conditional probabilities (see, e.g., (4.9)).

Accordingly, we now introduce restrictions on $\pi_{10}^{\bar{A}}$ and $\pi_{10}^{\bar{B}}$,

$$\pi_{10}^{\bar{A}} = 0 \ , \quad \pi_{10}^{\bar{B}} = 1 \ ; \tag{A.2}$$

and in that particular case, the parameter π_0 can be estimated by the observed proportion $p_U = p_{21}$.[41]

More generally, if we set $\pi_{10}^{\bar{A}}$ and $\pi_{10}^{\bar{B}}$ equal to two specified numerical values, then $\hat{\pi}_0 = p_{21}/\pi_{20}^{\bar{A}}\pi_{10}^{\bar{B}}$, and we obtain the explicit formulas for the maximum likelihood estimate of π_t, for $t = 1, 2, 3$,[42]

$$\begin{aligned}
\hat{\pi}_1 &= p_{11} - \hat{\pi}_0\pi_{10}^{\bar{A}}\pi_{10}^{\bar{B}} = p_{11} - p_{21}\Omega^{\bar{A}} \ , \\
\hat{\pi}_2 &= p_{12} - \hat{\pi}_0\pi_{10}^{\bar{A}}\pi_{20}^{\bar{B}} = p_{12} - p_{21}\Omega^{\bar{A}}/\Omega^{\bar{B}} \ , \\
\hat{\pi}_3 &= p_{22} - \hat{\pi}_0\pi_{20}^{\bar{A}}\pi_{20}^{\bar{B}} = p_{22} - p_{21}/\Omega^{\bar{B}} \ ,
\end{aligned} \tag{A.3}$$

where $\Omega^{\bar{A}}$ denotes the odds defined as in (7.1) and $\Omega^{\bar{B}}$ is defined similarly. (Compare (A.3) with (1.9).)

In this particular case, since set U consists only of one response pattern, the corresponding model of quasi-independence cannot be tested.

Consider next the case where there are three dichotomous items, say, A, B and C, and let (i, j, k) denote the response pattern in which items A, B and C are at levels i, j and k, respectively (for $i = 1, 2; j = 1, 2; k = 1, 2$). Corresponding to the three scale-type patterns in (A.1), we now have four such patterns,

$$(1, 1, 1), (1, 1, 2), (1, 2, 2), (2, 2, 2) \ . \tag{A.4}$$

Here set S consists of the four patterns in (A.4), and set U consists of the remaining four patterns in the $2 \times 2 \times 2$ table,

$$(1, 2, 1), (2, 1, 1), (2, 2, 1), (2, 1, 2) \ . \tag{A.5}$$

In this case, we obtain the explicit formulas for the maximum-likelihood estimates of the odds defined as in (7.1),[43]

$$\hat{\Omega}_{.0}^{\bar{A}} = p_{121}/p_{221} \ , \quad \hat{\Omega}_{.0}^{\bar{B}} = p_{211}/p_{221} \ , \quad \hat{\Omega}_{.0}^{\bar{C}} = p_{211}/p_{212} \ ; \tag{A.6}$$

and the corresponding estimated probabilities $\hat{\pi}_{10}^{\bar{A}}$, $\hat{\pi}_{10}^{\bar{B}}$, $\hat{\pi}_{10}^{\bar{C}}$ can be calculated directly from the estimated odds (see (7.3)). Similarly, we obtain the explicit formula for $\hat{\pi}_0$,[44]

---

[41] See related comments in Footnote 21.

[42] To avoid unnecessary complications, the specified numerical values of $\pi_{10}^{\bar{A}}$ and $\pi_{10}^{\bar{B}}$ should be such that none of the $\hat{\pi}_t$ in (A.3) are negative, and $\hat{\pi}_0 \leq 1$. This can be accomplished, e.g., with (A.2).

[43] For the sake of simplicity, we assume here that $p_{ijk} > 0$, for $(i, j, k) \in U$; or, more generally, that $p_{ijk} > 0$ for enough of the patterns (i, j, k) in set U so that the relevant parameters can be estimated. In this particular case, we need to assume that $p_{211} > 0$ and $p_{221} > 0$ (see (A.7) and (A.10)). For a related comment see Footnote 3.

[44] The $\hat{\pi}_0$ defined by (A.7) will satisfy the inequality $0 < \hat{\pi}_0 \leq 1$, if none of the $\hat{\pi}_t$ in (A.9) are negative. Similarly, the estimate $\hat{\pi}_0$ obtained from (1.8) will satisfy this inequality if none of the $\hat{\pi}_t$ in (1.9) are negative. This result for $\hat{\pi}_0$ holds true more generally when there are m dichotomous items ($m = 3, 4, \cdots$).

$$\hat{\pi}_0 = (p_{121} + p_{221})(p_{211} + p_{221})(p_{211} + p_{212})/p_{211}p_{221} \ . \quad (A.7)$$

Letting π_{ijk0}^{ABC} denote the probability that an individual will be in the 0th category *and* will be at level (i, j, k) on the joint variable (A, B, C), we see that

$$\hat{\pi}_{ijk0}^{ABC} = \hat{\pi}_0 \hat{\pi}_{i0}^{\bar{A}} \hat{\pi}_{j0}^{\bar{B}} \hat{\pi}_{k0}^{\bar{C}} \ , \quad (A.8)$$

and the formulas corresponding to (1.9) and (A.3) can be expressed[45]

$$\begin{aligned} \hat{\pi}_1 &= p_{111} - \hat{\pi}_{1110}^{ABC} \ , \quad \hat{\pi}_2 = p_{112} - \hat{\pi}_{1120}^{ABC} \ , \\ \hat{\pi}_3 &= p_{122} - \hat{\pi}_{1220}^{ABC} \ , \quad \hat{\pi}_4 = p_{222} - \hat{\pi}_{2220}^{ABC} \ . \end{aligned} \quad (A.9)$$

For the $\hat{\pi}_{ijk0}^{ABC}$ in (A.8), we obtain the explicit formulas[46]

$$\begin{aligned} \hat{\pi}_{ijk0}^{ABC} &= p_{ijk} \ , \quad \text{for} \quad (i, j, k) \in U \ , \\ \hat{\pi}_{1110}^{ABC} &= p_{121}p_{211}/p_{221} \ , \quad \hat{\pi}_{1120}^{ABC} = p_{121}p_{212}/p_{221} \ , \\ \hat{\pi}_{1220}^{ABC} &= p_{121}p_{212}/p_{211} \ , \quad \hat{\pi}_{2220}^{ABC} = p_{221}p_{212}/p_{211} \ . \end{aligned} \quad (A.10)$$

For each pattern (i, j, k) in set S, the preceding formulas show how the corresponding $\hat{\pi}_{ijk0}^{ABC}$ is determined in terms of the p_{ijk} for some patterns (i, j, k) in set U.

In this particular case, since there are zero degrees of freedom for testing the corresponding quasi-independence model (see (1.16) with $m = 3$), this model cannot be tested.[47] Thus, both in the case where $m = 3$ and $m = 2$ this model could not be tested; but all the parameters in the model are identifiable when $m = 3$, and they are not when $m = 2$ (unless additional restrictions are imposed).

[*Received August 1974. Revised June 1975.*]

REFERENCES

[1] Bishop, Y.M.M. and Fienberg, S.E., "Incomplete Two-Dimensional Contingency Tables," *Biometrics*, 25 (March 1969), 119–28.

[2] Goodman, Leo A., "The Analysis of Cross-Classified Data: Independence, Quasi-Independence, and Interactions in Contingency Tables With or Without Missing Entries," *Journal of the American Statistical Association*, 63 (December 1968), 1091–131.

[45] As with (1.9) and (A.3), we shall assume that none of the $\hat{\pi}_t$ in (A.9) are negative. If this is not the case, then a modification of the kind noted in Footnote 38 will be required here.

[46] The explicit formulas for the corresponding $\hat{\pi}_{ijk0}^{ABC}$ can be inserted in (A.9) to obtain explicit formulas for the $\hat{\pi}_t$ (for $t = 1, 2, 3, 4$).

[47] The corresponding chi-square values will be equal to zero, as can be seen from the first line of (A.10).

[3] ———, "On the Measurement of Social Mobility: An Index of Status Persistence," *American Sociological Review*, 34 (December 1969), 831–50.

[4] ———, "The Multivariate Analysis of Qualitative Data: Interactions Among Multiple Classifications," *Journal of the American Statistical Association*, 65 (March 1970), 225–56.

[5] ———, "Exploratory Latent Structure Analysis Using Both Identifiable and Unidentifiable Models," *Biometrika*, 61 Part 2, (1974), 215–31.

[6] ———, "The Analysis of Systems of Qualitative Variables When Some of the Variables Are Unobservable. Part I—A Modified Latent Structure Approach," *Americal Journal of Sociology*, 79 (March 1974), 1179–259.

[7] ———, "The Analysis of Systems of Qualitative Variables When Some of the Variables Are Unobservable. Part II—The Use of Modified Latent Distance Models," Unpublished manuscript.

[8] Guttman, Louis, "The Basis for Scalogram Analysis," in Samuel A. Stouffer *et al.*, eds., *Measurement and Prediction, Studies in Social Psychology in World War II, Vol. IV*, Princeton: University Press, 1950, 60–90.

[9] Hays, David G. and Borgatta, Edgar F., "An Empirical Comparison of Restricted and General Latent Distance Analysis," *Psychometrika*, 19 (December 1954), 271–9.

[10] Lazarsfeld, Paul F., "The Interpretation and Computation of Some Latent Structures," in Samuel A. Stouffer *et al.*, eds., *Measurement and Prediction, Studies in Social Psychology in World War II, Vol. IV*, Princeton: University Press, 1950, 413–72.

[11] ——— and Henry, Neil W., *Latent Structure Analysis*, Boston: Houghton-Mifflin, Co., 1968.

[12] McHugh, R.B., "Efficient Estimation and Local Identification in Latent Class Analysis," *Psychometrika*, 21 (December 1956), 331–47.

[13] Murray, J.R., "Statistical Models for Qualitative Data with Classification Errors," Ph.D. thesis, University of Chicago, 1971.

[14] Proctor, C.H., "A Probabilistic Formulation and Statistical Analysis of Guttman Scaling," *Psychometrika*, 35 (March 1970), 73–8.

[15] Schumacher, C.F., Maxson, G.R. and Martinek, H., "Tests for Creative Ability in Machine Design," Project ONR 458, Armed Services Technical Information Agency 21 284, 1953.

[16] Stouffer, Samuel A., "An Overview of the Contributions to Scaling and Scale Theory," in Samuel A. Stouffer *et al.*, eds., *Measurement and Prediction, Studies in Social Psychology in World War II, Vol. IV*, Princeton: Univeristy Press, 1950, 3–45.

[17] ——— and Toby, Jackson, "Role Conflict and Personality," *American Journal of Sociology*, 56 (March 1951), 395–406. Reprinted in Samuel A. Stouffer, *Social Research to Test Ideas*, New York: The Free Press, 1962. Reprinted in part in Matilda White Riley, *Sociological Research: A Case Approach*, New York: Harcourt, Brace and World, Inc., 1963.

Note: Footnote on page 363 was omitted, but reads as follows:

[1] The meaning of this model, and the interpretation of results obtained with it, will be discussed later. The methods which will be presented here for analyzing this model are somewhat similar to methods developed earlier (using the quasi-independence concept) to analyze the mover-stayer model (see, e.g., [3])

Leo A. Goodman is the Charles L. Hutchinson Distinguished Service Professor, Departments of Statistics and Sociology, University of Chicago, Chicago, Ill. 60637, and research associate at the Population Research Center of the University. This research was supported by Research Contract No. SOC 72-05228 A03 from the Division of the Social Sciences of the National Science Foundation. The author is indebted to D. Andrews, M. Burawoy, C. Clogg, O.D. Duncan, S. Haberman and P.F. Lazarsfeld for helpful comments.

Leo A. Goodman

Reprinted from: the JOURNAL OF THE AMERICAN STATISTICAL ASSOCIATION, December 1975, Vol. 70, No. 352.

Chapter 10

Exploratory Latent-Structure Analysis Using Both Identifiable and Unidentifiable Models

Summary

This paper considers a wide class of latent structure models. These models can serve as possible explanations of the observed relationships among a set of m manifest polytomous variables. The class of models considered here includes both models in which the parameters are identifiable and also models in which the parameters are not. For each of the models considered here, a relatively simple method is presented for calculating the maximum likelihood estimate of the frequencies in the m-way contingency table expected under the model, and for determining whether the parameters in the estimated model are identifiable. In addition, methods are presented for testing whether the model fits the observed data, and for replacing unidentifiable models that fit by identifiable models that fit. Some illustrative applications to data are also included.

Some key words: Contingency tables; Latent structure; Log linear models; Maximum likelihood estimation; Tests of fit.

1. Introduction

This paper deals with the relationships among m polytomous variables, i.e. with the analysis of an m-way contingency table. These m variables are manifest variables in that, for each observed individual in a sample, his class with respect to each of the m variables is observed. We also consider here polytomous variables that are latent in that an individual's class with respect to these variables is not observed. The classes of a latent variable will be called latent classes.

Consider first a 4-way contingency table which cross-classifies a sample of n individuals with respect to four manifest polytomous variables A, B, C and D. If there is, say, some latent dichotomous variable X, so that each of the n individuals is in one of the two latent classes with respect to this variable, and within the tth latent class the manifest variables (A, B, C, D) are mutually independent, then this two-class latent structure would serve as a simple explanation of the observed relationships among the variables in the 4-way contingency table for the n individuals. There is a direct generalization when the latent variable has T classes. We shall present some relatively simple methods for determining whether the observed relationships among the variables in the m-way contingency table can be explained by a T-class structure, or by various modifications and extensions of this latent structure.

To illustrate the methods we analyze Table 1, a 2^4 contingency table presented earlier by Stouffer & Toby (1951, 1962, 1963), which cross-classifies 216 respondents with respect to whether they tend towards universalistic values (+) or particularistic values (−) when confronted by each of four different situations of role conflict. The letters A, B, C and D in

Table 1. *Observed cross-classification of* 216 *respondents with respect to whether they tend toward universalistic* (+) *or particularistic* (−) *values in four situations of role conflict* (*A*, *B*, *C*, *D*)

A	*B*	*C*	*D*	Observed frequency	*A*	*B*	*C*	*D*	Observed frequency
+	+	+	+	42	−	+	+	+	1
+	+	+	−	23	−	+	+	−	4
+	+	−	+	6	−	+	−	+	1
+	+	−	−	25	−	+	−	−	6
+	−	+	+	6	−	−	+	+	2
+	−	+	−	24	−	−	+	−	9
+	−	−	+	7	−	−	−	+	2
+	−	−	−	38	−	−	−	−	20

Table 1 denote the dichotomous responses when confronted by the four different situations. In addition, a second illustrative example in Table 4 below will also be discussed briefly. Our analysis of these data leads to conclusions that are different from those presented earlier.

2. The latent class model unrestricted

Suppose that the manifest polytomous variables A, B, C and D consist of I, J, K and L classes, respectively. Let π_{ijkl} denote the probability that an individual will be at level (i, j, k, l) with respect to the joint variable (A, B, C, D) $(i = 1, \ldots, I; j = 1, \ldots, J; k = 1, \ldots, K; l = 1, \ldots, L)$. Suppose that there is a latent polytomous variable X, consisting of T classes, that can explain the relationships among the manifest variables (A, B, C, D). This means that π_{ijkl} can be expressed as follows:

$$\pi_{ijkl} = \sum_{t=1}^{T} \pi_{ijklt}^{ABCDX}, \tag{1}$$

where

$$\pi_{ijklt}^{ABCDX} = \pi_t^X \pi_{it}^{\bar{A}X} \pi_{jt}^{\bar{B}X} \pi_{kt}^{\bar{C}X} \pi_{lt}^{\bar{D}X} \tag{2}$$

denotes the probability that an individual will be at level (i, j, k, l, t) with respect to the joint variable (A, B, C, D, X). Here π_t^X denotes the probability that an individual will be at level t with respect to variable X; also $\pi_{it}^{\bar{A}X}$ denotes the conditional probability that an individual will be at level i with respect to variable A, given that he is at level t with respect to variable X, and finally $\pi_{jt}^{\bar{B}X}$, $\pi_{kt}^{\bar{C}X}$ and $\pi_{lt}^{\bar{D}X}$ denote similar conditional probabilities. Formula (1) states that the individuals can be classified into T mutually exclusive and exhaustive latent classes, and formula (2) states that within the tth latent class the manifest variables (A, B, C, D) are mutually independent $(t = 1, \ldots, T)$. The meaning of the latent polytomous variable X will be clarified further when particular examples are considered later.

The following elementary formulae, (3)–(8), are required to obtain the subsequent results:

$$\sum_{t=1}^{T} \pi_t^X = 1, \quad \sum_{i=1}^{I} \pi_{it}^{\bar{A}X} = 1, \quad \sum_{j=1}^{J} \pi_{jt}^{\bar{B}X} = 1, \quad \sum_{k=1}^{K} \pi_{kt}^{\bar{C}X} = 1, \; \sum_{l=1}^{L} \pi_{lt}^{\bar{D}X} = 1, \tag{3}$$

$$\pi_t^X = \sum_{i,j,k,l} \pi_{ijklt}^{ABCDX}, \tag{4}$$

$$\pi_t^X \pi_{it}^{\bar{A}X} = \sum_{j,k,l} \pi_{ijklt}^{ABCDX}. \tag{5}$$

Formulae similar to (5) can be obtained for π_t^X multiplied by $\pi_{jt}^{\bar{B}X}$, $\pi_{kt}^{\bar{C}X}$ and $\pi_{lt}^{\bar{D}X}$. In addition, we obtain

$$\pi_{ijklt}^{ABCD\bar{X}} = \pi_{ijklt}^{ABCDX}/\pi_{ijkl}, \tag{6}$$

where $\pi_{ijklt}^{ABCD\bar{X}}$ denotes the conditional probability that an individual is in latent class t, given that he was at level (i, j, k, l) with respect to the joint variable (A, B, C, D). Using (6), we can rewrite (4) and (5) as

$$\pi_t^X = \sum_{i,j,k,l} \pi_{ijkl}\pi_{ijklt}^{ABCD\bar{X}}, \tag{7}$$

$$\pi_{it}^{\bar{A}X} = (\sum_{j,k,l} \pi_{ijkl}\pi_{ijklt}^{ABCD\bar{X}})/\pi_t^X. \tag{8}$$

Formulae similar to (8) can be obtained for $\pi_{jt}^{\bar{B}X}$, $\pi_{kt}^{\bar{C}X}$ and $\pi_{lt}^{\bar{D}X}$. Without loss of generality, we can assume that $\pi_t^X > 0$; we also assume that $\pi_{ijkl} > 0$.

Let circumflexes denote the maximum likelihood estimates of the corresponding parameters in the latent-class model. From (1) and (2), we obtain

$$\hat{\pi}_{ijkl} = \sum_{t=1}^{T} \hat{\pi}_{ijklt}^{ABCDX}, \tag{9}$$

where

$$\hat{\pi}_{ijklt}^{ABCDX} = \hat{\pi}_t^X \hat{\pi}_{it}^{\bar{A}X} \hat{\pi}_{jt}^{\bar{B}X} \hat{\pi}_{kt}^{\bar{C}X} \hat{\pi}_{lt}^{\bar{D}X}; \tag{10}$$

and from (6) we obtain

$$\hat{\pi}_{ijklt}^{ABCD\bar{X}} = \hat{\pi}_{ijklt}^{ABCDX}/\hat{\pi}_{ijkl}. \tag{11}$$

If p_{ijkl} denotes the observed proportion of individuals at level (i, j, k, l) with respect to the joint variable (A, B, C, D), standard methods prove that the maximum likelihood estimates satisfy the following system of equations:

$$\hat{\pi}_t^X = \sum_{i,j,k,l} p_{ijkl}\hat{\pi}_{ijklt}^{ABCD\bar{X}}, \tag{12}$$

$$\hat{\pi}_{it}^{\bar{A}X} = (\sum_{j,k,l} p_{ijkl}\hat{\pi}_{ijklt}^{ABCD\bar{X}})/\hat{\pi}_t^X, \tag{13a}$$

$$\hat{\pi}_{jt}^{\bar{B}X} = (\sum_{i,k,l} p_{ijkl}\hat{\pi}_{ijklt}^{ABCD\bar{X}})/\hat{\pi}_t^X, \tag{13b}$$

$$\hat{\pi}_{kt}^{\bar{C}X} = (\sum_{i,j,l} p_{ijkl}\hat{\pi}_{ijklt}^{ABCD\bar{X}})/\hat{\pi}_t^X, \tag{13c}$$

$$\hat{\pi}_{lt}^{\bar{D}X} = (\sum_{i,j,k,} p_{ijkl}\hat{\pi}_{ijklt}^{ABCD\bar{X}})/\hat{\pi}_t^X. \tag{13d}$$

Compare (12) and (13a) with (7) and (8); recall that $\hat{\pi}_{ijklt}^{ABCD\bar{X}}$ in (12) and (13a)–(13d) was defined by (11), (10) and (9).

Let π denote the vector of parameters $(\pi_t^X, \pi_{it}^{\bar{A}X}, \pi_{jt}^{\bar{B}X}, \pi_{kt}^{\bar{C}X}, \pi_{lt}^{\bar{D}X})$ in the latent-class model, and let $\hat{\pi}$ denote the corresponding maximum likelihood estimate of the vector. To calculate $\hat{\pi}$, we apply the following iterative procedure. Start with an initial trial value for $\hat{\pi}$, say $\pi(0) = \{\pi_t^X(0), \pi_{it}^{\bar{A}X}(0), \pi_{jt}^{\bar{B}X}(0), \pi_{kt}^{\bar{C}X}(0), \pi_{lt}^{\bar{D}X}(0)\}$, which we discuss later. Then use formula (10) to obtain a trial value for $\hat{\pi}_{ijklt}^{ABCDX}$, replacing the terms on the right-hand side of (10) by the corresponding components in $\pi(0)$. We then use (9) to obtain a trial value for $\hat{\pi}_{ijkl}$, replacing the term on the right-hand side of (9) by the corresponding trial value; and we use (11) to obtain a trial value for $\hat{\pi}_{ijklt}^{ABCD\bar{X}}$, replacing the terms on the right-hand side of (11) by the corresponding trial values. Similarly, we use formula (12) to obtain a new trial value for $\hat{\pi}_t^X$, and (13*a*) to (13*d*) to obtain new trial values for $\hat{\pi}_{it}^{\bar{A}X}$, $\hat{\pi}_{jt}^{\bar{B}X}$, $\hat{\pi}_{kt}^{\bar{C}X}$ and $\hat{\pi}_{lt}^{\bar{D}X}$. Having thus obtained a new trial value for the vector $\hat{\pi}$, we repeat the procedure starting with the new trial value using in turn formulae (10), (9), (11), (12) and (13*a*)–(13*d*) to obtain the next trial value for $\hat{\pi}$. In this iterative procedure a latent class is deleted if the corresponding estimate tends to zero. The procedure will converge to a solution to the system of equations and to a corresponding likelihood. For some related but different results, see Haberman (1974).

From (12) and (13*a*)–(13*d*), we find that the components of $\hat{\pi}$ are such that

$$\sum_{t=1}^{T} \hat{\pi}_t^X = 1; \quad \sum_{i=1}^{I} \hat{\pi}_{it}^{\bar{A}X} = 1, \quad \sum_{j=1}^{J} \hat{\pi}_{jt}^{\bar{B}X} = 1, \quad \sum_{k=1}^{K} \hat{\pi}_{kt}^{\bar{C}X} = 1, \quad \sum_{l=1}^{L} \hat{\pi}_{lt}^{\bar{D}X} = 1. \tag{14}$$

Compare (14) with (3). Because of this, we can simplify the iterative procedure described above by estimating only $T-1$ of the π_t^X, say π_t^X for $t = 1, \ldots, T-1$, using (12) and by estimating only $I-1$ of the $\pi_{it}^{\bar{A}X}$, say, $\pi_{it}^{\bar{A}X}$ for $i = 1, \ldots, I-1$, using (13*a*), etc. The estimate of π_T^X can be obtained using (14); similarly for the estimate of $\pi_{It}^{\bar{A}X}$, etc.

The iterative procedure described above can be used when the manifest variables (A, B, C, D) are polytomous as well as in the special case when they are dichotomous. The earlier literature on maximum likelihood estimation in the latent class model (McHugh, 1956, 1958) dealt only with the special dichotomous case. In this case our procedure is easier to apply than are the formulae of McHugh.

We have concentrated on the case of four manifest variables (A, B, C, D). All the methods and results can be extended when there are m manifest variables ($m = 3, 4, \ldots$).

For the m-way contingency table, when the m manifest variables are dichotomous and $T \leqslant \frac{1}{2}(m+1)$, a determinantal method is available for calculating consistent estimates of the vector π of parameters in the latent class model under certain conditions; see Anderson (1954) and Lazarsfeld & Henry (1968, chapter 4).

The estimates so obtained are not asymptotically efficient (Anderson, 1959), except in the special case where $m = 3$ and $T = 2$. Even when the conditions specified in the earlier literature are satisfied, the determinantal method can yield estimates of π that are not permissible, e.g. where one or more of the components in the estimate of π lie outside the interval $[0, 1]$. When the components in this estimate all lie within the interval $(0, 1)$, this estimate can be used as the initial trial value for the maximum likelihood estimate $\hat{\pi}$ in the iterative procedure described earlier in this section.

When the components in the initial trial value for $\hat{\pi}$ all lie within the interval $(0, 1)$, the above iterative procedure will converge and yield a solution to (12) and (13*a*)–(13*d*). This solution will be either the maximum likelihood estimate $\hat{\pi}$, or some other solution to this system of equations, e.g. a terminal maximum, in which one or more of the components are 0 or 1. By trying various initial trial values for $\hat{\pi}$, we can compare the solutions obtained using the iterative procedure to see which solution minimizes the chi-squared statistic based upon the likelihood ratio

$$X^2 = 2 \sum_{i,j,k,l} f_{ijkl} \log (f_{ijkl}/\hat{F}_{ijkl}), \tag{15}$$

where

$$f_{ijkl} = np_{ijkl}, \quad \hat{F}_{ijkl} = n\hat{\pi}_{ijkl}, \tag{16}$$

with $\hat{\pi}_{ijkl}$ obtained from (9). The solution obtained which minimizes (15) yields the maximum likelihood estimate $\hat{\pi}$. For related comments, see Goodman (1974).

The procedure described above calculated $\hat{\pi}_{ijkl}$ from the vector $\hat{\pi}$, using (10) and (9). Having thus obtained $\hat{\pi}_{ijkl}$, we consider next whether the vector $\hat{\pi}$ is uniquely determined by the $\hat{\pi}_{ijkl}$. If $\hat{\pi}$ is uniquely determined, we say it is identifiable. If $\hat{\pi}$ is uniquely determined by the $\hat{\pi}_{ijkl}$ within some neighbourhood of π, we say it is locally identifiable. We now give a useful sufficient condition for local identifiability.

In the earlier literature, identifiability, or local identifiability, was defined with respect to the vector π of parameters in the model for the π_{ijkl} rather than with respect to the corresponding maximum likelihood estimate $\hat{\pi}$ determined by the $\hat{\pi}_{ijkl}$. The method which we shall now present can be used to study whether π is locally identifiable and/or whether $\hat{\pi}$ is locally identifiable. The former problem will be discussed first.

On combining (1) and (2), our model is

$$\pi_{ijkl} = \sum_{t=1}^{T} \pi_t^X \pi_{it}^{\bar{A}X} \pi_{jt}^{\bar{B}X} \pi_{kt}^{\bar{C}X} \pi_{lt}^{\bar{D}X}, \tag{17}$$

Formula (17) describes a set of $IJKL$ functions that transform the parameters

$$(\pi_t^X, \pi_{it}^{\bar{A}X}, \pi_{jt}^{\bar{B}X}, \pi_{kt}^{\bar{C}X}, \pi_{lt}^{\bar{D}X})$$

into the π_{ijkl}. Because of (3), we need consider only $T-1$ of the π_t^X, say,

$$\pi_t^X \quad (t = 1, \ldots, T-1),$$

only $I-1$ of the $\pi_{it}^{\bar{A}X}$, say, $\pi_{it}^{\bar{A}X}$ for $i = 1, \ldots, I-1$, etc. The value of π_T^X can be obtained using (3); similarly for the value of π_{It}^{AX}, etc. Thus we need consider only

$$T-1+(I+J+K+L-4)\,T = (I+J+K+L-3)\,T-1$$

parameters. We shall call this set of parameters a 'basic set'. Similarly, since

$$\sum_{i,j,k,l} \pi_{ijkl} = 1, \tag{18}$$

we need consider only $IJKL-1$ of the π_{ijkl}, say, π_{ijkl} for $(i,j,k,l) \neq (I,J,K,L)$. We shall also call this set of π_{ijkl} a basic set. When

$$IJKL < (I+J+K+L-3)\,T, \tag{19}$$

the number of parameters in the basic set exceeds the corresponding number of π_{ijkl}, and so the parameters will not be identifiable in this case.

Next suppose that (19) is not satisfied, i.e. that the number of parameters in the basic set does not exceed the corresponding number of π_{ijkl}. In this case, for each π_{ijkl} in the basic set, we calculate the derivative of the function π_{ijkl} described by (17) with respect to the parameters in the basic set, thus obtaining a matrix consisting of $IJKL-1$ rows and $(I+J+K+L-3)\,T-1$ columns. For example, in the column pertaining to the derivative with respect to π_t^X,

$$\frac{\partial \pi_{ijkl}}{\partial \pi_t^X} = \pi_{it}^{\bar{A}X} \pi_{jt}^{\bar{B}X} \pi_{kt}^{\bar{C}X} \pi_{lt}^{\bar{D}X} - \pi_{iT}^{\bar{A}X} \pi_{jT}^{\bar{B}X} \pi_{kT}^{\bar{C}X} \pi_{lT}^{\bar{D}X}, \tag{20}$$

for $t = 1, \ldots, T-1$; in the column pertaining to the derivative with respect to $\pi_{st}^{\bar{A}X}$,

$$\frac{\partial \pi_{ijkl}}{\partial \pi_{st}^{\bar{A}X}} = \begin{cases} \pi_t^X \pi_{jt}^{\bar{B}X} \pi_{kt}^{\bar{C}X} \pi_{lt}^{\bar{D}X} & (i = s), \\ -\pi_t^X \pi_{jt}^{\bar{B}X} \pi_{kt}^{\bar{C}X} \pi_{lt}^{\bar{D}X} & (i = I), \\ 0 & \text{otherwise}, \end{cases} \tag{21}$$

for $s = 1, \ldots, I-1$, etc. The second term on the right-hand side of (20), and the term on the second line on the right-hand side of (21), arise since

$$\pi_T^X = 1 - \sum_{t=1}^{T-1} \pi_t^X, \quad \pi_{It}^{\bar{A}X} = 1 - \sum_{i=1}^{I-1} \pi_{it}^{\bar{A}X};$$

see (3). By direct extension of a standard result about Jacobians, the parameters in the model will be locally identifiable if the rank of the matrix described above is equal to the number of columns, i.e. the number of parameters in the basic set. For a corresponding result in the special case where the variables are dichotomous, see McHugh (1956).

By replacing the π's by the corresponding $\hat{\pi}$'s in the preceding two paragraphs, the results can be applied to determine whether the maximum likelihood estimates of the parameters in the model are locally identifiable.

To test the null hypothesis that the T-class latent structure (17) is true, we can use the chi-squared statistic (15). Under this null hypothesis, the asymptotic distribution of the statistic (15) will be chi-squared with degrees of freedom

$$IJKL - 1 - \{(I+J+K+L-3)T - 1\} = IJKL - (I+J+K+L-3)T, \tag{22}$$

when the parameters in the latent structure are locally identifiable. When the parameters in the structure are not locally identifiable, various kinds of restrictions can be imposed upon the parameters in order to make them so. This would lead us to the analysis of restricted latent structures, rather than the unrestricted latent class model considered in the present section. We shall discuss various kinds of restricted structures later (§§4–6), but first we illustrate the application of the above techniques.

3. An example

Table 1 is concerned with whether respondents tend toward universalistic or particularistic values. Stouffer & Toby (1951, 1962, 1963) analyzed these data using a particular 5-class restricted latent structure, and they concluded that the underlying latent variable pertaining to universalistic versus particularistic values could be described by the 5 latent classes. In contrast to this conclusion, we shall show here that a much simpler model is congruent with these data.

Let H_1 denote the 2-class latent structure described in the preceding section, i.e. $T = 2$. To test the hypothesis that H_1 is true, we use the methods in that section to calculate the chi-squared statistic (15). It is 2·720 with 6 degrees of freedom. Thus this simple model fits the data very well indeed.

Table 2 gives both the likelihood ratio chi-squared (15) and the corresponding goodness-of-fit chi-squared for H_1 and for some other latent structures applied to Table 1; Table 3 gives the corresponding estimates obtained with the procedure introduced in §2. Having

Table 2. *Chi-squared values for some latent structures applied to Table* 1

Models	Number of latent classes	Degrees of freedom	Likelihood ratio chi-squared	Goodness-of-fit chi-squared
H_1	2	6	2·720	2·720
H_2, H_2', H_2''	3	2	0·387	0·423
H_3	4	0	0·000	0·000
$H_4, H_4', H_4'', H_4''', H_4^{iv}$	3	5	0·921	0·895
H_5	3	9	2·281	2·282
H_6	4	4	0·870	0·852
H_7	2	8	2·886	2·838
H_8	2	10	4·390	4·339
H_9	3	10	2·391	2·421

obtained such a good fit with H_1, we could stop with this model; but for purposes of comparison and illustration, we shall later consider the other models in Tables 2 and 3.

Model H_1 for Table 1 states that (*a*) there is a single latent dichotomous variable X pertaining to 'universalistic versus particularistic latent values', and (*b*) this latent variable alone can explain the observed relationships among all four manifest variables (A, B, C, D) in the table. From the estimated parameters for H_1 in Table 3, we see that, with respect to the joint manifest variable (A, B, C, D), the 'modal levels' are (1, 1, 1, 1) and (1, 2, 2, 2), for latent classes 1 and 2, respectively; and the latter latent class is modal, since $\hat{\pi}_2^X = 0{\cdot}721$. Thus, individuals in the modal latent class tend to be at manifest level 2, i.e. 'intrinsically' particularistic, except for the level on variable A, and individuals in the nonmodal latent class tend to be at manifest level 1, i.e. 'intrinsically' universalistic. The $\hat{\pi}_t^X$ for H_1 in Table 3 estimate the distribution of the latent variable X; and the other estimated parameters for this model can be used to estimate the effect of variable X upon each manifest variable (A, B, C, D).

The method used in the preceding paragraph to describe H_1 and its latent variable X can be applied in a similar way to the other models in Table 3; for some of these latent structures further insight into their meaning will be obtained by other means as well. We shall consider these latent structures here, in part in order to illustrate various problems that arise when determining whether the parameters in the model are identifiable, when moving from the simple 2-class model H_1 to the T-class models ($T = 3, 4, \ldots$), when moving from unrestricted to restricted latent structures, etc.

With respect to the estimated parameters in the 2-class model H_1, which are maximum likelihood estimates, the fact that they are locally identifiable can be established using either the method presented in the preceding section or a different determinantal method presented in the earlier literature. The latter method provides sufficient conditions for identifiability in the special case where the m manifest variables are dichotomous and $T \leqslant \frac{1}{2}(m+1)$ (see, e.g. Anderson, 1954), and these conditions were extended by Madansky (1960) to the dichotomous case where $T \leqslant 2^{\frac{1}{2}(m-1)}$. However, when considering the 3-class model H_2 in Table 3, the determinantal method in the earlier literature is not applicable, since in this case $T = 3 > 2^{\frac{3}{2}}$. Fortunately, we can apply to H_2 the method presented in the preceding section for studying local identifiability, and we find that H_2 has 14 parameters in its basic set, but the 15×14 matrix described there has rank 13. This model is not identifiable; models H_2' and H_2'' in Table 3 will produce the same estimated expected values $\hat{\pi}_{ijkl}$ as does H_2. Models H_2, H_2' and H_2'' produce the same $\hat{\pi}_{ijkl}$, but they provide different solutions

Table 3. *Estimated parameters in some latent structures applied to Table* 1

Model	Latent class t	$\hat{\pi}_t^X$	$\hat{\pi}_{1t}^{\bar{A}X}$	$\hat{\pi}_{1t}^{\bar{B}X}$	$\hat{\pi}_{1t}^{\bar{C}X}$	$\hat{\pi}_{1t}^{\bar{D}X}$
H_1	1	0·279	0·993	0·940	0·927	0·769
	2	0·721	0.714	0·330	0·354	0·132
H_2	1	0·208	0·997	0·973	0·986	0·884
	2	0·630	0·824	0·459	0·425	0·183
	3	0·162	0·404	0·052	0·256	0·067
H_2'	1	0·220	0·995	0·968	0·976	0·863
	2	0·672	0·806	0·428	0·407	0·170
	3	0·108	0·288	0·000	0·241	0·057
H_2''	1	0·193	0·998	0·980	1·000	0·913
	2	0·581	0·844	0·499	0·448	0·202
	3	0·226	0·481	0·095	0·269	0·075
H_3	1	0·226	0·999	0·965	0·970	0·852
	2	0·638	0·855	0·412	0·398	0·164
	3	0·102	0·189	0·104	0·148	0·034
	4	0·035	0·049	0·260	0·751	0·291
H_4	1	0·257	0·988	0·940	0·948	0·814
	2	0·627	0·812	0·401	0·364	0·136
	3	0·117	0·252	0·062	0·364	0·136
H_4'	1	0·257	0·988	0·940	0·948	0·814
	2	0·137	0·988	0·769	0·364	0·136
	3	0·607	0·664	0·253	0·364	0·136
H_4''	1	0·257	0·988	0·940	0·948	0·814
	2	0·103	0·988	0·940	0·364	0·136
	3	0·641	0·681	0·253	0·364	0·136
H_4'''	1	0·257	0·988	0·940	0·948	0·814
	2	0·091	1·000	1·000	0·364	0·136
	3	0·652	0·685	0·257	0·364	0·136
H_4^{iv}	1	0·257	0·988	0·940	0·948	0·814
	2	0·067	0·000	0·000	0·364	0·136
	3	0·676	0·796	0·383	0·364	0·136
H_5	1	0·175	1·000	1·000	1·000	1·000
	2	0·050	0·000	0·000	0·000	0·000
	3	0·775	0·796	0·420	0·437	0·175
H_6	1	0·253	0·991	0·953	0·945	0·811
	2	0·096	0·991	0·953	0·361	0·132
	3	0·009	0·684	0·256	0·945	0·811
	4	0·641	0·684	0·256	0·361	0·132
H_7	1	0·279	0·993	0·933	0·933	0·771
	2	0·721	0·714	0·342	0·342	0·132
H_8	1	0·231	0·986	0·986	0·986	0·841
	2	0·769	0·732	0·364	0·364	0·159
H_9	1	0·175	1·000	1·000	1·000	1·000
	2	0·050	0·000	0·000	0·000	0·000
	3	0·775	0·796	0·429	0·429	0·175

to the maximum likelihood equations; H_2' and H_2'' provide extreme solutions, whereas H_2 does not. Indeed, there is a one-dimensional continuum of models ranging between H_2' and H_2'' that will yield the same $\hat{\pi}_{ijkl}$. As $\hat{\pi}_1^X$ decreases from its value in H_2' to its value in H_2'', $\hat{\pi}_2^X$ decreases accordingly and all other estimated parameters increase. Note that $\hat{\pi}_{13}^{\bar{B}X} = 0$ in H_2', and $\hat{\pi}_{11}^{\bar{C}X} = 1$ in H_2''. If there is some need to select from among the various models that yield the same $\hat{\pi}_{ijkl}$, this can be done by the introduction of an *a priori* assumption of the kind considered in the following three paragraphs.

If we had assumed *a priori* that $\pi_{13}^{\bar{B}X} = 0$, and had estimated the other parameters in the model from the data, then there would have been 13, rather than 14, parameters in the basic set of estimated parameters, and they would have been identifiable. In the iterative procedure introduced in §2, if the initial trial value for $\hat{\pi}_{13}^{\bar{B}X}$ had been taken as zero, all subsequent values of $\hat{\pi}_{13}^{\bar{B}X}$ obtained by (13*b*) would also be zero. The iterative procedure can be applied directly in this case.

The remarks in the preceding paragraph can be applied also in the case where we assume *a priori* that $\pi_{11}^{\bar{C}X} = 1$, and then estimate the other parameters in the model from the data. More generally, if we assume *a priori* that a given set of conditional probabilities, e.g. $\pi_{j3}^{\bar{B}X}$ for $j = 1, \ldots, J$, or $\pi_{k1}^{\bar{C}X}$ for $k = 1, \ldots, K$, is equal to a specified set of zeros and ones which satisfy the corresponding condition (3), then the iterative procedure described herein can be applied directly in this case, simply by taking the assumed values as the initial trial values for the corresponding estimated parameters in the iterative procedure. Still more generally, if a given set of conditional probabilities, e.g. $\pi_{i1}^{\bar{A}X}$ for $i = 1, \ldots, I$, is assumed known, where the assumed values are not necessarily zeros and ones but they must satisfy the corresponding condition (3), then the iterative procedure would be modified by replacing the corresponding iterative calculation, e.g. (13*a*) for $\hat{\pi}_{i1}^{\bar{A}X}$, by the assumed values of the corresponding parameters at each iteration.

In the analysis of Table 1 using a 3-class latent structure, if we introduce a single restriction of the kind described above, namely that a given conditional probability is equal to a specified number in the closed interval defined by the corresponding values in H_2' and H_2'', say that $\pi_{13}^{\bar{B}X} = 0$ or that $\pi_{11}^{\bar{C}X} = 1$, then the chi-squared statistic (15) yields a value of 0·387 with 2 degrees of freedom as noted in Table 2. The number of degrees of freedom is equal to the number of π_{ijkl} in the basic set minus the corresponding number of estimated parameters in this case.

We have now discussed the case where $T = 2$ and $T = 3$. With respect to the case where $T = 4$, condition (19) is satisfied, since $16 < 5 \times 4$, and so the parameters in the 4-class model will not be identifiable. When the m manifest variables are dichotomous, condition (19) can be replaced by the condition that $2^m < (m+1)T$. Model H_3 in Table 3 is an example of such a model. In the analysis of Table 1, model H_3 yields a chi-squared value of zero, as noted in Table 2. Since H_3 was a 'super-saturated' model, i.e. the number of parameters in the basic set for this model exceeds the corresponding number of π_{ijkl}, the chi-squared statistic will have zero degrees of freedom.

From Table 3 we see that $\hat{\pi}_4^X$ in H_3 is relatively small, as $\hat{\pi}_4^X = 0{\cdot}035$. Since H_3 fits the data perfectly, it would be desirable to consider also 3-class models obtained by the deletion of latent class 4 from H_3. In the estimation procedure of §2 for 3-class models, as the components in the initial trial value for $\hat{\pi}$ we can use the estimated parameters in H_3 modified by the deletion of its fourth latent class, or by the absorption of this latent class in one or more of the other 3 classes. When this is done, we are led to 3-class models of the kind described earlier in this section, e.g. H_2, and also to 3-class models that provide terminal maxima rather than a global maximum.

Initial trial values in our estimation procedure for a 3-class model can be obtained either by (*a*) modifying the estimated parameters in 4-class models as indicated in the preceding paragraph, (*b*) modifying the estimated parameters in 2-class models, e.g. by inserting a third latent class in model H_1, (*c*) trial and error, or (*d*) using as initial values the estimated parameters in 3-class restricted latent structures of the kind discussed in §5, and removing the restrictions from the structures. The above remarks about initial values for 3-class models can be directly extended to T-class models for $T \geqslant 3$.

We have not yet discussed models H_4 to H_9 in Tables 2 and 3. These models will be discussed in §5 as particular examples of the kinds of restricted latent structures considered below.

4. Some restricted latent structures

As noted in §3, if a given set of conditional probabilities, e.g. $\pi_{i1}^{\bar{A}X}$ for $i = 1, \ldots, I$, is assumed known, then the estimation procedure introduced in §2 would be modified accordingly. Similarly, in the case where the set of probabilities π_t^X for $t = 1, \ldots, T$ is assumed known, this estimation procedure would be modified by using the known values of the π_t^X rather than $\hat{\pi}_t^X$ in the calculation of the $\hat{\pi}_{ijklt}^{ABCD\bar{X}}$ at each iteration, see (10), (9) and (11); but in order to ensure condition (14), the denominator on the right-hand side of (13*a*)–(13*d*) remains $\hat{\pi}_t^X$ defined by (12).

The estimation procedure introduced in §2 can also be modified in a straightforward way to accommodate the following kinds of T-class restricted latent structures.

(i) Models in which the following kind of condition is imposed upon the parameters:

$$\pi_{i1}^{\bar{A}X} = \pi_{i2}^{\bar{A}X} \quad (i = 1, \ldots, I). \tag{23}$$

(ii) More generally, models in which the T latent classes can be partitioned into α mutually exclusive and exhaustive subsets $\mathscr{T}_1^A, \ldots, \mathscr{T}_\alpha^A$, where $\alpha \leqslant T$, and/or into β mutually exclusive and exhaustive subsets $\mathscr{T}_1^B, \ldots, \mathscr{T}_\beta^B$, where $\beta \leqslant T$, such that

$$\pi_{it}^{\bar{A}X} = \pi_{it'}^{\bar{A}X} \quad (t, t' \in \mathscr{T}_a^A), \qquad \pi_{jt}^{\bar{B}X} = \pi_{jt'}^{\bar{B}X} \quad (t, t' \in \mathscr{T}_b^B), \tag{24}$$

where $a = 1, \ldots, \alpha$; $b = 1, \ldots, \beta$; $i = 1, \ldots, I$; $j = 1, \ldots, J$. Compare (24) with (23).

(iii) Models in which, in addition to condition (24), the following kind of condition is satisfied for certain specified pairs of subscripts, say (a, b), (a, a^*) and/or (b, b^*):

$$\pi_{it}^{\bar{A}X} = \pi_{jt'}^{\bar{B}X} \quad (t \in \mathscr{T}_a^B, \ t' \in \mathscr{T}_b^B);$$

$$\pi_{it}^{\bar{A}X} = \pi_{i^*t^*}^{\bar{A}X} \quad (t \in \mathscr{T}_a^A, \ t^* \in \mathscr{T}_{a^*}^A); \qquad \pi_{jt}^{\bar{B}X} = \pi_{j^*t^*}^{\bar{B}X} \quad (t \in \mathscr{T}_b^B, \ t^* \in \mathscr{T}_{b^*}^B), \tag{25}$$

where $i = 1, \ldots, I$, where there is a one-to-one correspondence between i and j, between i and i^*, and between j and j^*.

(iv) More generally, models in which the kinds of conditions described by (24) and (25), which were expressed in terms pertaining to variables A and B, are extended to other subsets of the m manifest variables in the m-way contingency table.

The above kinds of restricted latent structures are useful in the analysis of Table 1 and other contingency tables, e.g. Table 4 below, and they will be discussed further in §5 and examples given there.

To determine whether the estimated parameters in a restricted latent structure are locally identifiable, we can use a modified form of the method presented in § 2; see (20) and (21). For example, if restriction (24) is imposed, then the T columns pertaining to the derivative with respect to $\pi_{st}^{\bar{A}X}$ $(t = 1, \ldots, T)$, which we described by (21), would be replaced by α columns, where the ath column $(a = 1, \ldots, \alpha)$ is the sum of the corresponding columns obtained from (21) pertaining to the derivative with respect to $\pi_{st}^{\bar{A}X}$ for $t \in T_a^A$; and a similar kind of replacement would be made pertaining to the derivative with respect to $\pi_{st}^{\bar{B}X}$. If the T classes are partitioned into α, β, γ and δ subsets with respect to variables A, B, C and D, such that conditions corresponding to (24) are satisfied for each of these variables, then the number of columns in the matrix described by (20) and (21) will be reduced from $T-1+(I+J+K+L-4)T$ to $T-1+(I-1)\alpha+(J-1)\beta+(K-1)\gamma+(L-1)\delta$. The parameters in the restricted latent structure will be locally identifiable if the rank of the modified matrix is equal to its number of columns.

Under the null hypothesis that the restricted latent structure is true, if the parameters in the latent structure are locally identifiable, then the asymptotic distribution of the statistic (15) will be chi-squared with degrees of freedom equal to one less than the difference between the number of cells in the 4-way table, or m-way table, and the number of columns in the modified matrix described above. Compare this with (22); the number of degrees of freedom is equal to the number of π_{ijkl} in the basic set minus the corresponding number of independent parameters estimated under the model, when the parameters are locally identifiable.

We shall next describe some simple kinds of restrictions that would perforce make the parameters in the latent structure unidentifiable. Consider first the case where the T-class model is such that

$$\pi_{i1}^{\bar{A}X} = \pi_{i2}^{\bar{A}X}, \quad \pi_{j1}^{\bar{B}X} = \pi_{j2}^{\bar{B}X}, \quad \pi_{k1}^{\bar{C}X} = \pi_{k2}^{\bar{C}X}, \quad \pi_{l1}^{\bar{D}X} = \pi_{l2}^{\bar{D}X}. \tag{26}$$

In this case, formula (17) can be replaced by

$$\pi_{ijkl} = \sum_{t=2}^{T} \Theta_t^X \pi_{it}^{\bar{A}X} \pi_{jt}^{\bar{B}X} \pi_{kt}^{\bar{C}X} \pi_{lt}^{\bar{D}X}, \tag{27}$$

where

$$\Theta_t^X = \begin{cases} \pi_1^X + \pi_2^X & (t = 2), \\ \pi_t^X & (t = 3, \ldots, T). \end{cases} \tag{28}$$

Thus, we can collapse latent classes (1) and (2) to obtain an equivalent latent structure having $T-1$ classes rather than T classes; and the parameters π_1^X and π_2^X will not be identifiable unless additional restrictions, other than condition (3), are imposed upon them.

Consider next the case where the T-class model is such that

$$\pi_{j1}^{\bar{B}X} = \pi_{j2}^{\bar{B}X}, \quad \pi_{k1}^{\bar{C}X} = \pi_{k2}^{\bar{C}X}, \quad \pi_{l1}^{\bar{D}X} = \pi_{l2}^{\bar{D}X}. \tag{29}$$

In this case, formula (17) can be replaced by

$$\pi_{ijkl} = \sum_{t=2}^{T} \Theta_t^X \Theta_{it}^{\bar{A}X} \pi_{jt}^{\bar{B}X} \pi_{kt}^{\bar{C}X} \pi_{lt}^{\bar{D}X}, \tag{30}$$

where Θ_t^X is defined by (28) and

$$\Theta_{it}^{\bar{A}X} = \begin{cases} (\pi_1^X \pi_{i1}^{\bar{A}X} + \pi_2^X \pi_{i2}^{\bar{A}X})/\Theta_2^X & (t = 2), \\ \pi_{it}^{\bar{A}X} & (t = 3, \ldots, T). \end{cases} \tag{31}$$

Thus, here too we can collapse classes (1) and (2); and the parameters $\pi_1^X, \pi_2^X, \pi_{i1}^{\bar{A}X}, \pi_{i2}^{\bar{A}X}$ will not be identifiable unless additional restrictions, other than condition (3), are imposed upon them

Consider next the case where the T-class model is such that

$$\pi_{k1}^{\bar{C}X} = \pi_{k2}^{\bar{C}X}, \quad \pi_{l1}^{\bar{D}X} = \pi_{l2}^{\bar{D}X}. \tag{32}$$

In this case, (17) can be replaced by

$$\pi_{ijkl} = \sum_{t=1}^{T} \Theta_t^X \Theta_{it}^{\bar{A}X} \Theta_{jt}^{\bar{B}X} \pi_{kt}^{\bar{C}X} \pi_{lt}^{\bar{D}X}, \tag{33}$$

where the Θ's are equal to the corresponding π's for $t = 3, \ldots, T$; and where

$$\sum_{t=1}^{2} \Theta_t^X \Theta_{it}^{\bar{A}X} \Theta_{jt}^{\bar{B}X} = \sum_{t=1}^{2} \pi_t^X \pi_{it}^{\bar{A}X} \pi_{jt}^{\bar{B}X}, \tag{34}$$

for $i = 1, \ldots, I; j = 1, \ldots, J$. When $T > 2$, (34) imposes IJ restrictions on the Θ's; and because of (3) the number of Θ's that we need to consider is $2(I+J-2)+2 = 2(I+J-1)$. The $\Theta_{it}^{\bar{A}X}$ and $\Theta_{jt}^{\bar{B}X}$ are required to satisfy the same kind of condition as (3). Thus, the number of restrictions on the Θ's will be less than the number of Θ's that we need to consider whenever $IJ < 2(I+J-1)$, which will be the case whenever variable A or B is dichotomous or when both these variables are trichotomous. In these cases, the Θ's or the corresponding π's will not be identifiable unless additional conditions are imposed upon them.

The remarks in the preceding three paragraphs can be directly generalized to the m-way table. Conditions (26), (29) and (32) can be expressed more generally as follows: the corresponding conditional probabilities in latent classes 1 and 2 are equal for each of the m variables, see (26), or for $m-1$ of these variables, see (29), or for $m-2$ of these variables, see (32). Still more generally, we can consider the case where the corresponding conditional probabilities are equal in a given subset of the T latent classes, for the $m, m-1$, or $m-2$ variables.

5. Some applications of restricted models

We return now to the analysis of Table 1. Examination of the estimated parameters in Table 3 for model H_2, and/or H_2' and H_2'', would lead us to consider a number of restricted models of the kind presented in the preceding section. For example, if we impose the restriction that

$$\pi_{12}^{\bar{D}X} = \pi_{13}^{\bar{D}X}, \quad \pi_{12}^{\bar{C}X} = \pi_{13}^{\bar{C}X}, \tag{35}$$

then the chi-squared statistic (15) yields a value of 0·921. Model H_4 in Table 3 is an example of a restricted model that will yield this chi-squared value. Since (35) is the same kind of restriction as (32), model H_4 will not be identifiable. In this case, the number of Θ's that we need to consider in (34) exceeds by 2 the number of restrictions on the Θ's. If the matrix method presented in §4 for studying local identifiability is applied to H_4, we find that the 15×12 matrix described there has rank 10. Examination of the estimated parameters in H_4, and/or in some other models that yield the same $\hat{\pi}_{ijkl}$, would lead us to consider imposing the additional restriction that

$$\pi_{11}^{\bar{A}X} = \pi_{12}^{\bar{A}X}. \tag{36}$$

Model H_4' in Table 3 is an example of a model that satisfies (35) and (36). Examination of the estimated parameters in H_4' leads us to impose the additional restriction that

$$\pi_{11}^{\bar{B}X} = \pi_{12}^{\bar{B}X}, \tag{37}$$

which yields model H_4''. The estimated parameters in H_4'' are identifiable. For this model, the chi-squared (15) yields a value of 0·921 with 5 degrees of freedom. The $\hat{\pi}_{ijkl}$ for H_4, H_4' and H_4'' are the same.

Model H_4'' is a simple generalization of the 2-class model. In H_4'', latent class 2 is inserted between the other two classes. Latent class 2 is the same as latent class 1 with respect to two variables, A and B, and it is the same as latent class 3 with respect to the other two variables, C and D. Model H_4'' has one more parameter, and one less degree of freedom, than H_1.

If we had limited ourselves to models in which the parameters are identifiable, we could have begun our exploration with H_1, and then considered models of the H_4'' type, i.e. models in which latent class 2 is the same as latent class 1 with respect to two variables, e.g. A and B, or A and C, or A and D, and is the same as latent class 3 with respect to the other two variables.

The method presented in the earlier literature for obtaining consistent estimates of the parameters in H_1 can be extended as follows to obtain consistent estimates of the parameters in H_4'', which can be used as the components in the initial trial value for $\hat{\pi}$, under H_4'', in our maximum likelihood estimation procedure. Under H_4'', the latent structure for the 3-way marginal table $\{ABC\}$, obtained by ignoring variable D, can be expressed as

$$\pi_{ijk} = \sum_{t=1}^{3} \pi_t^X \pi_{it}^{\bar{A}X} \pi_{jt}^{\bar{B}X} \pi_{kt}^{\bar{C}X}, \tag{38}$$

where

$$\pi_{i1}^{\bar{A}X} = \pi_{i2}^{\bar{A}X}, \quad \pi_{j1}^{\bar{B}X} = \pi_{j2}^{\bar{B}X}. \tag{39}$$

Restriction (39) in the 3-way table is the same kind of restriction as (29) in the 4-way table; i.e. the corresponding conditional probabilities in latent classes 1 and 2 are equal for $m-1$ variables in the m-way table. Thus latent classes 1 and 2 can be collapsed to obtain a 2-class model for table $\{ABC\}$. By estimating the parameters in the 2-class model, we obtain consistent estimates of the following parameters in the 3-class model H_4'': π_3^X, $\pi_{13}^{\bar{C}X}$, and also $\pi_{1t}^{\bar{A}X}$ and $\pi_{1t}^{\bar{B}X}$ $(t = 1, 2, 3)$. A similar analysis of the 3-way marginal table $\{BCD\}$ yields consistent estimates of π_1^X, $\pi_{11}^{\bar{B}X}$, and also $\pi_{1t}^{\bar{C}X}$ and $\pi_{1t}^{\bar{D}X}$ $(t = 1, 2, 3)$. The consistent estimates thus obtained from $\{ABC\}$ and $\{BCD\}$ can be used to provide consistent estimates for all of the parameters in the 3-class model H_4'' for the 4-way table $\{ABCD\}$. Instead of using $\{ABC\}$ we could have used $\{ABD\}$; similarly, instead of $\{BCD\}$ we could have used $\{ACD\}$.

The method described in the preceding paragraph can be directly extended to more general restricted models for the m-way table (Goodman, 1974).

We arrived at the identifiable model H_4'' considered above by imposing restrictions (36) and (37) upon H_4. Other kinds of restrictions could have been imposed upon H_4 to obtain other identifiable models that would leave $\hat{\pi}_{ijkl}$ unchanged. See, for example, models H_4''' and H_4^{iv} in Table 3, which impose upon H_4 restrictions (40a) and (40b), respectively, for

$t = 2$. In addition to condition (35) imposed by H_4, the imposition of any one of the following restrictions would still leave $\hat{\pi}_{ijkl}$ unchanged

$$(\pi_{1t}^{\bar{A}X}, \pi_{1t}^{\bar{B}X}) = (1, 1) \quad (t = 2 \quad \text{or} \quad t = 3), \tag{40a}$$

$$(\pi_{1t}^{\bar{A}X}, \pi_{1t}^{\bar{B}X}) = (0, 0) \quad (t = 2 \quad \text{or} \quad t = 3), \tag{40b}$$

$$(\pi_{1t'}^{\bar{A}X}, \pi_{1t''}^{\bar{A}X}) = (1, 0) \quad ((t', t'') = (2, 3) \quad \text{or} \quad (3, 2)), \tag{40c}$$

$$(\pi_{1t'}^{\bar{B}X}, \pi_{1t''}^{\bar{B}X}) = (1, 0) \quad ((t', t'') = (2, 3) \quad \text{or} \quad (3, 2)). \tag{40d}$$

More generally, any set of Θ's that satisfy (34), and that are in the closed interval [0, 1], would leave $\hat{\pi}_{ijkl}$ unchanged; and these Θ's can be used by assuming that $\pi_{12}^{\bar{A}X}$ and $\pi_{12}^{\bar{B}X}$, or any two conditional probabilities from among the $\pi_{1t}^{\bar{A}X}$ and $\pi_{1t}^{\bar{B}X}$, for $t = 2, 3$, would be equal to the corresponding two Θ's.

Model H_4'' had four restrictions imposed upon it, see (35)–(37), and the removal of any one of the four restrictions, e.g. (37) as in H_4', would leave $\hat{\pi}_{ijkl}$ unchanged. Furthermore, the removal of two particular restrictions, namely (36) and (37) as in H_4, or the two restrictions included in (35) would also leave $\hat{\pi}_{ijkl}$ unchanged. If restrictions (36) and (37) are imposed but the two restrictions in (35) are not, then the imposition of any one of the sets of restrictions corresponding to (40a)–(40d), with A and B replaced by C and D in these formulae, would also leave $\hat{\pi}_{ijkl}$ unchanged. A generalization corresponding to the one at the end of the preceding paragraph can be made here too.

Examination of the estimated parameters in H_4^{iv} leads us to consider next model H_5 in Table 3, in which the following conditions are imposed upon the 3-class latent-class model

$$\pi_{11}^{\bar{A}X} = \pi_{11}^{\bar{B}X} = \pi_{11}^{\bar{C}X} = \pi_{11}^{\bar{D}X} = 1, \quad \pi_{12}^{\bar{A}X} = \pi_{12}^{\bar{B}X} = \pi_{12}^{\bar{C}X} = \pi_{12}^{\bar{D}X} = 0. \tag{41}$$

Examination of the estimated parameters in either H_2, H_2', H_2'' or H_4, or a more direct examination of Table 1 could also lead to consideration of a model equivalent to H_5, with labelling of latent classes 2 and 3 in H_5 interchanged. Model H_5 states that, with respect to the joint variable (A, B, C, D), individuals in latent classes 1 and 2 will be at levels (1, 1, 1, 1) and (2, 2, 2, 2), respectively, with probability one; and as usual the manifest variables (A, B, C, D) are mutually independent for the individuals in latent class 3. Thus, when levels (1, 1, 1, 1) and (2, 2, 2, 2) are deleted from the 4-way table $\{ABCD\}$, the manifest variables (A, B, C, D) will be quasi-independent under H_5; see, e.g., Goodman (1968). We could have arrived at H_5 by carrying out an analysis of quasi-independence in Table 1 with levels (1, 1, 1, 1) and (2, 2, 2, 2) deleted, followed by the insertion of latent classes 1 and 2 to account for the observed frequencies at levels (1, 1, 1, 1) and (2, 2, 2, 2), respectively.

We consider next model H_6 in Table 3, in which the following conditions are imposed upon the 4-class latent class model

$$\begin{aligned} &\pi_{11}^{\bar{A}X} = \pi_{12}^{\bar{A}X}, \quad \pi_{13}^{\bar{A}X} = \pi_{14}^{\bar{A}X}, \quad \pi_{11}^{\bar{B}X} = \pi_{12}^{\bar{B}X}, \quad \pi_{13}^{\bar{B}X} = \pi_{14}^{\bar{B}X}, \\ &\pi_{11}^{\bar{C}X} = \pi_{13}^{\bar{C}X}, \quad \pi_{12}^{\bar{C}X} = \pi_{14}^{\bar{C}X}, \quad \pi_{11}^{\bar{D}X} = \pi_{13}^{\bar{D}X}, \quad \pi_{12}^{\bar{D}X} = \pi_{14}^{\bar{D}X}. \end{aligned} \tag{42}$$

This model is an extension of H_1 and H_4''. In H_6, latent classes 2 and 3 are inserted between the two latent classes of H_1 which we now call, for convenience, latent classes 1 and 4, with latent class 2 the same as latent class1 with respect to two variables, A and B, and the same as latent class 4 with respect to the other two variables, C and D, and with latent class 3 the same as latent class 4 with respect to the former two variables, A and B, and the same as latent class 1 with respect to the latter two variables, C and D. Model H_6 has two more parameters and two less degrees of freedom than H_1, and one more parameter and one less degree of freedom than H_4''.

We can view latent variable X in H_6 as the joint latent variable (Y, Z), where latent variables Y and Z are dichotomous, and where latent level t in H_6 ($t = 1, 2, 3, 4$) describes the joint latent level (r, s) with respect to (Y, Z), with $(r, s) = (1, 1), (1, 2), (2, 1), (2,2)$ corresponding to $t = 1, 2, 3, 4$, respectively. Under H_6 in Table 3, variables A and B are affected by the level of latent variable Y but not Z, and variables C and D are affected by the level of latent variable Z but not Y. For further insight into the meaning of H_6 and its latent variables (Y, Z), see the corresponding interpretation of model II, a model of the H_6 type, in §6 below.

Having indicated above with H_6 how to obtain a latent structure containing two latent variables, Y and Z, we can use similar methods to obtain structures containing more than two latent variables. Model H_6, and models similar to it, will be found useful in the analysis of many m-way tables; see e.g., §6 below.

When we introduced H_6 above, it was first described as an extension of H_1, and we introduced H_4'' still earlier in a similar way. The various remarks following the introduction of H_4'' can be directly extended to H_6. As an example of a different identifiable model that yields the same $\hat{\pi}_{ijkl}$ as H_6, we can consider the 2-class model applied to the 3-way table consisting of variables A, B, and the joint variable (C, D), or variables C, D, and the joint variable (A, B), under certain conditions which we omit here to save space (Goodman, 1974).

Model H_6, and some of the other models considered earlier in the section, provide examples of latent structures that impose the kinds of conditions described by (24) expressed in terms pertaining to various subsets of the manifest variables. As examples of models that impose the kinds of conditions described by (25), we consider next H_7, H_8 and H_9 in Table 3, which impose conditions (43), (44) and (45), respectively, upon models H_1, H_7 and H_5, respectively.

$$\pi_{11}^{\bar{B}X} = \pi_{11}^{\bar{C}X}, \quad \pi_{12}^{\bar{B}X} = \pi_{12}^{\bar{C}X}, \tag{43}$$

$$\pi_{11}^{\bar{A}X} = \pi_{11}^{\bar{B}X}, \quad \pi_{11}^{\bar{D}X} = \pi_{22}^{\bar{D}X}, \tag{44}$$

$$\pi_{13}^{\bar{B}X} = \pi_{13}^{\bar{C}X}. \tag{45}$$

Models H_7 and H_8 can be interpreted in the same way as H_1, but there are fewer independent parameters in the former models. Similarly, model H_9 can be interpreted in the same way as H_5, but there are fewer independent parameters in H_9. Note that H_7 and H_9 state that both variables B and C are affected in the same way by the latent variable X; and H_8 states other things as well; see (44). From Table 2 we see that these models fit the data very well indeed.

6. Another example

Table 4 (Coleman, 1964, p. 171) cross-classifies schoolboys interviewed at two successive points in time. Variable A denotes self-perceived membership in the 'leading crowd', in it or out of it, at the time of the first interview, variable B denotes attitude concerning the leading crowd, favourable or unfavourable, expressed at the first interview and variables C and D denote the corresponding membership and attitude at the second interview. For brevity, we consider here only two models: (I) model H_1 in which there is one latent dichotomous variable, and (II) a model of the H_6 type in which there are two latent dichotomous variables, say V and W, where latent variable V can affect manifest variables A and C, and latent variable W can affect manifest variables B and D. The chi-squared values for models I and II are given in Table 5, together with the corresponding estimated parameters. The improvement in fit obtained with model II is dramatic.

Table 4. *Observed cross-classification of 3398 schoolboys, in interviews at two successive points in time, with respect to two dichotomous variables:* (1) *self-perceived membership in 'leading crowd', and* (2) *favourableness of attitude concerning the 'leading crowd'*

First interview		Second interview			
	Membership	+	+	−	−
	Attitude	+	−	+	−
Membership	Attitude				
+	+	458	140	110	49
+	−	171	182	56	87
−	+	184	75	531	281
−	−	85	97	338	554

Table 5. *Chi-squared values for two latent structures applied to Table 4 and the corresponding estimated parameters**

Model	Number of latent variables	Degrees of freedom	Likelihood ratio chi-squared	Goodness-of-fit chi-squared
I	1	6	249·502	251·171
II	2	4	1·270	1·281

Model	Latent class t	$\hat{\pi}_t^X$	$\hat{\pi}_{1t}^{\bar{A}X}$	$\hat{\pi}_{1t}^{\bar{B}X}$	$\hat{\pi}_{1t}^{\bar{C}X}$	$\hat{\pi}_{1t}^{DX}$
I	1	0·401	0·769	0·645	0·889	0·674
	2	0·599	0·101	0·467	0·090	0·499
II	1	0·272	0·754	0·806	0·910	0·832
	2	0·128	0·754	0·267	0·910	0·302
	3	0·231	0·111	0·806	0·076	0·832
	4	0·368	0·111	0·267	0·076	0·302

* For model II, the symbol X denotes the joint latent variable (V, W), and the four latent classes correspond to the four levels with respect to this joint variable; i.e. (1, 1), (1, 2), (2, 1) and (2, 2), respectively.

Model II for Table 4 states that (*a*) there are two latent dichotomous variables, V and W, pertaining to latent self-perceived membership in the leading crowd and latent attitude concerning the leading crowd, respectively; and (*b*) these two latent variables alone can explain the observed relationships among the manifest variables (A, B, C, D). The relationship between the two latent variables can be estimated using the $\hat{\pi}_t^X$ for model II in Table 5; the other estimated parameters for this model can be used to estimate the effects of each latent variable upon the corresponding manifest variables. For a comparison of this model with the models presented by the present author in his earlier article (1973) analyzing Table 4 above, see Goodman (1974).

The research was partly supported by the Division of the Social Sciences, National Science Foundation. For helpful comments the author is indebted to D. F. Andrews, O. D. Duncan, S. Haberman and A. Madansky.

References

Anderson, T. W. (1954). On estimation of parameters in latent structure analysis. *Psychometrika* **19**, 1–10.

Anderson, T. W. (1959). Some scaling models and estimation procedures in the latent class model. In *Probability and Statistics*, Ed. U. Grenander, pp. 9–38. New York: Wiley.

Coleman, J. S. (1964). *Introduction to Mathematical Sociology*. Glencoe, Illinois: Free Press.

Goodman, L. A. (1968). The analysis of cross-classified data: Independence, quasi-independence, and interactions in contingency tables with or without missing entries. *J. Am. Statist. Assoc.* **63**, 1091–131.

Goodman, L. A. (1973). The analysis of multidimensional contingency tables when some variables are posterior to others: A modified path analysis approach. *Biometrika* **60**, 179–92.

Goodman, L. A. (1974). The analysis of systems of qualitative variables when some of the variables are unobservable. I. A modified latent structure approach. *Am. J. Sociol.* **79**, 1179–259.

Haberman, S. H. (1974). Log-linear models for tables derived by indirect observation: Maximum likelihood equations. *Ann. Statist.* **2**. To appear.

Lazarsfeld, P. F. & Henry, N. W. (1968). *Latent Structure Analysis*. Boston: Houghton Mifflin.

Madansky, A. (1960). Determinantal methods in latent class analysis. *Psychometrika* **25**, 183–98.

McHugh, R. B. (1956). Efficient estimation and local identification in latent class analysis. *Psychometrika* **21**, 331–47.

McHugh, R. B. (1958). Note on 'Efficient estimation and local identification in latent class analysis'. *Psychometrika* **23**, 273–4.

Stouffer, S. A. & Toby, J. (1951). Role conflict and personality. *Am. J. Sociol.* **56**, 395–406. Reprinted in S. A. Stouffer (1962), *Social Research to Test Ideas* (New York: Free Press). Reprinted in part in R. M. Riley (1963), *Sociological Research: A Case Approach* (New York: Harcourt, Brace and World).

[*Received May* 1973. *Revised December* 1973]

Leo A. Goodman

Reprinted from: BIOMETRIKA (1974), Vol. 61, No. 2, pp. 215-231. Used with permission.

Part Five

Some Extensions to the Goodman System

The approach for the analysis of qualitative data presented in the preceding chapters includes methods and techniques that can be extended in many directions. In this section we include two such extensions.

Causal Systems of Qualitative Variables

Consider a logit model that specifies that the polytomous variables A, B, and C are causally prior to the polytomous variable D. Simple recursive models of this type were discussed in detail in part I. The model is recursive since the causal linkages run only one way (Duncan, 1976). Suppose that we also impose an additional recursive structure on the explanatory variables specifying that A and B are causally prior to C. We then have a recursive system containing two nested logit models. Similarly, we can conceive of more complex recursive systems containing any number of nested logit models.

In chapter 11 Goodman presents methods for generalizing the analysis of simple logit models to more complex systems of qualitative variables, both recursive and nonrecursive systems. For each hypothesized system he presents methods for (1) estimating the parameters in the system, (2) testing whether the hypothesized system fits the data, and (3) partitioning the test statistic into components that can be used to test individual logit models or other subsystems within the overall system. An application is included to illustrate the methods. This chapter develops and extends some of the concepts and methods introduced in chapter 6.

Multiple Populations

Researchers often obtain data from different populations (schools, regions, experimentals versus controls). The analysis of such data raises questions about the degree of homogeneity among the populations and the validity of pooling the data. In chapter 12 Goodman presents methods for analyzing the homogeneity and heterogeneity of data cross-classified for T populations. A threefold classification of models is introduced: (1) models that assume "complete homogeneity" among the tables, (2) models that allow "complete heterogeneity" among the tables, and (3) models that allow "partial heterogeneity." In this chapter Goodman also extends the stepwise procedures presented in chapter 5. "Guided" and "unguided" selection methods and "multidirectional" methods are introduced. To illustrate these methods he analyzes a set of two four-way tables.

Chapter 11

The Analysis of Multidimensional Contingency Tables When Some Variables Are Posterior to Others: A Modified Path Analysis Approach

Models and methods for analyzing the relations among a set of polytomous variables, when some of the variables are posterior to others, are presented. The techniques proposed here yield 'path diagrams' that are somewhat analogous to those used in path analysis. Earlier path analysis models are not suited to the case where the variables are polytomous, or dichotomous, but the models proposed herein are. For each of our models, methods are presented for (*a*) testing whether the model fits the data, (*b*) partitioning the test statistic into components that can be used to test submodels within the overall model, and (*c*) estimating the parameters in the model. An illustrative application is also included.

Some key words: Multidimensional contingency tables; Log linear models; Logistic models; System of simultaneous logistic models; Path analysis for polytomous variables; Maximum likelihood estimation: Tests of fit of models.

1. Introduction

For simplicity, consider first a 2^4 contingency table, where the four dimensions pertain to variables A, B, C and D. The usual terminology distinguishes between (*a*) response variables, and (*b*) explanatory or regressor variables, the variables of the second type being used to explain or predict variation in variables of the first type; see Cox (1970, pp. 1–2). In the present article, we shall consider, for example, (i) the case where variables A and B are used to explain variation in C, and variables A, B and C are used to explain variation in D; and (ii) the case where variables A, B and C are used to explain variation in D, and variables A, B and D are used to explain variation in C. Contrary to the usual way of analyzing a 2^4 table, where each variable is either a response or explanatory variable, in case (i) above variable C is both a response variable, whose variation is explained by A and B, and an explanatory variable, used to explain variation in D; in case (ii) variables C and D are both response and explanatory variables.

Case (i) will be of interest, for example, when variable C is measured posterior in time to variables A and B, and variable D is measured posterior in time to variable C; or when variables C and D are measured posterior in time to variables A and B, and variable C, together with A and B, is used to predict variation in D. Case (ii) will be of interest, for example, when variables C and D are measured posterior in time to variables A and B, and variable C, together with A and B, is used to predict variation in D, and variable D, together with A and B, is used to predict variation in C.

To illustrate the analysis of cases (i) and (ii), we shall consider Table 1, a 2^4 table presented by Coleman (1964, p. 171). This table cross-classifies schoolboys interviewed at two successive points in time, where variable A denotes membership in the leading crowd, in it or out of it, at the time of the first interview, B denotes attitude concerning the leading crowd, favourable or unfavourable, expressed at the first interview; C denotes membership in the leading crowd at the time of the second interview; and D denotes attitude concerning the leading crowd expressed at the second interview. In this table, variables A and B were measured prior in time to C and D, but our methods of analysis can be applied also when variables A and B are prior in some other sense, for example, when A and B are factors affecting C and D, or when they are explanatory or regressor variables.

Table 1. *Observed cross-classification of* 3398 *schoolboys, in interviews at two successive points in time, with respect to two dichotomous variables:* (1) *membership in 'leading crowd', and* (2) *favourableness of attitude concerning the 'leading crowd'*

First interview		Second interview			
	Membership	+	+	−	−
	Attitude	+	−	+	−
Membership	Attitude				
+	+	458	140	110	49
+	−	171	182	56	87
−	+	184	75	531	281
−	−	85	97	338	554

In analyzing a table of the same form as Table 1, *viz.* a turnover table, Anderson (1954) considered the conditional probability $\pi_{ijkl}^{AB\bar{C}\bar{D}}$ that the joint variable (C, D) is at level (k, l), given that the joint variable (A, B) is at level (i, j) $(i = 1, 2; j = 1, 2; k = 1, 2; l = 1, 2)$. He showed how to test the null hypothesis H that

$$\pi_{ijkl}^{AB\bar{C}\bar{D}} = \pi_{ik}^{A\bar{C}} \pi_{jl}^{B\bar{D}}, \tag{1}$$

where $\pi_{ik}^{A\bar{C}}$ is the conditional probability that variable C is at level k, given that variable A is at level i, and $\pi_{jl}^{B\bar{D}}$ is defined similarly. The null hypothesis H can be expressed in terms of the following hypotheses:

$$\pi_{ijk}^{AB\bar{C}} = \pi_{ik}^{A\bar{C}}, \tag{1a}$$

$$\pi_{ijkl}^{ABC\bar{D}} = \pi_{jl}^{B\bar{D}}, \tag{1b}$$

where $\pi_{ijk}^{AB\bar{C}}$ is the conditional probability that variable C is at level k, given that the joint variable (A, B) is at level (i, j), and $\pi_{ijkl}^{ABC\bar{D}}$ is the conditional probability that variable D is at level l, given that the joint variable (A, B, C) is at level (i, j, k). Letting H_a and H_b denote hypotheses $(1a)$ and $(1b)$, respectively, we see that H is equivalent to the hypothesis that both H_a and H_b are true, i.e. $H \equiv H_a \cap H_b$.

Under H_a, when variables A and B are used to explain variation in C, only variable A is actually used in the explanation. Under H_b, when variable A, B and C are used to explain variation in D, only variable B is actually used in the explanation. Thus, hypothesis H is a particular model that may be of interest in case (i) described earlier herein. In the present article, we shall present a wide class of models, which can be viewed as a generalization of hypothesis H, and which can be used in cases (i) and (ii) and more generally.

The class of models presented herein can also be viewed as a generalization of both the Markov-type models considered by Goodman (1970, pp. 240–1; 1971a) and S. Haberman, in his unpublished Ph.D. thesis, and the logistic models considered, for example, by Dyke & Patterson (1952), Cox & Lauh (1967), Bishop (1969) and Goodman (1970). For each of the models introduced, we shall show how to estimate the expected frequencies under the model, and we shall provide a chi-squared statistic for testing whether the model fits the data. This chi-squared statistic for the overall model can be partitioned into components that can be used to test separately the submodels within the overall model.

The models and methods presented herein are not limited to the 2^4 contingency table. They can be applied more generally to m-way contingency tables ($m = 3, 4, \ldots$) pertaining to polytomous variables.

2. Some simple examples

We shall first consider case (i) of the preceding section. For the 2^4 contingency table $\{ABCD\}$, let f_{ijkl} denote the observed frequency in cell (i,j,k,l) of the table ($i = 1, 2; j = 1, 2; k = 1, 2; l = 1, 2;$) and let f_{ijk}^{ABC} denote the observed frequency in cell (i,j,k) of the 2^3 marginal table $\{ABC\}$ formed from table $\{ABCD\}$ by ignoring variable D. Consider the conditional probability $\pi_{ijk}^{AB\bar{C}}$ that variable C is at level k, given that the joint variable (A, B) is at level (i,j); and the conditional probability $\pi_{ijkl}^{ABC\bar{D}}$ that variable D is at level l, given that the joint variable (A, B, C) is at level (i,j,k). Let $\Omega_{ij}^{AB\bar{C}}$ and $\Omega_{ijk}^{ABC\bar{D}}$ denote the corresponding conditional odds defined by

$$\Omega_{ij}^{AB\bar{C}} = \pi_{ij1}^{AB\bar{C}} / \pi_{ij2}^{AB\bar{C}}, \tag{2}$$

$$\Omega_{ijk}^{ABC\bar{D}} = \pi_{ijk1}^{ABC\bar{D}} / \pi_{ijk2}^{ABC\bar{D}}. \tag{3}$$

With respect to $\{ABC\}$, let H^{ABC} denote, say, the usual logistic model in which $\log \Omega_{ij}^{AB\bar{C}}$ is expressed as a linear function of the general mean and the main effects of variables A and B; and with respect to $\{ABCD\}$, let H^{ABCD} denote, say, the corresponding model in which $\log \Omega_{ijk}^{ABC\bar{D}}$ is expressed as a linear function of the general mean and the main effects of variables A, B and C. In other words, under H^{ABC}, we have

$$\Omega_{ij}^{AB\bar{C}} = \gamma^{\bar{C}} \gamma_i^{A\bar{C}} \gamma_j^{B\bar{C}}; \tag{4}$$

and under H^{ABCD}, we have

$$\Omega_{ijk}^{ABC\bar{D}} = \gamma^{\bar{D}} \gamma_i^{A\bar{D}} \gamma_j^{B\bar{D}} \gamma_k^{C\bar{D}}, \tag{5}$$

where

$$\prod_i \gamma_i^{A\bar{C}} = \prod_j \gamma_j^{B\bar{C}} = 1, \tag{6}$$

$$\prod_i \gamma_i^{A\bar{D}} = \prod_j \gamma_j^{B\bar{D}} = \prod_k \gamma_k^{C\bar{D}} = 1. \tag{7}$$

It will be simpler here to express the logistic models in the above multiplicative form via the odds, rather than in their usual additive form via the log odds. The parameters $\gamma_i^{A\bar{C}}$ and $\gamma_j^{B\bar{C}}$ in (4), and $\gamma_i^{A\bar{D}}$, $\gamma_j^{B\bar{D}}$ and $\gamma_k^{C\bar{D}}$ in (5), are the multiplicative main effects on $\Omega_{ij}^{AB\bar{C}}$ under H^{ABC} and on $\Omega_{ijk}^{ABC\bar{D}}$ under H^{ABCD}, respectively.

Let F_{ijk}^{ABC} denote the expected frequency in cell (i,j,k) of table $\{ABC\}$ under H^{ABC}, and let F_{ijkl}^{ABCD} denote the expected frequency in cell (i,j,k,l) of table $\{ABCD\}$ under H^{ABCD}. Under H^{ABC}, the observed frequencies in the 2^2 marginal table $\{AB\}$, formed from table $\{ABC\}$ by ignoring variable C, can be viewed as given; and under H^{ABCD}, the observed frequencies in table $\{ABC\}$ can be viewed as given. Model H^{ABC}, which we described by (4), can also be described in the following equivalent form:

$$F_{ijk}^{ABC} = \eta_{ij}^{AB} \tau_k^{C} \tau_{ik}^{AC} \tau_{jk}^{BC}, \tag{8}$$

where

$$\prod_k \tau_k^C = 1, \quad \prod_i \tau_{ik}^{AC} = \prod_k \tau_{ik}^{AC} = 1, \ldots \tag{9}$$

Models (4) and (8) are equivalent to the model of zero 3-factor interaction in $\{ABC\}$; see, for example, Bishop (1969) and Goodman (1970). Similarly, model H^{ABCD}, which we described by (5), can also be described in the following equivalent form:

$$F_{ijkl}^{ABCD} = \eta_{ijk}^{ABC} \tau_l^D \tau_{il}^{AD} \tau_{jl}^{BD} \tau_{kl}^{CD}, \tag{10}$$

where

$$\prod_l \tau_l^D = 1, \quad \prod_i \tau_{il}^{AD} = \prod_l \tau_{il}^{AD} = 1, \ldots \tag{11}$$

Let f_{ij}^{AB} denote the observed frequency in cell (i, j) of the 2^2 marginal table $\{AB\}$ formed from $\{ABC\}$ by ignoring variable C. Since table $\{AB\}$ can be viewed as given under model H^{ABC}, the η's in (8) are such that

$$\sum_k F_{ijk}^{ABC} = f_{ij}^{AB}. \tag{12}$$

Similarly, since table $\{ABC\}$ can be viewed as given under model H^{ABCD}, the η's in (10) are such that

$$\sum_l F_{ijkl}^{ABCD} = f_{ijk}^{ABC}. \tag{13}$$

Consider now the hypothesis H^* that both (4) and (5) are true, i.e.

$$H^* \equiv H^{ABC} \cap H^{ABCD}.$$

Letting F_{ijkl}^* denote the expected frequency in cell (i, j, k, l) under H^*, we find that

$$F_{ijkl}^* = F_{ijk}^{ABC} F_{ijkl}^{ABCD} / f_{ijk}^{ABC}; \tag{14}$$

see Appendix. From (8) and (10) we see that F_{ijkl}^* can also be expressed as

$$F_{ijkl}^* = \eta_{ij}^{AB} \tau_k^C \tau_{ik}^{AC} \tau_{jk}^{BC} \tau_l^D \tau_{il}^{AD} \tau_{jl}^{BD} \tau_{kl}^{CD} / \sigma_{ijk}^{ABC}, \tag{15}$$

where

$$\sigma_{ijk}^{ABC} = \sum_l \tau_l^D \tau_{il}^{AD} \tau_{jl}^{BD} \tau_{kl}^{CD}. \tag{16}$$

Although models H^{ABC} and H^{ABCD}, as described by (8) and (10), are examples of log linear models for contingency tables (see, for example, Goodman, 1970, Tables 3 and 4), model H^* for table $\{ABCD\}$ is not of this kind. As far as I know, models of the H^* type have not been considered in the earlier literature on the 4-way table.

To calculate the maximum likelihood estimate $\hat{F}_{ijkl}^*$ of F_{ijkl}^* under H^*, we use formula (14), with F_{ijk}^{ABC} and F_{ijkl}^{ABCD} replaced by their maximum likelihood estimates $\hat{F}_{ijk}^{ABC}$ and $\hat{F}_{ijkl}^{ABCD}$, respectively, which can be calculated by the usual iterative-scaling method (see, for example, Bishop, 1969 and Goodman, 1970) or by the usual methods of logit analysis. To test the null hypothesis H^*, we calculate the likelihood ratio chi-squared statistic

$$\chi^2(H^*) = 2 \sum_{i,j,k,l} f_{ijkl} \log (f_{ijkl} / \hat{F}_{ijkl}^*). \tag{17}$$

From (14) and (17), we see that $\chi^2(H^*)$ can be partitioned as

$$\chi^2(H^*) = \chi^2(H^{ABC}) + \chi^2(H^{ABCD}), \tag{18}$$

where

$$\chi^2(H^{ABC}) = 2 \sum_{i,j,k} f_{ijk}^{ABC} \log (f_{ijk}^{ABC} / \hat{F}_{ijk}^{ABC}), \tag{19}$$

$$\chi^2(H^{ABCD}) = 2 \sum_{i,j,k,l} f_{ijkl} \log (f_{ijkl} / \hat{F}_{ijkl}^{ABCD}). \tag{20}$$

The degrees of freedom corresponding to $\chi^2(H^*)$ will be the sum of the degrees of freedom corresponding to $\chi^2(H^{ABC})$ and $\chi^2(H^{ABCD})$.

The results presented above can be directly generalized to the case where model H^{ABC} for table $\{ABC\}$ is any 3-way log linear model, and where model H^{ABCD} for table $\{ABCD\}$ is any 4-way log linear model that includes table $\{ABC\}$ in the set of fitted marginals (see, for example, Goodman, 1970). In this case, with model H^* defined as $H^{ABC} \cap H^{ABCD}$, we can analyze both table $\{ABCD\}$ [see (14) and (17)] and the marginal table $\{ABC\}$.

The point of view presented above will help to clarify and simplify some earlier results by Birch (1963) for the 3-way table $\{ABC\}$. Considering the particular hypothesis H^{AB} that variables A and B are independent in the 2-way marginal table $\{AB\}$, Birch noted that H^{AB} could not be expressed conveniently in terms of the parameters of the 3-way log linear models. He then presented a different model for the 3-way table that could be used to express the hypothesis H^{AB}, and also the hypothesis $H' \equiv H^{AB} \cap H^{ABC}$, where H^{ABC} is the hypothesis of zero 3-factor interaction in table $\{ABC\}$. Birch's analysis (1963, pp. 229–32) can be simplified in the following way.

We shall first consider variables A and B as response variables, and then we shall use these variables to explain variation in variable C. Letting F_{ij}^{AB} denote the expected frequency in cell (i,j) of table $\{AB\}$ under H^{AB} and letting F_{ijk}^{ABC} be similarly defined for table $\{ABC\}$ under H^{ABC}, we see that F_{ijk}^{ABC} can be expressed by (8) and F_{ij}^{AB} can be written as

$$F_{ij}^{AB} = \eta\tau_i^A\tau_j^B, \tag{21}$$

where the τ's satisfy conditions of the form (9). Letting F'_{ijk} denote the expected frequency in cell (i,j,k) under H', we find, as in the Appendix, that

$$F'_{ijk} = F_{ij}^{AB}F_{ijk}^{ABC}/f_{ij}^{AB} = \eta\tau_i^A\tau_j^B\tau_k^C\tau_{ik}^{AC}\tau_{jk}^{BC}/\sigma_{ij}^{AB}, \tag{22}$$

where

$$\sigma_{ij}^{AB} = \sum_k \tau_k^C\tau_{ik}^{AC}\tau_{jk}^{BC}. \tag{23}$$

To calculate the maximum likelihood estimate $\hat{F}'_{ijk}$ of F'_{ijk} under H', we use formula (22), with F_{ij}^{AB} and F_{ijk}^{ABC} replaced by their maximum likelihood estimates $\hat{F}_{ij}^{AB}$ and $\hat{F}_{ijk}^{ABC}$, respectively. To test the null hypothesis H', we calculate the likelihood ratio chi-squared statistic

$$\chi^2(H') = 2 \sum_{i,j,k} f_{ijk} \log(f_{ijk}/\hat{F}'_{ijk}), \tag{24}$$

which can be partitioned into $\chi^2(H^{AB})$ and $\chi^2(H^{ABC})$.

The results in the preceding two paragraphs can be directly generalized to the case where model H^{AB} for table $\{AB\}$ is any 2-way log linear model, and where model H^{ABC} for table $\{ABC\}$ is any 3-way log linear model that includes table $\{AB\}$ in the set of fitted marginals. In addition to simplifying Birch's analysis, the results presented herein show how his results can be generalized.

3. A generalization

Let A^* now denote a set of a^* polytomous variables $(A_1, \ldots, A_{a^*})$, and let B^* and C^* be similarly defined as $(B_1, \ldots, B_{b^*})$ and $(C_1, \ldots, C_{c^*})$, respectively. Thus, A^* is an a^*-way variable, B^* is a b^*-way variable and C^* is a c^*-way variable. Let $i^* = (i_1, \ldots, i_{a^*})$ denote the level of variable A^* and let j^* and k^* be similarly defined for B^* and C^*, respectively. Let $f_{i^*j^*k^*}$ denote the observed frequency in cell (i^*, j^*, k^*) in the $(a^*+b^*+c^*)$-way table

$\{A^*B^*C^*\}$ and let $f_{i^*j^*}^{A^*B^*}$ and $f_{i^*}^{A^*}$ denote the corresponding frequencies in the (a^*+b^*)-way and a^*-way marginal tables $\{A^*B^*\}$ and $\{A^*\}$, respectively. For the a^*-way table $\{A^*\}$, let H^{A^*} denote any a^*-way log linear model; for the (a^*+b^*)-way table $\{A^*B^*\}$, let $H^{A^*B^*}$ denote any (a^*+b^*)-way log linear model that includes table $\{A^*\}$ in the set of fitted marginals; and for the $(a^*+b^*+c^*)$-way table $\{A^*B^*C^*\}$, let $H^{A^*B^*C^*}$ denote any $(a^*+b^*+c^*)$-way log linear model that includes table $\{A^*B^*\}$ in the set of fitted marginals. The expected frequencies are denoted by the corresponding F's. Let

$$H' \equiv H^{A^*} \cap H^{A^*B^*} \cap H^{A^*B^*C^*}.$$

Letting $F'_{i^*j^*k^*}$ denote the expected frequency in cell (i^*, j^*, k^*) under H', we find that

$$F'_{i^*j^*k^*} = F_{i^*}^{A^*} F_{i^*j^*}^{A^*B^*} F_{i^*j^*k^*}^{A^*B^*C^*} / (f_{i^*}^{A^*} f_{i^*j^*}^{A^*B^*}). \tag{25}$$

Formula (25) generalizes (14) and (22). To calculate the maximum likelihood estimate $\hat{F}'_{i^*j^*k^*}$ of $F'_{i^*j^*k^*}$ under H', we use formula (25) with the F's replaced by their maximum likelihood estimates. To test the null hypothesis H', we calculate $\chi^2(H')$ as in (24), which can be partitioned now into $\chi^2(H^{A^*})$, $\chi^2(H^{A^*B^*})$ and $\chi^2(H^{A^*B^*C^*})$.

4. An illustrative example

We shall now analyze Table 1 using some of the models presented in §2. When H^{ABC} is defined by (4) or (8), the likelihood ratio chi-squared $\chi^2(H^{ABC})$ is 0·02, with 1 degree of freedom. Model H^{ABC} is listed as H_1 in Table 2. In this table, for each model listed there, we give the set of fitted marginals, for example, $\{AB\}$, $\{AC\}$ and $\{BC\}$ for H^{ABC}, the degrees of freedom, and the likelihood ratio chi-squared. In addition, for those who prefer the goodness-of-fit chi-squared, we include this in Table 2, although we shall use the likelihood ratio rather than the goodness-of-fit chi-squared here.

Table 2. *Chi-squared values for some models pertaining to the 4-way table $\{ABCD\}$, Table 1, and the corresponding 3-way marginal table $\{ABC\}$*

Model*	Fitted marginals	Degrees of freedom	Likelihood ratio chi-squared	Goodness-of-fit chi-squared
H_1	$\{AB\}, \{AC\}, \{BC\}$	1	0·02	0·02
H_2	$\{AB\}, \{BC\}$	2	1005·12	979·18
H_3	$\{AB\}, \{AC\}$	2	27·18	27·23
H_4	$\{ABC\}, \{AD\}, \{BD\}, \{CD\}$	4	1·19	1·19
H_5	$\{ABC\}, \{BD\}, \{CD\}$	5	4·04	4·07
H_6	$\{ABC\}, \{AD\}, \{CD\}$	5	262·51	261·28
H_7	$\{ABC\}, \{AD\}, \{BD\}$	5	15·69	15·77
H_8	$\{ABC\}, \{CD\}$	6	267·46	265·47
H_9	$\{ABC\}, \{BD\}$	6	35·18	34·94
H_{10}	$\{AB\}, \{AC\}, \{BC\}, \{BD\}, \{CD\}$	6	4·06	4·10
H_{11}	$\{AB\}, \{AC\}, \{BD\}$	8	62·36	61·69

* Models H_1–H_3 pertain to table $\{ABC\}$ while models H_4–H_{11} pertain to table $\{ABCD\}$.

To test whether the main effect $\gamma^{A\bar{C}}$ in (4) is statistically significant, we consider the model obtained by setting $\tau^{AC} = 1$ in (8). This model is listed as H_2 in Table 2. Since

$$\chi^2(H_2) - \chi^2(H_1) = 1005{\cdot}10$$

with 1 degree of freedom, we reject the null hypothesis that $\tau^{AC} = 1$ in (8), i.e. $\gamma^{A\bar{C}} = 1$ in (4). Similarly, since $\chi^2(H_3) - \chi^2(H_1) = 27{\cdot}16$ with 1 degree of freedom, we also reject the null hypothesis that $\tau^{BC} = 1$ in (8), i.e. $\gamma^{B\bar{C}} = 1$ in (4).

Model H^{ABCD}, which we defined by (5) or (10), is listed as H_4 in Table 2. Since $\chi^2(H_4) = 1{\cdot}19$ with 4 degrees of freedom, this model fits the data. To test whether the main effect $\gamma^{A\bar{D}}$ in (5) is statistically significant, we consider the model obtained by setting $\tau^{AD} = 1$ in (10). This model is listed as H_5 in Table 2. Since $\chi^2(H_5) - \chi^2(H_4) = 2{\cdot}85$ with 1 degree of freedom, we do not reject the null hypothesis that $\gamma^{A\bar{D}} = 1$ at the 0·05 level. Similar tests for $\gamma^{B\bar{D}}$ and $\gamma^{C\bar{D}}$ in (5) indicate that these effects in (5), unlike the effect of $\gamma^{A\bar{D}}$, are statistically significant; see $H_6 - H_7$ in Table 2.

Since $\chi^2(H_5) = 4{\cdot}04$ with 5 degrees of freedom, model H_5 fits the data. This model can be expressed as

$$\Omega_{ijk}^{ABC\bar{D}} = \gamma^{\bar{D}}\gamma_j^{B\bar{D}}\gamma_k^{C\bar{D}}. \tag{26}$$

Tests of the main effects $\gamma^{B\bar{D}}$ and $\gamma^{C\bar{D}}$ in (26) indicate that they are statistically significant; see H_8 and H_9 in Table 2.

Having calculated $\hat{F}_{ijk}^{ABC}$ under H_1, i.e. under (4), and $\hat{F}_{ijkl}^{ABCD}$ under H_5, i.e. under (26), we can apply (14) to calculate $\hat{F}_{ijkl}^*$ under H^*, i.e. $H_1 \cap H_5$. The values of $\hat{F}_{ijkl}^*$ are given in Table 3. From (17) or (18), we find that $\chi^2(H^*) = 4{\cdot}06$ with 6 degrees of freedom. An alternative way of calculating $\hat{F}_{ijkl}^*$ and $\chi^2(H^*)$ is to take note of the fact that $H_1 \cap H_5$ is equivalent to the log linear model H_{10} in Table 2. Although $H_1 \cap H_4$ was not equivalent to any 4-way log linear model, $H_1 \cap H_5$ is.

Table 3. *Estimate of expected frequencies in Table 1 under model H^* described by equations* (4) *and* (26)

First interview: Membership	First interview: Attitude	Second interview: Membership + / Attitude +	Membership + / Attitude −	Membership − / Attitude +	Membership − / Attitude −
+	+	447·71	151·16	104·60	53·53
+	−	169·36	182·77	54·59	89·29
−	+	192·98	65·15	537·71	275·16
−	−	87·95	94·92	338·10	553·03

Before closing this section, let us return for a moment to the hypotheses H, H_a and H_b, defined by (1), (1*a*) and (1*b*), respectively. Hypothesis H_a, which pertains to table $\{ABC\}$, is equivalent to the hypothesis that variables B and C are conditionally independent, given the level of variable A. This hypothesis is listed as H_3 in Table 2; see, for example, Birch (1963) and Goodman (1970). Hypothesis H_b, which pertains to table $\{ABCD\}$, is equivalent to the hypothesis that variable D is conditionally independent of the joint variable (A, C), given the level of variable B. This hypothesis is listed as H_9 in Table 2. To calculate $\hat{F}_{ijkl}$ under H, i.e. $H_a \cap H_b$ or $H_3 \cap H_9$, we can apply (14), with H^{ABC} and H^{ABCD} denoting H_3 and H_9, respectively. To calculate $\chi^2(H)$, we can apply (17) or (18). An alternative way of calculating $\hat{F}_{ijkl}$ and $\chi^2(H)$ is to take note of the fact that H is equivalent to H_{11} in Table 2. From Table 2, we see that each of these hypotheses, viz. H_3, H_9 and H_{11}, is rejected by the data. Note that $\chi^2(H_{11})$ can be partitioned into $\chi^2(H_3)$ and $\chi^2(H_9)$.

We presented hypothesis H and the corresponding H_a and H_b in § 1 in order to indicate the kinds of model that were suggested in the earlier literature for the analysis of tables of the kind considered herein. We now see that hypothesis H is rejected by the data (Table 1) whereas the more general approach introduced herein leads to models that fit the data well, for example, $H_1 \cap H_4$ or $H_1 \cap H_5$.

5. Some path diagrams

Having calculated $\hat{F}_{ijk}^{ABC}$ under H_1 in Table 2 by the usual iterative-scaling methods, we can now use these estimates to calculate the maximum likelihood estimate $\hat{\tau}$ of each τ in (8); see, for example, Goodman (1970). Since

$$\gamma^{\bar{C}} = (\tau_1^C)^2, \quad \gamma_i^{A\bar{C}} = (\tau_{i1}^{AC})^2, \ldots, \tag{27}$$

each $\hat{\tau}$ can then be used to calculate the corresponding maximum likelihood estimate $\hat{\gamma}$ of the γ in (4). Similarly, using $\hat{F}_{ijkl}^{ABCD}$ under H_5, we can estimate the τ's in (10) with $\tau^{AD} = 1$, and the γ's in (26). These estimates are given in Table 4 together with the corresponding $\hat{\beta}$'s, where

$$\hat{\beta}^{\bar{C}} = \log \hat{\gamma}^{\bar{C}}, \quad \hat{\beta}^{A\bar{C}} = \log \hat{\gamma}^{A\bar{C}}, \ldots \tag{28}$$

The $\hat{\beta}$'s are the estimated parameters when (4) and (26) are expressed in additive form via natural logarithms.

A simple diagram describing the relationships among variables A, B, C and D, expressed by equations (4) and (26), is presented as Fig. 1. This diagram is somewhat analogous to the path diagrams used by Wright (1954); see also, for example, Tukey (1954), Turner & Stevens (1959) and Duncan (1966).

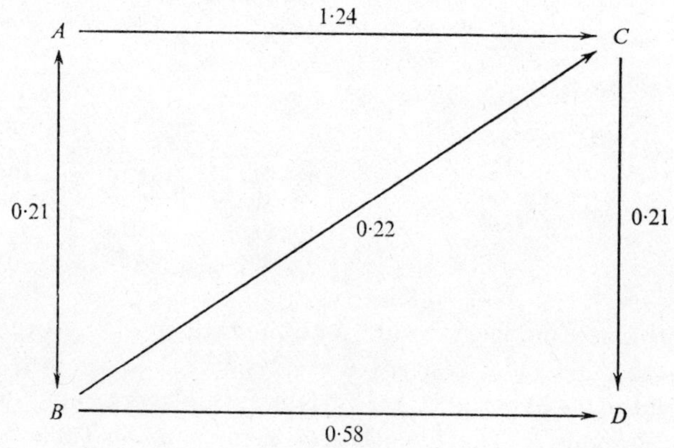

Fig. 1. Path diagram for the system of equations (4) and (26), applied to Table 1.

Each of the arrows pointing to C in Fig. 1 represents a main effect in (4), and each of the arrows pointing to D represents a main effect in (26). The numerical entries associated with these arrows are the corresponding $\hat{\beta}$'s in Table 4. In addition to providing a simple representation of the estimated main effects in (4) and (26), Fig. 1 also includes information about the relationship between variables A and B which is not included in equations (4) and (26). Note that variables A and B are prior to the other variables, and that the parameters $\gamma_i^{A\bar{B}}$ and $\gamma_j^{B\bar{A}}$ do not appear in (4) and (26). The information about the relationship between variables A and B, which is included in Fig. 1, was obtained in the following way.

For the 2-way marginal table $\{AB\}$, the saturated model can be expressed as

$$F_{ij}^{AB} = \eta\tau_i^A\tau_j^B\tau_{ij}^{AB}, \tag{29}$$

where the τ's satisfy conditions of the form (9). Compare (29) with (21). Let $\Omega_i^{A\bar{B}}$ and $\Omega_j^{B\bar{A}}$ denote the conditional odds defined by

$$\Omega_i^{A\bar{B}} = F_{i1}^{AB}/F_{i2}^{AB}, \quad \Omega_j^{B\bar{A}} = F_{1j}^{AB}/F_{2j}^{AB}. \tag{30}$$

The saturated model (29) can be expressed in any of the following equivalent forms:

$$\Omega_i^{A\bar{B}} = \gamma^{\bar{B}}\gamma_i^{A\bar{B}}, \tag{31}$$

or

$$\Omega_j^{B\bar{A}} = \gamma^{\bar{A}}\gamma_j^{B\bar{A}}, \tag{32}$$

or by the system of equations (31)–(32); the γ's satisfy conditions of the form (6). Note that

$$\gamma_1^{A\bar{B}} = \gamma_1^{B\bar{A}} = (\tau_{11}^{AB})^2. \tag{33}$$

The τ's in (29) can be estimated by the usual methods for the saturated 2-way log linear model (see, for example, Goodman, 1970), and each $\hat{\tau}$ can then be used to calculate the corresponding $\hat{\gamma}$, as in (27) and (33), and $\hat{\beta}$, as in (28). From (33), we see that $\hat{\beta}^{A\bar{B}} = \hat{\beta}^{B\bar{A}} = \hat{\beta}^{AB}$. The estimate $\hat{\beta}^{AB}$ is the numerical entry pertaining to the double-headed arrow in Fig. 1. Thus, in addition to including the $\hat{\beta}$'s corresponding to the main effects in (4) and (26), Fig. 1 also includes the $\hat{\beta}$ corresponding to the main effects in (31) and (32).

Table 4. *Estimate of the γ parameters in* (4) *and* (26), *and the corresponding β parameters*

Variable	γ Parameter	β Parameter
C	0·88	−0·13
AC	3·45	1·24
BC	1·24	0·22
D	1·35	0·30
BD	1·79	0·58
CD	1·23	0·21

When $\tau^{AB} = 1$ in (29), i.e. $\beta^{AB} = 0$, we have the usual null hypothesis that variables A and B are independent in the 2-way table $\{AB\}$; see (21). To test this null hypothesis, we calculate the usual likelihood ratio chi-squared, which is $\chi^2 = 35{\cdot}16$ with 1 degree of freedom for the data analyzed here. From this chi-squared test and the corresponding tests presented in the preceding section, we see that each of the numerical entries in Fig. 1 is statistically significant.

By the usual estimation methods for the saturated 2-way log linear model, we see that the standard deviation of $\hat{\beta}^{AB}$ can be estimated by

$$S = \tfrac{1}{2}\{\sum_{i,j}(f_{ij}+\tfrac{1}{2})^{-1}\}^{\frac{1}{2}}. \tag{34}$$

For the 3-way log linear model, the corresponding formula is

$$S = \tfrac{1}{4}\{\sum_{i,j,k}(f_{ijk}+\tfrac{1}{2})^{-1}\}^{\frac{1}{2}}, \tag{35}$$

and a similar formula is obtained for the 4-way table. The $\frac{1}{4}$ in (35) is replaced by $\frac{1}{8}$ for the 4-way table. For the data analyzed here, $S = 0{\cdot}036$ for table $\{AB\}$, $S = 0{\cdot}043$ for table

$\{ABC\}$ and $S = 0{\cdot}045$ for table $\{ABCD\}$. These estimated standard deviations apply to the $\hat{\beta}$'s calculated for the saturated models, for example, $\hat{\beta}^{AB}$ in table $\{AB\}$. The S value presented here for table $\{ABC\}$ is also an estimate of an upper bound for the standard deviation of the $\hat{\beta}$'s calculated for the unsaturated model for this 3-way table; see Table 4. A similar statement applies to the S value presented here for table $\{ABCD\}$.

Figure 1 is a path diagram corresponding to case (i) discussed in § 1. Note that, under (26), when variables A, B and C are used to explain variation in D, only variables B and C are actually used in the explanation. In Fig. 1, all arrows are single-headed except for the arrow between A and B, which pertains to (31) and (32). We shall next consider models that can be described by path diagrams in which double-headed arrows appear also among variables that are not prior to the others.

Consider model H_{10} in Table 2. Under this model, we see that the F_{ijkl} can be expressed as

$$F_{ijkl} = \eta \tau_i^A \tau_j^B \tau_k^C \tau_l^D \tau_{ij}^{AB} \tau_{ik}^{AC} \tau_{kj}^{BC} \tau_{jl}^{BD} \tau_{kl}^{CD}, \tag{36}$$

where the τ's satisfy conditions of the form (9). Let us now define the following conditional odds:

$$\Omega_{ijk}^{ABC\bar{D}} = F_{ijk1}/F_{ijk2}, \quad \Omega_{ijl}^{ABD\bar{C}} = F_{ij1l}/F_{ij2l}, \quad \Omega_{ikl}^{ACD\bar{B}} = F_{i1kl}/F_{i2kl}, \quad \Omega_{jkl}^{BCD\bar{A}} = F_{1jkl}/F_{2jkl}. \tag{37}$$

From (36) and (37), we see that

$$\Omega_{ijk}^{ABC\bar{D}} = \gamma^{\bar{D}} \gamma_j^{B\bar{D}} \gamma_k^{C\bar{D}}, \tag{38}$$

$$\Omega_{ijl}^{ABD\bar{C}} = \gamma^{\bar{C}} \gamma_i^{A\bar{C}} \gamma_j^{B\bar{C}} \gamma_l^{D\bar{C}}, \tag{39}$$

$$\Omega_{ikl}^{ACD\bar{B}} = \gamma^{\bar{B}} \gamma_i^{A\bar{B}} \gamma_k^{C\bar{B}} \gamma_l^{D\bar{B}}, \tag{40}$$

$$\Omega_{jkl}^{BCD\bar{A}} = \gamma^{\bar{A}} \gamma_j^{B\bar{A}} \gamma_k^{C\bar{A}}, \tag{41}$$

where the γ's satisfy conditions of the form (6). Note that the relationship between the γ's and τ's can be described by conditions of the form (27), and that

$$\gamma_1^{A\bar{B}} = \gamma_1^{B\bar{A}}, \quad \gamma_1^{A\bar{C}} = \gamma_1^{C\bar{A}}, \text{ etc.} \tag{42}$$

Under (38), the following γ's are set equal to one: $\gamma^{A\bar{D}}, \gamma^{AB\bar{D}}, \gamma^{AC\bar{D}}, \gamma^{BC\bar{D}}, \gamma^{ABC\bar{D}}$. Letting T^D denote the set of τ's corresponding to this set of γ's, we see that

$$T^D = \{\tau^{AD}, \tau^{ABD}, \tau^{ACD}, \tau^{BCD}, \tau^{ABCD}\}. \tag{43}$$

Under (39), the following γ's are set equal to one: $\gamma^{AB\bar{C}}$, $\gamma^{AD\bar{C}}$, $\gamma^{BD\bar{C}}$, $\gamma^{ABD\bar{C}}$. Letting T^C denote the set of τ's corresponding to this set of γ's, we see that

$$T^C = \{\tau^{ABC}, \tau^{ACD}, \tau^{BCD}, \tau^{ABCD}\}. \tag{44}$$

Since the union of the sets T^C and T^D is equivalent to the set of τ's that are taken as one under model (36), this model is equivalent to (38) and (39). More generally, model (36) is equivalent to any pair of equations from (38) to (41), except for the pair (39) and (40), any triplet of equations from (38) to (41), and the quartuplet (38) to (41).

After calculating $\hat{F}_{ijkl}$ under H_{10} by the usual iterative-scaling methods, we can then use these estimates to calculate the $\hat{\tau}$'s in (36), and each $\hat{\tau}$ can then be used to calculate the corresponding $\hat{\gamma}$ and $\hat{\beta}$. A simple diagram describing the relationships among variables A, B, C and D, expressed by (38) and (39), is presented as Fig. 2. The arrows pointing to C in Fig. 2 represent the main effects in (39), and the arrows pointing to D represent the main effects in (38). The main effects $\gamma_k^{C\bar{D}}$ in (38) and $\gamma_l^{D\bar{C}}$ in (39), which could be represented as two single-headed arrows, one from C to D and another from D to C, are represented in Fig. 2 by a double-headed arrow, since $\gamma_1^{C\bar{D}} = \gamma_1^{D\bar{C}}$, as in (42). In addition to including the $\hat{\beta}$'s corresponding to the main effects in (38) and (39), Fig. 2 also includes $\hat{\beta}^{AB}$ corresponding to the main effects in (31) and (32). Figure 2 is a path diagram corresponding to case (ii) discussed in §1.

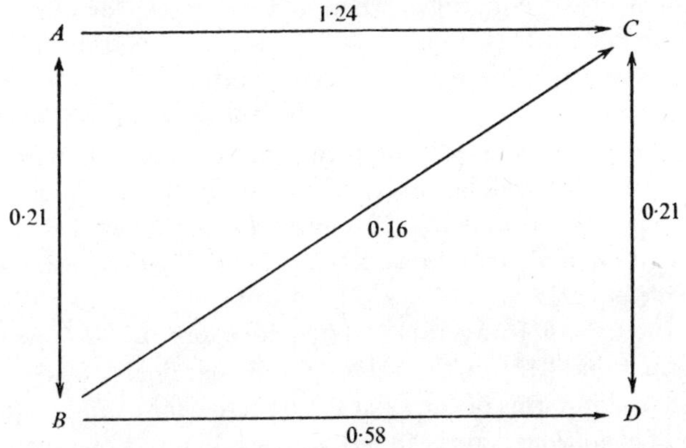

Fig. 2. Path diagram for the system of equations (38)–(39), applied to Table 1.

The reader may wonder why the $\hat{\beta}^{AB}$ in Fig. 2 pertains to (31) and (32), rather than to the corresponding γ in (40) or (41). Since variables A and B were viewed here as prior to the other variables, we considered models for the marginal table $\{AB\}$, for example, (21) and (29), and then models for tables $\{ABC\}$ and/or $\{ABCD\}$ given table $\{AB\}$, for example, (4) and (5) or (38) and (39). In cases where variables A and B are not prior to the other variables, equations of the form (40) and (41) will be of interest.

Since model H_{10}, which was equivalent to (38) and (39), was also equivalent to (4) and (26), the $\hat{\beta}^{BD}$ and $\hat{\beta}^{CD}$ in Fig. 2 are equal to the corresponding entries in Fig. 1. In addition, although $\hat{\beta}^{AC}$ was calculated from table $\{ABC\}$ in Fig. 1 and from table $\{ABCD\}$ in Fig. 2, these entries are equal to each other, since collapsing variable D in model (36) does not affect the τ^{AC} parameter. However, since collapsing variable D in model (36) does affect the B to C interaction in this model, due to the introduction of the multiplicative factor

$$\sigma_{jk}^{BC} = \sum_l \tau_l^D \tau_{jl}^{BD} \tau_{kl}^{CD}$$

in the corresponding model for the marginal table $\{ABC\}$, we see that $\hat{\beta}^{BC}$ in Fig. 1 differs from the corresponding quantity in Fig. 2. The B to C interaction in Fig. 1 is larger than the corresponding interaction in Fig. 2, because of the fact that $\hat{\beta}^{BD}$ and $\hat{\beta}^{CD}$ have the same sign, both are positive; for further details, see Goodman (1972a).

As we have already noted, Figs. 1 and 2 are path diagrams corresponding to cases (i) and (ii) discussed in §1. Now that we have indicated here how to analyze these cases, the interested reader can analyze other cases in a somewhat similar way. The results presented in the next section will facilitate the analysis of more complicated cases. For additional results, see Goodman (1972b).

6. Some generalizations

We return now to the notation of §3. For the a^*-way table $\{A^*\}$, under any a^*-way log linear model H^{A^*}, the $F^{A^*}_{i^*}$ can be expressed in terms of the τ parameters; see, for example, (29) or (36). Equivalently, the a^* sets of conditional odds, i.e. the Ω's [see, for example, (30) or (37)] can be expressed in terms of a system of a^* equations; see, for example, (31) and (32) or (38) to (41). For many a^*-way log linear models, the system of a^* equations can be reduced to a system of a'^* equations, where $a'^* < a^*$; see, for example, the comments following equations (38) to (41). Corresponding to each equation, there will be a set T of the τ's that are taken equal to one under that equation; see, for example, (43) and (44). The union of the T's for the a'^* equations describes the set of τ's that are taken equal to one under the system of a'^* equations; see, for example, comments following (43) and (44). The a^*-way log linear model H^{A^*} will be equivalent to the system of a'^* equations, if the set of τ's that are taken equal to one under H^{A^*} is equivalent to the set of τ's that are taken equal to one under the system of a'^* equations, having determined the latter set using the union of the T's for the a'^* equations.

Next consider the (a^*+b^*)-way table $\{AB\}$ under any (a^*+b^*)-way log linear model $H^{A^*B^*}$ that includes table $\{A^*\}$ in the set of fitted marginals. Under model $H^{A^*B^*}$, the $F^{A^*B^*}_{i^*j^*}$ can be expressed in terms of the τ parameters; see, for example, (8) or (10). Equivalently, the b^* sets of conditional odds can be expressed in terms of a system of b^* equations; see, for example, (4) or (5).

The above remarks can also be extended directly to the $(a^*+b^*+c^*)$-way table $\{A^*B^*C^*\}$ under any model $H^{A^*B^*C^*}$ that includes table $\{A^*B^*\}$ in the set of fitted marginals.

Let us return now to the special case of the 2^4 table $\{ABCD\}$. If a given equation for a set of conditional odds, for example, the $\Omega^{ABC\bar{D}}_{ijk}$ defined by (3) or (37), is multiplicative in the narrow sense that, say, $\Omega^{ABC\bar{D}}_{ijk}$ is expressed as a product of $\gamma^{\bar{D}}$ and the main effects (viz. $\gamma^{A\bar{D}}, \gamma^{B\bar{D}}, \gamma^{C\bar{D}}$), or a subset of these main effects [see, for example, (5) or (38)], then each main effect in this equation can be represented as a single-headed arrow in a simple diagram; see, for example, arrows pointing to D in Fig. 1. If a given model is equivalent to a system of such equations, then each of these equations can be represented by a set of arrows in this diagram; see, for example, the set of arrows pointing to C, and the set of arrows pointing to D, in Fig. 1. Because the effects satisfy condition (42), we replace two single-headed arrows between a given pair of variables in this diagram by a single double-headed arrow between these variables; see, for example, double-headed arrows in Fig. 2.

If a given equation for a set of conditional odds, for example, $\Omega^{ABC\bar{D}}_{ijk}$, includes one or more interaction effects, for example, $\gamma^{BC\bar{D}}$, the corresponding path diagram would not be quite as simple as those described above. Consider for a moment equation (38) modified by the inclusion of $\gamma^{BC\bar{D}}_{jk}$. In addition to the two arrows pointing to D in Fig. 1, representing $\hat{\beta}^{BD}$ and $\hat{\beta}^{CD}$, we would now insert a small circle or other appropriate symbol in Fig. 1 to represent $\hat{\beta}^{BC\bar{D}}$, with an arrow pointing to D from the circle. Consider next (38) and (39) modified by the inclusion of $\gamma^{BC\bar{D}}$ in (38) and $\gamma^{BD\bar{C}}$ in (39). In addition to the arrows pointing to C and D in Fig. 2, we would now insert the small circle introduced above, with arrows pointing to C and D from the circle. Each of the arrows from the small circle would have the same numerical value associated with it, viz. $\hat{\beta}^{BCD}$, as can be seen by a direct extension of (42) to the interaction effects.

An alternative way to represent equation (38) modified by the inclusion of $\gamma^{BC\bar{D}}$ would be to insert the following two arrows in Fig. 1; (*a*) an arrow from B to the middle of the arrow between C and D; and (*b*) an arrow from C to the middle of the arrow between B and D. The (*a*) arrow would indicate that the effect on variable D of being at a given level on variable C will depend also upon the level of variable B; and the (*b*) arrow would indicate that the effect on variable D of being at a given level on variable B will depend also upon the level of variable C. Similarly, we could represent (38) and (39), modified by the inclusion of $\gamma^{BC\bar{D}}$ in (38) and $\gamma^{BD\bar{C}}$ in (39), by inserting in Fig. 2 the *(a)* and *(b)* arrows and an arrow from D to the middle of the arrow between B and C. For a related suggestion, see Boudon (1968) and Lazarsfeld (1970). As in the preceding paragraph, each of the inserted arrows would have the same numerical value associated with it.

As we noted earlier, the models and methods presented herein can be directly extended to the m-way table pertaining to m polytomous variables. Even the path diagrams can be directly extended. However, in the more general case the numerical entry associated with each arrow in the diagram would be replaced, when the variables are polytomous, by a vector or matrix of numerical entries associated with each arrow in the diagram. Here each arrow is associated with the vector or matrix describing the corresponding effect in a given equation for a set of vectors of conditional odds; see, for example, Goodman (1971*b*).

This research was supported in part by the Division of the Social Sciences of the National Science Foundation. For helpful comments, the author is indebted to S. Haberman and S. Schooler.

Appendix

For table $\{ABCD\}$ in §2, let P_{ijk} be the probability that the joint variable (A, B, C) is at level (i, j, k); let $P_{ijk|l}$ be the conditional probability that variable D is at level l, given that (A, B, C) is at level (i, j, k); and let n be the sample size. Under H^{ABC} described by (8), we obtain

$$nP_{ijk} = F_{ijk}^{ABC}; \tag{A 1}$$

and under H^{ABCD} described by (5) or (10),

$$f_{ijk} P_{ijk|l} = F_{ijkl}^{ABCD}. \tag{A 2}$$

Under H^*, i.e. $H^{ABC} \cap H^{ABCD}$, we obtain

$$F_{ijkl}^{*} = nP_{ijk} P_{ijk|l} = F_{ijk}^{ABC} P_{ijk|l}, \tag{A 3}$$

which yields (14) using (A 2), and (15) using (8) and (10). By maximizing the likelihood under H^*, we find that

$$\hat{F}_{ijkl}^{*} = \hat{F}_{ijk}^{ABC} \hat{F}_{ijkl}^{ABCD} / f_{ijk}^{ABC}; \tag{A 4}$$

see comments preceding (17). Formulae (A 3) and (A 4) apply whether or not table $\{ABC\}$ is viewed as given under H^{ABCD}. With either view of $\{ABC\}$ under H^{ABCD}, model H^{ABCD} does not impose any restrictions on F^{ABC} under H^*, and H^{ABC} does not impose any restrictions on $P_{ijk|l}$. The above results can be extended directly to cover the cases pertaining to (22) and (25).

References

ANDERSON, T. W. (1954). Probability models for analyzing time changes in attitudes. In *Mathematical Thinking in the Social Sciences*, ed. P. F. Lazarsfeld, pp. 17–66. Glencoe, Illinois: Free Press.

BIRCH, M. W. (1963). Maximum likelihood in three-way contingency tables. *J. R. Statist. Soc. B* **25**, 220–33.

BISHOP, Y. M. M. (1969). Full contingency tables, logits, and split contingency tables. *Biometrics* **25**, 383–400.

BOUDON, R. (1968). A new look at correlation analysis. In *Methodology in Social Research*, eds. H. M. Blalock, Jr. and A. Blalock, pp. 199–235. New York: McGraw-Hill.

COLEMAN, J. S. (1964). *Introduction to Mathematical Sociology*. Glencoe, Illinois: Free Press.

COX, D. R. (1970). *The Analysis of Binary Data*. London: Methuen.

COX, D. R. & LAUH, E. (1967). A note on the graphical analysis of multidimensional contingency tables. *Technometrics* **9**, 481–8.

DUNCAN, O. D. (1966). Path analysis: Sociological examples. *Am. J. Sociology* **72**, 1–16.

DYKE, G. V. & PATTERSON, H. D. (1952). Analysis of factorial arrangements when the data are proportions. *Biometrics* **8**, 1–12.

GOODMAN, L. A. (1970). The multivariate analysis of qualitative data: Interactions among multiple classifications. *J. Am. Statist. Ass.* **65**, 226–56.

GOODMAN, L. A. (1971*a*). Partitioning of chi-square, analysis of marginal contingency tables, and estimation of expected frequencies in multidimensional contingency tables. *J. Am. Statist. Ass.* **66**, 339–44.

GOODMAN, L. A. (1971*b*). The analysis of multidimensional contingency tables: Stepwise procedures and direct estimation methods for building models for multiple classifications. *Technometrics* **13**, 33–61.

GOODMAN, L. A. (1972*a*). A general model for the analysis of surveys. *Am. J. Sociology* **77**, 1035–86.

GOODMAN, L. A. (1972*b*). Causal analysis of data from panel studies and other kinds of surveys. *Am. J. Sociology*, **77**, (to appear).

LAZARSFELD, P. F. (1970). A memoir in honor of Professor Wold. In *Scientists at Work*, ed. T. Dalenius *et al.*, pp. 78–103. Uppsala: Almqvist and Wiksell.

TUKEY, J. W. (1954). Causation, regression, and path analysis. In *Statistics and Mathematics in Biology*, ed. O. Kempthorne *et al.*, pp. 35–66. Ames, Iowa: Iowa State College Press.

TURNER, M. E. & STEVENS, C. D. (1959). The regression analysis of causal chains. *Biometrics* **15**, 236–58.

WRIGHT, S. (1954). The interpretation of multivariate systems. In *Statistics and Mathematics in Biology*, ed O. Kempthorne *et al.*, pp. 11–33. Ames, Iowa: Iowa State College Press.

[*Received January* 1972. *Revised May* 1972]

Leo A. Goodman

Reprinted from: BIOMETRIKA (1973), Vol. 60, No. 1, pp. 179-192. Used with permission.

Chapter 12

Guided and Unguided Methods for the Selection of Models for a Set of T Multidimensional Contingency Tables

The combined use of direct estimation (in saturated models) and indirect testing (of unsaturated models), proposed in [7, 8] for the analysis of a given multidimensional contingency table, is extended here to the analysis of a set of T multidimensional tables ($T \geq 2$). For this set of tables, a three-fold classification of models is introduced: (1) models that assume "complete homogeneity" among the tables; (2) models that allow "complete heterogeneity" among the tables; and (3) models that allow "partial heterogeneity." Stepwise procedures proposed in [8] for model selection are extended. "Guided" and "unguided" selection methods and "multidirectional" methods are introduced. For illustrative purposes, a set of two 4-way tables is analyzed.

1. INTRODUCTION

To illustrate the application of the methods and concepts that will be introduced herein, we shall reanalyze the set of two 4-way tables studied earlier in [3, 12, 13]. The data (Table 1) describe the cross-classification of students according to four dichotomized response variables, for a high IQ sample and a low IQ sample. The methods and concepts that will be presented here can be applied more generally to the case where the cross-classifications describe the observed joint-distribution with respect to m polytomous variables (where $m \geq 2$), and where there are T such tables (for $T \geq 2$).

Table 1 can be viewed as (a) two 4-way tables, or as (b) a single 5-way table (treating the "IQ variable" as the 5th dimension) in which the two parts of the table (i.e., the high IQ and low IQ parts) have fixed sample sizes. Each of these views of Table 1 leads to somewhat

Leo A. Goodman is the Charles L. Hutchinson Distinguished Service Professor, Departments of Statistics and Sociology, University of Chicago, Chicago, Ill. 60637, and research associate at the Population Research Center of the University. This research was supported in part by Research Contract No. NSF GS 1905 from the Division of the Social Sciences of the National Science Foundation. The author is indebted to C. Bingham, S. Haberman, G. Simon, and D. Wallace, for helpful comments.

different models for the analysis of the data. We shall discuss both these views in Sections 2 and 4.1.

In [7, Tables 3 and 4] we noted that there are 18 different hierarchical hypotheses for the 3-way table, 166 different hierarchical hypotheses for the 4-way table, and many more such hypotheses for the 5-way table. (These are hierarchical hypotheses about the log-linear model for the m-way table, for $m = 3$, 4, and 5, respectively; see [7].) When Table 1 herein is viewed as a 5-way table, each of the many hierarchical hypotheses about the

1. CROSS–CLASSIFICATION OF STUDENTS ACCORDING TO FOUR DICHOTOMIZED RESPONSE VARIABLES, FOR A HIGH IQ SAMPLE AND A LOW IQ SAMPLE

Item response				High IQ sample	Low IQ sample
A	B	C	D		
1	1	1	1	122	62
1	1	1	2	68	70
1	1	2	1	33	31
1	1	2	2	25	41
1	2	1	1	329	283
1	2	1	2	247	253
1	2	2	1	172	200
1	2	2	2	217	305
2	1	1	1	20	14
2	1	1	2	10	11
2	1	2	1	11	11
2	1	2	2	9	14
2	2	1	1	56	31
2	2	1	2	55	46
2	2	2	1	64	37
2	2	2	2	53	82

5-way table falls into one of four categories: (1) hypotheses that assume "complete homogeneity" between the two 4-way tables; (2) hypotheses that allow for "complete heterogeneity" between the two tables; (3) hypotheses that allow for "partial heterogeneity" between the two tables; (4) hypotheses that are inapplicable (because they can not incorporate the fact that the two 4-way tables have fixed sample sizes). The number of different hierarchical hypotheses in categories (1), (2), and (4) are 167, 165, and 167, respectively; there are many more such hypotheses in category (3). These hypotheses will be discussed in Section 3.

The preceding classification of hypotheses about the 5-way table will provide new interpretations for many of these hypotheses, in addition to providing useful models for the analysis of a set of T 4-way tables. (It will, of course, also provide a way of cataloguing the large number of hypotheses about the 5-way table.) More generally, a classification similar to the one in the preceding paragraph can be used to classify the hierarchical hypotheses about the $(m + 1)$-way table $(m \geq 1)$, and this classification will provide new interpretations for many of these hypotheses, in addition to providing useful models for the analysis of a set of T m-way tables. This will become more evident in Section 3.

In Section 4 we shall introduce extensions of the stepwise procedures proposed in [8] for the selection of models that fit the data in a given multidimensional contingency table. These extensions will facilitate application to *both* the analysis of a given multidimensional table and the analysis of T such tables. The extensions provide procedures for determining the sequences of models that should be examined, using the direct estimates in the saturated models (see, e.g., Tables 10 and 11 herein) as a partial guide, and they also provide procedures that do not use these estimates for guidance. In addition, we shall introduce here "multidirectional" stepwise procedures that can yield models that fit the data better than those obtained with the stepwise procedures proposed earlier.

As we have already noted, the methods presented herein are applicable to the case where the variables in the contingency table are polytomous and where $T \geq 2$. Some formulas that are particularly useful in the general case are included in Section 4.1.

The methods presented herein can be used to search for a concise description of the association among the m variables in each of T m-way tables, and to facilitate comparisons among these tables. These comparisons can serve various purposes. For example, they can lead to methods for discriminating among the corresponding m-way populations. If the level of the m variables is observed for a given individual (so that the particular m-way cell in which he falls is observed), but it is not known from which of the T populations he came, then the methods presented here can be used to facilitate his classification into one of these populations. For some comments on this particular problem, see Section 3.3.

2. SATURATED MODELS AND DIRECT ESTIMATION METHODS FOR A SET OF T TABLES

Consider the case where each of the T contingency tables is a 4-way $I \times J \times K \times L$ table, where the 4 dimensions pertain to variables A, B, C and D, respectively. For the tth table ($t = 1, 2, \cdots, T$) let f_{ijklt} denote the observed frequency in cell (i, j, k, l) of the table ($i = 1, 2, \cdots, I$; $j = 1, 2, \cdots, J$; $k = 1, 2, \cdots, K$; $l = 1, 2, \cdots, L$), when a sample of n_t observations is drawn from the corresponding tth population table, and let $F_{ijklt} = E\{f_{ijklt}\}$. Since $\sum_{i,j,k,l} f_{ijklt} = n_t$, we see that

$$\sum_{i,j,k,l} F_{ijklt} = n_t, \quad \text{for} \quad t = 1, 2, \cdots, T. \qquad (2.1)$$

Let $\xi_{ijklt} = \log F_{ijklt}$, where log refers to the natural logarithm throughout. (For simplicity, we assume $F_{ijklt} > 0$.) As in the analysis of variance (see, e.g., [1, 2, 7]), we can decompose ξ_{ijklt} as follows:

$$\begin{aligned}\xi_{ijklt} = \theta_t &+ \overset{A}{\lambda}_{it} + \overset{B}{\lambda}_{jt} + \overset{C}{\lambda}_{kt} + \overset{D}{\lambda}_{lt} + \overset{AB}{\lambda}_{ijt} + \overset{AC}{\lambda}_{ikt} \\ &+ \overset{AD}{\lambda}_{ilt} + \overset{BC}{\lambda}_{jkt} + \overset{BD}{\lambda}_{jlt} + \overset{CD}{\lambda}_{klt} + \overset{ABC}{\lambda}_{ijkt} \\ &+ \overset{ABD}{\lambda}_{ijlt} + \overset{ACD}{\lambda}_{iklt} + \overset{BCD}{\lambda}_{jklt} + \overset{ABCD}{\lambda}_{ijklt}, \qquad (2.2)\end{aligned}$$

where each λ_t sums to zero over any of its subscripts i, j, k, or l. For the tth table, the λ_t's represent the possible "effects" of the four variables on the ξ_{ijklt} (see, e.g., [8, p. 35]). Formula (2.2) describes the "saturated" model in which all possible "effects" are included in the tth table. From (2.1) and (2.2), we see that the θ_t satisfy the condition that

$$\{\exp \theta_t\}\{\textstyle\sum_{i,j,k,l} \exp (\overset{A}{\lambda}_{it} + \overset{B}{\lambda}_{jt} + \cdots + \overset{ABCD}{\lambda}_{ijklt})\} = n_t, \quad \text{for } t = 1, 2, \cdots, T. \tag{2.3}$$

Formula (2.2) provides a saturated model for each of the T 4-way tables, but it does not include explicit parameters that describe how the corresponding λ_t's in the T tables may be related to each other. We shall now introduce such parameters. Using the λ_t's in (2.2) we define the following λ's:

$$\overset{A}{\lambda}_i = \textstyle\sum_t \overset{A}{\lambda}_{it}/T, \quad \overset{AB}{\lambda}_{ij} = \sum_t \overset{AB}{\lambda}_{ijt}/T, \text{ etc.}$$

and

$$\overset{AE}{\lambda}_{it} = \overset{A}{\lambda}_{it} - \overset{A}{\lambda}_i, \quad \overset{ABE}{\lambda}_{ijt} = \overset{AB}{\lambda}_{ijt} - \overset{AB}{\lambda}_{ij}, \text{ etc.} \tag{2.4}$$

The superscript E can be viewed as pertaining to the additional "variable" (e.g., the "IQ variable" in Table 1) when the T 4-way tables are viewed as a 5-way table. Using (2.4) we can rewrite (2.2) as

$$\begin{aligned}\xi_{ijklt} = \theta_t &+ \overset{A}{\lambda}_i + \overset{AE}{\lambda}_{it} + \overset{B}{\lambda}_j \\ &+ \overset{BE}{\lambda}_{jt} + \cdots + \overset{AB}{\lambda}_{ij} + \overset{ABE}{\lambda}_{ijt} + \cdots + \overset{ABC}{\lambda}_{ijk} \\ &+ \overset{ABCE}{\lambda}_{ijkt} + \cdots + \overset{ABCD}{\lambda}_{ijkl} + \overset{ABCDE}{\lambda}_{ijklt},\end{aligned} \tag{2.5}$$

where each λ sums to zero over any of its subscripts i, j, k, l, or t. The λ's represent the possible "effects" of the five variables on the ξ_{ijklt}. (In the present context, we do *not* define a "main effect" λ pertaining to variable E, since (a) it would be a function of the θ_t rather than the λ_t's [used in (2.4)], and (b) it could not be treated in the same way as we shall treat the λ's in (2.5).)

Since formula (2.2) provides a separate saturated model for each of the T 4-way tables, the methods in [8, p. 38] for estimating the λ's in the saturated model for a given 4-way table can be applied directly to estimate the λ_t's in (2.2) for each of the T tables. These methods can also be extended to estimate the λ's in (2.5); see Section 4.1 herein.

3. UNSATURATED MODELS FOR THE SET OF T TABLES

Unsaturated models can be obtained by specifying that certain λ_t's in (2.2), or certain λ's in (2.5), are zero. Sections 3.1–3.3 provide examples of this method.

3.1 Models that Assume Complete Homogeneity

Consider the model in which all λ's in (2.5) that have the letter E among the superscripts are zero. This model states that the λ_t's in (2.2) are independent of t [see (2.4)]; i.e., that the corresponding T values of λ_t (for $t = 1, 2, \cdots, T$) are equal to each other. Under this model, the expected proportion F_{ijklt}/n_t of observations in cell (i, j, k, l) will also be independent of t; i.e.,

$$F_{ijklt}/n_t = P_{ijkl}, \quad \text{for} \quad t = 1, 2, \cdots, T. \tag{3.1}$$

Condition (3.1) states that the T tables are completely homogeneous. In other words, it states that variable E is independent of the joint variable $ABCD$.

The following models assume complete homogeneity:

M_0: All λ's in (2.5) are zero. This model states that, within the tth table ($t = 1, 2, \cdots, T$), all of the cells (i, j, k, l) are equiprobable. Condition (3.1) is satisfied for this model.

M_1: All λ's in (2.5) are zero except for the four 1-factor λ's; viz., λ_i^A, λ_j^B, λ_k^C, λ_l^D. This model states that, within the tth table ($t = 1, 2, \cdots, T$), variables A, B, C, D are mutually independent, *and* that (3.1) is satisfied.

M_2: All λ's in (2.5) are zero except for the four 1-factor λ's and the following six 2-factor λ's: λ_{ij}^{AB}, λ_{ik}^{AC}, λ_{il}^{AD}, λ_{jk}^{BC}, λ_{jl}^{BD}, λ_{kl}^{CD}. This model states that all 3-factor and 4-factor λ_t's in (2.2) are zero, *and* that (3.1) is satisfied.

M_3: All 4-factor and 5-factor λ's in (2.5) are zero, and so are all λ's that have the letter E among the superscripts. This model states that all 4-factor λ_t's in (2.2) are zero, *and* that (3.1) is satisfied.

M_4: All λ's in (2.5) that have the letter E among the superscripts are zero. This model states only that (3.1) is satisfied.

If the set of λ's in (2.5) that are zero under a given model includes all λ's that have the letter *E* among the

2. CHI-SQUARE VALUES FOR SOME MODELS THAT ASSUME COMPLETE HOMOGENEITY, APPLIED TO TABLE 1

Model	Fitted marginals	Degrees of freedom	Likelihood-ratio chi-square	Goodness-of-fit chi-square
M_0	{E}	30	2873.64	3255.17
M_1	{E},{A},{B},{C},{D}	26	188.36	200.43
M_2	{E},{AB},{AC},{AD},{BC},{BD},{CD}	20	74.46	73.01
M_3	{E},{ABC},{ABD},{ACD},{BCD}	16	69.81	68.86
M_4	{E},{ABCD}	15	68.37	67.63

3. ANALYSIS OF HETEROGENEITY AMONG TABLES AND ASSOCIATION WITHIN TABLES, APPLIED TO DATA IN TABLE 1

Source of variation	df	Chi-square	Numerical value
1. Total due to heterogeneity among the conditional tables {ABCD|E}, <u>and</u> to mutual dependence among A, B, C, D	26	$X^2(M_1)$	188.36*
1a. Due to heterogeneity among the conditional tables {ABCD|E}	15	$X^2(M_4)$	68.37*
1b. Due to mutual dependence among A, B, C, D in marginal table {ABCD}	11	$X^2(M_1\|M_4)$	119.99*
<u>Partition of (1b)</u>			
1b.1. Due to residual 4-factor interactions in {ABCD}	1	$X^2(M_3\|M_4)$	1.44
1b.2. Due to 3-factor interaction in {ABCD} (assuming 4-factor interaction is nil)	4	$X^2(M_2\|M_3)$	4.65
1b.3. Due to 2-factor interaction in {ABCD} (assuming 3-factor and 4-factor interactions are nil)	6	$X^2(M_1\|M_2)$	113.90*

NOTE: Asterisk denotes significance at the .05 level.

superscripts, then condition (3.1) will be satisfied. Models M_0–M_4 are examples of this. These five models are also examples of hierarchical hypotheses about (2.5). Applying the terminology and notation in [7], Table 2 describes, for each of the five models, the minimal set of marginals fitted under the model, the number of *df* used to test whether the model fits the data in Table 1 (where $I = J = K = L = T = 2$), and the chi-square value obtained in testing whether the model fits. (Table 2 gives both the likelihood-ratio chi square and the goodness-of-fit chi square, but our discussion will be limited to the former, which we denote by $X^2(M_i)$, for $i = 0, 1, \cdots, 4$.) When I, J, K, L, or T are not equal to two, see [7] for various ways to calculate the *df* pertaining to each model, and also for a description of the calculation of the chi squares.

If models of complete homogeneity are of interest, then M_4 should be tested first. If $X^2(M_4)$ is large, we need not test the other M_i (for $i = 0, 1, 2, 3$). [Note that $X^2(M_i)$ is monotonically non-increasing as i increases from 0 to 4.] On the other hand, if M_4 fits the data well, we could then analyze the marginal table obtained by ignoring variable E (i.e., by summing the corresponding T entries in cell (i, j, k, l) of the T tables). We can partition $X^2(M_i)$ (for $i = 0, 1, 2, 3$) into (a) $X^2(M_4)$ and (b) a chi square for testing a hypothesis corresponding to

4. CHI-SQUARE VALUES FOR SOME MODELS THAT ALLOW FOR COMPLETE HETEROGENEITY, APPLIED TO TABLE 1

Model	Fitted marginals	Degrees of freedom	Likelihood-ratio chi-square	Goodness-of-fit chi-square
M_1'	{AE},{BE},{CE},{DE}	22	130.48	132.46
M_2'	{ABE},{ACE},{ADE},{BCE},{BDE},{CDE}	10	11.56	11.67
M_3'	{ABCE},{ABDE},{ACDE},{BCDE}	2	1.84	1.85

M_i applied to the marginal table obtained by ignoring variable E (see [9]). The chi square for (b) above can be calculated directly from the marginal table, or as $X^2(M_i) - X^2(M_4)$ which we denote by $X^2(M_i | M_4)$. For example, Table 3 partitions $X^2(M_1)$ into components (1a) and (1b); with (1b) partitioned in turn into (1b.1), (1b.2) and (1b.3). Similarly, $X^2(M_2)$ can be partitioned into (1a), (1b.1) and (1b.2); and $X^2(M_3)$ can be partitioned into (1a) and (1b.1). These partitions shed further light on the meaning of the M_i (for $i = 1, 2, 3$).

If the marginal $\{E\}$ is included in the minimal set of marginals fitted under a given model (as in Table 2), then the model assumes complete homogeneity. If $\{E\}$ is not included in the minimal set, then either (a) at least one of the marginals in the minimal set is $\{EY\}$ where Y is a non-empty subset of the letters A, B, C, D (as in, e.g., Table 4), or (b) neither $\{E\}$ nor $\{EY\}$ is included in the minimal set. Case (a) will be discussed in Section 3.2–3.3; and case (b) describes models that are inappropriate in the present context since the total expected frequency (under the model) within the tth part of the 5-way table will not equal the fixed size n_t, except when n_t is independent of t. [In Table 1, n_t is independent of t, but our remark about the inappropriateness of case (b) models applies in the more general situation where n_t need not be independent of t.]

The analysis of Table 1 in [3] used four models that were expressed in different terms from the models presented here; but they are equivalent to M_1 and M_2 above and M_1' and M_2' below.

3.2 Models that Allow for Complete Heterogeneity

Consider now the following models, which are modifications of the corresponding M_i in Section 3.1 (for $i = 1, 2, 3$):

M_1': All λ_t's in (2.2) are zero except for the 1-factor λ_{it}^{A}, λ_{jt}^{B}, λ_{kt}^{C}, λ_{lt}^{D}. This model states that, within the tth table (i.e., given level t of the variable E), the variables A, B, C, D are mutually independent.

5. ANALYSIS OF CONDITIONAL ASSOCIATION IN TABLE 1

Source of variation	df	Chi-square	Numerical value
1. Total due to mutual dependence among A, B, C, D in the conditional table {ABCD|E}	22	$X^2(M_1')$	130.48*
1a. Due to residual 4-factor interaction in {ABCD|E}	2	$X^2(M_3')$	1.84
1b. Due to 3-factor interaction in {ABCD|E} (assuming 4-factor interaction is nil)	8	$X^2(M_2'\|M_3')$	9.72
1c. Due to 2-factor interaction in {ABCD|E} (assuming 3-factor and 4-factor interactions are nil)	12	$X^2(M_1'\|M_2')$	118.92*

NOTE: Asterisk denotes significance at the .05 level.

M_2': All 3-factor and 4-factor λ_t's in (2.2) are zero.

M_3': All 4-factor λ_t's in (2.2) are zero.

For the M_i', Table 4 gives the same kind of information presented in Table 2 for the M_i; and Table 5 partitions $X^2(M_1')$ into components (1a), (1b), and (1c). Similarly, $X^2(M_2')$ can be partitioned into (1a) and (1b). These partitions shed further light on the M_i', and they analyze further the association within the two 4-way tables (Table 1). The chi-square values in Tables 4 and 5 could have been obtained by analyzing each 4-way table separately, and then summing the corresponding chi squares for the two tables [7, Section 6.5].

Further light on the relationship between the M_i' and the corresponding M_i can be shed by partitions of the kind in Table 6. For example, we partition $X^2(M_2)$ here into components (1a) [i.e., $X^2(M_2')$] and (1b). [This partition provides a somewhat different view of M_2 than does the partition of M_2 by (1a), (1b.1), and (1b.2) of Table 3.] Each $X^2(M_i)$ (for $i = 1, 2, 3$) can be partitioned into $X^2(M_i')$ and a component corresponding to (1b) of Table 6.

6. ANALYSIS OF ASSOCIATION WITHIN TABLES AND HETEROGENEITY AMONG TABLES, APPLIED TO DATA IN TABLE 1

Source of variation	df	Chi-square	Numerical value
1. Total due to residual 3-factor and 4-factor interactions in the conditional tables {ABCD|E}, <u>and</u> to heterogeneity among the tables	20	$X^2(M_2)$	74.46*
1a. Due to residual 3-factor and 4-factor interactions in {ABCD|E}	10	$X^2(M_2')$	11.56
1b. Due to heterogeneity of corresponding 1-factor interactions and 2-factor interactions among the conditional tables {ABCD|E}	10	$X^2(M_2\|M_2')$	62.90*

NOTE: Asterisk denotes significance at the .05 level.

The M_i' are examples of a broader class of models that allow for complete heterogeneity. Any model is included in this class if each of the marginal tables in the minimal set of marginals fitted under the model includes the E dimension among its dimensions (as in Table 4), and the minimal set is not just $\{E\}$. Expressed in terms of the λ_t's of (2.2), for each model in the class, there is a specified (non-empty) set of λ_t's included in the model (the same set for $t = 1, 2, \cdots, T$), and no assumption is made as to whether the corresponding T values of λ_t are or are not equal to each other.

The class of models described above could be expanded further by not requiring that the set of λ_t's included in the model be the same for $t = 1, 2, \cdots, T$. For models of this kind, a separate analysis would be carried out for each of the T tables, using the procedures developed for the analysis of a given multidimensional contingency table, and the extensions described in Section 4 herein.

The class of models considered in the second paragraph preceding this one can be characterized in terms of the λ's of (2.5) as follows: For each model in this class, if a given λ that does not have the letter E among its superscripts is a member of the (non-empty) set of λ's that are included in the model, then the corresponding λ with the letter E added to its superscripts will also be a member of this set. On the other hand, the *expanded* class of models considered in the preceding paragraph can not be characterized in a simple way in terms of the λ's of (2.5). Both the class of models considered in the second paragraph preceding this one, and the expanded class considered in the preceding paragraph, can be characterized in terms of the λ_t's of (2.2).

3.3 Models that Allow for Partial Heterogeneity

Consider now the following models, which are modifications of M_4 in Section 3.1:

M_1'': All λ's in (2.5) that have an E among the superscripts are zero, except for the four 2-factor λ's: viz., λ_{it}^{AE}, λ_{jt}^{BE}, λ_{kt}^{CE}, λ_{lt}^{DE}.

M_2'': All λ's in (2.5) that have an E among the superscripts are zero, except for the four 2-factor λ's and the six 3-factor λ's; viz., λ_{ijt}^{ABE}, λ_{ikt}^{ACE}, λ_{ilt}^{ADE}, λ_{jkt}^{BCE}, λ_{jlt}^{BDE}, λ_{klt}^{CDE}. (In this model, the λ_{ijkt}^{ABCE}, λ_{ijlt}^{ABDE}, λ_{iklt}^{ACDE}, λ_{jklt}^{BCDE}, and λ_{ijklt}^{ABCDE} are zero.)

M_3'': Only λ_{ijklt}^{ABCDE} is zero. (Considering the λ's in (2.5) that have an E among the superscripts, the four 2-factor λ's, the six 3-factor λ's, and the four 4-factor λ's are included in the set of λ's that are not taken equal to zero in this model.)

Table 7 gives the results obtained when these models are applied to Table 1. Although M_4 of Section 3.1 assumes complete homogeneity [see (3.1)], we list it as M_0'' in Table 7, since modification of it led to M_1'', and then in turn to M_2'' and M_3''.

Table 8 partitions $X^2(M_0'')$ into components (1a)–(1d). Similarly, $X^2(M_1'')$ can be partitioned into (1b)–(1d); and $X^2(M_2'')$ can be partitioned into (1c)–(1d). These partitions analyze further the heterogeneity among the two 4-way tables (Table 1). The symbols (AE), (BE), etc. in Table 8 denote the corresponding λ^{AE}, λ^{BE}, etc.

Models M_i'' (for $i = 1, 2, 3$) allow for partial heterogeneity among the T tables by introducing into M_4 of Section 3.1 certain λ's from (2.5) that have an E among the superscripts. These models can also be described in

7. CHI-SQUARE VALUES FOR SOME MODELS THAT ALLOW FOR PARTIAL HETEROGENEITY APPLIED TO TABLE 1

Model	Fitted marginals	Degrees of freedom	Likelihood-ratio chi-square	Goodness-of-fit chi-square
$M_0''(=M_4)$	{ABCD},{E}	15	68.37	67.63
M_1''	{ABCD},{AE},{BE},{CE},{DE}	11	16.31	16.16
M_2''	{ABCD},{ABE},{ACE},{ADE},{BCE},{BDE},{CDE}	5	4.55	4.59
M_3''	{ABCD},{ABCE},{ABDE},{ACDE},{BCDE}	1	0.17	0.17

quite different terms. For example, when $T = 2$, M_1'' is equivalent to the model in which the logit pertaining to variable E is a sum of a general mean and the "main effects" of variables A, B, C, D on E (see, e.g., [2, 5, 7]); and when $T > 2$, we have a generalization thereof (see [7, 8]). Each M_i'' (for $i = 0, 1, 2, 3$) is equivalent to a

8. ANALYSIS OF HETEROGENEITY AMONG TABLES, APPLIED TO DATA IN TABLE 1

Source of variation	df	Chi-square	Numerical value
1. Total due to heterogeneity among the conditional tables {ABCD|E}	15	$X^2(M_0'')$	68.37*
1a. Due to (AE), (BE), (CE), (DE)	4	$X^2(M_0''\|M_1'')$	52.06*
1b. Due to (ABE), (ACE), (ADE), (BCE), (BDE), (CDE), in model that also includes 4 effects listed in (1a)	6	$X^2(M_1''\|M_2'')$	11.76+
1c. Due to (ABCE), (ABDE), (ACDE), (BCDE), in model that also includes 10 effects listed in (1a) and (1b)	4	$X^2(M_2''\|M_3'')$	4.38
1d. Due to (ABCDE), in model that also includes 14 effects listed in (1a), (1b), (1c)	1	$X^2(M_3'')$	0.17

NOTE: Asterisk denotes significance at the .05 level, and the symbol + denotes significance at the .10 level.

9. CHI-SQUARE VALUES FOR SOME MODELS THAT ALLOW FOR PARTIAL HETEROGENEITY AND/OR THAT ASSUME THAT CERTAIN INTERACTIONS WITHIN THE TABLES ARE NIL, APPLIED TO DATA IN TABLE 1

Model	Fitted marginals	Degrees of freedom	Likelihood-ratio chi-square	Goodness-of-fit chi-square
M_1''' $(=M_1)$	{A},{B},{C},{D},{E}	26	188.36	200.43
M_2'''	{AB},{AC},{AD},{AE},{BC}, {BD},{BE},{CD},{CE},{DE}	16	22.40	22.26
M_3'''	{ABC},{ABD},{ABE},{ACD},{ACE}, {ADE},{BCD},{BCE},{BDE},{CDE}	6	5.88	6.04
M_4''' $(=M_3'')$	{ABCD},{ABCE},{ABDE},{ACDE}, {BCDE}	1	0.17	0.17

corresponding model for the logit pertaining to variable E, which can be determined from the fitted marginals in Table 7 (see [7]). (Two models are called "equivalent" here if the maximum-likelihood estimates of the corresponding expected frequencies are identical under the two models.)

The M_i'' (for $i = 1, 2, 3$) are examples of a broader class of models that allow for partial heterogeneity. Any model is included in this class if the minimal set of fitted marginals consists of at least one marginal table (not $\{E\}$) that includes E among its dimensions and at least one marginal table that does not include E. Additional examples are given as M_i''' (for $i = 2, 3, 4$) in Table 9. Although M_1 of Section 3.1 is not in this class, we list it as M_1''' in Table 9, since modification of it can lead in turn to M_2''', M_3''', M_4'''.

In M_2''', all 3-, 4-, and 5-factor λ's in (2.5) are zero [i.e., all 3- and 4-factor λ_t's in (2.2) are zero, and the corresponding 2-factor λ_t's are homogeneous among the T tables]. In M_3''', all 4- and 5-factor λ's in (2.5) are zero [i.e., all 4-factor λ_t's in (2.2) are zero, and the corresponding 3-factor λ_t's are homogeneous]. In M_4''', the 5-factor λ in (2.5) is zero [i.e., the corresponding 4-factor λ_t's in (2.2) are homogeneous].

In contrast to the M_i''' (for $i = 1, 2, 3$), each of the M_i'' in Table 7 includes all of the λ_t's in (2.2). In M_1'', the corresponding 2-, 3-, and 4-factor λ_t's in (2.2) are homogeneous among the T tables; in M_2'', the corresponding 3- and 4-factor λ_t's are homogeneous; and in M_3'', the corresponding 4-factor λ_t's are homogeneous.

The M_i'' (for $i = 1, 2, 3$), and the other models introduced in this and the preceding section, will prove useful in, e.g., the classification (or discrimination) problem, where a given individual's class (i, j, k, l) with respect to variables (A, B, C, D) is observed, but his class t with respect to variable E (i.e., with respect to the T tables) is unknown and it is to be estimated on the basis of his (i, j, k, l) class. If, say, M_1'' fit the data in the set of T tables, then for the given individual whose class is (i, j, k, l), we can estimate his class with respect to variable E to be t^*, where t^* is such that

$$\hat{F}_{ijklt^*} = \max_t \hat{F}_{ijklt}, \tag{3.2}$$

10. STANDARDIZED VALUE OF THE MAIN EFFECTS AND INTERACTION EFFECTS AMONG THE FOUR VARIABLES IN THE TWO 4-WAY CONTINGENCY TABLES (TABLE 1)

Variable	High IQ sample Standardized value	Low IQ sample Standardized value
A	15.83	17.17
B	-16.98	-16.50
C	5.07	.88
D	2.75	- 2.65
AB	.10	- 2.17
AC	3.45	2.99
AD	- .22	.66
BC	3.18	2.54
BD	2.03	1.53
CD	1.58	2.18
ABC	1.02	.43
ABD	.20	- 1.78
ACD	.72	- .25
BCD	.61	- .44
ABCD	- 1.19	- .62

and where the $\hat{F}_{ijklt}$ are the estimated expected frequencies in cell (i, j, k, l) of the tth table $(t = 1, 2, \cdots, T)$ under M_1''. Use of (3.2) to determine t^* can be justified when the sample size n_t from the tth population is proportional to the proportion π_t of individuals in the tth population, and when we wish to minimize the expected frequency of mistakes in estimation (i.e., classification).

When n_t is not proportional to π_t, we would estimate the given individuals's class with respect to variable E to be t', where t' is such that

$$\pi_{t'}\hat{F}_{ijklt'}/n_{t'} = \max_t \pi_t\hat{F}_{ijklt}/n_t. \tag{3.3}$$

When the expected total cost of mistakes in classification (rather than the expected frequency of these mistakes) is to be minimized, criterion (3.3) can be directly extended to cover this more general situation.

The analysis of Table 1 in [12] used an unsaturated model that is different from but related to M_1''. The model in [12] is not included in the class of unsaturated models discussed in the present article; i.e., models obtained from (2.2) or (2.5). The methods used in [12] for this analysis are not as easy to apply as are the methods proposed herein.

4. METHODS FOR SELECTING MODELS THAT FIT THE DATA

4.1 Use of Direct Estimation Methods

In Section 2 we noted that the maximum likelihood estimate $\hat{\lambda}_t$ of each λ_t in (2.2) can be obtained by applying the methods in [8, p. 38] to each of the T 4-way tables. Thus, letting

$$y_{ijklt} = \log f_{ijklt} \tag{4.1}$$

(assuming $f_{ijklt} > 0$), we find, e.g., that

$$\hat{\lambda}_{it}^{A} = y_{i\cdots t} - y_{\cdots\cdot t}, \quad \hat{\lambda}_{ijt}^{AB} = y_{ij\cdot\cdot t} - y_{i\cdots t} - y_{\cdot j\cdot\cdot t} + y_{\cdots\cdot t}, \text{ etc.}, \tag{4.2}$$

where the dot denotes an average over the corresponding subscript. Each $\hat{\lambda}_t$ can be expressed as

$$\hat{\lambda}_t = \sum_{i,j,k,l} a_{ijkl} y_{ijklt}, \tag{4.3}$$

where the a_{ijkl} are constants that depend on which $\hat{\lambda}_t$ is being calculated, and where $\sum_{i,j,k,l} a_{ijkl} = 0$. The variance of $\hat{\lambda}_t$ can be estimated by

$$S^2_{\hat{\lambda}_t} = \sum_{i,j,k,l} a^2_{ijkl}/f_{ijklt}, \tag{4.4}$$

and the standardized value of λ_t can be calculated as $\hat{\lambda}_t/S_{\hat{\lambda}_t}$. Table 10 gives the standardized value of $\hat{\lambda}^A_{1t}$, $\hat{\lambda}^B_{1t}$, etc., when each f_{ijklt} in (4.1)–(4.4) is replaced by $f_{ijklt} + \frac{1}{2}$, as recommended in [7].

The maximum-likelihood estimate $\hat{\lambda}$ of each λ in (2.5) can be obtained from the $\hat{\lambda}_t$ using (2.4), or by applying the methods in [7] for the analysis of the 5-way table. Thus, e.g.,

$$\hat{\lambda}^A_i = \sum_t \hat{\lambda}^A_{it}/T = y_{i....} - y_{.....},$$

$$\hat{\lambda}^{AB}_{ij} = \sum_t \hat{\lambda}^{AB}_{ijt}/T = y_{ij...} - y_{i....} - y_{.j...} + y_{.....},$$

$$\hat{\lambda}^{AE}_{it} = \hat{\lambda}^A_{it} - \hat{\lambda}^A_i = y_{i...t} - y_{i....} - y_{....t} + y_{.....}, \text{ etc.} \tag{4.5}$$

Each $\hat{\lambda}$ can be expressed as a modified (4.3) and its variance can be estimated by a modified (4.4), with a_{ijkl} in (4.3)–(4.4) replaced by a_{ijklt}, and the summations taken over the index t as well. Table 11 gives the standardized values of $\hat{\lambda}^A_1$, $\hat{\lambda}^B_1$, etc., listed in decreasing order of their absolute values, first considering the 1-factor $\hat{\lambda}$'s, then the 2-factor $\hat{\lambda}$'s, etc.

The standardized values in Table 10 can be used as a partial guide in selecting unsaturated models of the kind described in Section 3.2, and the values in Table 11 can be used similarly in selecting models of the kind described in any of the subsections of Section 3. We shall illustrate the latter usage in Sections 4.3–4.5.

The variables considered in Tables 10 and 11 are all dichotomous. The a_{ijkl} used in (4.3) to obtain the $\hat{\lambda}_t$'s are particularly easy to calculate when variables A, B, C, D

are dichotomous, and similarly the a_{ijklt} used in the modified (4.3) to obtain the $\hat{\lambda}$'s are particularly easy to calculate when variables A, B, C, D, and E are dichotomous [7]. When these variables are not all dichotomous, the following results will be useful: Letting i_0, j_0, k_0, denote a given value of i, j, k, respectively, when $\hat{\lambda}$ in the modified (4.3) is taken as $\hat{\lambda}^A_{i_0}$, we see from (4.5) that

$$a_{ijklt} = [I\delta^i_{i_0} - 1]/(IJKLT), \qquad (4.6.a)$$

where δ denotes the Kronecker delta: $\delta^i_{i_0} = 1$ if $i = i_0$, and $\delta^i_{i_0} = 0$ otherwise. Similarly, when $\hat{\lambda}$ is taken as $\hat{\lambda}^{AB}_{i_0 j_0}$, we obtain

$$a_{ijklt} = [IJ\delta^i_{i_0}\delta^j_{j_0} - I\delta^i_{i_0} - J\delta^j_{j_0} + 1]/(IJKLT); \qquad (4.6.b)$$

and when $\hat{\lambda}$ is taken as $\hat{\lambda}^{ABC}_{i_0 j_0 k_0}$, we obtain

$$a_{ijklt} = [IJK\delta^i_{i_0}\delta^j_{j_0}\delta^k_{k_0} - IJ\delta^i_{i_0}\delta^j_{j_0} - IK\delta^i_{i_0}\delta^k_{k_0} - JK\delta^j_{j_0}\delta^k_{k_0} + I\delta^i_{i_0} + J\delta^j_{j_0} + K\delta^k_{k_0} - 1]/(IJKLT); \qquad (4.6.c)$$

etc. With formulas like (4.6.a, b, c) for the a_{ijklt}, we can express each $\hat{\lambda}$ using the modified (4.3), and we can then calculate $S^2_{\hat{\lambda}}$ using the modified (4.4). Formulas like (4.6.a, b, c) can also be obtained for the a_{ijkl} used in (4.3) to calculate the $\hat{\lambda}_t$'s and in (4.4) to calculate $S^2_{\hat{\lambda}_t}$.

Before closing this section, we include here some further comments on the case where the variables (A, B, C, D, E) are not all dichotomous. When some of these variables are trichotomous (and the others are dichotomous), the corresponding $\hat{\lambda}$'s can be replaced by linear and quadratic interaction effects pertaining to these variables, assuming that the levels of each trichotomous variable are equally spaced (see, e.g., [8]). (The $\hat{\lambda}$'s used herein do *not* require assumptions of this kind about the levels of the variables.) In some cases, use of the linear and quadratic interaction effects can facilitate the analysis based on the direct estimation method. For example, examination of the standardized values of the effects in [8, Table 2] would suggest that the quadratic effect between two particular variables (viz., variables S

11. STANDARDIZED VALUE OF THE MAIN EFFECTS AND INTERACTION EFFECTS AMONG THE FIVE VARIABLES OBTAINED BY TREATING THE TWO 4-WAY CONTINGENCY TABLES AS A SINGLE 5-WAY TABLE (TABLE 1)

Variable	Standardized value
B	-23.67
A	23.33
C	4.23
D	.09
E	-----
AC	4.55
BC	4.04
DE	3.82
CE	2.99
CD	2.66
BD	2.52
AB	- 1.45
AE	- .84
BE	- .45
AD	.31
ABE	1.60
ABD	- 1.11
ABC	1.03
ADE	- .62
BCE	.47
CDE	- .41
BDE	.36
ACE	.35
ACD	.34
BCD	.12
ABDE	1.39
ABCD	- 1.29
BCDE	.74
ACDE	.68
ABCE	.42
ABCDE	- .41

and T) could be deleted, but the corresponding linear interaction effect should not be. Unsaturated models in which some quadratic effects are deleted (but the corresponding linear effects are not) differ from the kinds of models considered in [8] and here. The latter class of models is wide enough to include models that provide a good fit for the data we have analyzed (both in [8] and here), but the researcher should be made aware of the possibility of expanding the models in the direction just indicated.

The remarks in the preceding paragraph can be directly extended to the case where the variables are not limited to trichotomous and/or dichotomous ones.

A. RELATIONSHIP AMONG MODELS IN TABLES 2, 4, 7, 9, WITH RESPECT TO THEIR IMPLICATIONS AND DEGREES OF FREEDOM

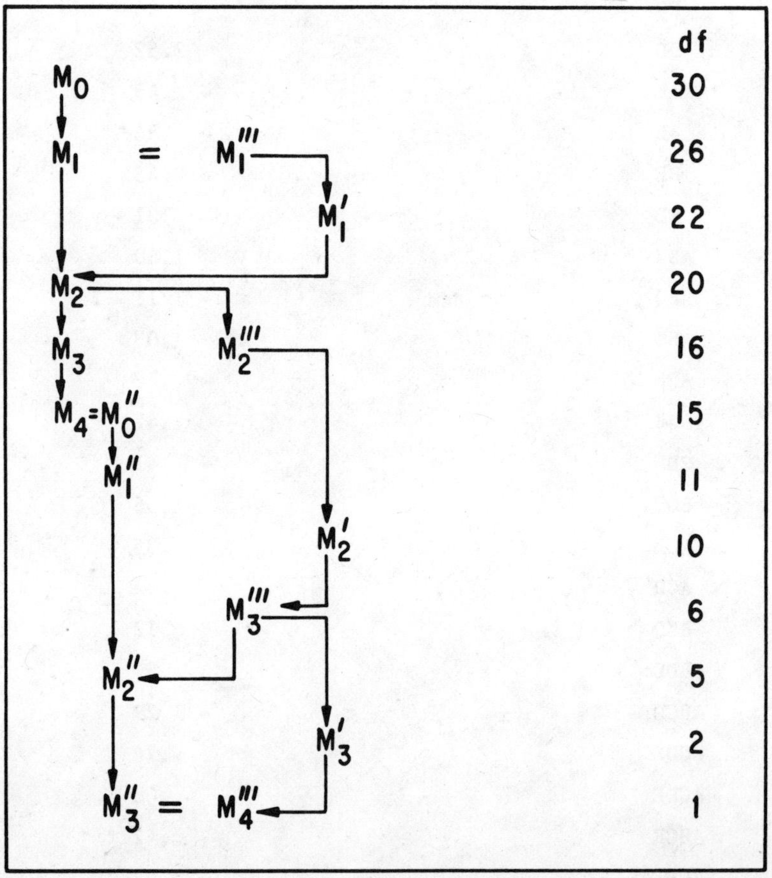

4.2 Guided and Unguided Procedures

The stepwise procedures in [8], and the extensions in Section 4.5 below, can be carried out with or without the aid of the standardized values obtained by the direct estimation method (see Table 10 and 11). We shall use the term "guided" to describe those procedures that use the results of the direct estimation method as a (partial) guide in determining the models that should be considered. The procedures used in Sec. 3 were "unguided." Various guided and unguided methods can be useful in selecting models.

For a given set of T tables, the models in Section 3 can be considered sequentially in various ways. For example, if M_4 of Table 2 did not fit the data well, we might then consider some of the models in Tables 4, 7, or 9. (Partitions of the kind in Tables 3, 5, 6, and 8 can also help in selecting models.) Tables 2, 4, and 7 view variables A, B, C, D, symmetrically; while Table 9 views A, B, C, D, E, symmetrically. If the latter view is taken as a starting point, we would begin with the models of Table 9.

If M_4 of Table 2 did not fit the data well, we might then proceed as indicated in the preceding paragraph; but if this model did fit the data, we might then proceed to consider M_3 of Table 2 (see related comments in Section 3.1). This procedure would use the results obtained with M_4 to help in determining the models that should be considered next. Such procedures are "unguided" in the strict sense used here (i.e., they do not use the results of the direct estimation method for guidance), but they are of course guided by the results obtained with M_4 and with the models examined at subsequent steps. The latter form of guidance uses the results obtained with the indirect testing procedure rather than with the direct estimation method. For simplicity, we shall use the term "guidance" here to refer to "direct-estimation guidance," though we could of course have referred also to "indirect-testing guidance."

To find models that fit the data, we need not limit consideration to the particular models listed in Tables 2, 4, 7, 9. Additional models can be obtained by filling in the "gaps" between the models in these tables. For example, in the gap between M_2''' and M_3''', we could insert M_2' (which introduces four 3-factor λ's into M_2''' and deletes six 3-factor λ's from M_3''') and other models that introduce λ's into M_2''' and delete λ's from M_3'''.

If a given model (say M^*) can be obtained by introducing λ's into another model (say M), then M implies M^*. We denote this by $M \Rightarrow M^*$; if M is true, then M^* will also be true. In this case, M is "simpler" than M^*. (Note, e.g., that $M_2''' \Rightarrow M_2' \Rightarrow M_3'''$.) If $M \Rightarrow M^*$, then when these models are fitted to a given set of data we find that

12. CHI-SQUARE VALUES FOR SOME MODELS PERTAINING TO TABLE 1

Model	Fitted marginals	Degrees of freedom	Likelihood-ratio chi-square	Goodness-of-fit chi-square
$H_1=M_3''=M_4'''$	{ABCD},{ABCE},{ABDE},{ACDE}, {BCDE}	1	0.17	0.17
$H_2=M_3'$	{ABCE},{ABDE},{ACDE},{BCDE}	2	1.84	1.85
$H_3=M_2''$	{ABCD},{ABE},{ACE},{ADE},{BCE}, {BDE},{CDE}	5	4.55	4.59
$H_4=M_3'''$	{ABC},{ABD},{ABE},{ACD},{ACE}, {ADE},{BCD},{BCE},{BDE},{CDE}	6	5.88	6.04
$H_5=M_2'$	{ABE},{ACE},{ADE},{BCE},{BDE}, {CDE}	10	11.56	11.67
$H_6=M_1''$	{ABCD},{AE},{BE},{CE},{DE}	11	16.31	16.16
$H_7=M_2'''$	{AB},{AC},{AD},{AE},{BC},{BD}, {BE},{CD},{CE},{DE}	16	22.40	22.26
H_8	{ABCD},{ABDE},{ACDE},{BCDE}	2	0.31	0.32
H_9	{ABCD},{ABDE},{ACDE},{BCE}	3	0.67	0.67
H_{10}	{ABCD},{ABDE},{ACE},{BCE},{CDE}	4	2.13	2.13
H_{11}	{ABCD},{ABDE},{BCE},{CDE}	5	2.13	2.13
H_{12}	{ABCD},{ABDE},{BCE}	6	2.37	2.37
H_{13}	{ABCD},{ABDE},{CE}	7	3.16	3.16
H_{14}	{ABDE},{ABC},{ACD},{BCD},{CE}	8	4.92	4.94
H_{15}	{ABDE},{ABC},{ACD},{CE}	9	5.32	5.35

Table 12 Continued

Model	Fitted marginals	Degrees of freedom	Likelihood-ratio chi-square	Goodness-of-fit chi-square
H_{16}	{ABDE},{ACD},{BC},{CE}	10	6.11	6.14
H_{17}	{ABD},{ABE},{ACD},{ADE},{BDE},{BC},{CE}	11	8.18	8.16
H_{18}	{ABE},{ACD},{ADE},{BDE},{BC},{CE}	12	10.11	10.00
H_{19}	{ABE},{ADE},{BDE},{AC},{BC},{CD},{CE}	13	12.59	12.64
H_{20}	{ACD},{ADE},{BDE},{BC},{CE}	14	13.95	14.73
H_{21}	{ADE},{BDE},{AC},{BC},{CD},{CE}	15	16.52	17.45
H_{22}	{ACD},{ADE},{BC},{BD},{CE}	16	19.21	19.01
H_{23}	{BDE},{AC},{AE},{BC},{CD},{CE}	17	21.08	21.06
H_{24}	{AC},{AE},{BC},{BD},{BE},{CD},{CE},{DE}	18	24.47	24.27
H_{25}	{AC},{AE},{BC},{BD},{CD},{CE},{DE}	19	26.24	25.88

$X^2(M) \geq X^2(M^*)$. We can test whether M is true by testing (a) whether M^* is true and (b) whether the particular λ's introduced into M to form M^* are zero (assuming M^* is true). This leads to the partitioning of $X^2(M)$ into $X^2(M^*)$ and $X^2(M) - X^2(M^*) = X^2(M \mid M^*)$, for testing (a) and (b), respectively.

Within each of Tables 2, 4, 7, 9, each model in the table implies the ones below it. Figure A shows how the models in these tables are related to each other in this respect, moving from the simplest model (M_0 with $30df$) to the least simple (M_3'' with $1df$). Note, e.g., that $M_1''' \Rightarrow M_1' \Rightarrow M_2 \Rightarrow M_2'''$ in Fig. A, and that M_2''' and M_3 are not comparable in this respect.

In addition to models that fit into the "gaps" between successive models in Table 2, 4, 7, 9, we might also consider models that do not do this. For example, model H_{16} of Table 12 fits the data well, but does not fit into such "gaps". (On the other hand, it does fit into the "gap" between M_2''' and M_4'''.)

Model H_{16} can be described in various ways (see [7, 8] and Section 3 herein). For example, it includes the 1- and 2-factor λ_t's, and λ_{ijlt}^{ABD} and λ_{iklt}^{ACD}; and the following λ_t's are homogeneous among the T tables: λ_{ikt}^{AC}, λ_{jkt}^{BC}, λ_{klt}^{CD}, λ_{iklt}^{ACD}. This model can also be interpreted as follows: It views each of the 4-way tables as a 3-way table consisting of variables B, C, and the joint variable AD; and it states that there is zero 3-factor interaction in each 3-way table, and that the 2-factor interactions between B and C and between C and AD are homogeneous among the T tables.

Additional interpretations of model H_{16} can be obtained by noting, e.g., that it is related to a model for the logit pertaining to variable C for each of the 4-way tables (see [7]); viz., the logit model in which the minimal set of fitted marginals is as follows: $\{ABDE\}$, $\{ACDE\}$, $\{BCE\}$. For each of the 4-way tables, this logit model expresses the logit pertaining to variable C as a sum of a "general mean" and the "main effects" of variables A, B, and D, and the "interaction effect" between variables A and D. Model H_{16} states both that the preceding logit model holds true *and* that the "main effects" and "interaction effect" in the logit model are homogeneous among the T tables, but the "general mean" may be heterogeneous among the T tables.

Corresponding to each number of *df* from 2 to 19, models H_8 to H_{25} in Table 12 were obtained by one or more of the selection procedures in [8], or by extensions of these procedures (see Section 4.3–4.5 below). Also included as H_1 to H_7 are models from Section 3 that were not grossly inadequate in fitting the data. Note the decrease in chi square when comparing H_2 to H_7 with the corresponding model among H_8 to H_{25} having the same number of *df*. As we did earlier with H_{16}, each model in Table 12 can be described in various terms (see [7, 8] and Section 3 herein).

The relationships among H_1 to H_7, with respect to their implications, can be obtained from Figure A. For H_8 to H_{19}, we note that $H_{19} \Rightarrow H_{18} \Rightarrow \cdots \Rightarrow H_8$. For H_{19} to H_{25}, the relationships among these models, with respect to their implications, are included in Figure B. (Model H'_{19} in Figure B will be introduced in Section 4.5 below.)

In the next two sections, we shall describe some of the ways in which guided methods can be used to extend the procedures in [8], and in the final section we introduce additional kinds of extensions. These extensions were used to obtain some of the models introduced in Table 12.

4.3 Guided Starting-Points for the Backward-Elimination Procedure

If we start the backward-elimination procedure (see [8]) with all λ's in (2.5) included in the model, we would

B. RELATIONSHIP AMONG MODELS H_{18} TO H_{25} IN TABLE 12, WITH RESPECT TO THEIR IMPLICATIONS AND DEGREES OF FREEDOM

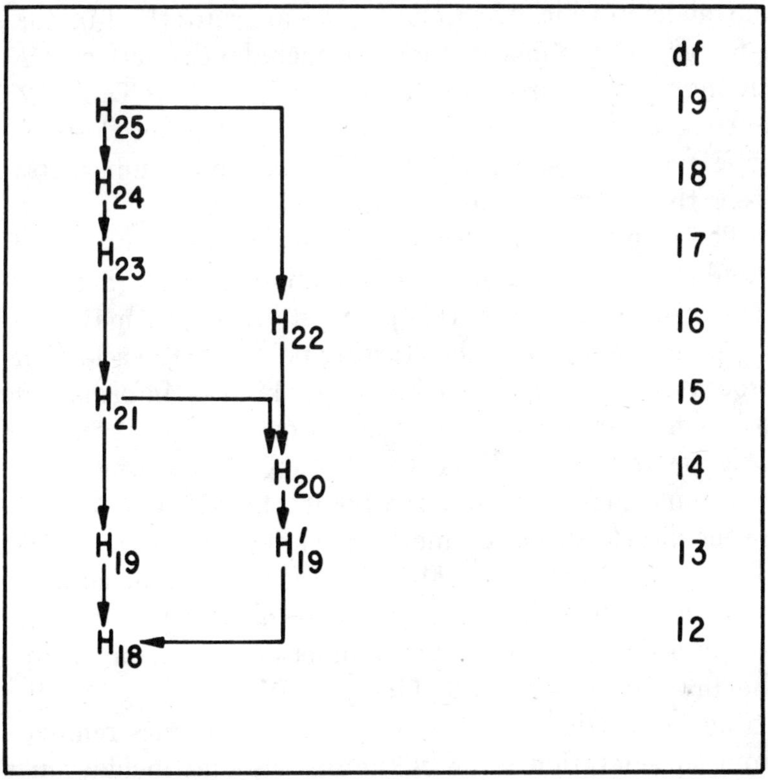

first eliminate λ^{ABCDE}, thereby obtaining H_1 of Table 12. At the next step, we would consider models in which λ^{ABCDE} and one 4-factor λ were eliminated (e.g., model H_2 of Table 12). Examination of the relative magnitudes of the 4-factor standardized values in Table 11 would suggest that the model in which λ^{ABCDE} and λ^{ABCE} are eliminated should provide the best fit, and indeed a comparison of the five models of this form confirms this. The best fitting model is H_8 in Table 12. Indeed, instead of starting backward-elimination with all λ's in (2.5) included in the model, examination of Table 11 might suggest that, for the analysis of Table 1, we could facilitate matters by starting the procedure with λ^{ABCDE} and λ^{ABCE} eliminated from the model.

Examination of the results obtained by the direct estimation method (see Table 11) will often suggest starting-points for the backward-elimination procedure that could facilitate matters greatly. For example, Table 11 would suggest that a suitable starting-point would be the model in which the following λ's are zero: the 3-factor, 4-factor, and 5-factor λ's that include the letters CE among their superscripts. This model is H_{13} of Table 12. For the analysis of Table 1, use of H_{13} (rather than H_1 or H_8) as the starting-point would reduce the number of steps that the selection procedure would take.

The results obtained by the direct estimation method should not be used in a simple-minded way. (See the cautionary remarks on this point in [7, 8].) It may also be worth noting that the starting-point for the selection procedure could be based upon the kind of information given in Tables 2 to 9, as well as on the results of the direct estimation method. In addition to their use in the determination of the starting-point, the results obtained by the direct estimation method (Table 11), and the kind of information given in Tables 2–9, can be used at various steps in the selection procedure (not only at the start of the procedure) to reduce the number of steps required by the procedure. With the kind of information given in Table 11 and in Tables 2–9, we can sometimes remove from consideration some of the models that include unnecessary λ's (and also some of the models that do not fit the data) without actually examining these models as we would normally do while proceeding step by step through the usual selection procedure.

4.4 Guided Starting-Points for Forward-Selection and the Stepwise Procedure

If we start forward-selection (or the stepwise procedure) with all λ's in (2.5) set equal to zero, we would begin with M_0 of Table 2, and would first consider which 1-factor λ's in (2.5) should be included in the model. If we start the procedure with all 2-, 3-, 4- and 5-factor λ's in (2.5) set equal to zero, we would begin with M_1 of Table 2, and would first consider which 2-factor λ's in (2.5) should be included in the model. Examination of the results obtained by the direct estimation method will often also suggest starting-points for this procedure that could facilitate matters. For example, Table 11 would suggest that at least the following six 2-factor λ's should be included in the model: λ^{AC}, λ^{BC}, λ^{DE}, λ^{CE}, λ^{CD}, λ^{BD}. The model that includes these λ's (and also the 1-factor λ's) can serve as a starting-point, and we would then consider which one of the remaining four 2-factor λ's should be inserted in the model at the next step. Comparison of the four models obtained by the insertion of a 2-factor λ indicates that the insertion of λ^{AE} provides the best fit. This model is H_{25} in Table 12. For the analysis of Table 1, use of the starting-point just noted, rather than M_0 or M_1, would reduce the number of steps that the procedures considered here would take.

The comments in the final paragraph of Section 4.3 apply to the procedures considered in the present section as well.

Before closing this section on forward-selection and stepwise procedures, it may be worth noting that, in addition to the reduction in the amount of calculation obtained with the use of guided starting-points, there are also other ways to reduce the amount of calculation required by these procedures. For example, to calculate the estimated expected frequencies under the various hierarchical models that are to be considered, the usual iterative-scaling method can be improved upon in the following way: It actually is not necessary to begin the iterative-scaling at the initial values of the estimated expected frequencies obtained when all λ's are set equal to zero (setting all initial values of the estimated expected frequencies equal to one, or to the ratio of the total frequencies and the number of cells in the table, or to any other constant). To calculate the estimated expected frequencies under a given hierarchical model (say, model H^{++}), we can begin the iterative-scaling using as the initial values the estimated expected frequencies under any other hierarchical model (say, model H^{+}) that consists of a subset of λ's in model H^{++}. When this subset is non-empty, we often obtain an improvement over the usual initial values (which are obtained by using the empty set for this subset—i.e., when all λ's are set equal to zero). Using this non-empty subset (i.e., using the estimated expected frequencies under model H^{+} as the initial values), we can often reduce the number of iterations required by the usual iterative-scaling method to calculate the estimated expected frequencies under model H^{++}. [If at least one of the marginals in the minimal set of marginals fitted under H^{+} is included in the minimal set of marginals fitted under H^{++} (and if the observed frequencies in the cells of this marginal are not all equal to each other), then an improvement will usually be obtained. On the other hand, if none of the marginals in the minimal set under H^{+} are included in the minimal set under H^{++}, then an improvement will not be obtained; but, in this case, we can often replace H^{+} by a different model (which is such that at least one of the marginals in its minimal set is also included in the minimal set under H^{++}) in order to obtain this improvement in the calculation of the estimated expected frequencies under H^{++}.]

When applying the forward-selection procedure, the estimated expected frequencies under the various models considered at a given step in the procedure can be calculated using as the initial values (in iterative-scaling) the estimated expected frequencies under models that were considered at preceding steps of the procedure. Similar kinds of improvements can also be obtained at some of the steps in the stepwise procedure.

4.5 Multidirectional Selection Procedures

Suppose, for the moment, that we arrived at H_{18} of Table 12. There are six marginals in the minimal set of marginals fitted under this model. We shall now consider which one of the six λ's in this model (corresponding to the six marginals) could be deleted by the backward-elimination procedure. Comparison of the six models obtained by the deletion of these λ's from H_{18} indicates that the deletion of λ^{ACD} yields the smallest chi square (viz., 12.59); and the deletion of λ^{ABE} yields the next smallest chi square (viz., 13.12). Model H_{19} of Table 12 is obtained when λ^{ACD} is deleted from H_{18}; and we shall let H'_{19} denote the model obtained when λ^{ABE} is deleted from H_{18}. Models H_{19} and H'_{19} have the same number of *df* (viz., 13); but there are seven marginals in the minimal set of marginals fitted under H_{19}, and only six marginals under H'_{19}. We next consider which one of the seven λ's in H_{19} (corresponding to the seven marginals), and which one of the six λ's in H'_{19} (corresponding to the six marginals) could be deleted. Comparison of the seven models obtained by the deletion of the λ's from H_{19} indicates that the deletion of λ^{ABE} yields the smallest chi square (viz., 15.72); whereas comparison of the six models obtained by the deletion of the λ's from H'_{19} indicates that the deletion of λ^{AB} yields the smallest chi square (viz., 13.95). The model obtained by the deletion of λ^{ABE} from H_{19}, and the model obtained by the deletion of λ^{AB} from H'_{19}, have the same number of *df* (viz., 14); but the latter model fits better. This model is H_{20} in Table 12 and in Figure B.

With the usual backward-elimination procedure, at a given step, we consider which one of the λ's in a given model (say, model H^*) can be eliminated, and we compare the models obtained by the deletion of the λ's from

H^*, selecting for further consideration the model (say, model H) that yields the smallest chi square (provided that the difference between the chi squares for H and H^* does not exceed a specified rejection level); but the model that yields the next smallest chi square (say, model H') is not considered further. The results just presented suggest that a modification of the usual procedure could yield models that fit better. In this modification, we consider further both H and H' (provided that the difference between the chi squares for H' and H^* also did not exceed a specified rejection level); comparing the models obtained by the deletion of the λ's from H with those obtained by the deletion of the λ's from H'. By this comparison, we would select for further consideration both the model (say, H'') that yields the smallest chi square and the model (say, H''') that yields the next smallest chi square [provided that the difference between the chi square for H'' and the chi square for the model from which H'' was obtained (viz., H or H') does not exceed a specified rejection level; and similarly for H'''].

The procedure proposed above selects for further consideration, at a given step, two models (say, H and H', or H'' and H''') rather than one. We shall refer to this as a "bidirectional" procedure. The "bidirectional" procedure could yield models that fit better than those obtained with the usual "unidirectional" procedure. If the former procedure is used at each step (following the initial step), the amount of calculation would be approximately doubled. Calculation could be reduced by applying the "bidirectional" procedure only at those steps where there is a small difference between the smallest and next smallest chi square, or where supplementary information (e.g., Table 11) would suggest that the second model (i.e., H' at one step, or H''' at the next step) will yield a better fit when one of the λ's is deleted from it than will the first model (i.e., H at one step, or H'' at the next step) when one of the λ's is deleted from it.

The preceding remarks can be directly generalized to obtain "multidirectional" procedures. In addition, the same kind of extension can be applied to forward-selection and to the stepwise-procedure in [8], for the analysis of a given table and for the analysis of T tables.

[*Received June 1971. Revised July 1972.*]

REFERENCES

[1] Birch, M.W., "Maximum-Likelihood in Three-Way Contingency Tables," *Journal of the Royal Statistical Society, Series B*, 25 (1963), 220–33.

[2] Bishop, Y.M.M., "Full Contingency Tables, Logits, and Split Contingency Tables," *Biometrics*, 25 (June 1969), 383–400.

[3] Bock, R. D., "Estimating Multinomial Response Relations," in R.C. Bose, ed., *Contributions to Statistics and Probability: Essays in Memory of S.N. Roy*, Chapel Hill, N. C.: University of North Carolina Press, 1969.

[4] Cox, D.R. and Snell, E.J., "A General Definition of Residuals," *Journal of the Royal Statistical Society, Series B*, 30 (1968), 248–65.

[5] Dyke, G.V. and Patterson, H.D., "Analysis of Factorial Arrangements When the Data are Proportions," *Biometrics*, 8 (March 1952), 1–12.

[6] Goodman, L.A., "Interactions in Multidimensional Contingency Tables," *Annals of Mathematical Statistics*, 35 (June 1964), 632–46.

[7] ———, "The Multivariate Analysis of Qualitative Data: Interactions Among Multiple Classifications," *Journal of the American Statistical Association*, 65 (March 1970), 226–56.

[8] ———, "The Analysis of Multidimensional Contingency Tables: Stepwise Procedures and Direct Estimation Methods for Building Models for Multiple Classifications," *Technometrics*, 13 (February 1971), 33–61.

[9] ———, "Partitioning of Chi-Square, Analysis of Marginal Contingency Tables, and Estimation of Expected Frequencies in Multidimensional Contingency Tables," *Journal of the American Statistical Association*, 66 (June 1971), 339–44.

[10] Haberman, S.J., "The General Log-Linear Model." Ph.D. thesis, Department of Statistics, University of Chicago, 1970.

[11] Lindley, D.V., "The Bayesian Analysis of Contingency Tables," *Annals of Mathematical Statistics*, 35 (December 1964), 1622–43.

[12] Martin, D.C. and Bradley, R.A., "Probability Models, Estimation, and Classification for Multivariate Dichotomous Populations," *Biometrics*, 28 (March 1972), 203–21.

[13] Solomon, H., "Classification Procedures Based on Dichotomous Response Vectors," in H. Solomon, ed., *Studies in Item Analysis and Prediction*, Stanford, California: Stanford University Press, 1961.

Leo A. Goodman

Appendix

Acquiring Computer Programs

All estimates and statistics reported for the log-linear models in parts I, II, and V can be reproduced exactly using the ECTA (Everyman's Contingency Table Analyzer) computer program. The source deck for this program is written in standard FORTRAN and the program consists of approximately 2000 cards. It is available with the EBCDIC character set (typically for IBM 360) or in BCD (usable with most other machines).

For further information write to

Professor Leo A. Goodman
Department of Statistics
University of Chicago
1118 East 58th Street
Chicago, IL 60637

Persons interested in the MLLSA (Maximum Likelihood Latent Structure Analysis) computer program should contact

Professor Clifford Clogg
504 Liberal Arts Tower
Pennsylvania State University
University Park, PA 16802

Index